How Far Are We from the Gauge Forces

THE SUBNUCLEAR SERIES

Series Editor: ANTONINO ZICHICHI, *European Physical Society, Geneva, Switzerland*

1. 1963 STRONG, ELECTROMAGNETIC, AND WEAK INTERACTIONS
2. 1964 SYMMETRIES IN ELEMENTARY PARTICLE PHYSICS
3. 1965 RECENT DEVELOPMENTS IN PARTICLE SYMMETRIES
4. 1966 STRONG AND WEAK INTERACTIONS
5. 1967 HADRONS AND THEIR INTERACTIONS
6. 1968 THEORY AND PHENOMENOLOGY IN PARTICLE PHYSICS
7. 1969 SUBNUCLEAR PHENOMENA
8. 1970 ELEMENTARY PROCESSES AT HIGH ENERGY
9. 1971 PROPERTIES OF THE FUNDAMENTAL INTERACTIONS
10. 1972 HIGHLIGHTS IN PARTICLE PHYSICS
11. 1973 LAWS OF HADRONIC MATTER
12. 1974 LEPTON AND HADRON STRUCTURE
13. 1975 NEW PHENOMENA IN SUBNUCLEAR PHYSICS
14. 1976 UNDERSTANDING THE FUNDAMENTAL CONSTITUENTS OF MATTER
15. 1977 THE WHYS OF SUBNUCLEAR PHYSICS
16. 1978 THE NEW ASPECTS OF SUBNUCLEAR PHYSICS
17. 1979 POINTLIKE STRUCTURES INSIDE AND OUTSIDE HADRONS
18. 1980 THE HIGH-ENERGY LIMIT
19. 1981 THE UNITY OF THE FUNDAMENTAL INTERACTIONS
20. 1982 GAUGE INTERACTIONS: Theory and Experiment
21. 1985 HOW FAR ARE WE FROM THE GAUGE FORCES

Volume 1 was published by W. A. Benjamin, Inc., New York; 2–8 and 11–12 by Academic Press, New York and London; 9–10 by Editrice Compositori, Bologna; 13–20 by Plenum Press, New York and London.

How Far Are We from the Gauge Forces

Edited by
Antonino Zichichi
European Physical Society
Geneva, Switzerland

PLENUM PRESS • NEW YORK AND LONDON

Library of Congress Cataloging in Publication Data

Main entry under title:

How far are we from the gauge forces.

(The Subnuclear series; v. 21)
"Proceedings of the twenty-first course of the International School of Subnuclear Physics, held August 3–14, 1983, in Erice, Trapani, Sicily, Italy."
Bibliography: p.
Includes index.
1. Particles (Nuclear physics)—Congresses. 2. Gauge fields (Physics)—Congresses. I. Zichichi, Antonino. II. International School of Subnuclear Physics (21st: 1983: Erice, Sicily) III. Series.

QC793.H69 1985 539.7′2 85-24412
ISBN 0-306-42203-4

Proceedings of the twenty-first Course of the International School of Subnuclear Physics, held August 3–14, 1983, in Erice, Trapani, Sicily, Italy

Printed in the United States of America

Eugene P. Wigner

PREFACE

During August 1983, a group of 89 physicists from 59 laboratories in 23 countries met in Erice for the 21st Course of the International School of Subnuclear Physics. The countries represented were Algeria, Australia, Austria, Canada, Czechoslovakia, the Federal Republic of Germany, Finland, France, Hungary, India, Israel, Italy, Japan, the Netherlands, South Africa, Spain, Sweden, Switzerland, Taiwan, Turkey, the United Kingdom, the United States of America, and Yugoslavia. The School was sponsored by the European Physical Society (EPS), the Italian Ministry of Education (MPI), the Italian Ministry of Scientific and Technological Research (MRST), the Sicilian Regional Government (ERS), and the Weizmann Institute of Science.

The programme of the School was mainly devoted to a review of the most significant results, both in theory and experiment, obtained in the field of the "electroweak" and of the "colour" forces of nature. The outcome of the Course was to present a clear picture of how far we are from the electronuclear formulation of these basic forces acting between quarks and leptons. And more generally, how far we are from the unification of all gauge forces of nature.

I hope the reader will enjoy this book as much as the students enjoyed attending the lectures and the discussion Sessions, which are the most attractive features of the School. Thanks to the work of the Scientific Secretaries, the discussions have been reproduced as faithfully as possible. At various stages of my work I have enjoyed the collaboration of many friends whose contributions have been extremely important for the School and are highly appreciated. I thank them most warmly. A final acknowledgement to all those who, in Erice, Bologna, Rome and Geneva, helped me on so many occasions and to whom I feel very much indebted.

Antonino Zichichi
November 1985
Geneva

CONTENTS

OPENING LECTURE

THEORETICAL LECTURES

SEMINARS ON SPECIALIZED TOPICS

REVIEW LECTURES

CELEBRATION OF E P WIGNER'S BIRTHDAY

CLOSING CEREMONY

ELEMENTARY PARTICLE PHYSICS TODAY

S. L. Glashow

Lyman Laboratory of Physics
Harvard University
Cambridge, Massachusetts 02138

The search for the ultimate constituents of matter and for the laws by which they combine has a somewhat periodic history, alternating between CHAOS, the discovery of ORDER, and the revelation of a new and confusing level of STRUCTURE. Begin with the chemical elements. Lavoisier was aware of some two dozen of them. Subsequent technical developments (like spectroscopy) led to the discovery and the synthesis of many more. Today, there are 108 known chemical elements, of which some twenty must be artificially produced. The time evolution of the number of known elements has been roughly linear for more than two centuries. Clearly, there are too many different varieties of atoms for them to be regarded as fundamental entities.

ORDER was established in 1869 with the development of the periodic table of the elements by Mendeleev. Not only did this system cogently display the regularities in chemical and physical properties of the elements, but it was predictive. The table had holes corresponding to elements yet undiscovered. Their chemical and physical properties were predicted. The discoveries of Gallium (1875), and Scandium (1879), and Germanium (1886) showed the power of the periodic table. Many scientists took the success of the periodic table as an indication of the composite nature of atoms.

Spectroscopy led to further evidence of atomic structure. The regularity of the Balmer series was recognized in 1885. But it was the discovery of the electron in 1895 and the atomic nucleus in 1911 that led us to the essentially correct view of STRUCTURE: the nuclear atom of Rutherford, Bohr, and Sommerfeld.

Thus, we were led to the next level of CHAOS, that of atomic nuclei. With the discovery of isotopes, it became clear that there were more species of the nucleus than of the atom. Nuclei could not be fundamental. The first tantalizing hint of nuclear ORDER is Prout's Law: that all atomic weights should be integer multiples of that of Hydrogen. (Indeed, it was in the test of this law that Lord Rayleigh performed the precise density measurements that led to the almost serendipitous discovery of Argon.) Prout's Law, though inexact, is remarkably well satisfied. However, it was Moseley's work that led to a more precise and predictive level of nuclear order. Moseley measured the characteristic X-rays of all the known elements from Aluminium to Gold. The systematic dependence of X-ray wavelength upon atomic weight revealed another integer law of the nuclei: atomic number, which Moseley correctly identified as the nuclear charge in multiples of the electronic charge. Again, there were predictive holes in Moseley's table corresponding to the then undiscovered elements Rhenium and Technetium. The stage was set for the discovery of the nuclear constituents.

With the discovery of the neutron in 1932 a new understanding of STRUCTURE was reached. Nucleons and electrons were the components of all matter. A new nuclear force was needed to bind nucleons together. The hypothetical "mesotrons" (now called pions), which mediate this force were discovered in cosmic rays in 1948. Yukawa won his Nobel Prize and the physics community rejoiced in its discovery of the ultimate constituents of matter. But, only a few years later the nuclear particles were to be subjected to yet a new variety of population explosion and a next level of CHAOS.

With the development of large particle accelerators and of detection systems, like the bubble chamber, many cousins of the pions and nucleons were discovered, as well as a host of strange particles. Many of the new particles were recognized to be excited states of the nucleon. Thus, energy level diagrams are common to atoms, to nuclei, and to the proton itself. This was a strong hint of the composite nature of hadrons. But, further progress to understand the structure of hadrons depended on the establishment of a new level of ORDER. This emerged with the development of the 8-fold way of Gell-Mann and Ne'eman. Hadrons were seen to appear as complete representations of the group SU(3), which was supposed to be an approximate symmetry group of the strong interactions. Baryons transformed according to one another of three representations: one-dimensional, eight-dimensional, or ten-dimensional. All mesons were either singlets or octets. One of the spectacular successes of this scheme was the prediction of the Ω^-, the tenth member of the $J = 3/2^+$ baryon decimet. With its discovery in 1964, it was clear to all that the eightfold-way was useful, valid, and predictive. But, why only certain representations, and why SU(3) among all groups?

Those questions were answered elegantly by Gell-Mann and Zweig with their invention of quarks, and with the mysterious quark rules: three quarks to a baryon, quark-antiquark to a meson, three antiquarks to an antibaryon, no other observable quark configurations, and spin-½ quarks to be treated as bosons. Only much later did a quantitative theory of strong interactions emerge in which these rules became comprehensible: today's gauge theory based on exact color SU(3) -- quantum chromodynamics.

Many experiments have identified quarks and gluons as the most fundamental known nuclear constituents. First and foremost, perhaps, are the deep-inelastic lepton scattering experiments initiated at SLAC in the 1960s. These results were incontrovertible evidence for the existence of pointlike charged particles (partons) within the nucleon. With the overwhelming success of the quark model in explaining hadron spectroscopy, it soon became evident that the parton is a quark. With the discovery of the J/ψ particle, a cleaner spectroscopic system emerged in which the quark model (supplemented with the charmed quark) proved itself to be a correct and predictive system. For example, the recently discovered $F = c\bar{s}$ meson at 1970 MeV lies almost precisely at its predicted (in 1975) mass of 1975 MeV. Finally, the observation and measurement of quark and gluon jets is as close to "seeing" these fundamental particles as we may get. We have reached the next (perhaps, the last?) level of STRUCTURE.

The number of truly fundamental particles is none too small for comfort: seventeen. The twelve fundamental fermions fit elegantly into today's periodic table of quarks and leptons. Of course, it may turn out that the number of fermion families (or, rows of the table) is larger than three, but not much larger since cosmology constrains the number of neutrinos to be not greater than four. Beyond these "matter particles" there are at least five varieties of "force particles": photons, gluons, $W^{\pm}$, Z^0, and the elusive Higgs boson. To date, fourteen of the particles have been "seen". All that remain are the top quark, the tau neutrino, and the Higgs boson.

Our theoretical framework: a gauge theory based upon $SU(3) \times SU(2) \times U(1)$, offers a complete and correct description of all confirmed phenomena at accessible energies. There are no generally established loose ends that require something extra in the theory. However, there are some intriguing indications. One is the existence of unexplained same-sign dileptons in high-energy neutrino interactions. This curious effect cannot be explained in terms of associated charm production, since the Ohio State triggered emulsion experiment shows neutrino production of $\sim$100 single charmed particles, but not one candidate for production of charmed particle pairs. A second potential problem is the sighting of apparent decays of Z^0 into a lepton pair and a photon. Such a process, if it is really there, admits of no conventional explanation. (More recently, in March 1984, other categories of anomalous events have been reported both by the UA(1) group and by the UA(2) group.) It would seem that the desert must begin to bloom in the very near future.

Permit me to conclude this talk with a list of six great outstanding problems of fundamental physics.

I. The Einstein Problem

Einstein devoted the last three decades of his life to the construction of a unified theory of gravitation and electromagnetism. His failure was due in part to his reluctance to accept quantum mechanics ("God does not play dice.") and his aversion to the existence of peculiarly nuclear forces. (He once argued that the nucleus is held together by gravitational forces.) Today, we have a unified theory of all forces except gravity, but still no quantum theory of gravity. While many of today's physicists devote themselves to this problem, as yet they produce more heat than light: unobservable particles and forces, and extra dimensions of space-time.

II. The Dirac Problem

This concerns the existence of very large numbers in our theory of physics, such as the ratio of the gravitational attraction and the electric repulsion of two electrons. In a "correct" theory, Dirac argues that no large dimensionless numbers can appear as fundamental parameters. They are not merely unsightly, but unnatural. To keep a parameter small, in the context of quantum field theory, requires careful fine-tuning of renormalized quantities. (Another possibility, advocated by Dirac, is the hypothesis that large numbers simply reflect the great age of our universe. Dirac conjectures that the gravitational constant is decreasing with time, $\dot{G}/G = 1/T$ where $T \approx 20$ billion years. Only very recently has experimental data ruled out this fascinating speculation.) Under the guise of "the gauge hierarchy problem", this question has the complete attention of many of today's practitioners.

III. The Rabi Problem

"Who ordered that?" said Isadore Rabi upon the identification of the muon as an obese electron. One family of quarks and leptons would be enough for the operation of spaceship Earth and its auxiliary solar power station. Why are there three families? Are there, perhaps, four? Why do all fermion families have exactly the same structure? Nanopoulos offers an "anthropic" partial answer to the first question. At least three families are needed if automatic CP conservation is to be avoided in the standard theory. CP violation was essential for the synthesis of baryons in the early universe: without it, intelligent life could never have evolved. Such arguments are satisfactory only to astrophysicists and followers of Velikovsky. Moreover, "anthropic" is not even an English word. The problem is not solved.

IV. The Cabibbo Problem

Nicola showed how the introduction of a single new parameter could explain much about the pattern of weak interaction decay phenomena. It was a grand accomplishment. Today, the standard theory requires the specification of nineteen fundamental dimensionless parameters. Surely, these are too many arbitrary constants to describe a truly fundamental theory. Most attempts to compute some of these parameters have led to the introduction of even more. For example, the grand unified theory reduces the number of gauge coupling constants from three to one. However, the number of independent Yukawa couplings is vastly increased.

V. The Dark Mass Problem

Traditionally, astronomers display the conceit that their studies of stars and galaxies reveal the nature of the bulk of the matter in our universe. Today, they are no longer so sure. Studies of the rotation curves of galaxies indicate that most of the matter in the universe is not in the form of luminous stars. Many of the galaxies that have been studied show an extensive halo of dark matter which is perhaps a factor of ten times more massive than the visible stellar system. What is the nature of this dark and dominant matter? In a last ditch attempt to defend their turf, many astronomers argue that the haloes consist of dead or dying stars. Yet, recent studies in the infrared performed by the IRAS satellite offer no support to this explanation. Particle physicists have suggested many possibilities: axions, massive neutrinos, nuggets of strange quark matter (nuclearites), or black holes. Here is one of the greatest uncertainties in science: the mass of the conjectured bodies which make up galactic haloes varies over some 70 orders of magnitude!

VI. The Problem of the Desert

Perhaps I am responsible for introducing into particle physics the concept of the Great Desert -- in which there are no more surprises in the mass range 10^2GeV to 10^{14} GeV. If this speculation is true, it would mark the death or decline of our study of the fundamental structure of matter. Over the centuries, our discipline has thrived on the frequent occurrence of surprises: the vacuum, oxygen, Brownian motion, argon, the electron, parity-violation, the tau lepton, *ad nauseum*. Generally, these surprises have led to a deeper understanding of nature: they are nature's clues in the Great Treasure Hunt of Science. Could it be that we have reached the end: that all we know is all we need to know? True, it seems technically possible that there is a desert. The low-energy SU(3) × SU(2) × U(1) theory works all too well. My own opinion is that we are now within the lull before a storm, soon to be engulfed in a new level of CHAOS. We must have the courage to persevere, to prove that our present smug theory is far from complete and probably wrong. It is our sacred duty to destroy the world view that we have labored so hard and so long to fashion.

This research was supported in part by the National Science Foundation under Grant No. PHY-82-15249

D I S C U S S I O N

CHAIRMAN : S.L. GLASHOW

Scientific Secretary : R.D. Ball

- *CATTO* :

The proton lifetime is now pushed up to 10^{32} years by IMB experiments. Is this lifetime compatible with SU(5) and other Grand Unified Theories, such as SO(10)?

- *GLASHOW* :

There is a disagreement between the predicted value of the proton lifetime in minimal SU(5) and the experiment. I think that the error in the theoretical prediction is understated, so although it is disappointing I do not think that the experimental situation is catastrophic even for naive SU(5). Marciano, who has done the most careful calculations for the proton lifetime in SU(5) has not done it in SO(10) because there is at least one additional parameter. There are two sets of gauge bosons which mediate proton decay and you can get different values for the proton lifetime depending on how you adjust these parameters. The example I gave complicating the fermion structure in naive SU(5) made dramatic changes either way in the predicted lifetime. Unfortunately it is very easy to stretch the proton lifetime out as long as you want - it is sad from an experimental point of view. Still and all, I believe that the lifetime of the proton is of the order of 10^{32} years.

- *BAGGER* :

Why does Professor Veltman believe that there are only three families ?

- *GLASHOW* :

Although I was not at Brighton and have not seen his paper, I understand that Veltman believes that an extra heavy lepton will lead to a contradiction with the measured ρ parameter of $SU(2) \times U(1)$. In order to say more, I would need to know just what Veltman has done.

- *DATÉ* :

The observed elementary particles have spin 0, ½ or 1. How does one understand the absence of higher spins ?

- *GLASHOW* :

Spin 0, ½ and 1 are the only renormalizable theories.

- *OLEJNÍK* :

There has been suggested the possibility that monopoles catalyze proton decay. Does the IMB limit on the decay $p \rightarrow \pi e^+$ put any limits on the characteristics of monopoles ?

- *GLASHOW* :

There is a preprint by the IMB people which I did not read giving a limit on the flux of monopoles assuming a certain cross-section for catalytic proton decay. The subject is a very rich and complicated one. Some people say that a much more sensitive detector than IMB for these things is neutrino stars, and some people say that if there is a sensible flux of monopoles then these monopoles will get trapped in neutron stars and cause too much X-ray emission by neutron stars. The only thing that concerns me is that there is absolutely no experimental indication for the existence of magnetic monopoles in any shape, form or manner, and consequently I am not interested in them. Of course, Grand Unified Theories do predict the existence of magnetic monopoles. A central cosmological problem, perhaps solved by inflation, is to explain why monopoles are not abundant in the universe.

- *RATOFF* :

What is the connection between the lifetime of the b-quark and the mass of the t-quark ?

- *GLASHOW* :

One thing that is known is the amount of CP violation and if you make the Kobayashi-Maskawa angles θ_2 and θ_3 very small then it is hard to get enough CP violation from the box diagram and it is clear that all the CP violation must come from the box diagram. However, it is still possible to recapture enough CP violation by making the mass of the t-quark large. Consequently, if you know that the b-quark lifetime is long and consequently that the angles are small, you know that the t-quark mass is big. Also folded in is the branching ratio between b-quarks going to the u versus b-quarks going into u. More about this in my next lecture.

- *RATOFF* :

Is there any theoretical appeal in having an almost diagonal Kobayashi-Maskawa matrix ?

- *GLASHOW* :

Several times today we have been talking about an almost diagonal matrix, but the Cabibbo angle is not really very small at all - it is $1/5$, and the other angles are much smaller. The CP violating phase is very large. There have been many thoughts along these lines in the past, most conspicuously by Harald Fritzsch. I do not think that what is called the "Fritzsch Ansatz" works any longer.

- *FRITZSCH* :

I do not believe it any more either. I have a suspicion that there are nevertheless strong relations between the angles and the masses and in particular that the angles are higher order electromagnetic radiative corrections. That is how I understand why all the angles are extremely small except for the first; it has to do with the fact that the u-quark is lighter than the d-quark.

- *GLASHOW* :

I wish Harald the best of luck. From my point of view calculating the masses of the quarks and the Kobayashi-Maskawa angles has been a central preoccupation of mine for the past three years and I have made absolutely zero progress.

- *MOUNT* :

What are the prospects for more precise estimates of decay rates for $p \to \pi^o e^+$ from, for example, SU(5) ? Does the present uncertainty reflect the errors in Λ_{QCD}, or uncertainty in how to perform the calculation ?

- GLASHOW :

There is a factor of ten or so coming from the uncertainty in the QCD parameter but there is a factor which I would estimate as a hundred coming from uncertainties of hadron physics. Do not forget that we have no simple explanation for the non-leptonic $\Delta I = \frac{1}{2}$ rule. Is the decay $p \to \pi^o e^+$ "allowed" like $K^o \to \pi^+\pi^-$, or is it "forbidden" like $K^+ \to \pi^+\pi^o$? At present, we have no way of knowing. Perhaps precise calculations of proton decay must await the development of more advanced lattice field theory calculations. QCD may be a perfect theory of strong interactions, but it does not yet permit the calculation of low-energy phenomenology.

There is a tremendous feeling of despair because our great hopes of being able to measure the detailed properties of proton decay have been dashed, because we have had the IBM group with 10 kilotonnes running for a year and they have no candidate. Now that means that if you do an experiment ten times better you might have a few candidates, but you will never have a thousand decays, never, and it is very, very depressing.

There is a possibility that because of supersymmetry or some other peculiar theory the proton decays predominately into $K\mu$ or something of this sort. There is not yet as good a limit on that decay mode because of neutrino processes which simulate this final state. Possibly in the next few years we will discover that the proton decays into $K\mu$ and we will accumulate a few dozen events. In this context it is interesting to note that with the 10 kilotonne operating American experiment, the next American device approved by the Department of Energy is a 1 kilotonne device.

- MARUYAMA :

If the Z^o or $W^{\pm}$ masses are heavier than expected from low energy experiments, how would you explain the difference ?

- GLASHOW :

If there is any difference in $\sin^2\theta_W$ between low energy neutrino physics and the $p\bar{p}$ collider experiment, it would be very interesting. The neutrino physics data has uncertainties from the quark model built into it. That means that given a choice between a measurement of $\sin^2\theta_W$ from the determination of the W mass and the neutrino scattering on hadrons I would simply say that the interpretation of the neutrino scattering in hadrons is wrong. That is why there is very strong pressure in America and at CERN to do careful measurements of neutrino scattering on electrons (which is extraordinarily difficult because the cross-section is 2000 times smaller). In this case there are no parton model uncertainties. Available data on neutrino scattering from hadrons is beset by uncertainties. As one example, consider the recently discovered result that the quark distribution in iron, or other large nuclei, is not the same as the quark distribution in nucleons. As another example, note that the UAI group claims that the uncertainty in $\sin^2\theta_W$ from neutrino-hadron data is ± .03, or 15%.

- *KLEVANSKY* :

Concerning the little and great deserts that occur between 10^2 - 10^6 eV and 10^{11} - 10^{23} eV : do you expect anything to be seen in them and why do you imagine they exist ?

- *GLASHOW* :

These are mysteries. We have no idea.

- *OGILVIE* :

How would you pick the energy and luminosity of the next big accelerator ?

- *GLASHOW* :

I would hope that the new physics, if it exists at all, exists at a few TeV, so that a 20 TeV machine would have more than enough energy to find something new. As for luminosity, I think that the Brookhaven people in arguing for the CBA have made it abundantly clear that high luminosity is essential. What we want is both high energy and high luminosity - a proton collider with at least 20 TeV total energy - our last opportunity to discover something really new. If there is nothing new at 20 TeV and 10^{33} luminosity then it's time to give up.

- *D'AMBROSIO* :

Why have you not mentioned the Higgs particles as a problem of SU(3) x SU(2) x U(1) ? Don't you think that they are quite unnatural in the standard model ?

- *GLASHOW* :

I did mention it in passing when I said that we are 82% towards finding all the particles and I mentioned that there were three particles missing. However, the Higgs particle is in a certain sense a very depressing particle because we do not know how much it weighs - it might weigh almost anything. But as soon as we have found it, as soon as we know its mass, then theory tells us all its properties in detail.

- *D'AMBROSIO* :

But don't you think that the Higgs boson is put in by hand, and is quite unnatural ?

- *GLASHOW* :

But the Higgs boson is the Higgs boson, and if Weinberg and Salam are right then its properties are defined. If they are wrong, then anything can happen, and that would be fine - that is what we are all waiting for, a contradiction. It would be nice to have a contradiction for a change. I vote for a large, complex and unexpected scalar meson sector.

- *D'AMBROSIO* :

But don't you think that this is an effective theory, and not the final theory ?

- *GLASHOW* :

How can we tell ? Make something better !

- *SIMIĆ* :

In what sense will eventual rigorous establishment of the triviality of φ^4-theory (or indeed any infrared stable theory) for which there are now indications, influence our way of looking at Weinberg-Salam or SU(5) ? Would such demonstrations necessarily mean that unified theory is just a phenomenological low energy theory ?

- *GLASHOW* :

I cannot answer that question. To me it is a miracle that φ^4-theory is trivial.

- *BATTISTON* :

In the last few years, papers have been published indicating that the t mass is just above the PETRA limit

- *GLASHOW* :

  Most of the published formulas, including mine, are wrong.

- *BATTISTON* :

  What are the constraints on this mass ?

- *GLASHOW* :

There are no constraints on the t-quark mass, except that it cannot be too heavy. where too heavy means more than a few hundred GeV. Physics answers some questions very well - like the mysteries of hadron spectroscopy - but we have no way of predicting the mass of a sixth quark. Yes, there have been various predictions based on random hypotheses; none of them were convincing at the beginning and in fact none of them came out to be true. We can no more compute the top mass than the muon mass. This is the mystery of the seventeen parameters of SU(3) x SU(2) x U(1).

DISCRETE MECHANICS

T. D. Lee

Columbia University
New York, N. Y. 10027

I. INTRODUCTION

At present, with QCD for the strong interaction, a unifying gauge theory for the electromagnetic and weak interactions, and the Einstein theory of gravitation (plus possible further unification of some or all of these interactions), we may feel that we have arrived at a closed system of physical laws. Nevertheless, the fact that any of the unifying gauge theoretical models requires about twenty parameters suggests that our understanding of these forces most likely remains phenomenological, far from fundamental. This uneasy feeling is reinforced by the enormous technical difficulty of quantizing gravity. Attempts to improve this situation have so far invariably led back to the original unsatisfactory state, with only a change in scale. The underlying cause may be traced to the foundation of our relativistic quantum local field theory. In such a theory only fields are observables, represented by operator functions embedded in a space-time continuum $x = (\vec{r}, t)$. Hence we can conceive of performing an infinite sequence of observations, at $x_1, x_2, \cdots,$ all lying infinitesimally close but with the result of the field measurements, say $\phi(x_1)$,

---

This research was supported in part by the U.S. Department of Energy.

$\phi(x_2)$, $\cdots$, fluctuating violently. The probability amplitude of such an unlikely event is proportional to e^{iA} with A as the corresponding action. Yet only the part of A that involves the gradient of the fields is effective in making such rapid fluctuations improbable. This then severely limits the candidates for an acceptable physical theory, and results in the present small set of renormalizable field theories: QED, QCD, Higgs, $\cdots$. All of these have been intensively investigated. Except for the variation in group algebra, we have virtually exhausted the possible theories within this limited repertoire.

Let us suppose that the concept of locality is only an approximate one. After all, it is not difficult to imagine that the present quantum local field theory should break down at a distance ℓ of the order of Planck's length. Once the notion of a fundamental length ℓ is accepted, one could contemplate that ℓ might even be much bigger (though about 10^{-16} cm in order to be consistent with present experimental observations). In any case, the existence of a fundamental length would break the present confinement of renormalizable field theories; it could naturally inhibit the possibility of making an infinite number of measurements that are infinitesimally close to each other in space-time.

Assuming that ℓ is the only scale in such a theory, at distances $r \gg \ell$ we may set $\ell/r \cong 0$. As a result, the theory becomes approximately scale-invariant. Scale-invariant theories are very few in number. We have the chiral symmetric theories for spin $\frac{1}{2}$, the gauge theories for spin 1 , etc. These are precisely the type of theories that are in use at present. They are almost all generalizations of the (classical) Maxwell equation which was originally derived from large-distance phenomena.

From this point of view, our present theories may well be approximations valid only at large distances. When we extrapolate these theories to small distances, new phenomenological constants are needed. This then accounts for the very large number of parameters that are currently required.

In this series of lectures, I wish to explore a new direction which may provide a fundamental length. I shall begin the discussion by regarding time as a dynamical variable. This will lead to the following two inter-related topics:

1. discrete quantum mechanics

and

2. random lattice field theories.

These subjects have recently been developed at Columbia[1-4] with the collaboration of N. H. Christ, R. Friedberg and H. C. Ren. In addition, there are also the recent papers by C. Itzykson and others.[5]

II. AN OVERVIEW: TIME AS A DYNAMICAL VARIABLE

As we all know, time has always been regarded as a continuous parameter in physics. Even in general relativity, although the metric is a dynamical variable, the continuous four-dimensional space that the metric is embedded in is not. This concept can be traced to Newtonian mechanics. The following quotation was written by James Clerk Maxwell[6] in 1877:

> Absolute, true and mathematical Time is conceived by Newton as flowing at a constant rate, unaffected by the speed or slowness of material things.

There is no evidence that this long-held view is in any kind of trouble. Nevertheless, we wish to explore an alternative: that time is a dynamical variable. Just like any other dynamical variable, time is affected by material things, and in fact it is part of matter.

Before introducing this new concept, let us review the role of time throughout all phases of mechanics. This is illustrated in the following table.

TABLE 1

	Usual Continuum Theory	
Classical Mechanics	$\vec{r}(t)$	dynamical variable
	t	parameter
Non-relativistic Quantum Mechanics	$\vec{r}(t)$	operator (observable)
	t	parameter
Relativistic Quantum Theory	field $\phi(\vec{r}, t)$	operator (observable)
	$\vec{r}$, t	parameters

In the usual continuum theory the position $\vec{r}(t)$ of a particle is a dynamical variable in classical mechanics, but the time t is a parameter. When we go over to the non-relativistic quantum mechanics, the observable $\vec{r}(t)$ becomes an operator while t remains a parameter. In the relativistic theory, $\vec{r}$ and t have to be treated on an equal basis. Two choices are open. Either regard t as an operator or $\vec{r}$ as a parameter. Our traditional course is to opt for the latter: only the fields are operators or observables. The space-time coordinates are merely parameters. An alternative route is to see whether we can regard t as an operator; this is then the essence of this new approach, which I call discrete mechanics.

Table 2 summarizes the fundamental concepts of the discrete mechanics.

TABLE 2

	Discrete Theory	
Classical Mechanics	$\vec{r}$, t	both dynamical variables
Non-relativistic Quantum Mechanics	$\vec{r}$, t	both operators (observables)
Relativistic Quantum Theory	$\vec{r}$, t, ϕ	all operators (observables)

Thus, in the discrete version of relativistic quantum theory, the space-time position, as well as the field, is considered a dynamical variable. For example, in a collision experiment of, say, $e^+e^- \to \mu^+\mu^-$ at LEP region P4 in 1988, the precise location and time of the collision should be regarded as part of the measurement, on the same footing as the electric field, magnetic field, ... In order to incorporate such a view, let us start with the discrete theory in its classical form.

III. CLASSICAL MECHANICS

Consider the example of a non-relativistic point particle of unit mass moving in a potential $V(\vec{r})$. In Table 3, which appears on the next page, we give the familiar formulation of classical continuum mechanics in the left column with A_c = action. The corresponding discrete version is given in the right column.

Let us consider a large time interval T. A fundamental postulate of discrete mechanics is that within such a time interval a particle can only assume N space-time positions

$$(\vec{r}_n\ ,\ t_n)$$

with $n = 1, 2, \cdots, N$. The ratio

$$\frac{N}{T} \equiv \rho$$

is a fundamental constant of the theory. For convenience and without loss of generality we have arranged $t_1, t_2, \cdots, t_N$ in ascending order. The discrete action A_d is then given in Table 3. Unlike the continuum case (where only $\vec{r}(t)$ is the dynamical variable) we regard $\vec{r}_n$ and t_n both as dynamical variables. Consequently there are two sets of equations:

$$\frac{\partial A_d}{\partial \vec{r}_n} = 0 \tag{3.1}$$

which gives the discrete version of Newton's law, and in addition

$$\frac{\partial A_d}{\partial t_n} = 0 \tag{3.2}$$

which yields the energy conservation. There are altogether 4N unknowns:

$$\vec{r}_n = (x_n\ ,\ y_n\ ,\ z_n) \quad \text{and} \quad t_n\ .$$

Exactly the same number of equations is supplied by (3.1) and (3.2). In continuum mechanics conservation of energy is a consequence of Newton's equation. This is not so in discrete mechanics.

Had we treated time merely as a predetermined discrete parameter, the system would then violate time-translational invariance, and lead to energy non-conservation. Historically this has always been the difficulty encountered in any attempt to treat time as a discrete parameter. Here, by viewing $\vec{r}_n$ and t_n as dynamical variables we bypass this problem.

TABLE 3

Continuum Mechanics	Discrete Mechanics
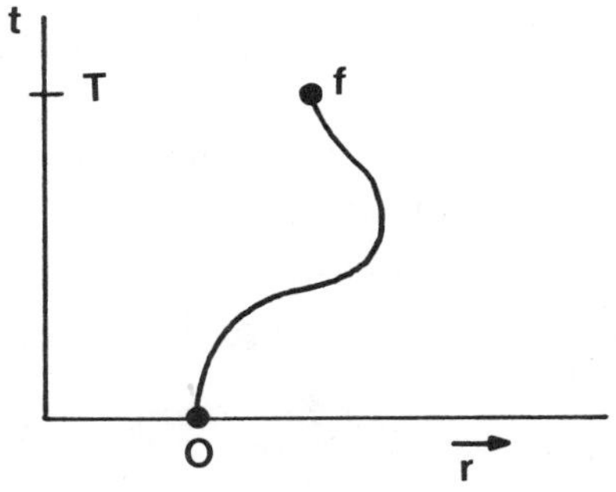	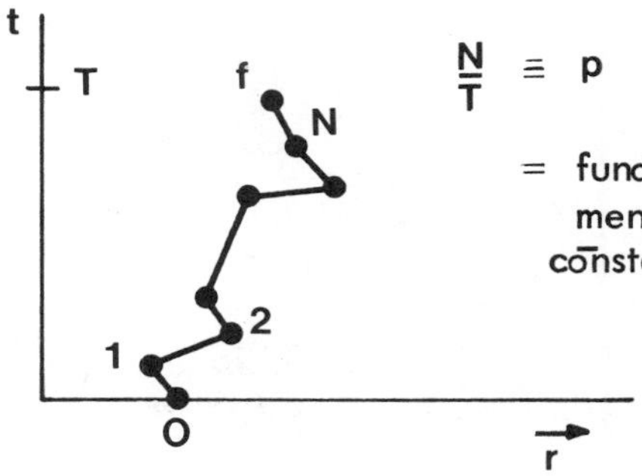 $\frac{N}{T} \equiv p$ = fundamental constant
$A_c = \int_0^T (\frac{1}{2}\dot{\vec{r}}^2 - V)\, dt$	$A_d = \sum_n \left\{\frac{1}{2}\vec{v}_n^2 - \frac{1}{2}\left[V(\vec{r}_n) + V(\vec{r}_{n-1})\right]\right\} \cdot (t_n - t_{n-1})$ $\vec{v}_n = \frac{\vec{r}_n - \vec{r}_{n-1}}{t_n - t_{n-1}}$
fix $\vec{r}(0) = \vec{r}_0$ $\vec{r}(T) = \vec{r}_f$	fix $(\vec{r}_n, t_n) = \begin{cases} (\vec{r}_0, 0) & \text{when } n = 0 \\ (\vec{r}_f, T) & \text{when } n = N+1 \end{cases}$
$\frac{\delta A_c}{\delta \vec{r}(t)} = 0$ gives $\ddot{\vec{r}} = -\nabla V$	$\frac{\partial A_d}{\partial \vec{r}_n} = 0$ gives $\frac{\vec{v}_{n+1} - \vec{v}_n}{\frac{1}{2}(t_{n+1} - t_{n-1})} = -\vec{\nabla} V(\vec{r}_n)$
$\vec{r}(t)$ = dynamical variable	$\frac{\partial A_d}{\partial t_n} = 0$ gives
t = parameter	$E_n \equiv \frac{1}{2}v_n^2 + \frac{1}{2}\left[V(\vec{r}_n) + V(\vec{r}_{n-1})\right]$ $= E_{n+1}$

Example 1 For a free particle $V = 0$, Eq. (3.1) gives $\vec{v}_n =$ constant, which also satisfies (3.2). The trajectory of the particle is always a straight line, the same as in the continuum version. Therefore Newton's first law remains unaltered.

Example 2 In the case that the particle is subject to a constant force,

$$\vec{\nabla} V = \text{constant},$$

it is easy to see that the solution of (3.1) and (3.2) determines that $(\vec{r}_1, t_1)$, $(\vec{r}_2, t_2), \cdots, (\vec{r}_N, t_N)$ all lie on a parabola, similar to the continuum case. In addition, the time intervals $t_2 - t_1$, $t_3 - t_2$, $\cdots$ are all equal.

Example 3 For a harmonic oscillator $V = \frac{1}{2}\omega^2 x^2$, Eqs (3.1) and (3.2) can be solved numerically. In Figure 1 on the next page we give the result for $\omega = 1$, $N = 40$ and $T = 6$ (which is near 2π). The ordinate is x and the abscissa t for $n = 1, 2, \cdots, 40$. At each point (x_n, t_n) the discrete action A_d is stationary with respect to both x_n and t_n. The former gives the equality between momentum-change and impulse, whereas the latter gives energy conservation. Thus both the position x_n and the time t_n respond to the potential. Both depend on the dynamics.

The consecutive time spacing $\tau_n \equiv t_n - t_{n-1}$ is not uniform and is plotted in Figure 2 for n running from 1 to 40. The horizontal line indicates the average τ_n ($= T/N = 6/40$). Note that the τ_n for $n = 8, 17, 28$ and 36 are much larger than the average. As we can see, τ_n is almost a periodic function in n, but with a frequency $\simeq 4\omega$, which can also be established analytically when N/T is very large.

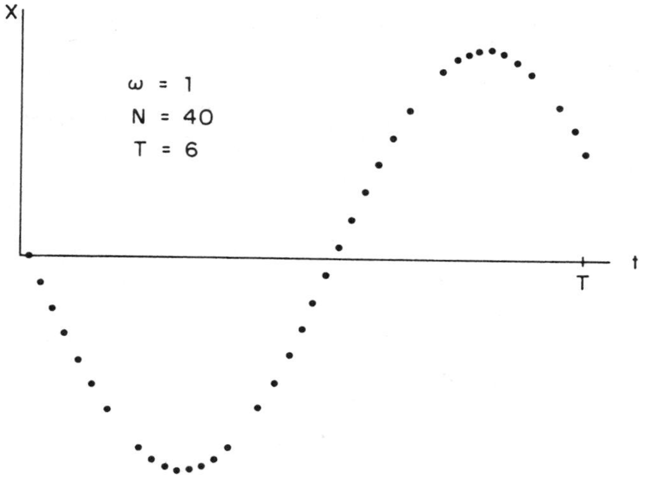

Figure 1. Numerical solution for a harmonic oscillator potential $V = \frac{1}{2}\omega x^2$ in discrete mechanics.

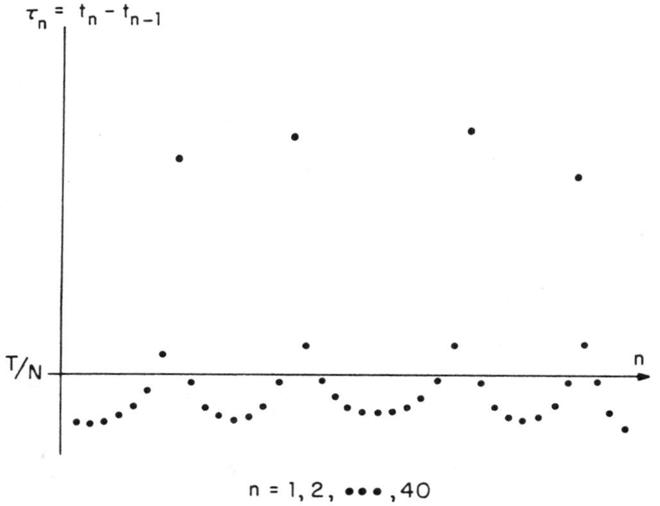

Figure 2. Plot of the consecutive time-spacings $\tau_n = t_n - t_{n-1}$ for n running from 1 to 40. The horizontal line indicates the average τ_n (= T/N = 6/40).

IV. NON-RELATIVISTIC QUANTUM MECHANICS

4.1. General Formulation

Extension of the above classical system of a point particle to quantum mechanics is straightforward, as shown in the following table.

TABLE 4

Non-relativistic Continuum Quantum Mechanics	Non-relativistic Discrete Quantum Mechanics
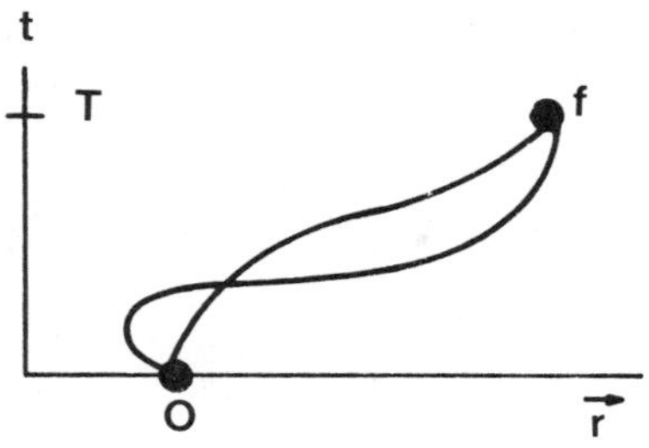	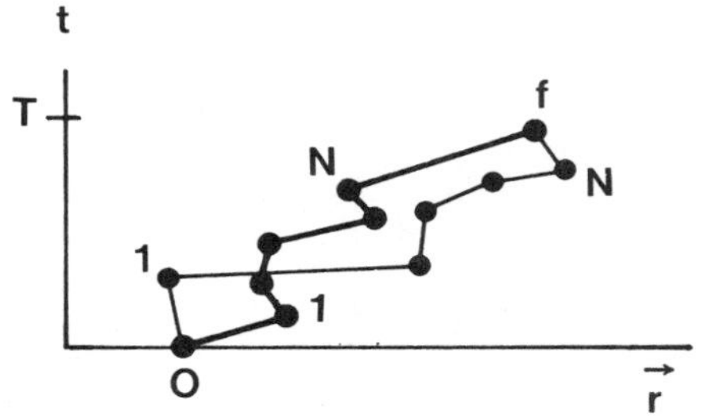
$H_0 = -\frac{1}{2}\nabla^2 + V \qquad (\hbar = m = 1)$	$\frac{N}{T} = \rho \equiv \frac{1}{\ell}$ = fundamental constant
$e^{-iH_0T} = \int e^{iA_c}\,[d\vec{r}(t)]$	$G(T) \equiv \int e^{iA_d}\, J \prod_n d^3r_n\, dt_n$
$e^{-H_0T} = \int e^{-\mathcal{A}_c}\,[d\vec{r}(t)]$	$\mathcal{G}(T) \equiv \int e^{-\mathcal{A}_d}\, J \prod_n d^3r_n\, dt_n$
where	where
$\begin{matrix} A_c \\ \mathcal{A}_c \end{matrix} = \int \left(\frac{1}{2}\dot{\vec{r}}^{\,2} \mp V\right) dt$	$\begin{matrix} A_d \\ \mathcal{A}_d \end{matrix} = \sum_n \frac{(\vec{r}_n - \vec{r}_{n-1})^2}{2\epsilon_n} \mp \frac{1}{2}\epsilon_n\left[V(r_n) + V(r_{n-1})\right]$
$[d\vec{r}(t)] \equiv \lim_{\substack{T = N\epsilon \\ \epsilon \to 0}} J \prod_n d^3r_n$	
$J \propto (1/\epsilon)^{3N/2}$	$J \propto \prod_n (1/\epsilon_n)^{-3/2}$

In the left column of Table 4 we give the usual continuum Feynman path integration formulation. For the Green's function e^{-iH_0T}, each path carries an amplitude e^{iA_c} where A_c is the continuum action. Because t is just a parameter, the integration is only over the continuous path $[d\vec{r}(t)]$, which is defined to be

$$\lim_{\substack{T=N\epsilon \\ \epsilon\to 0}} J \prod_n d^3 r_n$$

where

$$J = \left(\frac{1}{2i\pi\epsilon}\right)^{3N/2} \tag{4.1}$$

The limit $\epsilon \to 0$ is necessary because in Feynman's formulation the time interval T is divided into N equal intervals of ϵ each. Thus energy is not conserved if $\epsilon \neq 0$. For large T, the matrix element e^{-iH_0T} is oscillatory. It is more convenient to introduce the analytic continuation $T \to -iT$; in that case the Green's function becomes

$$e^{-H_0T} = \int e^{-\mathcal{A}_c} [d\vec{r}(t)] \tag{4.2}$$

where $\mathcal{A}_c$ is the energy integrated over the path $\vec{r}(t)$:

$$\mathcal{A}_c = \int \left(\tfrac{1}{2}\dot{\vec{r}}^{\,2} + V\right) dt \quad . \tag{4.3}$$

The new Green's function (4.2) is identical to that used in statistical physics and is well-behaved at large T.

In the discrete version, the continuum Green's function e^{-iH_0T} is replaced by G(T) and e^{-H_0T} by $\mathcal{G}(T)$, both of which are given in Table 4. Because t_n are now dynamical variables just like $\vec{r}_n$, the path integrations are over

$$\prod_n d^3 r_n \, dt_n$$

where n varies from 1 to N with

$$\frac{N}{T} = \rho \equiv \frac{1}{\ell} = \text{fundamental constant.} \tag{4.4}$$

For clarity, we first let each t_n vary independently from 0 to T. It is convenient to denote their spacings by a set of positive time intervals $\epsilon_1, \epsilon_2, \cdots, \epsilon_N$ so that through a permutation the t_n become chronologically ordered

$$t_1, t_2, t_3, \cdots \rightarrow \epsilon_1, \epsilon_1 + \epsilon_2, \epsilon_1 + \epsilon_2 + \epsilon_3, \cdots . \tag{4.5}$$

In the $(\vec{r}, t)$ plane, the consecutive points that define the discrete path become

$$(\vec{r}_1, \epsilon_1), (\vec{r}_2, \epsilon_1 + \epsilon_2), (\vec{r}_3, \epsilon_1 + \epsilon_2 + \epsilon_3), \cdots . \tag{4.6}$$

Each discrete path carries an amplitude e^{iA_d} or $e^{-\mathcal{A}_d}$ given in Table 4; the path of stationary phase corresponds to the one described by classical discrete mechanics. In the continuum case it is well known that the amplitudes e^{iA_c} and $e^{-\mathcal{A}_c}$ are related to the transformation matrix elements

$$\prod_t \langle \vec{r}(t + dt) \mid \vec{r}(t) \rangle , \tag{4.7}$$

which describes the fact that the operator $\vec{r}_{op}(t)$ at time t does not commute with $\vec{r}_{op}(t + dt)$. Here the discrete amplitudes e^{iA_d} and $e^{-\mathcal{A}_d}$ can be viewed as representing the corresponding product of matrix elements

$$\prod_n \langle \vec{r}_{n+1}, t_{n+1} \mid \vec{r}_n, t_n \rangle \tag{4.8}$$

where $t_1, t_2, \cdots$ are assumed to be arranged in a chronological order through the permutation (4.5).

The discreteness in the new mechanics refers to the discrete number of measurements. Each measurement determines the position of the particle $\vec{r}_n$ and the time t_n. Both $\vec{r}_n$ and t_n can take on any of the continuous eigenvalues of the operators $(\vec{r}_n)_{op}$ and $(t_n)_{op}$, which however do not commute with $(\vec{r}_{n+1})_{op}$ and $(t_{n+1})_{op}$, as indicated by (4.8).

In discrete mechanics there is no Hamiltonian or Lagrangian, but only action. What is the physical meaning of these Green's functions? As we shall establish in the next section, for $T \gg \ell$, if we neglect terms $O(e^{-T/\ell})$ or $O(e^{-iT/\ell})$ but keep $O(\ell/T)$, $O(\ell^2/T^2)$, etc., then*

$$G(T) \sim e^{-iHT}$$

and

$$\mathcal{G}(T) \sim e^{-HT} \tag{4.9}$$

where

$$H = H^\dagger = H_0 + \frac{1}{8}\ell^2\left[\nabla^2 - V,\ [\nabla^2, V]\right] + O(\ell^3) \tag{4.10}$$

and

$$H_0 = -\tfrac{1}{2}\nabla^2 + V \tag{4.11}$$

is the Hamiltonian in the continuum theory.

According to (4.9), $G(T)$ and $\mathcal{G}(T)$ are the Green's functions of a "Schroedinger" equation

$$-\frac{1}{i}\,\frac{\partial\psi}{\partial T} = H\psi \tag{4.12}$$

or its analytic continuation

$$-\frac{\partial\psi}{\partial T} = H\psi \tag{4.13}$$

where the Hermitian $H = H^\dagger$ plays the role of an effective Hamiltonian. Consequently, the usual physical interpretation of the continuum quantum mechanics can be carried over to the discrete version. For example, the eigenvalues of H are the energy levels of the system and the limit $T \to \infty$ of e^{-iHT} is the S-matrix, which is unitary.

It is not difficult to carry out high-order corrections in ℓ. The details will be given in the next section. For a free particle $V = 0$, the effective Hamiltonian H is identical to H_0; this is then the quantum mechanical version of the classical result: the motion of a free particle is unaltered when the

---

* The precise mathematical meaning of $\sim$ is given by (4.39) and (4.59) below.

continuum mechanics is replaced by the discrete mechanics.

In the usual continuum quantum mechanics, at any time t we have the following commutation relation between the position operator $x_{op}(t)$ and its conjugate momentum operator $p_{op}(t)$:

$$[p_{op}(t),\ x_{op}(t)] = -i\hbar \ , \tag{4.14}$$

which gives the familiar uncertainty relation

$$\Delta p(t) \cdot \Delta x(t) \geqslant \frac{\hbar}{2} \ . \tag{4.15}$$

Since t is a continuous parameter, as t varies we have an <u>infinite</u> number of (4.15). On the other hand, although the commutator between t and its derivative operator $i\hbar\frac{\partial}{\partial t}$ is also $-i\hbar$, because there is only one t we have just <u>one</u> uncertainty relation of the type

$$\Delta E \cdot \Delta t \geqslant \frac{\hbar}{2} \ . \tag{4.16}$$

This asymmetry is removed in discrete mechanics. There are N relations each of

$$\begin{aligned} &\Delta p_n \cdot \Delta x_n \geqslant \frac{\hbar}{2} \\ \text{and} \quad &\Delta E_n \cdot \Delta t_n \geqslant \frac{\hbar}{2} \end{aligned} \tag{4.17}$$

where p_n and E_n refer to the eigenvalues of $-i\hbar\,\partial/\partial x_n$ and $i\hbar\,\partial/\partial t_n$. The eigenvalue of the effective Hamiltonian H, given by (4.10), represents the coarse grain average of E_n.

4.2. Effective Hamiltonian

We begin with the discrete action $\mathcal{A}_d$ given in Table 4:

$$\mathcal{A}_d = \sum_{n=1}^{N+1} \left\{ \frac{(\vec{r}_n - \vec{r}_{n-1})^2}{2\epsilon_n} + \frac{\epsilon_n}{2}\left[V(\vec{r}_n) + V(\vec{r}_{n-1})\right] \right\} \tag{4.18}$$

where $\vec{r}_{N+1} = \vec{r}_f$, ϵ_n are real > 0, and satisfy

$$\sum_{n=1}^{N+1} \epsilon_n = T \ . \tag{4.19}$$

In the $(\vec{r}, t)$ space, the N discrete points that define the path are given by (4.6). As before, let the original t-coordinates of these N points be $t_1, t_2, \cdots, t_n$, each of which varies independently between 0 and T. The two sequences $\epsilon_1, \epsilon_1 + \epsilon_2, \cdots$ and $t_1, t_2, \cdots$ are related by the permutation (4.5).

For a given N, the Green's function $\mathcal{G}_N(T)$ is given by *

$$\mathcal{G}_N(T) \equiv \int e^{-\mathcal{A}_d} \, J \prod_{n=1}^{N} d^3 r_n \, dt_n \tag{4.20}$$

where

$$J = \prod_{n=1}^{N+1} \left(\frac{1}{2\pi \epsilon_n} \right)^{3/2} \ . \tag{4.21}$$

Instead of assuming a definite N, we shall keep only T fixed, but sum over N with a probability distribution proportional to

$$\frac{1}{N!} \left(\frac{1}{\ell}\right)^N e^{-T/\ell} \ . \tag{4.22}$$

This technique is familiar in statistical mechanics, corresponding to the transition of a canonical ensemble to a grand canonical ensemble.

In place of the matrix element $\langle \vec{r}_f \,|\, e^{-TH_0} \,|\, \vec{r}_0 \rangle$ in the continuum theory, the Green's function in the discrete mechanics is now given by the following sum over all discrete paths leading from $\vec{r}_0$ at $t = 0$ to $\vec{r}_{N+1} = \vec{r}_f$ at $t = T$:

$$\langle \vec{r}_f \,|\, \mathcal{G}(T, \ell) \,|\, \vec{r}_0 \rangle \equiv \sum_{N=0}^{\infty} \frac{1}{N!} \left(\frac{1}{\ell}\right)^N e^{-T/\ell} \, \mathcal{G}_N(T)$$

$$= \sum_{N=0}^{\infty} \frac{1}{N!} \left(\frac{1}{\ell}\right)^N e^{-T/\ell} \int J \prod_{n=1}^{N} d\tau_n \, d^3 r_n \, e^{-\mathcal{A}_d} \tag{4.23}$$

* We add a subscript N to the Green's function defined in Table 4.

where ℓ^{-1} plays the same role as the fugacity in a grand partition function in statistical mechanics.

Recalling that

$$\langle \vec{r}' \,|\, e^{\frac{1}{2}\epsilon\nabla^2} \,|\, \vec{r} \rangle = \left(\frac{1}{2\pi\epsilon}\right)^{3/2} e^{-\frac{1}{2\epsilon}(\vec{r}' - \vec{r})^2} , \tag{4.24}$$

we see that the matrix elements of

$$U(\epsilon) \equiv e^{-\frac{1}{2}\epsilon V} e^{\frac{1}{2}\epsilon\nabla^2} e^{-\frac{1}{2}\epsilon V} \tag{4.25}$$

are

$$\langle \vec{r}' \,|\, U(\epsilon) \,|\, \vec{r} \rangle = \left(\frac{1}{2\pi\epsilon}\right)^{3/2} e^{-\frac{1}{2\epsilon}(\vec{r}' - \vec{r})^2 - \frac{1}{2}\epsilon\left[V(\vec{r}') + V(\vec{r})\right]} . \tag{4.26}$$

Hence, (4.23) becomes

$$\langle \vec{r}_f \,|\, \mathcal{G}(T, \ell) \,|\, \vec{r}_0 \rangle = \sum_{N=0}^{\infty} \frac{1}{N!} \left(\frac{1}{\ell}\right)^N e^{-T/\ell} \int \prod_{n=1}^{N} dt_n \cdot \langle \vec{r}_f \,|\, U(\epsilon_{N+1})\, U(\epsilon_N) \cdots U(\epsilon_1) \,|\, \vec{r}_0 \rangle . \tag{4.27}$$

Because each t_n varies independently from 0 to T, we have

$$\int \prod_{n=1}^{N} dt_n = T^N .$$

The time-spacings ϵ_n are real and positive, related to the t_n by (4.5). Thus $\int \prod_{n=1}^{N} d\epsilon_n = T^N/N!$. Hence,

$$\int \prod_n dt_n \cdots = N! \int \prod d\epsilon_n \cdots$$

and (4.27) can be written as

$$\mathcal{G}(T, \ell) = \sum_{N=0}^{\infty} \left(\frac{1}{\ell}\right)^N e^{-T/\ell} \int \prod_{n=1}^{N} d\epsilon_n\, U(\epsilon_{N+1})\, U(\epsilon_N) \cdots U(\epsilon_1) . \tag{4.28}$$

We note that for a free particle $V = 0$, the continuum Hamiltonian, given by (4.11) is $H_0 = -\frac{1}{2}\nabla^2$; hence, $U(\epsilon) = e^{\frac{1}{2}\epsilon\nabla^2}$ and therefore $\mathcal{G}(T, \ell) = e^{-TH_0}$, the same as the continuum case.

For a general V , we multiply the right hand side of (4.28) by

$$1 = \int d\epsilon_{N+1}\, \delta\Big(\sum_{1}^{N+1} \epsilon_n - T\Big) = \int_{-i\infty}^{i\infty} \frac{dz}{2\pi i} \int_0^{\infty} d\epsilon_{N+1}\, e^{Tz - \sum_1^{N+1} \epsilon_n z} \qquad (4.29)$$

which takes care of the constraint (4.19), and write $\exp(-T/\ell)$ as $\exp(-\sum_1^{N+1} \epsilon_n/\ell)$. Hence,

$$\mathcal{G}(T, \ell) = \int_{-i\infty}^{i\infty} \frac{dz}{2\pi i}\, e^{Tz} \sum_{N=0}^{\infty} \Big(\frac{M(z)}{\ell}\Big)^N M(z) \qquad (4.30)$$

where

$$M(z) = \int_0^{\infty} d\epsilon\, e^{-\epsilon(\frac{1}{\ell}+z)}\, U(\epsilon) \quad . \qquad (4.31)$$

Defining

$$\overline{H}(z) \equiv M^{-1}(z) - z - \frac{1}{\ell} \quad , \qquad (4.32)$$

we have

$$\mathcal{G}(T, \ell) = \int_{-i\infty}^{i\infty} \frac{dz}{2\pi i} \frac{e^{Tz}}{z + \overline{H}(z)} \quad . \qquad (4.33)$$

Anticipating that the average ϵ_n is $O(\ell)$ which is small, we expand the deviation of $U(\epsilon_n)$ from $e^{-\epsilon_n H_0}$ in powers of ϵ_n . Using (4.11) and (4.25), and writing

$$K \equiv -\tfrac{1}{2}\nabla^2 \quad , \qquad H_0 = K + V$$

and

$$U(\epsilon) = e^{-\frac{1}{2}\epsilon V}\, e^{-\epsilon K}\, e^{-\frac{1}{2}\epsilon V} \quad , \qquad (4.34)$$

we find that the expansion

$$U(\epsilon) - e^{-\epsilon H_0} = \sum_1^{\infty} \frac{\epsilon^n}{n!} h_n \qquad (4.35)$$

is given by

$$h_1 = h_2 = 0 \quad ,$$

$$h_3 = -\tfrac{1}{2}\Big[K + \frac{V}{2}, [K, V]\Big] \quad , \qquad (4.36)$$

$$h_4 = \tfrac{1}{2}\big[K^2 + (K+V)^2, [K,V]\big] + \tfrac{1}{2}\big[K, [K, V^2]\big] - [K,V]^2 \quad ,$$

etc. Combining (4.31), (4.32) and (4.35), we obtain the power series of $\overline{H}(z)$

$$\overline{H}(z) = H_0 + \ell^2\,\overline{H}_2(z) + \ell^3\,\overline{H}_3(z) + \ell^4\,\overline{H}_4(z) + \cdots \tag{4.37}$$

where

$$\begin{aligned}\overline{H}_2(z) &= -h_3 \ ,\\ \overline{H}_3(z) &= 2z\,h_3 - h_4 - h_3H_0 - H_0h_3 \ ,\\ \overline{H}_4(z) &= -3z^2h_3 + 3z(h_4 + h_3H_0 + H_0h_3)\\ &\qquad - (h_5 + H_0h_4 + h_4H_0 + H_0h_3H_0) \ .\end{aligned} \tag{4.38}$$

Substituting (4.37) into (4.33), we shall establish the following asymptotic expression for $\mathcal{G}(T, \ell)$ when T is large:

<u>Theorem</u> $$\mathcal{G}(T, \ell) = \xi^\dagger e^{-TH}\,\xi + O(e^{-T/\ell}) \ , \tag{4.39}$$

where in powers of ℓ

$$\xi = 1 - \ell^3h_3 - 3\,\ell^4(H_0h_3 + h_3H_0 + \tfrac{1}{2}h_4) + O(\ell^5) = \xi^\dagger \tag{4.40}$$

and

$$\begin{aligned}H = H_0 &- \ell^2h_3 - \ell^3(h_4 + 2h_3H_0 + 2H_0h_3)\\ &-\ell^4(h_5 + \tfrac{5}{2}h_4H_0 + \tfrac{5}{2}H_0h_4 + 3h_3H_0^2 + 3H_0^2h_3 + 4H_0h_3H_0)\\ &+ O(\ell^5)\end{aligned} \tag{4.41}$$

with $H_0 = -\frac{1}{2}\nabla^2 + V$ as before, and h_3, h_4, $\cdots$ given by (4.35) and (4.36).

<u>Proof</u> Let us assume that the potential $V(\vec{r})$ is bounded from below. Without any loss of generality, we may set* $V(\vec{r}) \geqslant 0$.

<u>Lemma</u> In the complex z-plane, if $\mathrm{Re}\, z > 0$ the determinant $|\, z + \overline{H}(z)\, | \neq 0$.

---

*A shift $V(\vec{r}) \to V(\vec{r}) + v$ where v is a constant leads, in accordance with (4.25)-(4.27), to $U(\epsilon) \to U(\epsilon)\, e^{-\epsilon v}$ and $\mathcal{G}(T, \ell) \to \mathcal{G}(T, \ell)\, e^{-Tv}$.

Take any normalized state vector ψ. We have $0 < \psi^\dagger e^{-\epsilon K} \psi \leqslant 1$ where $K = -\frac{1}{2}\nabla^2$, as before. Thus, the expectation value of $U(\epsilon) = e^{-\frac{1}{2}\epsilon V} e^{-\epsilon K} e^{-\frac{1}{2}\epsilon V}$ has the same bounds: $0 < \psi^\dagger U(\epsilon)\psi \leqslant 1$. From (4.31) we see that

$$\psi^\dagger M(z)\psi = \int_0^\infty d\epsilon\, e^{-\epsilon(\frac{1}{\ell}+z)} \psi^\dagger U(\epsilon)\psi \ ;$$

its magnitude satisifes

$$\begin{aligned} |\psi^\dagger M(z)\psi| &\leqslant \int_0^\infty d\epsilon\, |e^{-\epsilon(\frac{1}{\ell}+z)} \psi^\dagger U(\epsilon)\psi| \\ &= \int_0^\infty d\epsilon\, e^{-\epsilon(\frac{1}{\ell}+\mathrm{Re}\, z)} |\psi^\dagger U(\epsilon)\psi| \\ &\leqslant \frac{\ell}{1+\ell\, \mathrm{Re}\, z} \ . \end{aligned} \tag{4.42}$$

Hence when $\mathrm{Re}\, z > 0$, we find

$$|\psi^\dagger M(z)\psi| < \ell \ ,$$

which implies that the magnitude of every eigenvalue of $M(z)$ is less than ℓ, and therefore that of $M^{-1}(z)$ is greater than ℓ^{-1}; consequently the determinant $|M^{-1}(z) - \ell^{-1}| \neq 0$. From (4.32) we have $z + \overline{H}(z) = M^{-1}(z) - \ell^{-1}$ and the lemma is proved.

We now turn to the proof of the theorem.

In the complex z-plane, the roots of $|z + \overline{H}(z)| = 0$ all lie on the half plane $\mathrm{Re}\, z \leqslant 0$. For small ℓ, those roots that are near $|z| = O(\ell^{-1})$ or larger give an $\leqslant O(e^{-T/\ell})$ contribution to $\mathcal{G}(T, \ell)$; they can be ignored when $T \to \infty$. To find the roots that are $O(1)$, we can expand $\overline{H}(z)$ in powers of ℓ.

(i) Neglecting $O(\ell^3)$, (4.37) becomes

$$\overline{H}(z) = H_0 - \ell^2 h_3 \tag{4.43}$$

which is independent of z. Hence, in accordance with (4.33),

$$\mathscr{G}(T, \ell) = \int_{-i\infty}^{i\infty} \frac{dz}{2\pi i} \frac{e^{Tz}}{z + H_0 - \ell^2 h_3} = e^{-TH}$$

where

$$H = H_0 - \ell^2 h_3 + O(\ell^3)$$

in agreement with (4.41).

(ii) Neglecting $O(\ell^4)$, $\overline{H}(z)$ is a linear function of z:

$$\overline{H}(z) = \alpha + \beta z \tag{4.44}$$

where

$$\alpha = H_0 - \ell^2 h_3 - \ell^3 (h_4 + h_3 H_0 + H_0 h_3) + O(\ell^4) \tag{4.45}$$

and

$$\beta = 2\ell^3 h_3 + O(\ell^4) \ . \tag{4.46}$$

Thus,

$$z + \overline{H}(z) = (1+\beta)^{\frac{1}{2}} (\bar{\alpha} + z)(1+\beta)^{\frac{1}{2}} + O(\ell^4) \tag{4.47}$$

where

$$\bar{\alpha} = (1+\beta)^{-\frac{1}{2}} \alpha (1+\beta)^{-\frac{1}{2}} \ . \tag{4.48}$$

Substituting (4.47) into (4.33), we obtain

$$\mathscr{G}(T, \ell) = \xi^{\dagger} e^{-TH} \xi$$

where

$$\xi = (1+\beta)^{-\frac{1}{2}} = 1 - \ell^3 h_3 + O(\ell^4)$$

and

$$H = \bar{\alpha} + O(\ell^4) = H_0 - \ell^2 h_3 - \ell^3 (h_4 + 2h_3 H_0 + 2H_0 h_3) + O(\ell^4)$$

in agreement with (4.40) and (4.41).

Similar considerations can be extended to higher order corrections in ℓ.

Comments Consider a time interval $T \gg \ell$. Neglect $O(e^{-T/\ell})$ but keep $O(\ell/T)$, $O(\ell^2/T^2)$, $O(\ell^3/T^3)$, $\cdots$. According to (4.39)

$$\mathscr{G}(T, \ell) = \xi^{\dagger} e^{-TH} \xi \ .$$

Analytically continue $T \to iT$ to the "real" time, we have $\mathscr{G}(T, \ell) \to G(T, \ell)$

where $$G(T, \ell) = \xi^{\dagger} e^{-iTH} \xi \quad . \tag{4.49}$$

Regard H as the (effective) Hamiltonian. The corresponding Schroedinger equation with T as the time is

$$-\frac{1}{i}\frac{\partial \psi}{\partial t} = H\psi \quad . \tag{4.50}$$

Given $\psi = \psi(0)$ at $T = 0$, we find

$$\psi(T) = e^{-iTH}\psi(0) \quad . \tag{4.51}$$

The transition amplitude from $\psi(0) = \psi_a$ to

$$\psi(T) = \psi_b$$

is

$$M_{ba}(T) = \psi_b^{\dagger} e^{-iHt} \psi_a \quad . \tag{4.52}$$

Define

$$\psi(T) = \xi\,\phi(T) \,, \quad \text{or} \quad \phi(T) = \xi^{-1}\psi(T) \tag{4.53}$$

then

$$-\frac{1}{i}\frac{\partial \phi}{\partial T} = \mathcal{H}\phi \tag{4.54}$$

where

$$\mathcal{H} = \xi^{-1} H \xi \quad . \tag{4.55}$$

The same transition amplitude $M_{ba}(T)$ can also be written as

$$M_{ba}(T) = \phi_b^{\dagger}\, G(T, \ell)\, \phi_a \tag{4.56}$$

where, in accordance with (4.53),

$$\psi_b = \xi\,\phi_b \quad \text{and} \quad \psi_a = \xi\,\phi_a \tag{4.57}$$

and $G(T, \ell)$ is given by (4.49).

The ϕ's are state-vectors in a Hilbert space whose metric is

$$g = \xi^\dagger \xi$$

which is positive. From (4.53) we see that

$$\frac{\partial}{\partial T} (\phi^\dagger g \phi) = 0$$

for $\phi = \phi(T)$ satisfying (4.54). Under the transformation

$$\phi \to \psi = \xi \phi \tag{4.58}$$

we have the metric

$$g = \xi^\dagger \xi \to g = 1$$

and correspondingly

$$G(T, \ell) \to e^{-iTH} \tag{4.59}$$

which is the Green's function of the Schroedinger equation $H\psi = i\ \partial\psi/\partial T$.

4.3. Applications

When we make the transition from continuum to discrete, it is of interest to study the lowest order correction to the energy levels in the usual continuum theory. Let $|>$ be an eigenstate of the continuum Hamiltonian $H_0 = -\frac{1}{2m}\nabla^2 + V$:

$$H_0 \,|> = E_0 \,|> . \tag{4.60}$$

From (4.36) and (4.41), we have

$$H = H_0 + \tfrac{1}{2}\ell^2 \left[K + \frac{V}{2}, \ [K, V]\right] + O(\ell^3)$$

where $K = -\frac{1}{2m}\nabla^2$ is the usual kinetic energy operator. Because $K + \frac{V}{2} = H_0 - \frac{V}{2}$ and, on account of (4.60),

$$< | \left[H_0, [K, V]\right] | > = 0 . \tag{4.61}$$

To $O(\ell^2)$ the corresponding energy in the discrete quantum mechanics for a non-relativistic particle of mass m moving in a potential V is given by ($\hbar = 1$)

$$E = \langle | H | \rangle + O(\ell^3)$$
$$= E_0 - \frac{1}{4m} \ell^2 \langle | (\nabla V)^2 | \rangle + O(\ell^3) \ . \qquad (4.62)$$

Examples (i) In the case of a one-dimensional harmonic oscillator $V = \frac{1}{2} m \omega^2 x^2$, $\nabla V = m \omega^2 x$ we have $E_0 = (n + \frac{1}{2}) \omega$ and

$$E = (1 - \tfrac{1}{4} \ell^2 \omega^2)(n + \tfrac{1}{2}) \omega + O(\ell^3) \qquad (4.63)$$

where $n = 0, 1, 2, \cdots$.

(ii) Let V be the Coulomb potential experienced by an electron of mass m_e at a distance r from a nucleus of charge Z distributed uniformly within a sphere of radius R :

$$V = \begin{cases} -Z\alpha / r & \text{for } r \geqslant R \\ -\frac{1}{2} Z\alpha \left[3 - (\frac{r}{R})^2 \right] / R & \text{for } r \leqslant R \end{cases} \qquad (4.64)$$

where $\alpha \cong 1/137$ is the fine structure constant. The additional energy shift ΔE in the discrete mechanics may be estimated by using the non-relativistic wave function $\psi(\vec{r}) = \langle \vec{r} | \rangle$. We find

$$\Delta E = - \frac{\ell^2}{4m_e} \langle | (\nabla V)^2 | \rangle \cong - \frac{6\pi}{5m_e R} | Z\alpha \ell \psi(0) |^2 . \qquad (4.65)$$

For the 2s orbit in hydrogen

$$| \psi(0) |^2 = (8\pi a^3)^{-1} \qquad (4.66)$$

where $a = (m_e \alpha)^{-1}$ is the Bohr radius. Setting the agreement between the experimental observation and the theoretical QED calculation of this level to an accuracy ϵ Rydberg, we obtain

$$\ell m_e < \left[\frac{10}{3} (137)^3 \epsilon m_e R \right]^{\frac{1}{2}} \qquad (4.67)$$

which, for $R \sim .8$ fermi and[7] $\epsilon \sim 10^{-11}$, gives

$$\ell < 1.6 \times 10^{-14} \text{ cm} . \tag{4.68}$$

Better limits ($\ell < 10^{-16}$ cm) can be set by using results from the hyperfine splitting,[8] e^+e^- collisions[9] and the forward dispersion relations[10] in πp.

4.4. Jacobian and Supersymmetry

Let us consider a free particle of unit mass. From (4.18) and (4.20) we see that

$$<\vec{r}_f | \mathcal{J}_N | \vec{r}_0> = \int \prod_1^N d^3r_n \, dt_n \, J \, e^{-<\vec{r}_f | \mathcal{A}_d | \vec{r}_0>} \tag{4.69}$$

where

$$J = \prod_1^{N+1} \left(\frac{1}{2\pi \epsilon_n}\right)^{3/2} \tag{4.70}$$

and

$$<\vec{r}_f | \mathcal{A}_d | \vec{r}_0> = \sum_1^{N+1} \frac{1}{2\epsilon_n} (\vec{r}_n - \vec{r}_{n-1})^2 \tag{4.71}$$

with $\vec{r}_{N+1} = \vec{r}_f$ and $\sum_1^{N+1} \epsilon_n = T$.

The Jacobian J is also related to $\mathcal{A}_d$ by

$$J = \left(\frac{1}{2\pi T}\right)^{3/2} \Big/ \int \prod_1^N d^3r_n \, e^{-<0 | \mathcal{A}_d | 0>} \tag{4.72}$$

in which $<0 | \mathcal{A}_d | 0>$ is the same matrix element (4.71) but with $\vec{r}_0 = \vec{r}_f = 0$. Thus, apart from an overall multiplicative factor $(2\pi T)^{-3/2}$, the dependence of J on ϵ_n is entirely contained in

$$\left(\int \prod_1^N d^3r_n \, e^{-<0 | \mathcal{A}_d | 0>} \right)^{-1} . \tag{4.73}$$

The appearance of J is reminiscent of a similar circumstance in statistical mechanics. It was known in classical statistical mechanics that if the ensemble average is over only the coordinate q_n-space, then there is an additional Jacobian in the integral

$$\int J \prod dq_n \cdots \quad . \tag{4.74}$$

On the other hand, if the integration is over the space of both the coordinate q_n and its conjugate momentum p_n then the corresponding Jacobian is 1 ; (4.74) becomes simply

$$\int \prod dq_n dp_n \cdots \quad . \tag{4.75}$$

The true significance was not fully understood until quantum mechanics. Here, the appearance of a multiplicative factor in the integration suggests that we should consider the integration to be performed in a bigger space in which the multiplicative factor is a constant. Only by viewing from the bigger space can we justify the label "Jacobian" for J in the sub-space.

Let us stack the components (x_n, y_n, z_n) of all $\vec{r}_n$ together forming a $\mathcal{N} = 3N$ component column vector:

$$\phi = \begin{pmatrix} x_1 \\ y_1 \\ z_1 \\ \vdots \\ z_N \end{pmatrix} \quad . \tag{4.76}$$

The matrix element $< 0 \mid \mathcal{A}_d \mid 0 >$ can be written as

$$< 0 \mid \mathcal{A}_d \mid 0 > \; = \; \tfrac{1}{2} \tilde{\phi} M \phi \tag{4.77}$$

where $\tilde{\phi}$ is the transpose of ϕ and M is an $\mathcal{N} \times \mathcal{N}$ real symmetric matrix. We introduce $2\mathcal{N}$ elements of a Grassmann algebra: $z_1, \cdots, z_{\mathcal{N}}$ and $\bar{z}_1, \cdots, \bar{z}_{\mathcal{N}}$ which satisfy

$$\{ z_i, z_j \} = \{ z_i, \bar{z}_j \} = \{ \bar{z}_i, \bar{z}_j \} = 0 \tag{4.78}$$

for $i, j = 1, \cdots, \mathcal{N}$. Define

$$< 0 \mid \mathcal{A} \mid 0 > \; \equiv \; \tfrac{1}{2} \tilde{\phi} M \phi + \bar{\psi} M^{\frac{1}{2}} \psi \; = \; < 0 \mid \mathcal{A}_d \mid 0 > + \bar{\psi} M^{\frac{1}{2}} \psi \tag{4.79}$$

where

$$\psi = \begin{pmatrix} z_1 \\ \vdots \\ z_{\mathcal{N}} \end{pmatrix} \quad \text{and} \quad \overline{\psi} = (\bar{z}_1 \cdots \bar{z}_{\mathcal{N}}) \ . \tag{4.80}$$

The function $\mathcal{A}$ belongs to a general class of supersymmetrical functions considered by Parisi and Sourlas.[11] It is invariant under the supersymmetric transformation

$$\begin{aligned} \delta\phi_i &= \overline{\psi}_i \, \varepsilon + \bar{\varepsilon} \, \psi_i \ , \\ \delta\psi &= -\varepsilon \, M^{\frac{1}{2}} \phi \\ \text{and} \qquad \delta\overline{\psi} &= -\tilde{\phi} \, M^{\frac{1}{2}} \bar{\varepsilon} \end{aligned} \tag{4.81}$$

where ε and $\bar{\varepsilon}$ are anticommuting parameters.

One can readily verify that

$$\int \prod_1^{\mathcal{N}} d\phi_i \, d\bar{z}_i \, dz_i \, e^{-<0|\mathcal{A}|0>} = (2\pi)^{\frac{1}{2}\mathcal{N}} \ . \tag{4.82}$$

Thus, apart from the factor $(2\pi)^{-\frac{1}{2}\mathcal{N}} (2\pi T)^{-3/2}$, (4.69) is identical to

$$\int \prod_n dt_n \, d^3r_n \prod_i d\bar{z}_i \, dz_i \, e^{-<\vec{r}_f|\mathcal{A}|\vec{r}_i>} \ , \tag{4.83}$$

where, similar to (4.79)

$$<\vec{r}_f|\mathcal{A}|\vec{r}_0> = <\vec{r}_f|\mathcal{A}_d|\vec{r}_0> + \overline{\psi} M^{\frac{1}{2}} \psi \ ; \tag{4.84}$$

i.e.,

$$<\vec{r}_f|\mathcal{G}_N|\vec{r}_0> = (2\pi)^{-\frac{1}{2}\mathcal{N}} (2\pi T)^{-3/2} \int \prod_n dt_n \, d^3r_n \prod_i d\bar{z}_i \, dz_i \, e^{-<\vec{r}_f|\mathcal{A}|\vec{r}_0>} . \tag{4.85}$$

Similar considerations can also be extended to relativistic field theories. The simplifications derived strongly suggest that supersymmetry should play an important and perhaps more basic role in discrete mechanics.

V. SPIN 0 FIELD

5.1. General Discussion

As an example of the relativistic quantum field theory, we first discuss the case of a massless scalar field ϕ interacting with an arbitrary external current j. The comparison between the new discrete theory and the usual continuum formalism is given in Table 5.

TABLE 5

Continuum theory	Discrete theory
$e^{-TH_0} = \int e^{-A_c}\, J \prod_x d\phi(x)$	$\mathcal{G}(T) = \int e^{-A_d}\, J \prod_{i=1}^{N} d^4x_i\, d\phi_i$
$\phi(x)$ = dynamical variable	ϕ_i and x_i are both dynamical variables $(i = 1, 2, \cdots, N)$
x = 4-dimensional euclidean coordinate (parameter)	$\dfrac{N}{\text{volume}} = \rho = \left(\dfrac{1}{\ell}\right)^4$ = fundamental constant
$A_c = \int \left[\frac{1}{2}\left(\dfrac{\partial\phi}{\partial x_\mu}\right)^2 - j\phi\right] d^4x$	$A_d = \sum_{\ell_{ij}} \frac{1}{2}\lambda_{ij}(\phi_i - \phi_j)^2 - \sum_i j_i\phi_i$
Equation of motion: $-\dfrac{\partial^2\phi}{\partial x_\mu^2} = j(x)$	Equation of motion: $\sum_{j \text{ linked to } i} \lambda_{ij}(\phi_i - \phi_j) = j_i$
Laplace equation	Kirchoff law

For simplicity, we give here only the euclidean version. (See section VIII for the Minkowski version.) In the continuum theory, the field $\phi(x)$ is the dynamical variable; the space-time coordinate x is just a parameter. Hence, the path integration extends only to $d\phi(x)$. In contrast, the corresponding Green's function in the discrete theory consists of integrations over d^4x_i as well as $d\phi_i$. Because of the d^4x_i integration, it is possible to have both the 4-dimensional rotational symmetry and the translational invariance.

In the discrete formulation, given a 4-dimensional volume

$$\Omega = L^3 T ;$$

we postulate that there can be at most N measurements; each determines the space-time position x_i of the observation and the value of the field ϕ_i at x_i. The ratio

$$\rho = \frac{N}{\Omega} \equiv \left(\frac{1}{\ell}\right)^4 \tag{5.1}$$

is a fundamental constant of the theory. For a given set $\{x_i, \phi_i\}$ where $i = 1, 2, \cdots, N$, the action A_d is identical to that of a random lattice. In order to construct A_d, one must solve the following two problems:

(i) Given N points in a volume Ω, it is desirable to couple only nearby points, simulating the local character of the corresponding density of the continuum action. How can we do that?

(ii) Assume that an algorithm is given, so that only neighboring pairs of sites, say i and j, are coupled. Each pair gives a link ℓ_{ij} and contributes to the action a term $(\phi_i - \phi_j)^2$ multiplied by a weight-factor λ_{ij}. The sum over all links

$$\tfrac{1}{2} \sum_{\ell_{ij}} \lambda_{ij} (\phi_i - \phi_j)^2 \tag{5.2}$$

replaces the integral

$$\tfrac{1}{2} \int \left(\frac{\partial\phi}{\partial x_\mu}\right)^2 d^4x \tag{5.3}$$

in the continuum theory. Because different links ℓ_{ij} have different lengths and orientations, it is reasonable that they should carry different weights λ_{ij}. What would be the best choice for these weight functions?

We observe that if ϕ_i is identified as the "electric potential" and

$$\lambda_{ij}^{-1} = \text{electric resistance between } i \text{ and } j \ , \tag{5.4}$$

then the equation of motion

$$\sum_{\substack{j \text{ linked} \\ \text{to } i}} \lambda_{ij}(\phi_i - \phi_j) = j_i \tag{5.5}$$

is identical to the Kirchoff law of an electric circuit, with j_i as the external current entering at point i. In this analog problem, we would like to determine these resistances λ_{ij}^{-1} so that the electric potential ϕ_i gives the "best" discrete approximation to the corresponding continuum solution $\phi(x)$ of the Laplace equation

$$- \frac{\partial^2 \phi}{\partial x_\mu^2} = j(x) \ . \tag{5.6}$$

The answers to these two problems are given below.

5.2. Simplitial Decomposition in the Euclidean Space

Consider a D-dimensional euclidean space. Given an arbitrary distribution of N points (called lattice sites, or simply sites) in a finite volume Ω, our first task is to decompose Ω into non-overlapping simplices whose vertices are those sites. To avoid complications due to the boundary, we may assume Ω to be a D-dimensional rectangular volume with the standard periodic boundary condition. For $D = 2$, the simplices are triangles; for $D = 3$, they are tetrahedra. In general, a D-simplex consists of $D + 1$ vertices. Between any pair of vertices, we draw a straight line which forms a link. Hence; there are altogether

$$C_2^{D+1} = \tfrac{1}{2}(D+1)\ D$$

different links per simplex. Regarding the links as sides of triangles, we can form

$$C_3^{D+1} = \frac{1}{3!}(D+1)\ D(D-1)$$

different triangles per simplex. Likewise, since the triangles can be viewed as surfaces of tetrahedra, there are

$$C_4^{D+1} = \frac{1}{4!}(D+1)\ D(D-1)(D-2)$$

different tetrahedra per simplex; etc.

For our applications; we wish to draw links only between nearby sites. This is related to the problem of partitioning Ω into non-overlapping simplices. Let us introduce the following concept of clusters of neighboring sites: consider an arbitrary group of $D+1$ random lattice sites. They lie on the surface of a hypersphere in D dimensions, which will be referred to as their circumscribed sphere. If the inside of that sphere is free of all lattice sites, then this group of $D+1$ sites is called a cluster of neighbors. We form a D-simplex for every cluster, using its sites as vertices.

We note that for any $D+1$ infinitesimal volumes $d^D r_1$, $d^D r_2$, $\cdots$, $d^D r_{D+1}$ in Ω, the probability that each volume element should contain a lattice site is

$$\rho^{D+1} \prod_{i=1}^{D+1} d^D r_i \quad ;$$

the probability that there is no lattice point inside their circumscribed sphere is given by the Poisson formula

$$\exp(-v\rho) \quad , \tag{5.7}$$

where ρ is the site-density, v is the volume of the circumscribed sphere, related to its radius R by

$$v = \begin{cases} (2\pi)^{D/2}\ R^D/D!! & \text{if } D \text{ is even,} \\ 2(2\pi)^{(D-1)/2}\ R^D/D!! & \text{if } D \text{ is odd.} \end{cases}$$

Hence, the probability that $d^D r_1$, $d^D r_2$, $\cdots$, $d^D r_{D+1}$ should form a cluster and therefore be the vertices of a D-simplex is

$$\rho^{D+1} \prod_{i=1}^{D+1} d^D r_i \ \exp(-v\rho) \quad . \tag{5.8}$$

The factor $\exp(-v\rho)$ insures that the vertices of the simplex cannot be too far apart.

Let us examine, out of the N given sites, all C_{D+1}^{N} combinations of possible groupings of $D+1$ sites. Whenever a group fulfills the above condition of a cluster, a D-simplex is formed. The simplices thus constructed define our basic random lattice, which has the following properties:

Theorem (i) There is no overlap between the volumes of any two simplices, and (ii) the volume-sum of all simplices is Ω . [The proof of the theorem is given in the next section.]

In Figure 3, we give an example of a two-dimensional random lattice.

From (5.8), we can compute the various kinematic properties of the random lattice. For $n > m$, let

$$N_{n/m} \equiv \text{average number of } n\text{-simplices per } m\text{-simplex;}$$

i.e., each m-simplex is on the average shared by $N_{n/m}$ n-simplices. For example, each site (0-simplex) is shared by an average of $N_{1/0}$ links (1-simplices) and each link by $N_{2/1}$ triangles (2-simplices), etc. These averages are given in Table 6.

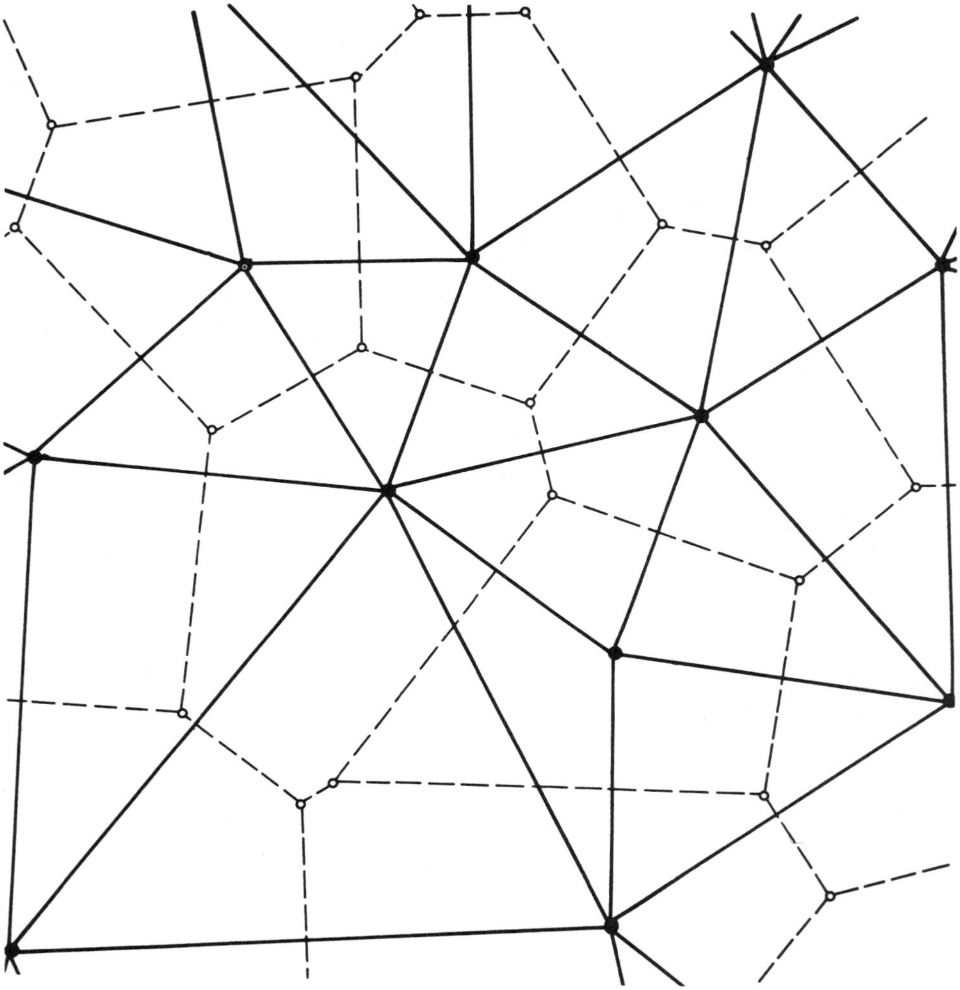

Figure 3. An example of a two-dimensional random lattice (solid lines) and its dual (dashed lines). Dots are the lattice sites, and open circles are the corners of the dual cells. [See section 5.5 for the definition of the dual lattice.]

TABLE 6

		D = 2	D = 3	D = 4
$N_{1/0}$	=	6	$2\left(1+\frac{24}{35}\pi^2\right)$	$\frac{340}{9}$
$N_{2/0}$	=	6	$\frac{144}{35}\pi^2$	$\frac{590}{3}$
$N_{3/0}$	=		$\frac{96}{35}\pi^2$	$\frac{2860}{9}$
$N_{4/0}$	=			$\frac{1430}{9}$
$N_{2/1}$	=	2	$\frac{144\pi^2}{35+24\pi^2}$	$\frac{177}{17}$
$N_{3/1}$	=		$\frac{144\pi^2}{35+24\pi^2}$	$\frac{429}{17}$
$N_{4/1}$	=			$\frac{286}{17}$
$N_{3/2}$	=		2	$\frac{286}{59}$
$N_{4/2}$	=			$\frac{286}{59}$
$N_{4/3}$	=			2

We note that

$$N_{D/(D-1)} = 2 ,$$

which is the consequence of the fact that in D-dimensions, each (D-1) - simplex is shared by two D-simplices. In addition from the same table we have

$$N_{D/(D-2)} = N_{(D-1)/(D-2)} = \begin{cases} 6 & \text{when } D = 2 \\ 5.2276 & \text{when } D = 3 \\ 4.8475 & \text{when } D = 4 \end{cases}$$

which becomes 4 when $D \to \infty$.

When $D = 4$,

$$N_{1/0} = \frac{340}{9} \cong 37.8$$

and (5.9)

$$N_{2/0} = \frac{590}{3} \cong 197 ,$$

whereas for a regular "cubic" lattice, $N_{1/0} = 8$ and $N_{2/0} = 24$.

5.3. Proof of the Basic Theorem on the Linking Algorithm

To prove the basic simplitial decomposition theorem stated in the previous section, start with N random lattice sites in a D-dimensional euclidean volume Ω with periodic boundary conditions. Let sites $\{ a_1 , a_2 , \cdots , a_{D+1} \}$ and $\{ b_1 , b_2 , \cdots , b_{D+1} \}$ form two clusters with S_a and S_b as their respective circumscribed hyperspheres. By definition, S_a passes through $a_1 , a_2 , \cdots , a_{D+1}$, and S_b through $b_1 , b_2 , \cdots , b_{D+1}$; furthermore, the insides of both S_a and S_b are free of any lattice sites. Hence, S_a cannot be completely inside S_b , nor vice versa. Consequently, S_a and S_b must either be totally separated from each other, which makes (i) an obvious conclusion, or intersect each other.* In the latter case, the intersection I

* The possibility that two different clusters share the same circumsphere (i.e., S_a and S_b are identical) is of zero measure, and is therefore ignored.

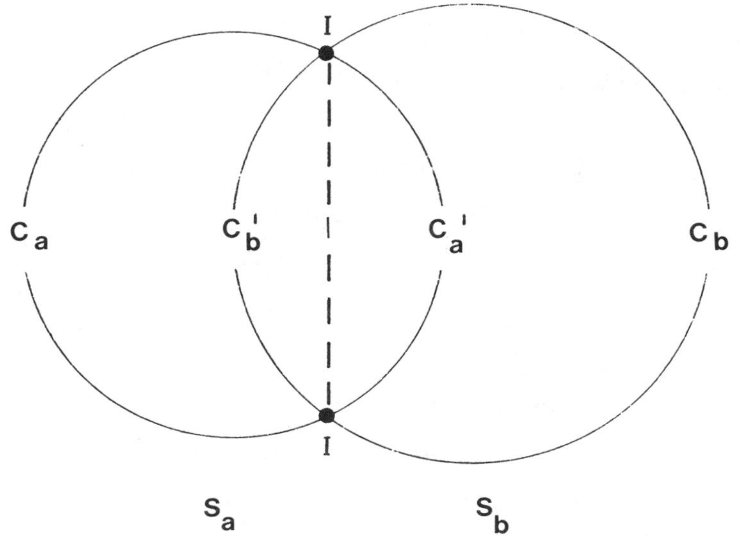

Figure 4. The intersection I of two hyperspheres S_a and S_b cuts each sphere S_i into two caps C_i and C_i' where $i = a$ or b.

of these surfaces S_a and S_b defines a $(D-2)$-dimensional sphere. As shown in Figure 4, I cuts S_a into two separate caps C_a and C_a' with C_a' lying entirely inside S_b. For definiteness, we exclude the border I from the caps C_a and C_a' so that the sets C_a, C_a' and I are disjoint. The sites in $\{a_i\}$ can be on C_a and I, but not on C_a'; otherwise, some of the a_i would be inside the other sphere S_b, in violation of the assumption. Likewise, I separates S_b into two caps C_b and C_b' with C_b' inside S_a. [Again I is not included in C_b and C_b'.] Hence, none of the b_i can be on C_b'. From $\{a_i\}$ and $\{b_i\}$ we now construct two D-simplices, S_a and S_b. These two simplices must lie on the opposite sides of the $(D-1)$-dimensional subspace that contains I. Proposition (i) now follows.

We note that S_a and S_b may share some common boundary. If the sets

$\{a_i\}$ and $\{b_i\}$ have a single site in common, then there is a common vertex between S_a and S_b; if there are two sites in common, then they share a common line, etc. The maximum contact between these two simplices is for $\{a_i\}$ and $\{b_i\}$ to have D sites in common; in that case, S_a and S_b share a common (D-1)-simplex, say s, whose vertices are all on I.

We now turn to the proof of (ii). Every D-simplex is bounded by (D-1)-simplices, each of which has two sides. We shall show that every such (D-1)-simplex is shared by two D-simplices, one on each side. [In two dimensions, this means that every link is shared by two triangles; in three dimensions, every triangle is shared by two tetrahedra, etc.] Assuming that this is not true, then there must be a D-simplex, say τ_a, which has on its surface a (D-1)-simplex, say s, that is not shared by any other D-simplices. Let S_a be the circumscribed sphere of τ_a, and I that of s. We denote by S_b any hypersphere that intersects S_a at I but, as shown in Figure 1, we require the centers of S_a and S_b to lie on opposite sides of s. Consider now the family of all such S_b. By assumption, none of the S_b can contain any lattice points except those on I (i.e., on s). This is clearly false, since S_b can be arbitrarily large and Ω is finite with a periodic boundary condition. [A convenient way to visualize this situation is to consider an array of identical volumes Ω, each containing exactly the same site distribution. These volumes are placed side by side to form an infinite periodic rectangular lattice whose unit cell is Ω. It is then easy to see that the radius of S_b can be arbitrarily large.]

From any D-simplex, we can go through its boundary and reach other D-simplices. By repeating this process, we must eventually cover a volume Ω' which is within Ω and has no boundary. Consequently, $\Omega' = \Omega$, and the proof of the theorem is complete.

A simple example of $D = 2$ and $N = 3$ is given in Figure 5.

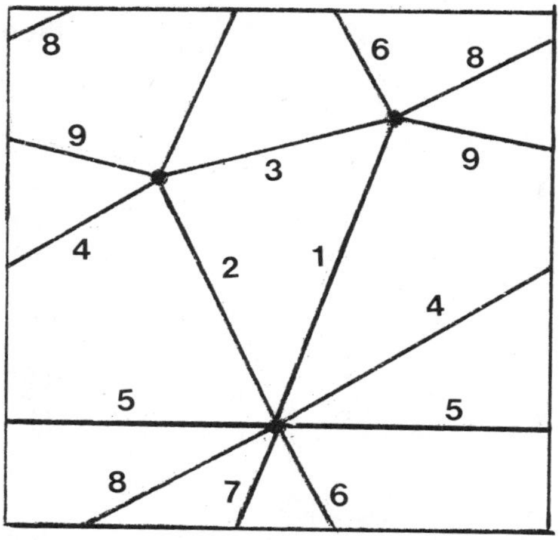

Figure 5. Ω is a two-dimensional square with a periodic boundary condition. There are three sites, nine links and six triangles. The links are marked 1 through 9 .

5.4. Generalization of the Theorem

Consider again the problem of N lattice sites randomly distributed in a D-dimensional "rectangular" euclidean volume Ω with periodic boundary conditions. Let $x_1, x_2, \cdots, x_D$ be the cartesian coordinates, and $x_\mu(a)$ the coordinates of the ath site with $\mu = 1, 2, \cdots, D$.

Recall the definition of a "cluster" given in section 5.2. For any group of $D+1$ sites, say $a_1, a_2, \cdots, a_{D+1}$, let S be the $D-1$ dimensional spherical surface (called circumscribed sphere) that passes through these $(D+1)$-sites. To construct S, we may follow a somewhat formal procedure which will however make our subsequent generalization easier. Take an arbitrary point c (which does not have to be a site). With c as the center, consider the sequence of all spherical surfaces $S_R(c)$ of different radii R. Analytically, $S_R(c)$ is represented by

$$\sum_{\mu=1}^{D} [x_\mu - x_\mu(c)]^2 = R^2 \tag{5.10}$$

where $x_\mu(c)$ is the μth component of the position vector of c. By varying R and c, and requiring the $S_R(c)$ to pass through all these $D+1$ sites $a_1, a_2, \cdots, a_{D+1}$, we can find the circumscribed sphere S. According to the definition given in section 5.2, if the interior of S is empty of lattice sites, then the set $\{a_1, a_2, \cdots, a_{D+1}\}$ is called a cluster (of neighbors).

This definition of clusters will now be generalized. Instead of the spherical surfaces $S_R(c)$, we consider the sequence of convex surfaces $\mathcal{S}_R(c)$ of a given shape (with respect to the fixed cartesian coordinate-axes), with c as the center and R as its linear scale. To be specific, we may introduce the standard spherical coordinates r, $\theta_1, \theta_2, \cdots, \theta_{D-2}, \phi$ by

$$\begin{aligned} x_1 - x_1(c) &= r\cos\theta_1 , \\ x_2 - x_2(c) &= r\sin\theta_1 \cos\theta_2 , \\ &\cdots \\ x_{D-1} - x_{D-1}(c) &= r\sin\theta_1 \sin\theta_2 \cdots \sin\theta_{D-2}\cos\phi \\ \text{and} \quad x_D - x_D(c) &= r\sin\theta_1 \sin\theta_2 \cdots \sin\theta_{D-2}\sin\phi \end{aligned} \tag{5.11}$$

where $\theta_1, \theta_2, \cdots, \theta_{D-2}$ vary independently from 0 to π and ϕ from $-\pi$ to π. Instead of (5.10) [which is $r = R$], $\mathcal{S}_R(c)$ is now represented by

$$r = f(\theta_1, \theta_2, \cdots, \theta_{D-2}, \phi)\; R \tag{5.12}$$

where f is any given convex function. The following are two examples in $D = 2$:

(i) $\mathcal{S}_R(c)$ are squares whose sides are parallel to the x_1 and x_2 axes. Equation (5.12) becomes

$$| x_+ - x_+(c) | + | x_- - x_-(c) | = R \tag{5.13}$$

where

$$x_{\pm} \equiv \frac{1}{\sqrt{2}} (x_1 \pm x_2)$$

and

$$x_{\pm}(c) = \frac{1}{\sqrt{2}} (x_1(c) \pm x_2(c)) \quad . \tag{5.14}$$

(ii) $\mathcal{S}_R(c)$ are ellipses given by

$$r^2 = (\cos^2 \phi + a^2 \sin^2 \phi)^{-1} R^2 \tag{5.15}$$

where a is a fixed constant,

$$x_1 - x_1(c) = r \cos \phi \qquad \text{and} \qquad x_2 - x_2(c) = r \sin \phi \quad .$$

As R changes, (5.15) gives a sequence of ellipses of varying sizes but the same shape.

Given a convex shape function $f(\theta_1, \theta_2, \cdots, \theta_{D-2}, \phi)$, for an arbitrary set of $D+1$ sites, say $a_1, a_2, \cdots, a_{D+1}$, if by varying c and R we can find a surface $\mathcal{S}_R(c)$ that passes through these $D+1$ sites, then that particular $\mathcal{S}_R(c)$ is called the circumscribed surface $\mathcal{S}_a$ of the set $\{a_i\}$. These $D+1$ sites form a cluster* (of neighbors) if, in addition, the interior of $\mathcal{S}_a$ is free of lattice sites. For our problem of N random sites in a volume Ω, we consider all possible groups of $D+1$ sites. Whenever a group is a cluster, we form a simplex using the sites as its vertices. The simplices thus formed satisfy the following theorem.

---

* It is possible that a set of $D+1$ sites $\{a_i\}$ may not have a circumscribed surface, in which case these sites do not form a cluster. For example, in two dimensions if $\mathcal{S}_R(c)$ are the squares given by (5.13) and $\{a_i\}$ consists of three sites all lying on, say, the straight line $x_1 = x_2$, then it is not possible to have a square $\mathcal{S}_R(c)$ passing through these three sites.

Theorem (i) There is no overlap between the volumes of any two simplices, and (ii) the volume-sum of all simplices is Ω .

Proof In section 5.3, the entire proof rests only on the convexity of the spherical surfaces. Hence, identical arguments also apply to the present case.

Comments For any given site-distribution, the simplitial decomposition is not unique, depending on the particular convex function f in (5.12). When f is not a constant (i.e., not spherically symmetric), rotational symmetry may be restored by integrating over the orientation of f . Nevertheless the criterion of using spheres as the circumscribed surface is clearly the simplest choice; it also enjoys the maximum symmetry in a euclidean space. Hence, in Chapters V, VI and VII, throughout our discussions of relativistic theories in the euclidean space we shall always use spheres as the circumscribed surfaces. Different criteria will be adopted only when we discuss gravity in Chapter VIII.

5.5. Weights and the Laplacian

We now return to the discrete action A_d given in Table 5:

$$A_d = \frac{1}{2} \sum_{\ell_{ij}} \lambda_{ij}(\phi_i - \phi_j)^2 - \sum_i j_i \phi_i \quad . \tag{5.16}$$

As discussed in section 5.1, the field equation is obtained by setting $\partial A_d / \partial \phi_i = 0$, which leads to the "Kirchoff Law"

$$\sum_{\substack{j \text{ linked} \\ \text{to } i}} \lambda_{ij}(\phi_i - \phi_j) = j_i \quad . \tag{5.17}$$

Its continuum analog is the Laplace equation

$$- \frac{\partial^2 \phi}{\partial x_\mu^{\ 2}} = j \quad . \tag{5.18}$$

In (5.17), why can't λ_{ij} be simply a constant, independent of the link ℓ_{ij} ?

Consider, e.g., a one-dimensional random distribution of N sites at positions x_n (with $x_1 < x_2 < \cdots < x_{n-1} < x_n < \cdots$). Assign a field variable ϕ_n to each x_n. Most of us will agree that the approximation

$$dx \frac{d^2\phi}{dx^2} \sim \frac{\phi_{n+1} - \phi_n}{x_{n+1} - x_n} - \frac{\phi_n - \phi_{n-1}}{x_n - x_{n-1}} \tag{5.19}$$

is better than

$$dx \frac{d^2\phi}{dx^2} \sim \lambda(\phi_{n+1} - 2\phi_n + \phi_{n-1}) \tag{5.20}$$

where λ is a constant. This is because when $\phi(x)$ = linear function of x, or correspondingly

$$\phi_n = \text{linear function of } x_n ,$$

both $\frac{d^2\phi}{dx^2}$ and the right hand side of (5.19) are zero, but not the discrete approximation (5.20). This obvious criterion can be readily generalized.

In a multi-dimensional space when $\phi(\vec{r})$ is a linear function of the position vector $\vec{r}$, the Laplace equation is

$$\frac{\partial^2\phi}{\partial x_\mu^2} = 0$$

where x_μ is the μ-component of $\vec{r}$. We require the weights λ_{ij} in the discrete case to satisfy

$$\sum_{\substack{j \text{ linked} \\ \text{to } i}} \lambda_{ij}(\phi_i - \phi_j) = 0 \tag{5.21}$$

at all i for an arbitrary distribution of the site-position vectors $\vec{r}_1, \vec{r}_2, \cdots, \vec{r}_n$ when ϕ_i is a linear function of $\vec{r}_i$; i.e.,

$$\phi_i = a + \vec{b} \cdot \vec{r}_i \tag{5.22}$$

where a and $\vec{b}$ are constants.

To determine λ_{ij}, it is useful to introduce the dual lattice. Let i be an arbitrary site of a D-dimensional random lattice ($i = 1, 2, \cdots, N$). Take any point p in Ω. We say p belongs to i if i is the nearest site to p. The dual to site i is the volume ω_i consisting of all p belonging to i. Each ω_i is a convex polyhedron (called a cell) in D dimensions. The surface of a cell consists of a number of $(D-1)$-dimensional faces, formed by points that belong to two or more sites. The intersections of these $(D-1)$-dimensional faces are $(D-2)$-dimensional faces, which consist of points each belonging to three or more sites, etc. The corners (i.e., vertices) of these cells are single points, each belonging to, and only to, $D+1$ sites. Let c be one of these corners. From our construction we see that c is equidistant from all the $D+1$ sites to which it belongs. Using c as the center, draw a hypersphere through these $D+1$ sites. Since no site is nearer to c than the aforementioned $D+1$ sites, there cannot be any lattice site inside the hypersphere. Hence, these $D+1$ sites satisfy the definition of a cluster, and form a D-simplex, whose dual is the corner c.

The division of Ω into cells gives the dual of our random lattice. Each link in the random lattice is perpendicular to (but need not intersect) a $(D-1)$-dimensional face in its dual; each triangle is perpendicular to a $(D-2)$-dimensional face, etc. This gives a one-to-one correspondence between an n-simplex in the random lattice and a $(D-n)$-dimensional face in its dual. While each D-simplex has a fixed structure (e.g., fixed number of vertices and faces, etc.), this is not so for the cells. For this reason, as we shall discuss, it is often more convenient to formulate the action in terms of simplices, instead of cells. The dual cells are called Voronoi polyhedra in the literature.[12]

In Figure 3 we give a two-dimensional example of a random lattice and its dual.

We set the weights λ_{ij} between any pair of sites i and j to be

$$\lambda_{ij} = \begin{cases} 0 & \text{if } i \text{ and } j \text{ not linked} \\ s_{ij}/\ell_{ij} & \text{if linked} \end{cases} \tag{5.23}$$

Figure 6. The dotted line indicates the boundary of the Voronoi cell, dual to the lattice site i. The dual to the link ℓ_{ij} is s_{ij}.

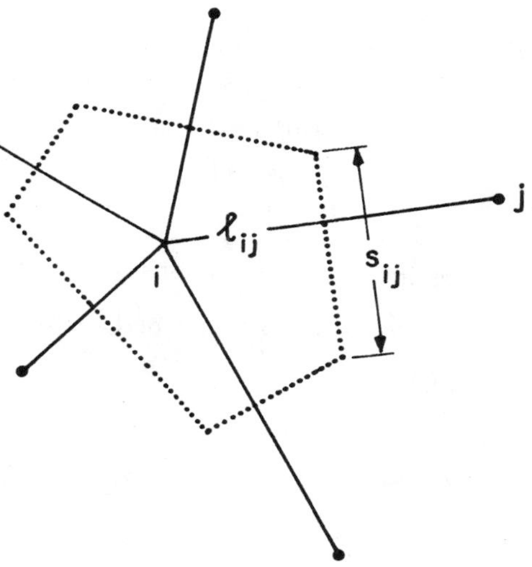

where ℓ_{ij} is the length of the link and s_{ij} is the "volume" of the corresponding $(D-1)$-dimensional surface in the dual lattice. When $D = 2$, s_{ij} is the border-length of the dual cell. [See Figure 6.]

Theorem. The λ_{ij} given by (5.23) satisfies

$$\sum_{\substack{j \text{ linked} \\ \text{to } i}} \ell_{ij}^{\mu} \lambda_{ij} = 0 \tag{5.24}$$

at all sites i and

$$L^{\mu\nu} \equiv \sum_{i,j} \ell_{ij}^{\mu} \ell_{ij}^{\nu} \lambda_{ij} = 2\Omega\delta^{\mu\nu} \tag{5.25}$$

where ℓ_{ij}^{μ} is the μ-component of $\vec{\ell}_{ij} = \vec{r}_i - \vec{r}_j$ and $\delta^{\mu\nu} = 0$ if $\mu \neq \nu$ and 1 otherwise.

Proof When $D = 2$, s_{ij} is simply the border length in the dual lattice, and $\vec{\ell}_{ij} \lambda_{ij}$ is a vector of magnitude s_{ij} in the direction of $\vec{\ell}_{ij}$. A 90° rotation turns it into $\vec{s}_{ij}$. From Figure 6 we see clearly that the sum $\sum \vec{s}_{ij}$ over all j (keeping i fixed) is a null vector, and therefore (5.24) follows.

In any dimension D, two or larger, we may take any constant vector $\vec{V}$. From Gauss' theorem it follows that for any dual cell ω_i

$$0 = \int_{\omega_i} \vec{\nabla} \cdot \vec{V} \, d^D r = \int \vec{V} \cdot d\vec{S} \, . \tag{5.26}$$

Because the (D-1)-dimensional polyhedra s_{ij} (with i fixed but j varying) form the boundary of ω_i, and because $\hat{\ell}_{ij} \equiv \vec{\ell}_{ij}/\ell_{ij}$ is the normal vector of s_{ij}, (5.26) becomes

$$0 = \vec{V} \cdot \sum_j \hat{\ell}_{ij} \, s_{ij}$$

which implies the first part of the theorem.

To prove (5.25), we again apply Gauss' theorem, but to the following integral:

$$\omega_i \, \delta^{\mu\nu} = \int_{\omega_i} \frac{\partial}{\partial x^\nu} (\vec{r} - \vec{r}_i)^\mu \, d^D r = \int (\vec{r} - \vec{r}_i)^\mu \, dS^\nu$$

$$= \sum_j r_{ic}^{\ \mu} \, \hat{\ell}_{ij}^{\ \nu} \, s_{ij} \tag{5.27}$$

where r^ν is the ν-component of $\vec{r}$ and $r_{ic}^{\ \mu}$ is the μ-component of $\vec{r}_{ic} \equiv \vec{r}_c - \vec{r}_i$. The vector $\vec{r}_c$ is given by

$$\vec{r}_c \equiv \frac{1}{s_{ij}} \int_{s_{ij}} \vec{r} \, d^{D-1} r \, , \tag{5.28}$$

and is the position vector of the center of mass c of s_{ij}. [See Figure 7.]

Figure 7. c is the center of the polyhedron s_{ij}, dual to the link ℓ_{ij}.

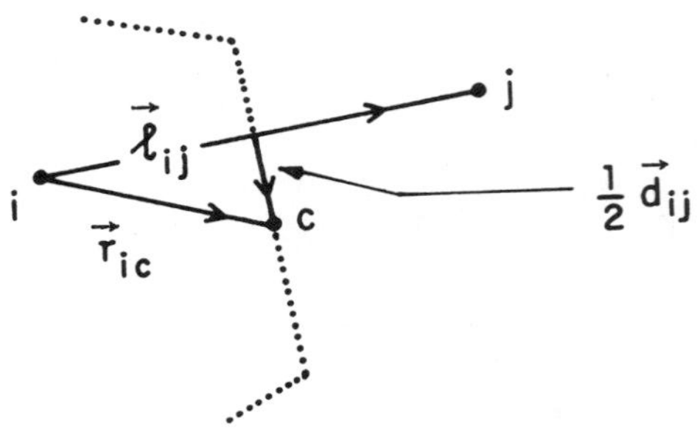

The link-vector ℓ_{ij} is related to $\vec{r}_{ic}$ and $\vec{r}_{jc}$ by

$$\vec{\ell}_{ij} = \vec{r}_{ic} - \vec{r}_{jc} \quad . \tag{5.29}$$

Define

$$\vec{d}_{ij} \equiv \vec{r}_{ic} + \vec{r}_{jc} \quad . \tag{5.30}$$

Equation (5.27) can then be written as

$$\omega_i \delta^{\mu\nu} = \frac{1}{2} \sum_j (\vec{\ell}_{ij} + \vec{d}_{ij})^\mu \hat{\ell}_{ij}^{\ \nu} s_{ij} \quad .$$

Summing over all dual cells ω_i and noting that $\vec{d}_{ij}$ and s_{ij} are even under the exchange of i and j but $\vec{\ell}_{ij}$ and $\hat{\ell}_{ij}$ are odd, we derive

$$\Omega \delta^{\mu\nu} = \sum_i \omega_i \delta^{\mu\nu} = \frac{1}{2} \sum_{i,j} \ell_{ij}^{\ \mu} \hat{\ell}_{ij}^{\ \nu} s_{ij} \tag{5.31}$$

which establishes (5.25) and completes the proof of the theorem.

From the identities (5.24) and (5.25), we see that when, according to (5.22), ϕ_i is a linear function of $\vec{r}_i$

$$\sum_j \lambda_{ij}(\phi_i - \phi_j) = \vec{b} \cdot \sum_j \lambda_{ij} \vec{\ell}_{ij} = 0$$

at all i , and the sum over all sites i and j

$$\frac{1}{2} \sum_{\ell_{ij}} \lambda_{ij}(\phi_i - \phi_j)^2 = \frac{1}{4} \sum_{i,j} \lambda_{ij}(\phi_i - \phi_j)^2 = \frac{1}{2} \vec{b}^{\,2} \Omega \ .$$

The former satisfies condition (5.21)-(5.22) for $\sum_j \lambda_{ij}(\phi_i - \phi_j)$ to be a good discrete version of the Laplace operator, and the latter makes the lattice action (5.2) identical to the continuum value (5.3) for an arbitrary lattice before the limit $\ell_{ij} \to 0$, provided that $\phi(\vec{r})$ is a linear function of $\vec{r}$.

In addition, we can show that if the external current is non-zero only at site 0

$$j_i = \delta_{i0} \ ,$$

then at large distance r_i from 0 the solution ϕ_i becomes the same as that in a continuum; e.g., for $D = 4$ the solution of the Kirchoff equation satisfies

$$\phi_i \rightarrow \frac{1}{4\pi^2 r_i^2} \tag{5.32}$$

when $r_i \rightarrow \infty$. In the corresponding continuum case we have

$$\phi = (4\pi^2 r^2)^{-1} \tag{5.33}$$

which is the solution of

$$-\frac{\partial^2 \phi}{\partial x_\mu^2} = \delta^4(\vec{r}) \ . \tag{5.34}$$

This then insures that the long wave-length limit of the quantum propagator in the discrete theory is the same as the continuum propagator k^{-2}.

The explicit solution of $\phi_i \equiv \phi(r_i)$ for a random lattice of 4^4 sites in $D = 4$ for a site-density $\rho = 1$ is plotted in Figure 8. The average of $\phi(r_i)$ over the position-distribution of lattice sites gives the propagator in the discrete theory. As we can see from Figure 8, even without the average, the result for such a small sample is already relatively smooth, changing from

$$\frac{1}{4\pi^2 r^2} - .00195 \tag{5.35}$$

at large r to a constant $\cong .0165$ at $r = 0$. The constant $-.00195$ in (5.35) is due to the finite size $N = 4^4$ in the sample; it should approach zero and $\phi \rightarrow (4\pi^2 r^2)^{-1}$ as $N \rightarrow \infty$.

Another pleasant feature is that when $\vec{r}_i$ changes, the classical lattice solution ϕ_i and the corresponding action are continuous in $\vec{r}_1, \vec{r}_2, \cdots, \vec{r}_N$ [proved in Ref. 1]. The applicability of continuity concept and the existence of exact equalities, such as (5.24) and (5.25), make it possible to use analytic method in the discrete theory.

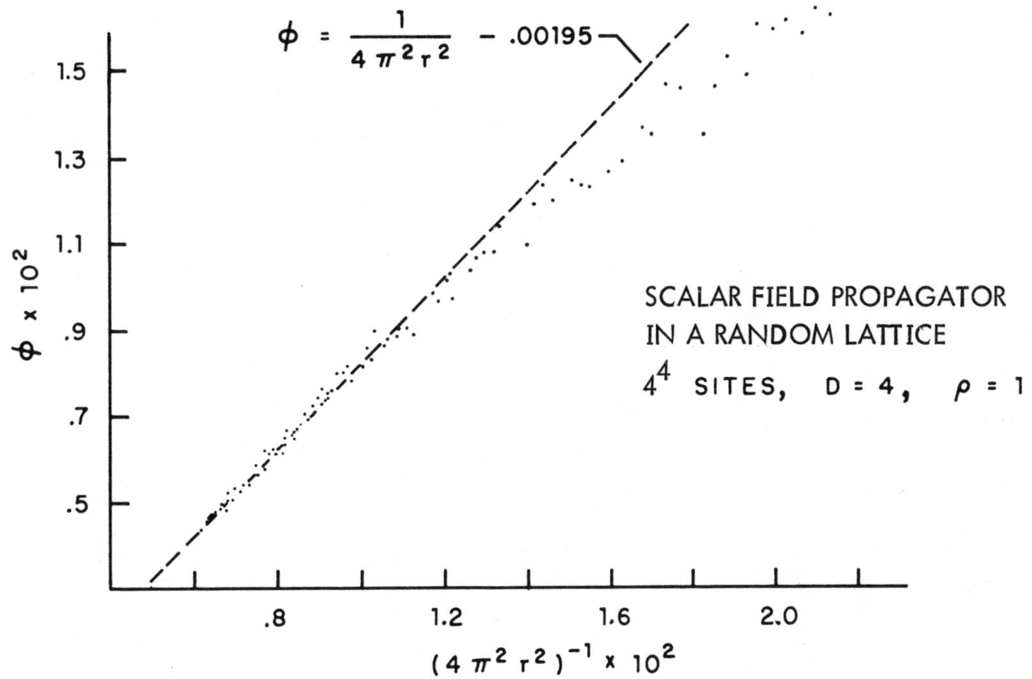

Figure 8. Scalar field propagator in a random lattice.

VI. SPIN - ½ FIELD

The continuum action in a D-dimensional euclidean volume Ω is a hermitian bilinear form given by*

$$A_c(\Omega) = \int_\Omega (\psi^\dagger \gamma^0 \gamma^\mu \frac{\partial}{\partial x^\mu} \psi + m\psi^\dagger \gamma^0 \psi)\, d^D r \tag{6.1}$$

where the dagger denotes hermitian conjugation and $\gamma^0, \gamma^1, \cdots, \gamma^D$ are anti-commuting hermitian matrices whose squares are the unit matrix. The corresponding lattice action is

$$A_d(\Omega) = \frac{1}{2} \sum_{i,j} \psi_i^\dagger \gamma^0 \gamma^\mu \ell_{ij}^\mu \lambda_{ij} \psi_j + m \sum_i \omega_i \psi_i^\dagger \gamma^0 \psi_i \tag{6.2}$$

where λ_{ij} is given by (5.23) and ω_i is the volume of the D-dimensional polyhedron dual to i. Because of (5.24), it follows that

$$\psi_j = \text{a constant spinor} \tag{6.3}$$

satisfies the massless field equation

$$\sum_j \gamma^\mu \ell_{ij}^\mu \lambda_{ij} \psi_j = 0 \tag{6.4}$$

at all sites i.

Next, we consider the function

$$\psi_j = \chi\, e^{i\vec{p}\cdot\vec{r}_j} \tag{6.5}$$

where χ is a constant spinor; the corresponding lattice action is

$$A_d(\Omega) = \frac{1}{2}(\chi^\dagger \gamma^0 \gamma^\mu \chi) \sum_{i,j} \ell_{ij}^\mu \lambda_{ij}\, e^{i\vec{p}\cdot\vec{\ell}_{ij}} + m\chi^\dagger \gamma^0 \chi\, \Omega \ .$$

For small $\vec{p}$, $e^{i\vec{p}\cdot\vec{\ell}_{ij}} = 1 + i\,\vec{p}\cdot\vec{\ell}_{ij} - \frac{1}{2}(\vec{p}\cdot\vec{\ell}_{ij})^2 + \cdots$. By using (5.25) and

* We exclude other hermitian choices such as $A_c(\Omega) = \int_\Omega (i\,\psi^\dagger \gamma^\mu \frac{\partial}{\partial x^\mu} \psi + m\psi^\dagger \psi)\, d^D r$ by requiring the field equation to be $(\gamma^\mu \frac{\partial}{\partial x^\mu} + m)\,\psi = 0$.

neglecting $O(\vec{p}^{\,3})$, we see that

$$A_d(\Omega) = \left[i(\chi^\dagger \gamma^0 \gamma^\mu \chi)\, p^\mu + m\chi^\dagger \gamma^0 \chi \right] \Omega$$

which has the same form as in the continuum case.

The lattice action (6.2) may be resolved in terms of its eigenvectors $\psi_i(n)$ and eigenvalues $E(n)$:

$$\tfrac{1}{2} \sum_j \gamma^\mu \ell_{ij}^{\ \mu} \lambda_{ij} \psi_j(n) + m\omega_i \psi_i(n) = E(n)\, \omega_i \gamma^0 \psi_i(n) . \qquad (6.6)$$

We write

$$A_d(\Omega) = \sum_{n,i} E(n) \omega_i \psi_i(n)^\dagger \psi_i(n) .$$

It is easy to verify that (6.5) is not a solution of the eigenvector equation (6.6). Hence, the low-lying spectrum $E(n)$ of the lattice action is more complicated than the continuum case.

For the zero-mass case, it can be shown that in any dimension D, besides the constant spinor solution (6.3) there is <u>another</u> zero-energy solution which has a rather complicated dependence on $\vec{r}_i$. The full implications of this additional solution and its relation to chiral anomaly are yet to be investigated.

VII. GAUGE THEORY

7.1. General Discussion

As in the conventional treatment[13] of lattice gauge theory, we introduce gauge variables by assigning a group element $U(i,j)$ to a link connecting the points (i.e., sites) i and j in the lattice. For an $SU(N)$ gauge theory, we can set $U(i,j)$ to be an arbitrary $N \times N$ unitary matrix with determinant 1. It is convenient to define both $U(i,j)$ and $U(j,i)$ with $U(j,i) = U(i,j)^{-1}$. Next, with each elementary triangle of vertices i,j and k we associate a group element $U_{ijk} = U(i,j)\, U(j,k)\, U(k,i)$. Recall that in our construction of the random lattice we identify a "cluster" of $D+1$ vertices with the property that the $D-1$ dimensional sphere, on whose surface the cluster of $D+1$ points lies, has no lattice points in its interior. Each pair of points in such a cluster is joined by a link. An elementary triangle Δ_{ijk} is then formed of any three points i, j and k in the same cluster and the links between them.

The discrete action for the lattice is defined using these group variables U_{ijk}:

$$A_d = \frac{1}{g^2} \sum_{\Delta_{ijk}} \kappa_{ijk} f(U_{ijk}) \; , \tag{7.1}$$

where g is the coupling constant, and the sum extends over all triangles Δ_{ijk}. The coefficients κ_{ijk} are given by, similar to (5.23),

$$\kappa_{ijk} = \tau_{ijk}/\Delta_{ijk} \tag{7.2}$$

where Δ_{ijk} is the area of the plaquette and τ_{ijk} the $(D-2)$-dimensional volume of its dual. Just as in (5.24) and (5.25) we can prove (Ref. 1) that

$$\sum_k \kappa_{ijk} \Delta_{ijk}^{\mu\nu} = 0 \tag{7.3}$$

for all sites i and j that are linked, and

$$T^{\mu\nu\rho\sigma} \equiv \frac{1}{4} \sum_{\Delta_{ijk}} \kappa_{ijk} \Delta_{ijk}^{\mu\nu} \Delta_{ijk}^{\rho\sigma} = \frac{1}{8} (\delta^{\mu\rho} \delta^{\nu\sigma} - \delta^{\mu\sigma} \delta^{\nu\rho}) \, \Omega \; . \tag{7.4}$$

The function $f(U)$ obeys

$$f(u U u^{-1}) = f(U) \tag{7.5}$$

for all group elements u and U; the simplest choice is

$$f(U) = \tfrac{1}{2} \operatorname{tr}(I - U) + \text{c.c.} = \operatorname{Re}\left[\operatorname{tr}(I - U)\right] \; , \tag{7.6}$$

with I = unit matrix and, for the $SU(N)$ theory, U = an arbitrary $N \times N$ unitary matrix of unit determinant, in accordance with the conventional Wilson formalism.

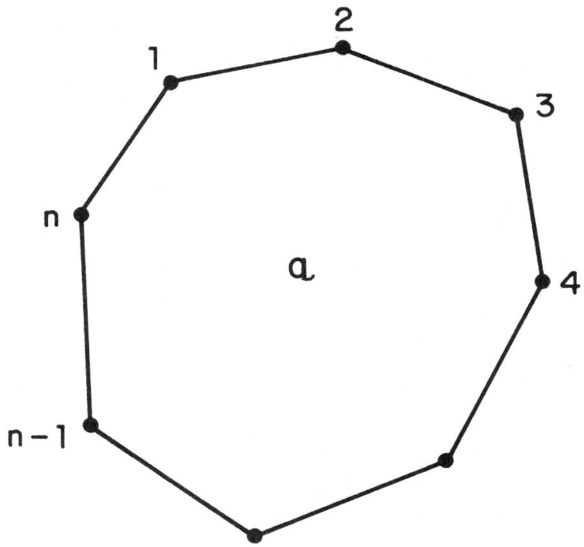

Figure 9. Wilson loop on a random lattice.

Next, consider a large loop of area $\mathcal{a}$ bounded by links ℓ_{12}, ℓ_{23}, $\cdots$, as shown in Figure 9. [The linear dimension of the loop is assumed to be $O(\mathcal{a}^{\frac{1}{2}})$ in any direction.] Define

$$\exp(-T\mathcal{a}) \equiv Z^{-1} \int [d^D r_i]\,[dU(i,j)]\; e^{-A_d/g^2}\, W_L \tag{7.7}$$

where

$$Z \equiv \int [d^D r_i]\,[dU(i,j)]\; e^{-A_d/g^2}\,, \tag{7.8}$$

T is the string tension, A_d is the action given by (7.1) and

$$W_L = \text{trace}\; U(1,2)\; U(2,3)\; \cdots\,, \tag{7.9}$$

with the product extending over the boundary of $\mathcal{a}$.

7.2. Strong Coupling Limit: String Tension, String Thickness and Glueball Spectrum

In the strong coupling limit $(g \to \infty)$, one can show that the theory becomes identical to a relativistic string model. For $D = 4$, we find the string tension T is given by

$$T = 2\left(\frac{\pi}{\sqrt{3}}\right)^{\frac{1}{2}} (\rho \ln g^2)^{\frac{1}{2}} \tag{7.10}$$

with ρ = site-density. Furthermore, the thickness d of the string is

$$d^2 = \frac{1}{2\pi T} \ln \mathcal{a} \ . \tag{7.11}$$

Both (7.10) and (7.11) are derived in Ref. 1. The string thickness (7.11) in a random lattice is identical to that given by a relativistic string model.[14] Thus, unlike a rigid "cubic" lattice the strong coupling limit of the random lattice gauge theory is already quite physical, since it carries the phenomenological virtues of a relativistic string model. From (7.10) we see that quarks are confined in the strong coupling limit.

As will be shown, the result of the numerical calculation indicates that there is no phase transition in a random lattice non-Abelian gauge theory. Consequently, when we change the coupling from strong to weak, quarks should remain confined.

The string tension concerns calculations for a given two-dimensional figure (Wilson loop). At the next level of complexity, we consider the glueball propagator between two plaquettes, separated by a large time interval T. Let Δ and Δ' be these two plaquettes. Then in the strong coupling limit we find that these two plaquettes are connected by tetrahedra of the form given in Figure 10.

When $T \to \infty$, the corresponding propagator has a leading behavior which can be expressed as a superposition of different angular momentum J states, each having an asymptotic behavior

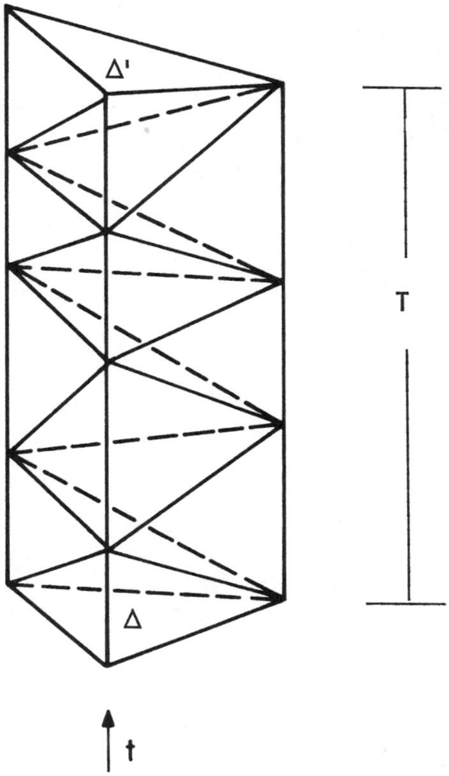

Figure 10. Glueball propagator in the strong coupling limit.

$$\sim e^{-M_J T} \qquad (7.12)$$

where M_J is the lowest glueball mass of angular momentum J. By subtracting away the $e^{-M_J T}$ term, the remaining part has an asymptotic behavior $\sim e^{-M_{J*} T}$, where M_{J*} is the second lowest glueball mass of angular momentum J, etc.

For low angular momentum states, we find that in the strong coupling limit[15] the glueball rotational levels are identical to that of a rigid body of moment of inertia I_1, I_2 and I_3 where

$$I_1 = .631\, \rho^{1/4} (\ln g^2)^{5/4}$$

and

$$I_2 = I_3 = .341\, \rho^{1/4} (\ln g^2)^{5/4} \quad .$$

Let M_0 be the ground state mass with $J = 0$ and parity $+$. The low-lying excitation energy $M_J - M_0$ is given by

J	Parity	$M_J - M_0$
1	-	$\frac{1}{2}(I_1^{-1} + I_2^{-1})$
1*	-	I_2^{-1}
2	+	$2 I_1^{-1} + I_2^{-1}$
2*	+	$\frac{1}{2} I_1^{-1} + \frac{5}{2} I_2^{-1}$
2**	+	$3 I_2^{-1}$

Qualitatively, the three $J = 2$ levels are consistent with the recent experimental results of S. Lindenbaum.[16]

For high angular momentum we have in the strong coupling limit[15]

$$M_J \propto \sqrt{J} \tag{7.13}$$

so that the system exhibits the typical Regge behavior of a relativistic string in rotation. The proportional constant is exactly that corresponding to a double string (since quark number = 0).

7.3. Numerical Results[4]

Numerical programs for a random lattice gauge theory were set up by Fried-

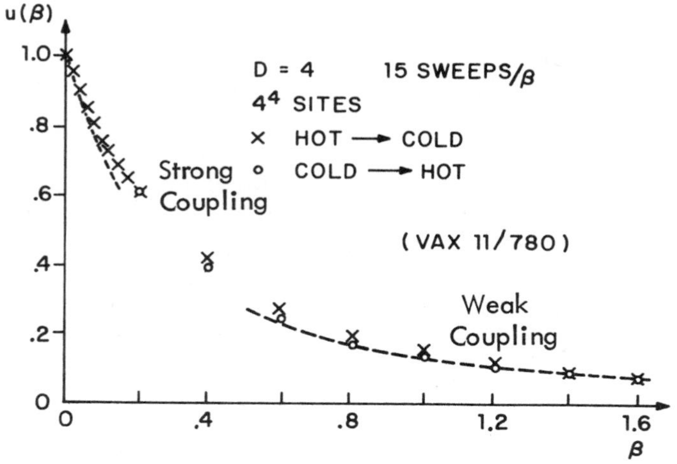

(a)

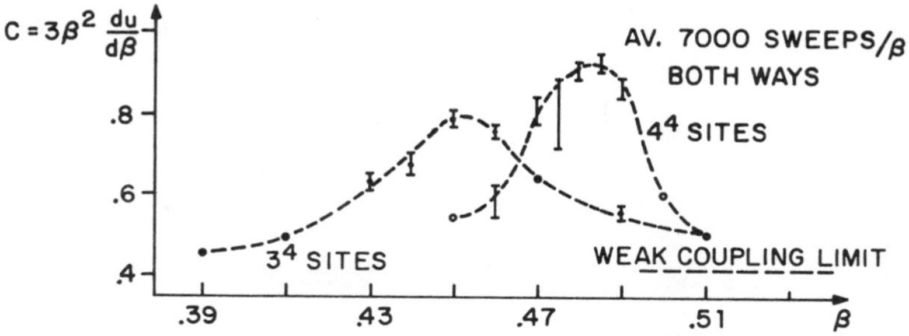

(b)

Figure 11. Numerical results for a four-dimensional U(1) random lattice field theory:

a) Average plaquette energy vs. $\beta = 1/g^2$,

b) Specific heat vs. β . (H. C. Ren)

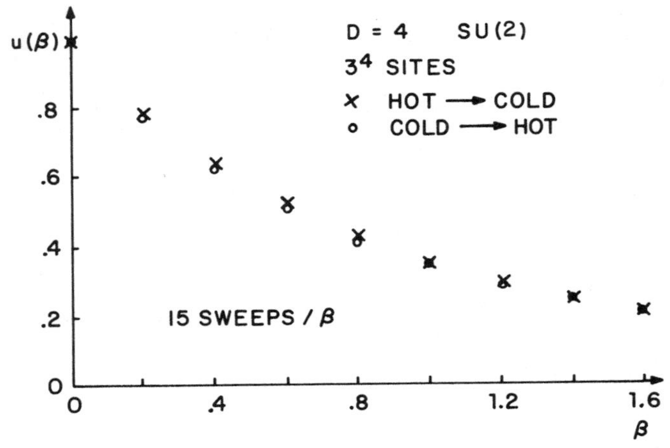

(a)

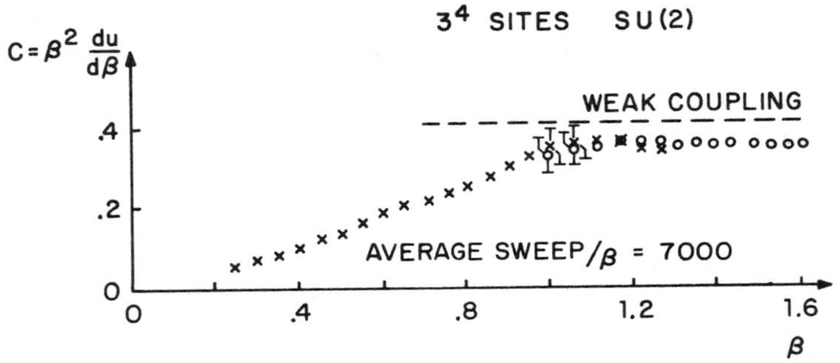

(b)

Figure 12. Numerical results for a four-dimensional SU(2) random lattice field theory:

a) Average plaquette energy vs. β ,

b) Specific heat vs. β . The error bars indicated are applicable to all points. (H. C. Ren) berg and Ren at Columbia; the computations were carried out by Ren. In Figures 11a–11b we give the average plaquette energy u and specific heat C vs. $\beta = 1/g^2$ for the U(1) theory.

The corresponding plots for an SU(2) theory are given in Figures 12a–12b. We see that the specific heat has a peak in the U(1) theory, but not in the SU(2) theory. For U(1), the peak becomes steeper when the number of lattice sites increases, suggesting that there is a phase transition. On the other hand, the specific heat curve for SU(2) has no peak, indicating that the passage from strong to weak coupling is a smooth one. Consequently, while both theories are confined in the strong coupling limit, the weak coupling limit is consistent with deconfinement in the U(1) theory (QED), but not in a non-Abelian gauge theory. Extension to SU(3) and also calculations on Wilson string tension are in progress.

In contrast, we given in Figure 13 the recent numerical calculation by N. H. Christ and A. Terrano[17] for the SU(3) gauge theory on a regular lattice. As we can see, there is a sharp peak in the specific heat, suggesting that the transition from strong to weak in a regular lattice is by no means smooth, unlike that in a random lattice.

VIII. QUANTUM GRAVITY

8.1. General Discussion

The technical difficulties of quantizing gravity are well known. Among them the two most prominent are

(i) nonrenormalizability

and (ii) energy not positive definite.

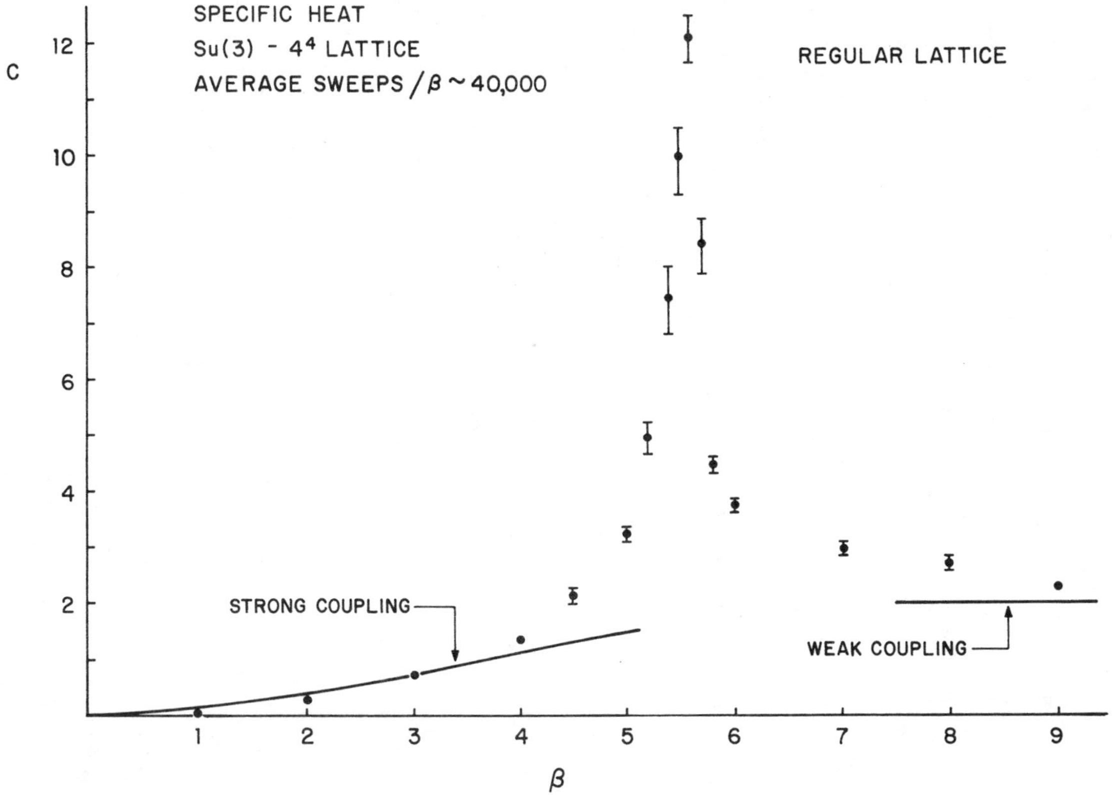

Figure 13. Specific heat vs. β for a four-dimensional SU(3) theory on a "cubic" lattice. (N. H. Christ and A. Terrano)

In connection with (ii), the usual Einstein equation in the euclidean space *

$$G_{\mu\nu} = R_{\mu\nu} - \tfrac{1}{2} g_{\mu\nu} R = 0 \qquad (8.1)$$

gives solutions that are saddle points of the continuum action

$$A_c = \int \sqrt{|g|}\ R\, d^4x \ ; \qquad (8.2)$$

this makes quantization in the euclidean space extremely inconvenient.

Our strategy is to formulate the discrete theory of gravity directly in the minkowski space. In the next section, we shall show how a random lattice can be set up in the minkowski space. By keeping the density of lattice sites $\rho = (1/\ell)^4$ finite, we avoid the ultraviolet divergence. Because we work directly in the minkowski space, the path integrals become finite. The latter follows from the elementary observation that an integral such as

$$\int_{-\infty}^{\infty} \int_{-\infty}^{\infty} e^{i(ax^2 + by^2)}\, dx\, dy$$

---

* Throughout this chapter, our conventions for the continuum theory are:

$$ds^2 = g_{\mu\nu}\, dx^\mu\, dx^\nu$$

where the metric for a <u>flat</u> euclidean space is

$$(g_{\mu\nu}) = (\delta_{\mu\nu}) = \begin{pmatrix} 1 & & & \\ & 1 & & \\ & & 1 & \\ & & & 1 \end{pmatrix} \qquad (8.3)$$

and for a <u>flat</u> minkowski space is

$$(g_{\mu\nu}) = (\eta_{\mu\nu}) = \begin{pmatrix} -1 & & & \\ & 1 & & \\ & & 1 & \\ & & & 1 \end{pmatrix} . \qquad (8.4)$$

The terms "euclidean space" and "minkowski space" are for <u>arbitrary</u> $g_{\mu\nu}$, but with $|g| \equiv \det(g_{\mu\nu}) > 0$ and < 0 respectively.

exists for any real constants a and b independently of their signs, while the corresponding integral

$$\int_{-\infty}^{\infty} \int_{-\infty}^{\infty} e^{-(ax^2 + by^2)} \, dx \, dy$$

exists only for a and b both positive. [A saddle point of the exponent at the origin $x = y = 0$ would have opposite signs for a and b.]

8.2. Random Lattice in a Minkowski Space[18]

Let x^μ be the Cartesian coordinates in a D-dimensional flat minkowski space where the superscript $\mu = 0, 1, 2, \cdots, D-1$ denotes the space-time component with $x^0 =$ time t. The metric $\eta_{\mu\nu}$ is given by (8.4). The invariant volume element is

$$d^D x = dx^0 dx^1 \cdots dx^{D-1} , \tag{8.5}$$

the same as in a flat euclidean space. Consider a random distribution of N lattice sites in a volume Ω. Clearly, N, Ω and the site-density

$$\rho \equiv \frac{N}{\Omega} \tag{8.6}$$

are all Lorentz invariant.

The immediate problem with setting up a relativistic field theory on a random lattice in a minkowski space is the concept of "neighbors". In order to simulate, e.g., the d'Alambertian operator (or other derivatives) in a continuum theory, the corresponding discrete operator should involve only a few nearby sites (similar to the relation between (5.21) and the Laplace equation in the euclidean space). To be specific, let us consider the example of a free massless scalar field: there should be an expression like (5.21)

$$\sum_{\substack{j \text{ linked} \\ \text{to } i}} \lambda_{ij}(\phi_i - \phi_j) = 0 \tag{8.7}$$

which would approach the d'Alambertian equation

$$-\frac{\partial^2\phi}{(\partial x^0)^2} + \frac{\partial^2\phi}{(\partial x^1)^2} + \cdots + \frac{\partial^2\phi}{(\partial x^{D-1})^2} = 0 \qquad (8.8)$$

when the site-density $\rho \to \infty$. In order to construct (8.7) we must first have an appropriate linking algorithm. In the minkowski space, the (distance)^2 between any two points p and q is given by

$$\ell_{pq}^{\ 2} \equiv -\left[x^0(p) - x^0(q)\right]^2 + \left[x^1(p) - x^1(q)\right]^2 + \cdots + \left[x^{D-1}(p) - x^{D-1}(q)\right]^2 \qquad (8.9)$$

where $x^\mu(p)$ and $x^\mu(q)$ are the μth coordinates of p and q. A small $|\ell_{pq}|$ clearly does not imply that $x^\mu(p)$ and $x^\mu(q)$ are near to each other. Yet, in the d'Alambertian (8.8), we only deal with the differences between ϕ at points that are infinitesimally close to each other. Therefore, the linking algorithm in a minkowski space cannot be based simply on the invariant distance ℓ_{pq}, in contrast to the situation in a euclidean space.

In a given random distribution of N lattice sites, we can envisage each site carrying its light cone. Between any two points p and q (which may or may not be the lattice sites), let us define

$$n_{pq} \equiv \text{number of light cones crossed by the straight line } \overline{pq}\,. \qquad (8.10)$$

Under a Lorentz transformation, a light cone remains a light cone, and a straight line remains a straight line. If $\overline{pq}$ crosses the light cone of, say, the ith site in one Lorentz frame, this crossing must also be observed in any other Lorentz frame. Consequently, n_{pq} is invariant. As we shall see, this number n_{pq} can be used to establish the concept of "neighbors" in a minkowski space.

Take the example of two points p and p' lying on each other's light cone. The invariant distance $\ell_{pp'}$ is always zero. However, $n_{pp'}$ varies depending on the relative positions of p and p'. A large $n_{pp'}$ indicates p and p' are far apart. The use of a number (not a distance) to distinguish the nearness or apartness of two points should be familiar to any New Yorker. In New York, for example, from Lincoln Center one has to travel about 55 blocks to Columbia University, but only about 10 blocks to Rockefeller University. Even though the east-west blocks are longer than the north-south ones, since 55 is a much larger number than 10 one may conclude that Lincoln Center is nearer to Rockefeller than to Columbia. In the following, we shall take the example of a two-dimensional minkowski space to illustrate how the number n_{pq} can be used for linking the random lattice sites.

When $D = 2$, the coordinates x^0 and x^1 will be written simply as t and x. Define

$$x_+ \equiv \frac{1}{\sqrt{2}} (x + t)$$

and

$$x_- \equiv \frac{1}{\sqrt{2}} (x - t) \quad . \tag{8.11}$$

The light cone of any point consists of two lines, one parallel to the x_+ axis and the other to the x_- axis. Under a Lorentz transformation

$$\begin{pmatrix} t \\ x \end{pmatrix} \rightarrow \begin{pmatrix} \cosh\theta & \sinh\theta \\ \sinh\theta & \cosh\theta \end{pmatrix} \begin{pmatrix} t \\ x \end{pmatrix} , \tag{8.12}$$

both x_+ and x_- undergo a scale transformation

$$x_+ \rightarrow e^{\theta} x_+ \qquad \text{and} \qquad x_- \rightarrow e^{-\theta} x_- , \tag{8.13}$$

keeping their product $x_+ x_-$ invariant.

There is an alternative way of deriving n_{pq}, which is defined by (8.10). Instead of "moving" from p to q along the straight line $\overline{pq}$, one starts from p, travels along its light cone (say the one that is parallel to the x_+ axis) towards q, crossing n_+ light cones (of other sites) until one reaches the light cone of q; then one makes a turn and travels along the light cone of q, crossing another n_- light cones until one reaches q. It is easy to show that (see Figure 14)

$$n_{pq} = n_+ + n_- \quad . \tag{8.14}$$

Under a Lorentz transformation, the $x_\pm$ - coordinates of p, q and all the lattice sites undergo the same scale transformation (8.13); therefore, n_+ and n_- are separately Lorentz invariant.

For convenience, let us distribute the N lattice sites in a box of volume $\Omega = L^2$, which is bounded by the light cones of two points, labeled B and X. These two points are symmetrically placed with respect to the origin O, so that all three points B, O and X lie on one straight line, with O in the middle. The boundary of the box consists of four lines given by

$$x_+ = \frac{L}{2} e^{\alpha} , \qquad x_- = \frac{L}{2} e^{-\alpha} ,$$
$$x_+ = -\frac{L}{2} e^{\alpha} \quad \text{and} \quad x_- = -\frac{L}{2} e^{-\alpha} , \tag{8.15}$$

as shown in Figure 14. The box is therefore "rectangular" in shape, with sides $L_+ \equiv Le^{\alpha}$ and $L_- \equiv Le^{-\alpha}$; its volume is $L_+ L_- = L^2$. The parameter α defines the shape of the box. Under the Lorentz transformation (8.12), the volume of the box is invariant, but the shape changes* with

$$\alpha \to \alpha + \theta . \tag{8.16}$$

---

* In a discrete field theory on a flat minkowski-space random lattice, besides integrating over the positions of all lattice sites i we also integrate over the parameter α which defines the box shape. From Figure 14, we see that B is confined to the hyperbola $\ell_{BO}^2 = -\frac{1}{2}L^2 =$ constant, where ℓ_{BO} is the minkowski distance between B and the origin O; the shape of the box is in turn uniquely determined by the position of B on that hyperbola. We may regard B as an extra-lattice site. The integration over the box shape becomes, then, simply an integration over the position of the site B, subject to the constraint that ℓ_{BO}^2 is fixed.

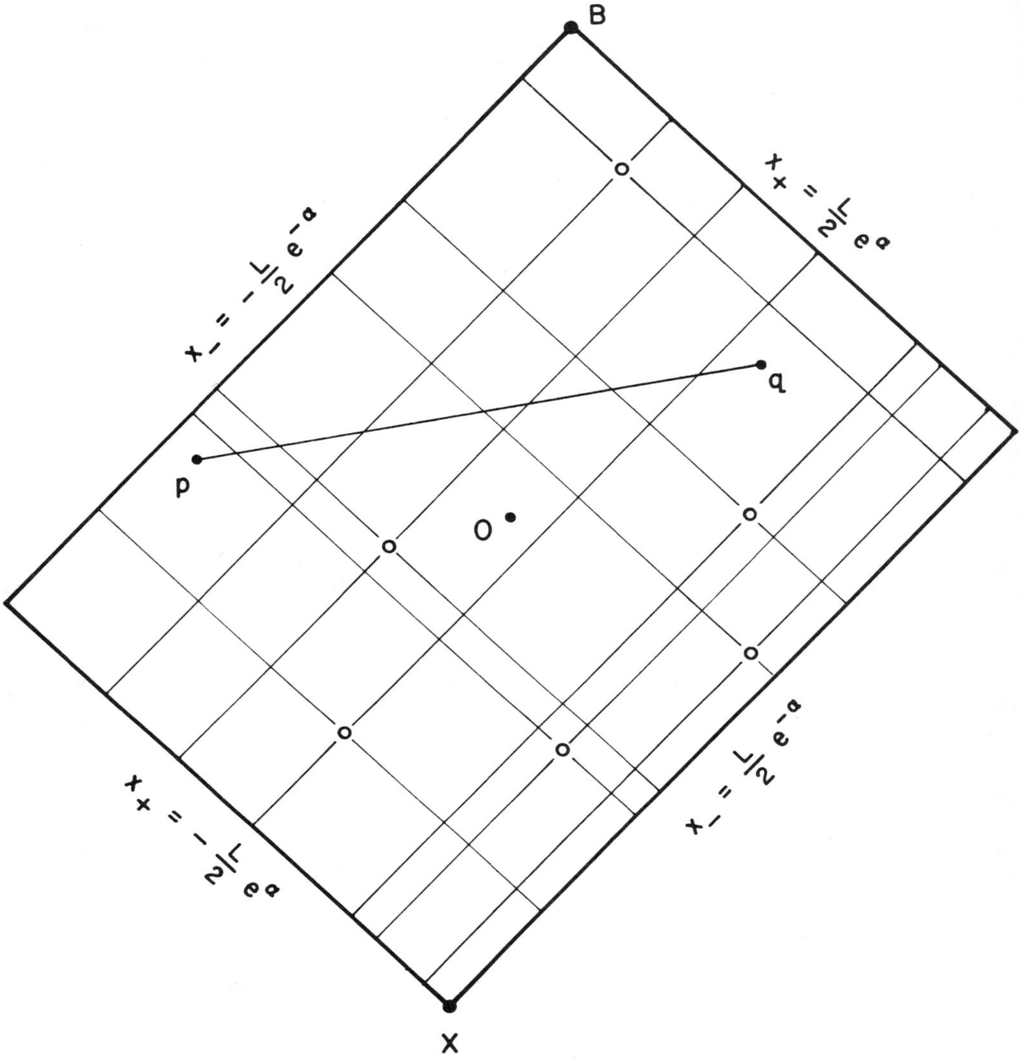

Figure 14. In this example of $N = 6$ sites (represented by circles) in a $D = 2$ box of volume L^2 (bounded by the light cones of B and X), we have $n_{pq} = 7$, $n_+ = 4$ and $n_- = 3$, confirming (8.14). The origin O is the mid-point of the straight line $\overline{BX}$; therefore, the minkowski lengths squared between BO and OX are $\ell^2_{BO} = \ell^2_{OX} = -L^2/2$

By definition the number n_{pq} is an integer which is somewhat awkward to use. As we shall see, in the limit of $N \to \infty$ and $L \to \infty$, but at a fixed site-density

$$\rho = \frac{N}{L^2} = \text{constant}, \tag{8.17}$$

a continuous parameter ν_{pq} can be derived which is more suitable for our purpose. We define

$$\nu_{pq} \equiv n_{pq}/N^{\frac{1}{2}}$$

and

$$\nu_{\pm} \equiv n_{\pm}/N^{\frac{1}{2}} \quad ; \tag{8.18}$$

(8.14) becomes

$$\nu_{pq} = \nu_{+} + \nu_{-} \quad . \tag{8.19}$$

Let

$$\Delta_{+} \equiv x_{+}(p) - x_{+}(q)$$

and

$$\Delta_{-} \equiv x_{-}(p) - x_{-}(q) \tag{8.20}$$

where $x_{\pm}(p)$ and $x_{\pm}(q)$ are the $x_{\pm}$-coordinates of p and q. Since the total number of light cones crossed by the entire length of the box side $L_{+} = L e^{\alpha}$ is N, the average number of n_{+} is

$$< n_{+} > = \frac{|\Delta_{+}|}{L_{+}} N = \frac{|\Delta_{+}|}{L} N e^{-\alpha} \quad ,$$

with $< >$ denoting the average. Correspondingly,

$$<\nu_{+}> = \frac{< n_{+} >}{N^{\frac{1}{2}}} = \frac{|\Delta_{+}|}{L} N^{\frac{1}{2}} e^{-\alpha} = \rho^{\frac{1}{2}} |\Delta_{+}| e^{-\alpha} \quad . \tag{8.21}$$

Likewise, we have

$$<\nu_{-}> = \rho^{\frac{1}{2}} |\Delta_{-}| e^{\alpha} \quad . \tag{8.22}$$

For a random distribution of lattice sites, the fluctuation in $n_\pm$ is given by

$$< n_\pm^2 > - < n_\pm >^2 = < n_\pm > .$$

In the limit of $N \to \infty$ at a constant ρ, we have

$$< \nu_\pm^2 > - < \nu_\pm >^2 = \frac{< n_\pm >}{N} = \frac{< \nu_\pm >}{N^{\frac{1}{2}}} \to 0 . \qquad (8.23)$$

Thus, we can neglect the fluctuation in $\nu_\pm$, and regard $\nu_\pm$ simply as continuous parameters, equal to $\rho^{\frac{1}{2}} \mid \Delta_\pm \mid e^{\mp\alpha}$, which can vary from 0 to ∞. From (8.18)-(8.21) it follows that

$$\nu_{pq} = \rho^{\frac{1}{2}} \left[\mid x_+(p) - x_+(q) \mid e^{-\alpha} + \mid x_-(p) - x_-(q) \mid e^{\alpha} \right]. \qquad (8.24)$$

Under a Lorentz transformation, $x_+ \to x_+ e^\theta$, $x_- \to x_- e^{-\theta}$, $\alpha \to \alpha + \theta$; therefore ν_{pq} is invariant, as expected.

We shall now follow the steps given in section 5.4 for the generalization of the linking algorithm. Take an arbitrary point c as the center. Consider the sequence of curves $\mathcal{S}_R(c)$ with varying R, each of which consists of all points p that satisfy

$$\nu_{pc} = \rho^{\frac{1}{2}} R . \qquad (8.25)$$

From (8.24) we see that $\mathcal{S}_R(c)$ are parallelopipes, which fulfill the convexity requirement of (5.12). As in section 5.4, take any group of $D + 1 = 3$ sites. If by varying c and R, we can find an $\mathcal{S}_R(c)$ passing through these three sites and if, in addition, the interior of that $\mathcal{S}_R(c)$ is empty, then these three sites form a cluster of neighbors; they should be linked to each other and become the vertices of a triangle. The theorem proved in section 5.4 insures that by going through all groups of three sites, the triangles thus formed do not intersect each other, and together they fill the entire volume $\Omega = L^2$. Under a Lorentz transformation, sites move and the box-shape changes. However, if two sites are linked in one frame, then they are also linked in any other frame.

For a "square" box (i.e., $\alpha = 0$), (8.24) becomes

$$\nu_{pq} = \rho^{\frac{1}{2}} \left[|x_+(p) - x_+(q)| + |x_-(p) - x_-(q)| \right] . \qquad (8.26)$$

The curves $\mathcal{S}_R(c)$ are squares whose sides are parallel to the x and t axes. By using (8.25)-(8.26) for the construction of the circumscribed curves $\mathcal{S}_R(c)$, we can decompose a random lattice in the minkowski space (inside a "square" box) into triangles. An explicit example is given in Figure 15a. In Figure 15b, we give the lattice for the same site distribution in the euclidean space, by using circles as the circumscribed curves. It is of interest to compare these two decompositions.

Comments

1. The method discussed in this section can be readily generalized to $D > 2$. The details of such a minkowski space random lattice field theory will be given elsewhere.[18]

2. At large distances the field propagator in the euclidean space is proportional to r^{2-D} which, for $D > 2$, approaches 0 as $r \to \infty$. Hence for a large euclidean volume, the effect of the boundary can be neglected.

3. The situation is different in the minkowski space. From (8.24)-(8.25) we see that the corresponding convex function f used in (5.12) for simplitial decompositions depends explicitly on the shape parameter α of the box. Lorentz invariance is restored only after the average of the shape parameter.

8.3. Discrete Action

First take a random lattice in a flat D-dimensional minkowski space. Label

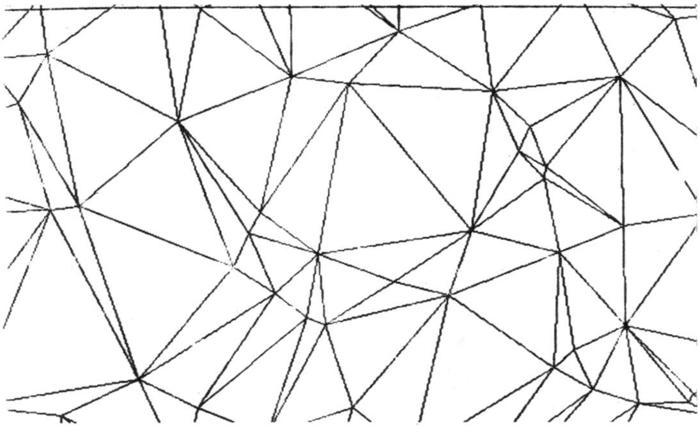

(a)

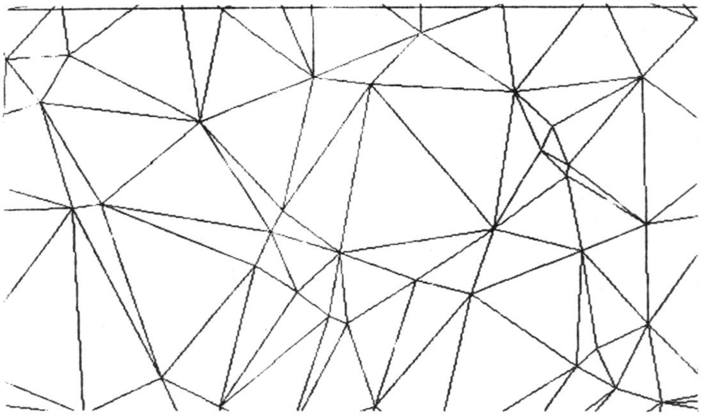

(b)

Figure 15. (a) A two-dimensional minkowski space random lattice (inside a square box).
(b) Same lattice-site distribution, but in a euclidean space.

each site by $i = 1, 2, \cdots$, and each D-simplex by $\mathring{\tau}(a)$ with $a = 1, 2, \cdots$. Let $\mathring{x}^\mu(i)$ be the μth-coordinate of the ith site. For every linked pair of sites i and j, the (link-length)^2 is given by the usual expression

$$\mathring{\ell}_{ij}^{\,2} = \eta_{\mu\nu} \left[\mathring{x}(i) - \mathring{x}(j)\right]^\mu \left[\mathring{x}(i) - \mathring{x}(j)\right]^\nu \tag{8.27}$$

where $\eta_{\mu\nu}$ is defined by (8.4).

Consider now an arbitrary variation*

$$\mathring{\ell}_{ij}^{\,2} \rightarrow \ell_{ij}^{\,2} \quad . \tag{8.28}$$

Correspondingly, each D-simplex changes from

$$\mathring{\tau}(a) \rightarrow \tau(a) \tag{8.29}$$

with the same vertices, but different link-lengths. Following Regge,[19, 20] we assume that the <u>interior</u> of each D-simplex <u>remains flat.</u> Hence, inside each simplex $\tau(a)$, we may set up an arbitrary Lorentz frame $\Sigma(a)$. A vector ϕ^μ viewed in $\Sigma(a)$ can be written as a column matrix

$$\phi(a) \equiv \begin{pmatrix} \phi^0(a) \\ \phi^1(a) \\ \vdots \\ \phi^{D-1}(a) \end{pmatrix} \quad . \tag{8.30}$$

Under a Lorentz transformation of $\Sigma(a)$,

---

* For each plaquette Δ_{ijk} in the lattice, the new lengths of its three sides ℓ_{ij}, ℓ_{jk} and ℓ_{ki} are assumed to satisfy the triangular inequality in a flat minkowski space. Likewise, for each tetrahedron, the length-squared of its six sides ℓ_{ij}^2 satisfies the tetrahedral inequalities, etc.

$$\phi(a) \rightarrow u(a)\ \phi(a) \tag{8.31}$$

where, by definition, the real $D \times D$ matrix $u(a)$ satisfies

$$\tilde{u}(a)\ \eta\ u(a) = \eta \tag{8.32}$$

with $\sim$ denoting the transpose and

$$\eta = \begin{pmatrix} -1 & 0 & 0 & 0 \\ 0 & 1 & 0 & 0 \\ 0 & 0 & 1 & 0 \\ 0 & 0 & 0 & 1 \end{pmatrix} , \tag{8.33}$$

as before.

Each D-simplex, say $\tau(a)$, has $D+1$ vertices and $\frac{1}{2}(D+1)D$ links; viewed in $\Sigma(a)$, apart from the absolute position of its center-of-mass, all the properties of $\tau(a)$ such as shape, volume, orientation, etc. are completely determined by the D independent link-vectors: $\ell_{12}^{\ \mu}$, $\ell_{13}^{\ \mu}$, $\cdots$, $\ell_{1\,D+1}^{\ \mu}$ where $1, 2, \cdots, D+1$ refer to the vertices of $\tau(a)$. Under a Lorentz transformation of $\Sigma(a)$, these vectors transform according to (8.31). From these vectors, we can form a total of $\frac{1}{2}(D+1)D$ independent invariants which may be represented by all the $(\text{link-length})^2$ of $\tau(a)$.

So far, each $\tau(a)$ is separate from all other D-simplices. Suppose that in the original flat minkowski space, two D-simplices, say $\mathring{\tau}(a)$ and $\mathring{\tau}(b)$, share a common $(D-1)$-simplex $\mathring{s}$ whose vertices are, say, sites $1, 2, \cdots, D$. Under the link-length variation (8.28), these two D-simplices transform from

$$\mathring{\tau}(a) \rightarrow \tau(a) \quad \text{and} \quad \mathring{\tau}(b) \rightarrow \tau(b) \ ; \tag{8.34}$$

viewed in $\Sigma(a)$ we have

$$\mathring{s} \rightarrow s(a) \ , \tag{8.35}$$

with each of its $\frac{1}{2}D(D-1)$ link-vectors changing from

$$\overset{\circ}{\ell}{}_{ij}^{\mu} \rightarrow \ell_{ij}^{\mu}(a) \tag{8.36}$$

where i and $j = 1, 2, \cdots, D$. Likewise, viewed in $\Sigma(b)$ we have

$$\overset{\circ}{s} \rightarrow s(b) \tag{8.37}$$

and

$$\overset{\circ}{\ell}{}_{ij}^{\mu} \rightarrow \ell_{ij}^{\mu}(b) \quad . \tag{8.38}$$

Since in accordance with (8.28)

$$\ell_{ij}(a)^2 = \ell_{ij}(b)^2 = \ell_{ij}^{\;2} \tag{8.39}$$

for all i and j from $1, 2, \cdots, D$, there exists a Lorentz transformation $U(b,a)$ which can transform the set $\{\ell_{ij}^{\mu}(a)\}$ to $\{\ell_{ij}^{\mu}(b)\}$, and therefore $s(a)$ to $s(b)$; i.e.,

$$\ell_{ij}^{\mu}(b) = U(b, a)_{\nu}^{\;\mu}\, \ell_{ij}^{\nu}(a) \tag{8.40}$$

for all i and $j = 1, 2, \cdots, D$. Under Lorentz transformations of both reference frames $\Sigma(a)$ and $\Sigma(b)$, we have on account of (8.31)

$$U(b, a) \rightarrow u(b)\, U(b, a)\, u(a)^{-1} \tag{8.41}$$

where $u(a)$, $u(b)$ and $U(b, a)$, being Lorentz transformations, all satisfy (8.32).

Next we discuss the parallel transport of any vector ϕ. We start with ϕ inside a D-simplex $\tau(a)$. Viewed in $\Sigma(a)$, we write $\phi = \phi(a)$, in accordance with (8.30). Because the interior of $\Sigma(a)$ is flat, the parallel transport of ϕ within $\tau(a)$ is trivial. When we transport ϕ from $\tau(a)$ to one of its adjacent D-simplices, say $\tau(b)$, ϕ becomes $\phi(b)$ [when viewed in $\Sigma(b)$], where $\phi(b)$ is <u>defined</u> to be

$$\phi^{\mu}(b) = U(b, a)_{\nu}^{\;\mu}\, \phi^{\nu}(a) \quad . \tag{8.42}$$

Clearly, $U(b, a)$ does not depend on which point in the interior of $\tau(a)$ we start from, nor on which point in the interior of $\tau(B)$ we end.

Continuing this way, we can parallel transport any vector ϕ along a general path P. It is convenient to think first of the path P in the original flat minkowski lattice. We assume P to be a continuous path starting from the interior of any D-simplex, say $\mathring{\tau}(a_1)$, passing successively through a sequence of D-simplices $\mathring{\tau}(a_2)$, $\mathring{\tau}(a_3)$, $\cdots$, and finally arriving at the interior of $\mathring{\tau}(a_n)$. In going from one D-simplex, say $\mathring{\tau}(a_i)$, to the next one $\mathring{\tau}(a_{i+1})$, the path P always passes through their common $(D-1)$-simplex (i.e., P never intersects any $(D-2)$-simplex, or $(D-3)$-simplex, etc.). After the link-length variation (8.28), P becomes a path from $\tau(a_1)$ through $\tau(a_2)$, $\tau(a_3)$, $\cdots$ to $\tau(a_n)$, the parallel transport of any vector ϕ along P is given by

$$\phi^{\mu}(a_n) \;=\; U(P)_{\nu}^{\ \mu}\, \phi^{\nu}(a_1) \qquad (8.43)$$

where

$$U(P) \;=\; U(a_n\,,\, a_{n-1}) \cdots U(a_2\,,\, a_1)\;. \qquad (8.44)$$

When $a_n = a_1$, then P is a closed path C, in which case (8.44) becomes

$$U(C) \;=\; U(a_1\,,\, a_{n-1})\, U(a_{n-1}\,,\, a_{n-2}) \cdots U(a_2\,,\, a_1)\;. \qquad (8.45)$$

Two different paths P (and therefore also two different C) that pass through the same sequence of $\tau(a_1)$, $\tau(a_2)$, $\cdots$ have the same $U(P)$ (or $U(C)$). Under arbitrary Lorentz transformations of $\Sigma(a_1)$, $\Sigma(a_2)$, $\cdots$ we have from (8.41)

$$U(P) \;\rightarrow\; u(a_n)\, U(P)\, u(a_1)^{-1}$$

and

$$U(C) \;\rightarrow\; u(a_1)\, U(C)\, u(a_1)^{-1}\;. \qquad (8.46)$$

The departure of $U(C)$ from the unit matrix implies that the space is curved. Note that C is a closed path passing from one D-simplex to the next D-simplex by going through their common $D-1$ simplex. Since the interiors of all D-simplices are flat and since $U(C)$ is independent of the precise passage of

C through these (D - 1) - simplices, the curvature must rest solely in the (D - 2) - simplices that C encircles. In any D - dimensional space, a (D - 2) - simplex, say γ, is shared by several D - simplices, say

$$\tau(a_1),\ \tau(a_2),\ \cdots,\ \tau(a_m) \tag{8.47}$$

in a sequential order; γ is also shared by an equal number of (D - 1) - simplices which are the common surfaces of these D - simplices, (8.47). For each γ, we introduce a closed path $C(\gamma)$ which encircles γ by following the order (8.47) and ending in $\tau(a_1)$. [As before, it is easiest to visualize the closed path $C(\gamma)$ as first drawn in the original flat minkowski space, before the variation (8.28).] Define

$$U_\gamma \equiv U(C(\gamma)) = U(a_1, a_m)\, U(a_m, a_{m-1}) \cdots U(a_2, a_1) \tag{8.48}$$

which depends only on γ and the reference system $\Sigma(a_1)$. When $D = 2$, γ is a site. When $D = 3$, γ is a link represented by a vector ℓ^μ; when $D = 4$, γ is a plaquette $\Delta^{\mu\nu}$, etc. In general, γ can be represented by a tensor of rank $D - 2$:

$$\gamma^{\mu_1 \mu_2 \cdots \mu_{D-2}}, \tag{8.49}$$

whose components are referred to the reference frame $\Sigma(a_1)$.

Given γ and $\tau(a_1)$ in (8.47), there remains the question of the sense of $C(\gamma)$, defined by the sequence (8.47) versus its inverse order. Without any loss of generality, we can set $C(\gamma)$ to be a planar figure; the two-dimensional area that it encloses, when viewed in $\Sigma(a_1)$ and set in the order (8.47), can be represented by a tensor of second rank $C(\gamma)^{\alpha\beta}$. We require that the sense of the curve $C(\gamma)$ and the tensor γ satisfy the usual right-hand relation; i.e.,

$$\epsilon_{\mu_1 \cdots \mu_{D-2}\alpha\beta}\, C(\gamma)^{\alpha\beta}\, \gamma^{\mu_1\mu_2 \cdots \mu_{D-2}} \tag{8.50}$$

is positive for all γ , where

$$\epsilon_{\mu_1 \cdots \mu_D} = \begin{cases} 1 & \text{if } \mu_1 \cdots \mu_D \text{ is an even permutation of } 1,2,\cdots, \\ -1 & \text{if } \mu_1 \cdots \mu_D \text{ is an odd permutation of } 1,2,\cdots, \\ 0 & \text{otherwise} . \end{cases} \tag{8.51}$$

Let $J^{\rho\sigma} = -J^{\sigma\rho}$ be the D-dimensional real matrix representation of the generator of the Lorentz group

$$\eta^{\sigma\nu} x^\rho \frac{\partial}{\partial x^\nu} - \eta^{\rho\nu} x^\sigma \frac{\partial}{\partial x^\nu} \tag{8.52}$$

where x^ρ is the coordinate of points inside $\tau(a_1)$, with respect to the frame $\Sigma(a_1)$. When $D = 4$,

$$J^{01} = -J^{10} = \begin{pmatrix} 0 & 1 & 0 & 0 \\ 1 & 0 & 0 & 0 \\ 0 & 0 & 0 & 0 \\ 0 & 0 & 0 & 0 \end{pmatrix} ,$$

$$J^{12} = -J^{21} = \begin{pmatrix} 0 & 0 & 0 & 0 \\ 0 & 0 & 1 & 0 \\ 0 & -1 & 0 & 0 \\ 0 & 0 & 0 & 0 \end{pmatrix} , \tag{8.53}$$

etc.

The action of the discrete gravity is defined to be, apart from a proportionality constant,

$$\sum_\gamma \text{trace} \left(\epsilon_{\rho\sigma\mu_1 \cdots \mu_{D-2}} J^{\rho\sigma} \gamma^{\mu_1 \cdots \mu_{D-2}} U_\gamma \right) . \tag{8.54}$$

Because of the trace, the above expression is independent of the frame $\Sigma(a_1)$. Also, if the link-variation (8.28) is merely due to a coordinate change from $\overset{\circ}{x}(i) \to \overset{\circ}{x}(i)'$ with $\overset{\circ}{\ell}_{ij}{}^\mu \to \ell_{ij}{}^\mu \equiv [\overset{\circ}{x}(i)' - \overset{\circ}{x}(j)']^\mu$, then $U_\gamma = 1$ and (8.54) remains zero. The invariances of the action under arbitrary independent Lorentz transformations of all $\Sigma(a)$ and under arbitrary independent translations (when U_γ is near 1) of all sites are the discrete equivalent of Einstein's arbitrary coordinate transformations in the continuum theory.

We note that when $D = 4$ the discrete QCD action in a flat minkowski space (i.e., without gravity) is, similar to (7.6),

$$A_{QCD} = \frac{1}{g^2} \sum_{\Delta_{ijk}} \lambda_{ijk} \ \mathrm{Re} \left[\mathrm{trace} \ (1 - u_{ij} u_{jk} u_{ki}) \right] \qquad (8.55)$$

with u_{ij} the SU(3) matrix associated with link u_{ij} and

$$\lambda_{ijk} = \frac{\text{area of the dual of } \Delta_{ijk}}{\text{area of } \Delta_{ijk}} \quad ; \qquad (8.56)$$

the corresponding generating function is

$$Z_{QCD} = \int e^{i A_{QCD}} \left[d^4 x_i \right] \left[du_{ij} \right] \quad . \qquad (8.57)$$

When we include gravity, the action A and the generating function Z become

$$A = A_{gravity} + A_{QCD} \qquad (8.58)$$

and

$$Z = \int e^{iA} \left[d\ell_{ij}^{\ 2} \right] \left[du_{ij} \right] \qquad (8.59)$$

where

$$A_{gravity} = \frac{1}{G} \sum_{\Delta_{ijk}} \mathrm{trace} \left(\epsilon_{\rho\sigma\mu\nu} \ J^{\rho\sigma} \ \Delta_{ijk}^{\mu\nu} \ U_{\Delta_{ijk}} \right) , \qquad (8.60)$$

$J^{\rho\sigma}$ and $U_{\Delta_{ijk}}$ are both 4×4 real matrices defined by (8.53) and (8.48); A_{QCD} remains given by (8.55), but with the area Δ_{ijk} and its dual-area calculated according to the new length-squares, given by (8.28). The integration $\left[d^4 x_i \right]$ in (8.57) is now replaced by $\left[d\ell_{ij}^{\ 2} \right]$ in (8.59). [When some of the $\ell_{ij}^{\ 2}$ become too large, the lattice has to be re-linked.] In both (8.57) and (8.60) we keep the total number of lattice sites N and the sum Ω of all flat simplitial volumes fixed; their ratio $N/\Omega = \ell^{-4}$ determines the fundamental length ℓ in the discrete theory.

Comments

1. The coupling g^2 of QCD is dimensionless, but the gravitational coupling G is of dimension $(\text{length})^2$. We have

$$G \sim (\text{Planck length})^2 \, . \tag{8.61}$$

2. In the weak coupling limit, $g^2 \to 0$ and $G \to 0$ the discrete action approaches the usual continuum theory.

3. Only the simplest forms of the discrete actions for QCD and gravity are given in (8.55) and (8.60). If one wishes, one may consider more complicated invariant actions; these would correspond to different discrete theories, although with the same continuum limit, if it exists. [For example, inside the trace we may add higher order polynomials in $u_{\Delta_{ijk}}$ for QCD and $U_{\Delta_{ijk}}$ for gravity.] Further details will be given elsewhere.[21]

IX. CONCLUDING REMARKS

In this series of lectures, I have tried to show how by viewing time as a dynamical variable, a fundamental length ℓ can be introduced which removes all ultra-violet divergences and thereby makes quantization of gravity possible. Because all symmetries, unitary and space-time transformations, are respected, even for renormalizable theories such as QCD the discrete mechanics gives new insight into their solutions. At the minimum this represents an advance in theoretical techniques; at the maximum it may change our basic concept of space-time.

Of course, we do not know how nature really works, and there is no evidence that the discrete mechanics is more appropriate than the usual continuum mechanics. However, once a theoretical possibility of such scope is open, it seems difficult not to face the intellectual challenge. Even at the present stage of enormous imperfection, I find aspects of this new mechanics appealing. I hope, during the past four days, you have shared with me some of my fun in developing this program, and some of my excitement in anticipating its future.

References

1. N. H. Christ, R. Friedberg and T. D. Lee, Nucl. Phys. B202, 89, B210, [FS6], 310, 337 (1982).
2. T. D. Lee, Phys. Lett. 122B, 217 (1983); CU-TP-266 (to be published in the proceedings of the Shelter Island II Conference).
3. R. Friedberg and T. D. Lee, Nucl. Phys. B225 [FS9], 1 (1983).
4. H. C. Ren, CU-TP-272 (Nucl. Phys.); R. Friedberg and H. C. Ren, CU-TP-268 (Nucl. Phys.).
5. C. Itzykson, in Proceedings of the Trieste Workshop, eds. R. Lengo et al. (Singapore, World Scientific, 1983); E. J. Gardner, C. Itzykson and B. Derrida (J. Phys. A); C. Itzykson, "Fields on a Random Lattice," Saclay DPh. G/SPT/83/148.
6. J. C. Maxwell, in A Treatise on Electricity and Magnetism (New York, Dover Publications, 1954).
7. S. R. Lundeen and F. M. Pipkin, Phys. Rev. Lett. 46, 232 (1981).
8. G. Feinberg (private communication); L. Bracci, G. Fiorentini, G. Mezzorani and P. Quarati, Phys. Lett. 133B, 231 (1983).
9. B. Adeva et al., Phys. Rev. Lett. 48, 967 (1982); Proceedings of the 1981 International Symposium on Lepton and Photon Interactions at High Energies, ed. W. Pfeil (Bonn, Physikalisches Institut, 1981).
10. L. A. Fajardo et al., Phys. Rev. D24, 46 (1981); S. J. Lindenbaum, in Particle and Interaction Physics at High Energies (Oxford, Oxford University Press, 1973).
11. G. Parisi and N. Sourlas, Nucl. Phys. B206, 321 (1982).
12. J. M. Ziman, Models of Disorder (New York, Cambridge University Press, 1979); J. L. Meijering, Philips Research Report 8, 270 (1953).
13. K. Wilson, Phys. Rev. D10, 2455 (1974).
14. M. Luscher, Nucl. Phys. B180 [FS2], 1 (1981).
15. N. H. Christ, R. Friedberg and T. D. Lee (to be published).
16. S. Lindenbaum, this proceeding.
17. N. H. Christ and A. E. Terrano, CU-TP-259 (Nuclear Methods and Instrumentation).
18. R. Friedberg, T. D. Lee and H. C. Ren (to be published).

19. T. Regge, Nuovo Cimento 19, 558 (1961). Our treatment is based extensively on Regge's idea.

20. M. Rocek and R. M. Williams, Phys. Lett. 104B, 3. (1981), in Quantum Structure of Space and Time, eds. M. J. Duff and C. J. Isham (Cambridge, Cambridge University Press, 1981).

21. G. Feinberg, R. Friedberg, T. D. Lee and H. C. Ren (to be published).

D I S C U S S I O N S

CHAIRMAN : T.D. LEE

Scientific Secretary: S. Catto

DISCUSSION 1

- *WIGNER* :

I was very impressed by, and largely in agreement with, your whole very interesting presentation. I was, however, missing one point which I consider to be important. It is based on the fact that an observer cannot ascertain the situation which prevails over space simultaneously with his observation. At best he can recognise the situation on his negative light-cone. This leads to the suggestion to introduce, instead of the time t, the variable $\tau = t + r/c$. The equations in terms of this τ become a bit complicated but the question prevails whether this change of variables is necessary ?

- *LEE* :

Yes, I assume that you are considering the classical version, not the quantum mechanical one. There is a difference between the two. In the usual continuum mechanics the classical equation of motion is second order in time whereas a quantum mechanical one is a first order differential equation in time. If you consider the quantum mechanical version, then it would be much more reasonable to start with negative infinity on the light-cone.

- *WIGNER* :

I am afraid my remark does apply principally to quantum mechanics. I believe we should try to modify our equations so that they use the variable τ instead of t.

- *AMI* :

The discrete time units vary along the path. Does that mean we have to measure them at discrete space-time points ?

- *LEE* :

The discreteness is a discrete number of measurements that you can make. The result is a discrete set of numbers which correspond to the results of the measurements.

- *AMI* :

But each of them do not necessarily satisfy the energy conservation law, or do they ?

- *LEE* :

If they are solutions of the equations, then by definition they satisfy energy conservation, since half of the equations are the equations of energy conservation. Going back to your first question, you do not have the liberty to choose space-time points arbitrarily. If you do, then you do not have the conservation of energy because then you would not have infinitesimal time translation operation in your scheme, without which energy cannot be conserved.

- *AMI* :

In you formalism relative positions of adjacent times vary arbitrarily whereas in Feynman's path integral formalism they are chronologically ordered. Then, what happens to causality ?

- *LEE* :

Since you have a one-dimensional distribution you can always by permutation re-arrange time variables in a chronologically increasing order. In the relativistic version which we have not touched yet, space and time measurements will just result in N points. The re-arrangement will not be so meaningful nor is it necessary. Feynman's version is diffrent because t is just a parameter, so you parametrize in a constant division, then go the limit $N \to \infty$. In our version t is a response to the measurement.

- *CATTO* :

Could you please tell us how does one determine the fundamental length ?

- *LEE* :

The $1/\rho$ is a constant of the theory. It of course affects the energy levels, forward dispersion relations, e^+e^- collisions, ...etc. So any deviations from the continuum theory will give you an indication of how big or how small it is. At present we do not know of any violation, unless you want to take Pipkin's experiment very seriously where a deviation from the QED predictions for the hydrogen spectrum is found which happens to agree in sign with the deviations predicted by our approach. This way you get $\ell \approx 1.6 \times 10^{-14}$cm. Conversely if ℓ is really that big, then it could have been measured from the Lamb shift experiment. From our present knowledge of forward dispersion relations, e^+e^- annihilations, ... etc., probably ℓ cannot be that big; it must be less than 10^{-15}cm.

- *BAGGER* :

I would like to know how canonical quantization works in this picture ?

- *LEE* :

Canonical quantization is more complicated in discrete mechanics. One may view this as an advantage, because ordinary canonical quantization introduces an artificial separation between space and time.

- *BAGGER* :

Does the density ρ appear in the canonical commutation relations ?

- *LEE* :

Yes, it does. The canonical commutation relations are more complicated than usual.

- *NEMESCHANSKY* :

How do you define the canonical momenta ? What is the canonical momentum of the dynamical variable t and what is the commutation relation r_n, t_n ?

- *LEE* :

The canonical momenta for r_n and t_n are defined as $\frac{\partial}{\partial r_n}$ and $\frac{\partial}{\partial t_n}$;

r and t commute at any given point: $[r_n, t_n] = 0$. However, $(r_n)_{op}$ and $(t_n)_{op}$ do not commute with $(r_{n+1})_{op}$ and $(t_{n+1})_{op}$. Their eigenstates are related by

$$\langle r_n, t_n \mid r_{n+1}, t_{n+1} \rangle \sim e^{iA}$$

- *NEMESCHANSKY* :

Do you get angular momentum conservation ?

- *LEE* :

Since discrete mechanics is rotationally symmetric, angular momentum conservation can be derived explicitly.

- Question :

In your formalism you treat position and time on an equal footing. However, in Schrödinger equation you have first derivative in time and second derivative in position. What happens to Schrödinger equation in your formalism ?

- *LEE* :

If you keep all the powers of ℓ/T to arbitrary power but neglect things which are exponentially small, then the Green's function $G(\tau)$ satisfies the Schrödinger equation with H as the Hamiltonian and T as the time. So this represents the effective Schrödinger equation.

- *ZEPPENFELD* :

If you take the limit $N \to \infty$ in your harmonic oscillator example, I expect the distribution of space points to approach a linear combination of sines and cosines. Now what happens to the distribution of the time differences $t_n - t_{n-1}$? Does it also approach a stable curve ?

- *LEE* :

You can work it out analytically by a systematic expansion. Consider N to be very, very large, with T fixed. So density ρ becomes very large also. You can start with the zeroth-order mean solution and expand in powers of ℓ, where ℓ is the ratio T/N, which becomes very small when N becomes large. In zeroth-order on gets the familiar sin and cos solutions. To the next power in ℓ, you get the distribution of time spacings which fluctuates with $\frac{1}{4}$ x period of the harmonic oscillator. This information is completely lost in the usual continuum case.

- *ZEPPENFLED* :

If you use another discretization of the harmonic oscillator, do you still get this result ?

- *LEE* :

To the lowest order, nothing changes. Differences will only show up at higher powers of ℓ.

- *ZEPPENFELD* :

So, is there any physical meaning to the Δt distribution ?

- *LEE* :

Of course, if you change your potential you will also change the Δt distribution to the first order. So, it just reflects properties of the potential.

- *BALLOCCHI* :

I believe that the idea of considering time as a dynamical variable is very exciting. Can you briefly mention the new results you obtain in describing the real world and that you obtain by means of other existing theories ?

- *LEE* :

This theory reduces the existing convention theory to an approximation. As you will see in the case of quantization of gravity the conventional method fails.

- *HEISE* :

Doesn't the assumption of the dynamical time variable lead to a more complicated formulation of space-time structure ?

- *LEE* :

I am afraid that complication is a subjective judgment.

- *LAMARCHE* :

Do you personally believe that the discrete length will be the inverse of Planck's mass ?

- *LEE* :

I consider the Planck length (10^{-33}cm) to be a lower bound for ℓ. On the other hand the existing experiments put an upper bound of 10^{-15}cm. So, ℓ has to lie somewhere in this interval.

- *BERNSTEIN* :

In the non-relativistic case do all the inertial observers measure the same ρ ?

- *LEE* :

Yes, ρ is a constant of motion. But actually it is not difficult to make ρ a dynamical variable. I may get to that later on .

- *BERNSTEIN* :

Is CPT conserved in discrete mechanics ?

- *LEE* :

Yes, if you stick to the Lorentz group.

- *HOU* :

So, due to Lorentz invariance, there is one fundamental constant ρ, not four or more ?

- *LEE* :

It better be only one.

- *KLEVANSKY* :

What is meant by the term "quasi local" in your theory of discrete mechanics ?

- *LEE* :

Averages over distances larger than the fundamental length ℓ.

- *KLEVANSKY* :

Is their special significance the fact that the energy conservation relation now arises independently of the Newton's law relation ?

- *LEE* :

Only in that it arises in a new way through the fact that time is now a dynamical variable.

- *OHTA* :

What is the reason for introducing a fundamental length scale and discretizing time ?

- *LEE* :

First, I think it is a possibility. The present continuum theory has serious problems. Of course not for the classical theory, nor for usual quantum mechanics, nor for QED or QCD. But if you try to quantize gravity, you do not know how to do that. Presumably gravity is a fundamental theory of nature so you have to quantize it. Second, in the present theory you have some 20 parameters. Within the context of local quantum field theories we only have a limited number of possible theories, namely, the renormalizable ones. Within the context of renormalizable theories, the only thing which we have not exhaustively studied is the size of the gauge group, but we have already gone through very large groups like SU(10), etc., and the number of parameters does not seem to decrease. It seems, therefore, useful to explore other possibilities beyond the present set of renormalizable theories.

- *TAVANI* :

When you neglect terms that are exponentially small in the expansion of $G(\tau)$, retaining only the inverse powers of the density of measurements you obtain a not very beautiful non-local theory. How could you get rid of this non-locality ?

- *LEE* :

In this approximation, the non-locality is only in space, not in time. The situation is not different from that in the usual nonrelativistic quantum mechanics. The usual unitarity is only with respect to time, not with respect to space.

- *DATÉ* :

It seems that to make a transition from the continuum mechanics to the discrete mechanics one introduces a larger number of generalized co-ordinates e.g. 1-particle in 3-dimension has three generalized co-ordinates in the continuum mechanics while one has 4N number of generalized co-ordinates $\vec{x}_1, \ldots, \vec{x}_N, t_1, \ldots, t_N$, in the discrete mechanics. After this essential modification one repeats the standard steps of proposing a variational principle, and a discretized lagrangian. A transition to discrete quantum mechanics may be made in the canonical way. One then may expect the usual Schrödinger equation to be replaced by many Schrödinger equations, e.g., $i\frac{\partial}{\partial t_n} |\psi\rangle = H_n |\psi\rangle$, $n = 1, \ldots, N$. Is this true ?

- *LEE* :

No. In making the passage to quantum mechanics one considers the path integral approach as more fundamental, modifies it by introducing integrations over the $t_1, \ldots, t_N$'s. That defines the quantum mechanics. A modified (new) Schrödinger-type equation can be derived from this definition.

- *DATÉ* :

Is the dynamics supposed to determine the fundamental scale ?

- *LEE* :

Dynamics and observations together determine the fundamental scale.

- *DATÉ* :

Suppose one considers a particular model system and computes all the relevant observables. Do the results depend upon $\rho = N/T$, or on N and T separately ?

- *LEE* :

Given the total time T over which the system is "observed", since $N = T/\ell$, the results depend upon the ratio $\rho = \frac{N}{T}$.

- *D'AMBROSIO* :

The discrete mechanics has to reproduce the correct limit in the continuum limit. I can understand how this happens in case of linear equations, but maybe in the case of non-linear equations the fluctuations can be very large.

- *LEE* :

Yes, but only for a zero-measure set of points.

- *BURGESS* :

Is your modified Hamiltonian H positive definite ?

- *LEE* :

I think H has a lower bound (i.e., can be made positive by a shift) if the corresponding H in the usual continuum problem does.

DISCUSSION 2 (Scientific Secretaries: G. Heise and M. Tavani)

- *KLEVANSKY* :

In the massless spin-0 field case, can the identities involving the length ℓ_{ij} and λ_{ij} be simply derived ? What do they mean physically ?

- *LEE* :

I did not have time this morning to derive them but I will do it now. Let ℓ_{ij} be a link in Euclidean space and $\lambda_{ij} = \frac{S_{ij}}{\ell_{ij}}$ where S_{ij} is the dual "volume"; if you were in two dimensions the S_{ij} would be also a length, in three dimensions this would be an area, in four dimensions this would be 3-dimensional volume. The formula $\sum_j \vec{\ell}_{ij}\lambda_{ij} = 0$ is valid for an arbitrary random distribution of points.

Consider a vector $\vec{E}$ in a D-dimensional space, you can think it like an electric field, that is a constant. Thus, by definition

$$\int \vec{\nabla}.\vec{E}d^D_r = 0$$

furthermore, by Gauss theorm and by the definition of our quantities :

$$\int \vec{\nabla}.\vec{E}d^D_r = \int \vec{E}.d\vec{S} = \vec{E}.\sum_j \hat{\ell}_{ij}S_{ij}$$

and this gives you the first identity.

This identity incidentally proves that

$$\sum_j \lambda_{ij} (\phi_i - \phi_j)$$

is a good Laplacian. In fact if the potential Φ is a linear function in r (when $j_i = 0$) :

$$\Phi_i = \Phi_o + \vec{E}.\vec{r}_i$$

thus $$(\phi_i - \phi_j) = \vec{E}\cdot(\vec{r}_i - \vec{r}_j).$$

Because $\vec{r}_i - \vec{r}_j = \vec{\ell}_{ij}$ we get

$$\sum_{ij} \vec{\ell}_{ij} \lambda_{ij} = 0 .$$

To prove the other identity

$$\sum_{ij} \ell^\mu_{ij} \ell^\nu_{ij} \lambda_{ij} = 2\Omega\delta^{\mu\nu}$$

we have to consider the next step of complexity.

Consider the quantity

$$\frac{\partial(\vec{r} - \vec{r})^\mu}{\partial r^\nu} = \delta^{\mu\nu}$$

the integration over D-dimensional space gives the volume of the dual cell ω_i *)

$$\omega_i \delta^{\mu\nu} = \int \frac{\partial(\vec{r} - \vec{r}_i)^\mu}{\partial x^\nu} d^D r$$

by Gauss theorem

$$\int \frac{\partial(\vec{r} - \vec{r}_i)^\mu d^D r}{\partial x^\nu} = \int (\vec{r} - \vec{r}_i)^\mu dS^\nu$$

consider the last integration on the surface. $\hat{\lambda}^\nu_{ij}$ is a constant

*) N.H. Christ, R. Friedberg and T.D. Lee. Nucl.Phys.B.202,89(1982)

vector located on the surface of the cell; $\vec{r}$ is the position vector of a point on the surface and $\vec{r}_i$ is constant. So $\vec{r}$ integrated on the surface is exactly the center of mass of the surface $\vec{r}_c$ and thus

$$\int (\vec{r} - \vec{r}_i)^{\mu} dS^{\nu} = \sum_j (\vec{r}_c - \vec{r}_i)^{\mu} \hat{\ell}^{\nu}_{ij} \, S_{ij}.$$

From these relations, after some manipulation we can derive

$$\sum_i \omega_i \delta^{\mu\nu} = \Omega \delta^{\mu\nu} = \frac{1}{2} \sum_{ij} \ell^{\mu}_{ij} \ell^{\nu}_{ij} \, \lambda_{ij} \; .$$

This result can be generalized not only to links but to plaquettes, to tetrahedra, etc. This enables one to go on to higher spin fields.

- *AMI* :

In the spin zero field case do you have a theoretical framework from which you can obtain analytical results ?

- *LEE* :

Yes, this was developed by Friedberg and me in the same way as in the non-realistic case in terms of a cluster expansion and corrections can be systematically computed for small ℓ.

- *POSCHMANN* :

In your example of the harmonic oscillator the spacing between two space time points was fixed. It was a solution of the equations. Why do you consider now random distribution of points ?

- *LEE :*

The previous harmonic oscillator example was considered in terms of classical mechanics in which the solution is determined. In the quantum case the position of your points is the result of your measurement; it is not fixed and it can vary. The result of these measurements can be associated with a quantum probability amplitude. In the integration over the distribution of these points, the integrand is identical to that of a random lattice.

- *OGILVIE :*

You have shown how to calculate matrix elements using the discrete mechanics version of the path integral. Also you have shown us how to derive an effective Hamiltonian operator. Is there a conventional Hilbert space formalism for discrete quantum mechanics derived, for example, using Wightman reconstruction theorem. If so, how does it differ from conventional quantum mechanics ?

- *LEE :*

G(T) is an operator and it connects any function at initial time $\psi(0, \vec{r}_o)$ into $\psi(T, \vec{r}_j)$; in this case this matrix is well defined in the conventional Hilbert space. In addition G(T) becomes a unitary matrix if T becomes large if you neglect terms of the order $O(e^{-T/e})$, keeping only $(\ell/T)^n$ terms.
(Friedberg - T.D. Lee Theorm)

- *LEURER :*

What is the role of the Jacobian in your functional integration?

- *LEE :*

The Jacobian plays a role in the next order correction in our integration. This is true in the familiar WKB case also; the variation of phase gives the classical path and the Jacobian gives the next order correction.

- *LEURER* :

Is the choice of links in a random lattice unique ?

- *LEE* :

Yes, it is unique if you want to maintain rotational symmetry.

- *VAN DEN DOEL* :

What about fermion fields in a random lattice ?

- *LEE* :

The introduction of spin $\frac{1}{2}$ fields in a lattice has its own characteristics and it is not completely solved yet.

It is not difficult to write the discrete version of Dirac equation

$$\sum_{\substack{j \\ j \text{ linked to } i}} \gamma^{\mu} \ell^{\mu}_{ij} \lambda_{ij} \psi_j = 0$$

The trouble with fermion theory is that the discrete action

$$A_d = \sum_{\ell_{ij}} \psi_i \gamma^{\mu} \ell^{\mu}_{ij} \lambda_{ij} \psi_j$$

is not a good action owing to the absence of chiral anomaly. In a lattice either you have chiral symmetry or you do not. If there is chiral symmetry then there is no chiral anomaly. In a regular lattice you have 2^D zero energy modes, where D is the dimension, but in the random lattice the number of zero modes is always 2, irrespective of dimension of space. But 2 is sufficient to not giving you the chiral anomaly.

- *BURGESS* :

What is the role played by the anomaly in continuum field theory and in a lattice theory ?

- *LEE* :

The continuum field theory has a chiral anomaly given by the well-known triangle diagram. In the discrete theory there is no anomaly because everything is well-defined.

- *BURGESS* :

In field theory you have a variety of Green's functions, such as retarded, advanced and Feynman ones. In your discrete theory, where the field depends on discrete variables, how do you formulate the analogous boundary value problem ?

- *LEE* :

In the continuum field theory the Green's functionals satisfy a second order differential equation. The choice of Green's functions depend on the position of the singularity in the complex k^2-plane.

In the discrete case we do not as yet know the complete analytical solution for the propagator.

- *TAVANI* :

We know that different actions on the lattice can have the same correct continuum limit. That is the degree of arbitrariness of your action in your lattice.?

- *LEE* :

This is analogous to the transition of classical mechanics to quantum mechanics. Knowing the classical Hamiltonian, we only know the quantum Hamiltonian in the limit $\hbar = 0$. Likewise, knowing the continuum action we only know the discrete action in the limit $\ell = 0$.

- *CATTO* :

How do you get the normalization of your Green function to obtain the right Jacobian ?

- *LEE* :

By requiring $G(T) \rightarrow$ unitary matrix when $T >> \ell$.

DISCUSSION 3 (Scientific Secretaries: R.H. Bernstein, S.P. Klevansky)

- *OGILVIE* :

Can you calculate Λ_{QCD} as a function of the fundamental density ρ of the random lattice ?

- *LEE* :

Yes. This is equivalent to computing 1-loop diagrams in the weak coupling limit for the random lattice; this is analytically difficult, but should present no problem numerically. I want to point out that in the strong coupling limit it is not necessary to calculate Λ_{QCD}, because we can analytically compute the string tension and the glueball structure and then use the scale determined from the string tension.

- *OGILVIE* :

I am concerned about the amount of computer time involved in making accurate numerical calculations on the random lattice because of the high connectivity of the lattice. What can you tell me about this ?

- *LEE* :

The amount of time needed for the random lattice is not much more than is needed for the regular lattice; it takes a little longer for the random lattice because you have to set up the links, but after this each sweep takes about the same time. However, the amount of time taken is not the key issue: the advantage of the random lattice is that it respects Poincaré invariance at each step.

Of course, we still have more calculations to do: this is still at an early stage, and we have not had the time to complete most of the interesting calculations.

- *BURGESS* :

In continuum theories there are phenomena like instantons and monopoles which rely on the topology of the field configurations and spacetime. Do these phenomena arise in the discrete version, and if so, how ?

- *LEE* :

Yes, but we have not resolved the full extent of topological solitons in the discrete case (except in the continuum limit). For the non - Abelian gauge theory on a lattice, monopoles and other topological soliton solutions do exist. However, I believe they are only metastable. In the limit $\ell \to 0$ they become stable.

- *HOU* :

How do you get quantum numbers such as spin and mass for glueballs on a lattice ?

- *LEE* :

The calculation is as follows: you start with two plaquettes and do a calculation similar to that of the Wilson loop. In the Wilson loop calculation you evaluate

$$\frac{1}{Z}\int \left[dr_i\right]\left[du_L\right] W_L e^{-A/g^2}$$

Here you evaluate a similar quantity, using two plaquettes $\overset{3}{\triangle}_{1\;\;2}$ and $\overset{3'}{\triangle}_{1'\;\;2'}$. You calculate the connected graphs, shown symbolically as :

$$\frac{1}{Z}\int \left[dr_i\right]\left[du_L\right] e^{-A/g^2} \;\overset{3}{\triangle}_{1\;\;2}\;\; \overset{3'}{\triangle}_{1'\;\;2'} \;.$$

The first plaquette 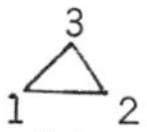

is characterized not only by U_{12}, U_{23} and U_{31}, but also by its orientation; one can attach a coordinate system to the plaquettes and define Euler angles α,β,γ and α',β',γ' for each plaquette respectively. The graph is now a function of the spacing between the two plaquettes, which give the time T, and their relative orientation. First integrate over the group elements; now you have the propagator from one plaquette to the other as a function of the Euler angles. Let us fix the orientation of one plaquette because only the relative orientation is of importance. Now we multiply by $D^J{}_{mm'}(\alpha\beta\gamma)$ to analyse the system in terms of the various angular momentum states. When you take the limit at large T, the dominant term is the e^{-m_jT} where m_j refers to the lowest state of angular momentum j.

- *TAVANI* :

In the random lattice, the metric depends on the potential. In the non-relativistic case we recover the usual independence only in the asymptotic region and when the potential is zero. What happens in the relativistic case ?

- *LEE* :

This is an unsolved and fascinating problem; the field is just too new.

- *TAVANI* :

In continuum QCD we have at least three different mass scales: the confinement scale, the $U_A(1)$ symmetry breaking mass, and the chiral symmetry breaking scale. Is there some hope of understanding these scales and of finding relationships between them ?

- *LEE* :

I hope so. For confinement, I think we can calculate the answer in a few years. The large distance behavior we can understand; to explain the phenomenon of finding <u>small</u> masses at short distances will require new ideas.

- *VAN DER SPUY* :

Is there a strong CP or CT problem on the random lattice ?

- *LEE* :

No, both are conserved.

- *VAN DER SPUY* :

Are there instantons in this theory ?

- *LEE* :

In the discrete theory it is not clear whether there are instantons or not. It may well be that there is a leak because the topology is not enclosed. There may be an absolute vacuum, but this has not been worked out yet.

- *VAN DER SPUY* :

Do you expect axions to exist ?

- *LEE* :

That is not my problem.

- *VAN DER SPUY* :

Do you regard the Higgs mechanism for giving fermions and bosons mass as natural ?

- *LEE* :

No, I think it is contrived.

- *HWANG* :

You have shown the correspondence between the strong coupling limit and the relativistic string. However, the relativistic string can only be quantized in 26 dimensions. Does this problem arise in your strong coupling limit ?

- *LEE* :

The correspondence between the strong coupling limit and the relativistic string is only approximate through the first order correction term in a series expansion. The complete theory which is quantized is <u>not</u> the string theory so the problem does not arise.

- *OHTA* :

Our favourite theories tell us all low energy theories are renormalizable, but in your approach no special meanings are given to such theories. Can you explain in your approach why only renormalizable theories are relevant in the low energy world ?

- *LEE* :

Yes. Assuming the theory has a fundamental length ℓ, at a large distance $r >> \ell$, one may set $\ell \to 0$ and therefore the theory has no scale. This automatically leads to renormalizable theories like QED and QCD. The part of QCD that contains a scale, quark masses and so forth, is the part we do not understand.

- *VAN DEN DOEL* :

Can the weak interactions be put on the lattice ?

- *LEE* :

Yes.

- *VAN DEN DOEL* :

Isn't there a chiral symmetry problem ?

- *LEE* :

Yes if one wants a deep understanding, but no if one just wants to do computations. Phenomenologically it is quite easy to break the chiral symmetry "by hand on a lattice.

- *LEURER* :

You mentioned that the phase transition in the regular lattice is due to the loss of rotational invariance.

- *LEE* :

At least the Λ-like behavior of the specific heat curve and the regularity of the lattice seem to have a close relationship.

- *LEURER* :

Why do you have a phase transition in U(1) for both the regular and random lattices ?

- *LEE* :

You must have on because in the strong coupling limit the theory is confined in both lattices and you know that continuum QED (i.e. weak coupling) is not confirmed, so there must be a phase transition.

- *BALL* :

Does the discrete dynamics of the pure gauge theory depend on the representation of the group and not just its Lie Algebra as in the continuum case ?

- *LEE* :

Yes, because different elements of different groups have different dynamics.

- *BALL* :

Do you know, for instance, whether SO(3) still has a phase transition in the random lattice case ?

- *LEE* :

SO(3) has not been computed yet. I expect there will be no phase transition.

DISCUSSION 4 (Scientific Secretaries: G. Hou, and M. Ogilvie)

- *KLEVANSKY* :

You showed us a comparison of two random lattices; one generated by choosing your three sites to pass through a circle, the other by the square method. Is the one method preferred to the other and if so, why ?

- *LEE* :

In the Euclidean case, it is better to use the circle method because one would like the linking algorithm as invariant as possible. In Minkowski space one works with parallelopipeds.

- *BURGESS* :

Suppose one used a lattice defined by analytic continuation from Euclidean space rather than using the Minkowski lattice, would you get different physical results ?

- *LEE* :

This question is a deep one and I do not know the answer. If one uses different lattices with different definitions of linking one might get different results. I don't know yet if there is a criterion one could use to choose between the different possibilities.

- *AMI* :

How would you choose a coordinate system in each simplex ?

- *LEE* :

The choice of coordinates is arbitrary. One wants the final answer to be independent of the choice made, which is why one takes the trace in the action.

- *WIGNER* :

If you divide 4-space into a 4-dimensional simplex and measure only distances between points can you get all the invariants of the curvature ?

- *LEE* :

For this discrete geometry yes, it is sufficient to know the lengths of each link. From these one can get all the invariants.

- *VAN DEN DOEL* :

There is an infinite number of choices for the lattice action, reducing to Hilbert action in the continuum limit. Quantum gravity is non-renormalizable so all these actions will lead to very different theories, how are we to choose a particular action ?

- *LEE* :

By requiring the discrete action to be invariant under the general coordinate transformation, a unique action does arise.

- *VAN DEN DOEL* :

Do you think supergravity can be put on a lattice ?

- *LEE* :

Yes, sure.

- *VAN DEN DOEL* :

I don't see how.

- *LEE* :

It will be a challenge for you.

- *SIMIĆ* :

Doesn't your theory explicitly violate dilation invariance ?

- *LEE* :

Yes, of course. A fundamental length scale has been introduced.

- *OGILVIE* :

In your random lattice theory of gravity is the "gauge volume" infinite, as in conventional quantum gravity.

- *LEE* :

When ℓ becomes very large, one must re-link the points, thereby avoiding the infinite volume difficulty.

- *BALL* :

Could you show us what Einstein's equations look like on the lattice ?

- *LEE* :

For that I have to refer you to a forthcoming paper by Friedberg and myself.

LATTICE CALCULATIONS IN GAUGE THEORY

Claudio Rebbi

Physics Department
Brookhaven National Laboratory
Upton, NY 11973

1. INTRODUCTION

In 1975 Kenneth Wilson gave in Erice a series of lectures[1], where he showed how the lattice regularization of quantum gauge theories could provide a clue to the understanding of strong coupling phenomena and, ultimately, of hadronic dynamics. The expectations raised by Wilson's pioneering work[1-2] have been largely fulfilled. The lattice formulation, coupled to powerful numerical techniques, has allowed the derivation of a variety of important results, mainly for the theory of strong interactions (Quantum Chromo Dynamics or QCD), which could not have been attained by more conventional methods relying on perturbative expansions. We are now at a stage where calculations of the spectrum of hadrons, entirely from first principles, are being performed. While the outcomes of different investigations sometimes reveal discrepancies, a consequence of current computational limitations, the overall trend of the results is very encouraging. It appears likely that, with computer developments lying just ahead of us and, undoubtedly, with refinements in our numerical techniques, within a few years we shall be able to calculate the properties of QCD much as it is possible to investigate atomic physics starting from the Schrödinger equation.

In two lectures it would not be possible to give a comprehensive review of lattice gauge theories and their applications. Too much work has been done in this field to be summarized in two hours. Thus, rather than attempt a global exposition, I shall concentrate on a few topics. After a very concise recapitulation of some basic facts, I am planning to discuss the capabilities and limitations of current computational techniques, to conclude with an outline of what is presently one of the most important endeavors, namely the calculation of the spectrum in a theory of quarks and gluons. More complete accounts of lattice gauge theories may be found in several reviews and extensive lecture notes (see, for instance, Refs. 3 and 4).

2. BASICS

The main motivation for defining a quantum field theory on the lattice is to achieve a regularization (of ultraviolet divergences) which does not rely on a perturbative expansion. Computations at strong and intermediate coupling become possible. In the case of gauge field theories the lattice regularization preserves explicit gauge invariance.

Most frequently the lattice is hypercubic in Euclidean space time [but other geometries and also random lattices have been considered. See Prof. Lee's lectures at this school.] Let a be the lattice spacing. Coordinates x^μ (restricted to integer multiples of a) will be used to denote the lattice points and increments of a in the μ direction will be denoted by $\hat{\mu}$. Matter fields ψ_x and $\bar{\psi}_x$ are defined at the vertices of the lattice, while finite elements $U_x{}^\mu$ of the gauge group are assigned to the oriented links (from x to $x + \hat{\mu}$) of the lattice. The above constitute the dynamical variables.

The action S consists of a pure gauge part S_G and a matter part S_M coupling together gauge and matter dynamical variables. Different choices for S_G or S_M are possible, corresponding to different schemes of regularization. For the SU(3) gauge group of Quantum Chromo Dynamics a widely used form for S_G is Wilson's

$$S_G = \frac{6}{g^2} \sum (1 - \frac{1}{3} \mathrm{ReTr} U_\square), \tag{2.1}$$

where $\square$ denotes a generic plaquette of the lattice, i.e. an elementary square with vertices in x, $x + \hat{\mu}$, $x + \hat{\mu} + \hat{\nu}$ and $x + \hat{\nu}$, and correspondingly

$$U_\square = U_x^{\nu\dagger}\, U_{x+\hat{\nu}}^{\mu\dagger}\, U_{x+\hat{\mu}}^{\nu}\, U_x^{\mu} \quad . \tag{2.2}$$

g is the unrenormalized coupling constant. The matter action will be considered later; here we just mention that couplings between matter variables at different sites are made gauge invariant by the inclusion of the appropriate gauge variables, e.g.

$$\bar{\psi}_{x+\hat{\mu}}\, U_x^{\mu}\, \psi_x \quad . \tag{2.3}$$

Quantum expectation values of observables are defined as weighted averages over all values of the dynamical variables

$$\langle \mathcal{O} \rangle = Z^{-1} \int \prod_{x,\mu} dU_x{}^{\mu} \prod_x (d\bar{\psi}_x d\psi_x)\, \mathcal{O}\, (U,\bar{\psi},\psi) e^{-S} \quad , \tag{2.4}$$

$$Z = \int \prod_{x,\mu} dU_x{}^{\mu} \prod_x (d\bar{\psi}_x d\psi_x) e^{-S} \quad , \tag{2.5}$$

the integrals being invariant group integrals for the gauge variables and integrals over the elements of a Grassmann algebra for fermionic matter fields. If the lattice is initially restricted to a finite volume V, Eqs. (2.4) and (2.5) are mathematically well defined. The limit $V \to \infty$ and the analytic continuation $t \to it$ back to Minkowskian space-time (if needed) are to be performed at the end of the calculations.

There is a strong formal analogy between the above definition of the quantum expectation values and the definition of thermodynamical averages in statistical mechanics. The quantity $6/g^2$ in Eq. (2.1) plays the role of the Boltzmann factor $1/kT$, while the unnormalized action

$$E = \Sigma_{\square}(1 - \frac{1}{3}\,\mathrm{ReTr}\,U_{\square}) \equiv \sum_{\square} E_{\square} \qquad (2.6)$$

plays the role of the internal energy. In the analogy strong coupling corresponds to high temperature, weak coupling to low temperature. The notation β for $6/g^2$ (in SU(3); with other factors at numerator for different gauge groups) is frequently used.

Finally, and this is a most important point, the lattice regularization must be removed by the process of renormalization to recover a continuum theory. This can be done if the theory possesses a critical point g_0 such that all correlation lengths (in lattice units) become infinite as $g \to g_0$. If the critical point has suitable scaling properties, reducing the lattice spacing while $g \to g_0$ according to a well defined functional relationship $a = a(g)$ will make all physical quantities tend to finite limits as $a \to 0$. For asymptotically free theories one expects $g_0 = 0$ and the functional form of the equation $a = a(g)$ can be determined from perturbative arguments. For SU(3) it is

$$a(g) = \frac{1}{\Lambda}\left(\frac{11g^2}{16\pi^2}\right)^{-51/121} e^{-8\pi^2/11g^2}(1 + O(g^2)) =$$

$$= \frac{1}{\Lambda}\left(\frac{33}{8\pi^2\beta}\right)^{-51/121} e^{-(4\pi^2/33)\beta}(1 + O(\beta^{-1})) \; . \qquad (2.7)$$

Λ, the lattice scale parameter, is introduced in the above equation to provide correct dimensional units but has no direct physical significance. However, the continuum limits of all observables can be expressed in terms of Λ and, by eliminating Λ from the expressions of two or more observables, meaningful relationships among these can be established.

By Monte Carlo (MC) numerical calculations Λ can be related to physical observables, such as meson and baryon masses, or to quantities which have a well defined phenomenological value, such as the string tension. This, in turn, allows one to associate to every value of g (or of $\beta = 6/g^2$) a correspondingmagnitude for the lattice spacing a. Using early MC determinations of the string tension[5] one would find

$$\begin{aligned} a &\approx .53 \text{ fm} \quad \text{for} \quad \beta = 5 \\ a &\approx .31 \text{ fm} \quad \text{for} \quad \beta = 5.5 \\ a &\approx .17 \text{ fm} \quad \text{for} \quad \beta = 6 \\ a &\approx .10 \text{ fm} \quad \text{for} \quad \beta = 6.5 \\ a &\approx .06 \text{ fm} \quad \text{for} \quad \beta = 7 \quad . \end{aligned} \tag{2.8}$$

Recent results indicate that these values may be in error by as much as a factor of 2. Substantial amounts of careful numeriical work will be needed to resolve the discrepancies. However, the trend indicated by the numbers in Eq. (2.8) is the important thing to keep in mind. The relevant scale for hadron physics is of the order of 1 fm. In a very small window of values of β the lattice spacing varies from half a hadronic site to a small fraction of it. A very small lattice spacing would of course allow a very refined analysis of hadronic structure, but then the number of points in the 4-dimensional lattice needed to contain the hadron becomes prohibitively high, given the numerical nature of the only calculations which can be presently performed. Thus, in all practical calculations, one is limited to a small window of values of β, precisely the domain where the lattice spacing is hopefully fine enough to give a reasonable resolution of a hadron yet coarse enough that a lattice of not more than 10 to 20 sites across may contain the hadron well.

3. CALCULATIONAL TECHNIQUES AND LIMITATIONS

Wilson's lattice formulation allows the application to quantized field theories of computational techniques, which have been successful in the study of thermodynamical systems and which, as has been stated before, can not be implemented in the framework of a perturbative regularization. These computational methods are strong coupling expansions, mean field expansions and numerical simulations. Strong coupling and mean field techniques are illustrated in a recent review article[6] and I shall say very little about them. The general idea of strong coupling methods is to expand integrands like

$$e^{-\beta\sum_{\Box}(1 - 1/3 \, \mathrm{Re\,Tr}\, U_{\Box})} , \qquad (3.1)$$

which appear in the expressions for the observables

$$\langle \mathcal{O} \rangle = Z^{-1} \int \prod_{x,\mu} dU_x{}^{\mu} \, \mathcal{O}(U) \, e^{-\beta\sum_{\Box}(1-1/3 \, \mathrm{Re\,Tr}\, U_{\Box})} \qquad (3.2)$$

(we assume here for simplicity a pure gauge system), in powers of β, i.e., in powers of the inverse bare coupling constant square:

$$\langle \mathcal{O} \rangle = Z^{-1} \int \prod_{x,\mu} dU_x{}^{\mu} \, \mathcal{O}(U) [1+\beta\sum_{\Box}(1-1/3 \, \mathrm{Re\,Tr}\, U_{\Box}) + \ldots]. \qquad (3.3)$$

The terms in the expansion are integrals over the gauge group variables $U_x{}^{\mu}$ of polynomials in the same variables. These integrals can be calculated by group theoretical methods and represented in terms of strong coupling diagrams.

Monte Carlo simulations proceed instead by statistical sampling of the sum over configurations which defines the expectation value of observables. The method requires a finite lattice. Then the link variables $U_x{}^{\mu}$ are finite in number and are all stored in the memory of a computer. Their collection defines a configuration C of the system. A suitable algorithm modifies the variables $U_x{}^\mu$ one by one according to a definite stochastic procedure. A sequence of very many configurations

$$\rightarrow C_i \rightarrow C_{i+1} \rightarrow C_{i+2} \rightarrow \qquad (3.4)$$

is thus generated. The algorithm is such that the probability of encountering a configuration C in the sequence tends, as $i \rightarrow \infty$, to a distribution proportional to $e^{-S(C)}$. Then the expectation values of the observables can be approximated by

$$\langle \mathcal{O} \rangle \approx \frac{1}{n} \sum_{i=i_0+1}^{i=i_0+n} \mathcal{O}(C_i) \quad , \qquad (3.5)$$

where $\mathcal{O}(C_i)$ represents the value taken by the observable in the configuration C_i, i_0 is the number of steps one allows for the sequence to reach statistical equilibrium, and n is the total number of configurations included in the sampling. n should be large enough to assure a sufficiently precise determination of $\langle \mathcal{O} \rangle$.

What are the limitations of these non-perturbative techniques? Let us recall that the goal of the calculations is to determine observable quantities of the continuum quantum field theory. Thus, one wants to calculate the observables on the lattice, but at a value of β ($\beta \equiv 6/g^2$ in SU(3)), considered directly or reached through extrapolation, sufficiently large so that scaling toward the continuum limit has already taken place. Actually, one would even like to calculate at different values of β, all in the scaling domain, to make sure that the observables scale properly toward the continuum limit, and to calculate with different forms of the lattice action, to verify that the continuum limit is indeed universal.

In the SU(3) theory with Wilson's action the onset of scaling occurs at $\beta \approx 6$ where a, the lattice spacing, equals .17 fm (or less). A lattice of 10 points per direction corresponds then to a space-time volume of $\approx (1.7\ \text{fm})^4$, definitely a rather minimal size to explore properties of hadrons. Yet such a lattice contains 40,000 links, and therefore to register a configuration 720,000 real numbers (the real and imaginary parts of the elements of 40,000 3×3 complex matrices) must be stored in the memory of the computer (some saving in storage can be achieved, at the expense of a few more computations in the Monte Carlo algorithm). Moreover, the upgrading of one of the link variables (i.e., the replacement of a definite $U_x{}^\mu$ with a new $U_x{}^{\mu'}$ in the step from C_i to C_{i+1}) implies the performance of more than 4000 arithematic operations. This must be done for all links of the lattice, completing one Monte Carlo iteration, and then over and over again, sometimes for thousands of iterations. Clearly, while MC simulations can be done at any value of β, the serious limitation is the one placed on the size of the lattice. As discussed at the end of Section 2, through the interconnection between β (or g) and a, the limitation on the size of the lattice becomes effectively a limitation on how deeply one can investigate the scaling domain.

The problem with the strong coupling expansions is that their domain of validity is for small β. Their predictions must be extrapolated to large values of β (large enough to be in the scaling domain) and, even barring singularities which might act as obstacles to the extrapolation, terms of high order must be included on the expansion to obtain sufficient accuracy. In a sense, the difficulty in producing terms of high order can also be seen as a size limitation. Strong coupling diagrams are related to geometrical arrangements of plaquettes, which involve larger and larger extensions of the lattice the higher the complexity and order of the diagram become. Thus, expansions restricted to relatively small order effectively probe only small lattice volumes and, although Monte Carlo simulations and strong coupling expansions are completely different calculational techniques, the same kind of size problems appear to limit the accuracy of both.

Monte Carlo simulations where the gauge group is SU(2) require between 4.5 and 18 times less memory storage than their SU(3) counterparts, according to whether one considers as gauge group the full SU(2) or approximates its manifold with its maximal non-Abelian finite subgroup (such approximation works very well for SU(2)[7-9], but no analogous discretization appears to work for SU(3)[10]). The CP time required for the upgrade of a link variable is reduced, with respect to the SU(3) case, by comparable amounts. Thus calculations for SU(2) gauge theories have been done on larger lattices and/or with better statistics than for SU(3). (From the conceptual point of view, the justification for considering an SU(2) gauge theory is that, because of its non-Abelian character, it should embody many of the physical properties expected in the more realistic SU(3) model).

I have selected the following figures to illustrate some of the results obtained by Monte Carlo simulations of the SU(2) gauge system. Figures 1 and 2 (from Ref. 9) exhibit the string tension $\kappa(I,\beta)$ in dimensionless unit as function of separation (I lattice spacings) and coupling parameter ($\beta \equiv 4/g^2$ in SU(2)). The string tension in physical units would be given by

$$\sigma = \kappa\, a^{-2} \quad . \tag{3.6}$$

As a matter of fact, the true string tension is the constant force between two static sources at infinite separation. Thus $\kappa(I,\beta)$ should be thought as giving the force at finite separation rather than the string tension. It is in the nature of the numerical calculation that because of problems of statistics and of finite size effects the force κ can be measured only at

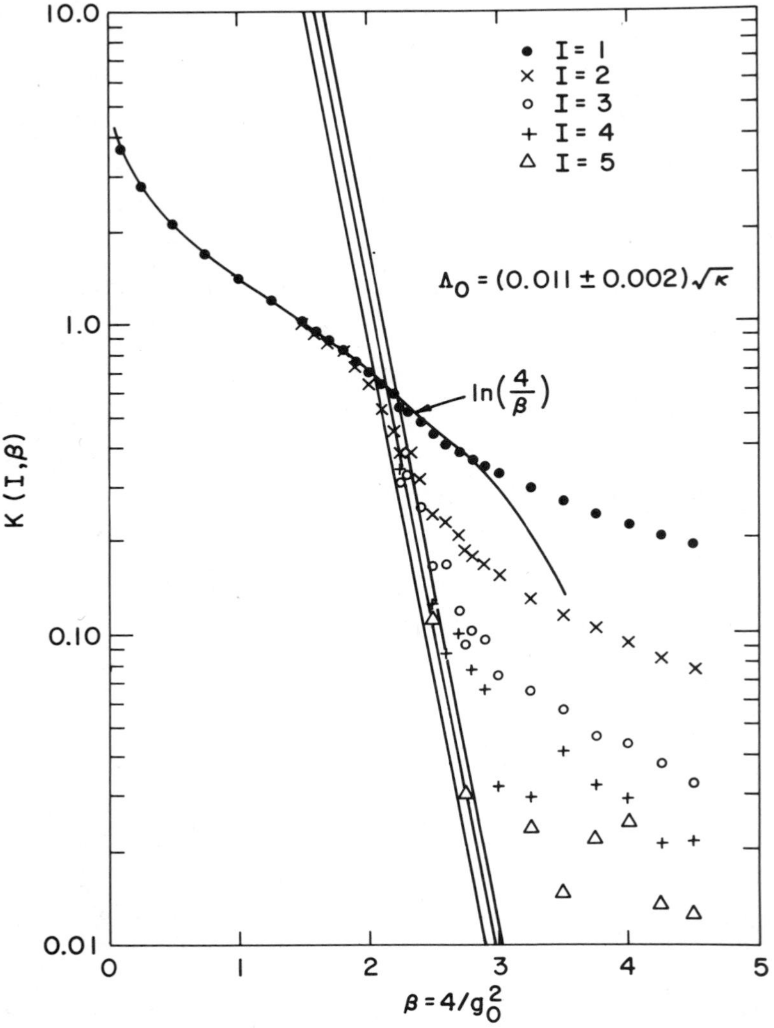

Fig. 1. Monte Carlo determination of the string tension in the SU92) guage theory.

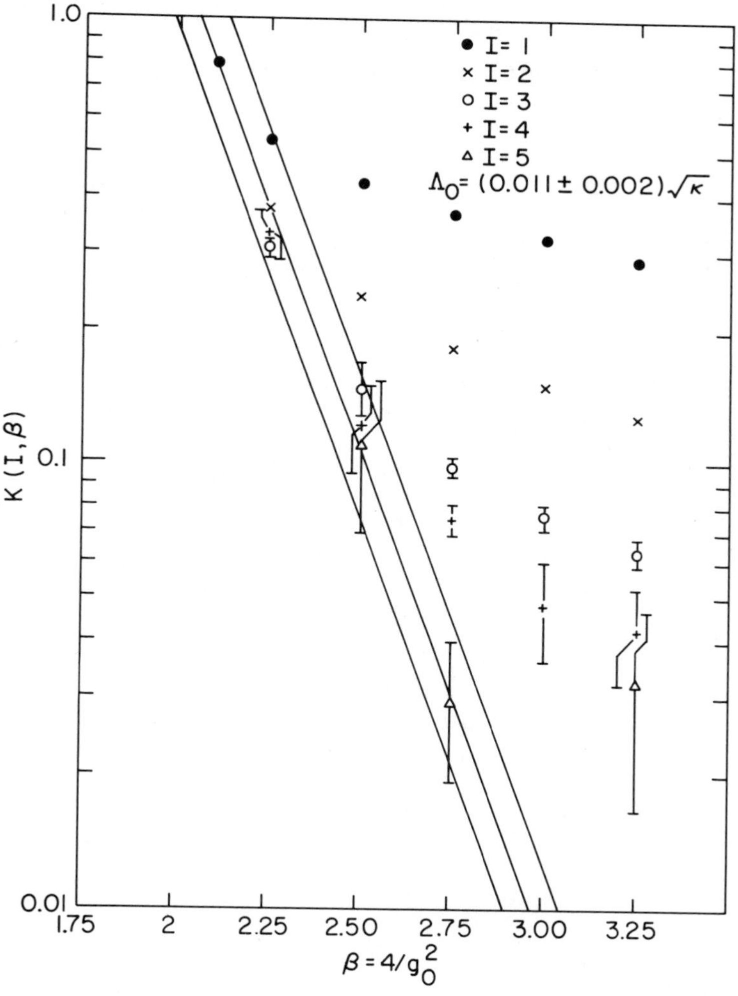

Fig. 2. An expansion of central part of Fig. 1, showing also statistical errors.

limited separations. Nevertheless, the results clearly indicate that for strong coupling (low β, the solid line beginning at the left of the figure represents the lowest order term in the strong coupling expansion) the force saturates already at one lattice spacing, which agrees with the notion that for strong coupling the lattice is very coarse and a separation of one lattice spacing represents already a large distance. As β becomes larger, $\kappa(I,\beta)$ clearly decreases with I, but there are still signs of saturation for $I \leq 5$ in the intermediate β region. The limiting curve, i.e. the envelope of the $\kappa(I,\beta)$ curves, agrees well with the expected scaling behavior, given by

$$\kappa(\infty,\beta) = \sigma\, a^2(\beta) \quad ,. \tag{3.7}$$

$$a(\beta) = \frac{1}{\Lambda}\left(\frac{11}{6\pi^2\beta}\right)^{-51/121} e^{-(3\pi^2/11)\beta}(1 + O(\beta^{-1})) \quad \text{(for SU(2))}, \tag{3.8}$$

and represented by the three parallel lines (central fit plus estimate of the error) in the figures. Scaling toward the continuum appears to set in at $\beta \approx 2.2$. Using the Monte Carlo results and the physical value of the string tension one can determine Λ; then Eq. (3.8) would give

$a \approx .43$ fm at $\beta = 2$

$a \approx .20$ fm at $\beta = 2.3$

$a \approx .09$ fm at $\beta = 2.6$.

We see again that the lattice spacing varies very rapidly in a narrow region of β. Indeed, the value $a \approx .43$ fm at $\beta = 2$ represents an extrapolation to a point where scaling has really not yet taken place. On the other hand, at $\beta = 2.6$ the value of the lattice spacing is already so small that a separation of 5 lattice units can barely test the asymptotic behavior of the force.

A determination of the potential between two static sources in the SU(2) theory is presented in Fig. 3 (from Ref. 11).

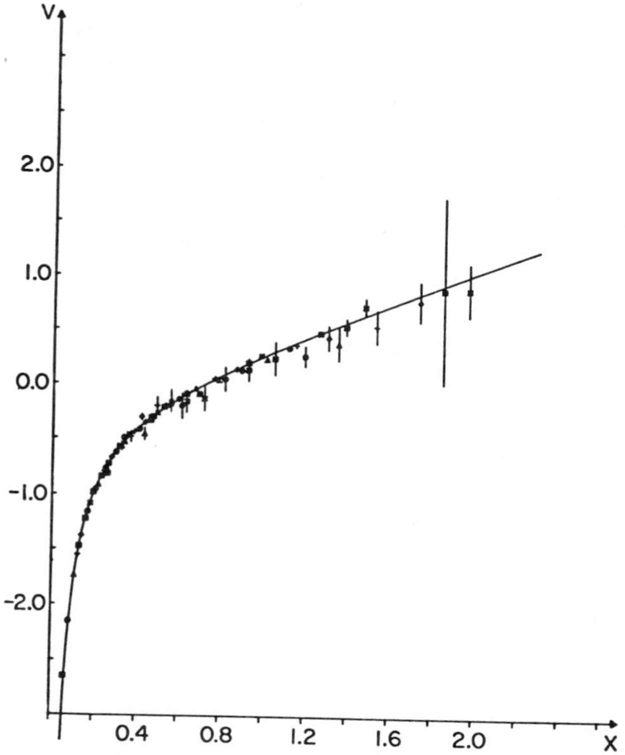

Fig. 3. Monte Carlo determination of the potential between two static charges in the SU(2) guage theory. Using the phenomenological value of the string tension to set the scale, units are of approximately 500 MeV for V and 0.4 fm for x.

Results obtained at different values of lattice separation and β are combined in a single graph, using Eq. (3.8). Scaling appears to work very well (points obtained at different values of β all lie on the same curve). The continuous curve represents a fit to the potential in terms of a function which interpolates between the asymptotic freedom behavior at small separations and a linear behavior at large distances. V and x are given in units of a mass parameter μ defined as $\mu = \Lambda/0.012$. μ was chosen as the average value of $\sqrt{\sigma}$ from the determinations of Ref. 12 and 9. The slope of the straight portion of the potential curve is, however, slightly less than one. This indicates a value for $\sqrt{\sigma}$ somehow smaller than μ. $\sqrt{\sigma} = \Lambda/0.013$ (as in Ref. 12) appears consistent with the results for the potential. Such margins of uncertainty can be expected from the numerical nature of the calculations and the limitations of size and statistics.

Figures 4 and 5 (from Ref. 13) illustrate the restoration of

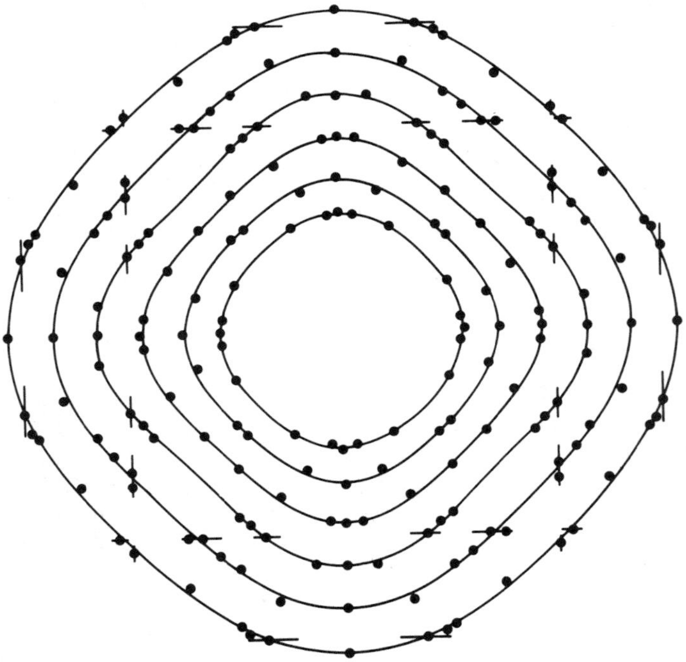

Fig. 4. Equipotential lines, interpolated through the lattice, in the SU(2) guage theory at $\beta = 2$.

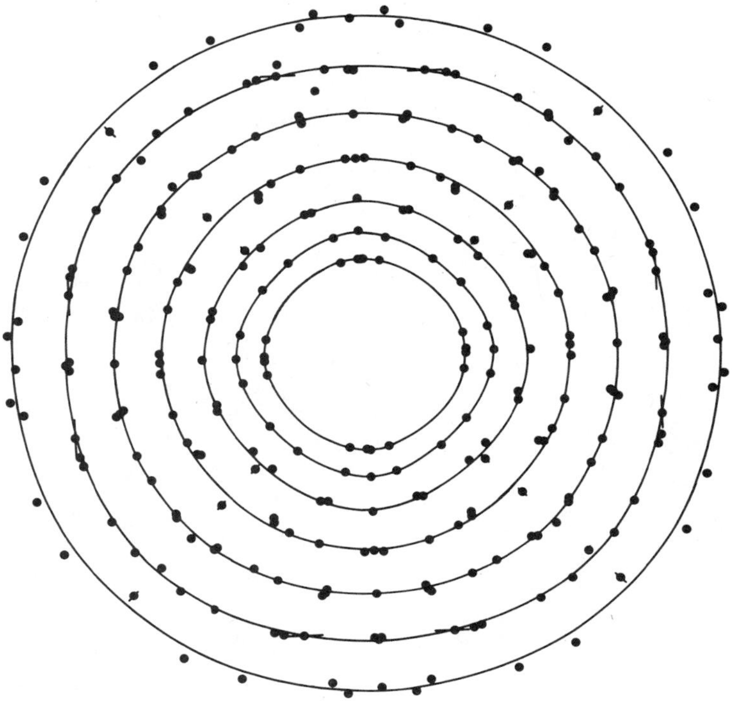

Fig. 5. Same as in Fig. 4, but at $\beta = 2.3$.

rotational invariance as scaling sets in in the SU(2) gauge theory. The points represent positions, interpolated in-between lattice sites, where the potential between two static sources (one fixed at the origin, the other moveable) takes constant values. These equipotential points are then further fitted by curves of the form $\rho(\delta) = \rho_0 + \Delta\rho \cos 4\theta$ (in polar coordinates), which thus represent equipotential lines interpolated through the lattice. The graph in Fig. 4 shows the equipotential lines at $\beta = 2$, just before the beginning of scaling. The distortion of the lines due to the lattice is quite apparent. The same type of calculation repeated at $\beta = 2.3$, reveals equipotential lines of a shape consistent with the spherical symmetry of the continuum theory. Thus, as scaling sets in, rotational symmetry also appears to be recovered. Figs. 6 and 7 (from Ref. 14) illustrate

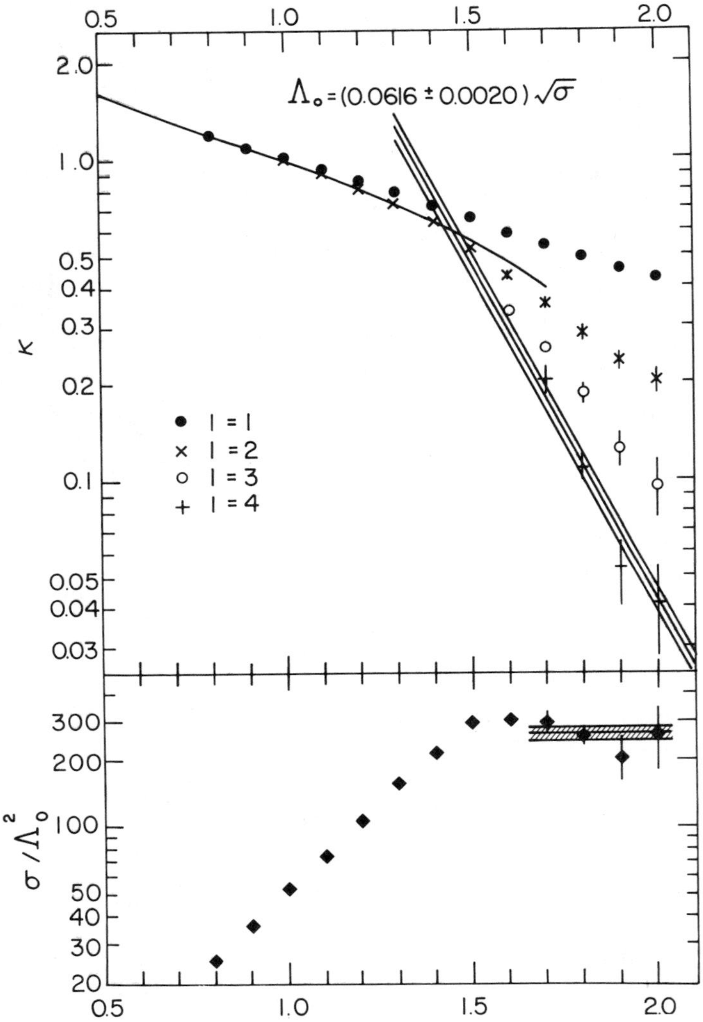

Fig. 6. Monte Carlo determination of the string tension in an SU(2) model based on Manton's form of the action.

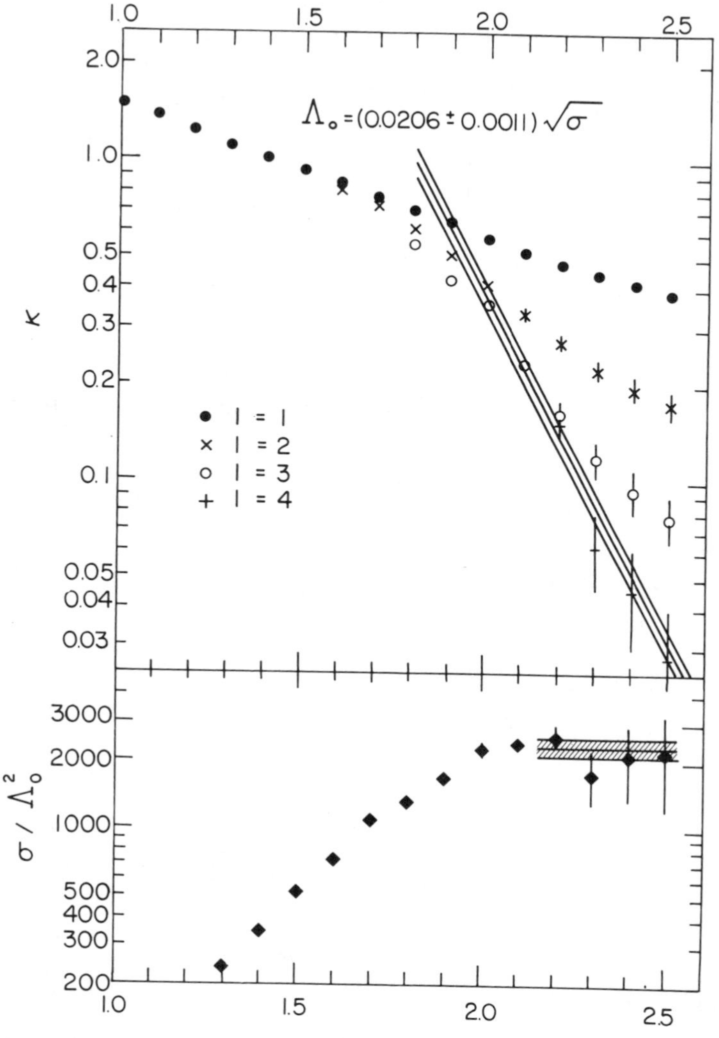

Fig. 7. Monte Carlo determination of the string tension in an SU(2) model based on the heat-kernel action.

determinations of the string tension in SU(2) models where the action, while still being based on the sum over individual plaquettes, has a different functional form than $S = \beta \sum_{\square}(1-1/2 \text{ Tr } U_{\square})$. The top part of the graph exhibits the string tension κ in lattice units as function of separation and β (see the discussion accompanying Figs. 1 and 2), while the bottom part gives the asymptotic value of κ (the one obtained in correspondence to the largest meaningful separation) rescaled by $[\Lambda a(\beta)]^{-2}$. $\kappa(\Lambda a)^{-2}$ should tend to the constant value σ/Λ^2 if the theory scales properly. Figures 6 and 7 show that scaling seems to occur independently of the specific form of the action. Only the relationship between $\sqrt{\sigma}$ and Λ (this last quantity, we recall, does not have a direct physical meaning) is action-dependent. Further studies[15–16] indicate that ratios of physical observables are independent of the form of the action, thus confirming universality.

Finally, going back to previous considerations on the validity of strong coupling expansions, in Fig. 8 a comparison is presented between results obtained by strong coupling and Monte Carlo techniques. The dotted lines represent the Monte Carlo determination of the scaling form of the string tension in the SU(2) system. The other lines represent results of strong coupling expansions, carried out to the highest and next to highest available orders, either as direct series (dashed-dotted lines) in powers of β (plus the leading term, going as $\ln(\beta/4)$), or after some resummations have been performed (solid and dashed lines). No attempt was made to incorporate in the strong coupling expansions information about the scaling behavior; rather one wishes to see whether indications of scaling emerge from them. While the results of the strong coupling expansions are consistent with the scaling behavior for large β, as determined by Monte Carlo simulations, their domain of validity seems to stop short of scaling.

After this discussion of Monte Carlo results obtained for SU(2) models, I would like to return to the more realistic SU(3) systems with some computational considerations. An efficient code, making use of the discrete approximation for the SU(2) group, can perform one MC iteration on a 16^4 lattice in about 13 secs; this corresponds to $\approx 50\,\mu s$ for the upgrading of one $U_x{}^\mu$

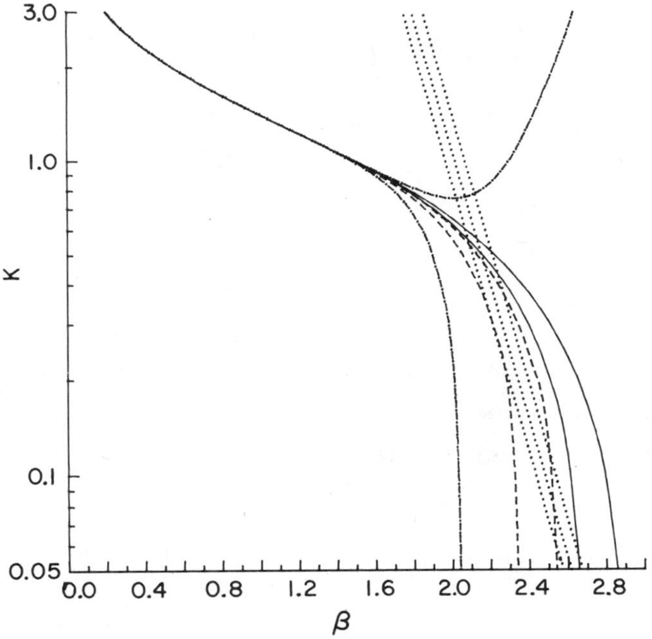

Fig. 8. Comparison of strong coupling and Monte Carlo results for the string tension in the SU(2) theory.

link variable. The simulation of an SU(3) system of 8^4 sites on the same computer requires approximately 16 secs per iteration, i.e. ≈ 1ms for the upgrading of a single link. The results presented for SU(2) systems indicate that the lattice sizes and statistics considered in the various Monte Carlo simulations may be large enough to produce reliable numerical information on the continuum limit. One might expect that computations of similar scope applied to SU(3) models would produce results of comparable accuracy. This may be the case; however, in spite of the larger computational demands presented by the simulation of SU(3) models, up to now essentially comparable resources have gone into the analysis of SU(2) and SU(3) systems. Interesting results have been obtained for the SU(3) theory. But one should not be surprised if, with the extension of calculations to larger lattices and greater statistics, discrepancies with respect to earlier results manifest themselves. One needs being able to perform simulations on SU(3) models of the same, or larger, magnitudes as can be done for SU(2). This requires the use of vectorized computers, parallel processors or special purpose computers, which can exploit the organized nature of Monte Carlo calculations so as to increase very substantially the computational speed. A few groups have already been working along these directions and link upgrade times in the SU(3) theory as low as 40 μs are being achieved[17]. As more and more powerful computational resources become available, we may be confident that reliable numerical results will become established for the SU(3) gauge theory as well.

4. CALCULATIONS OF MASSES IN THE HADRONIC SPECTRUM

Attempts to determine masses and other properties of the hadronic spectrum typically proceed through the numerical evaluation of two-point functions. Let us consider a theory with several quark flavors, specified by an index i. The matter field action will be a lattice transcription of Dirac's continuum action. There are some problems in formulating Dirac's equation on the lattice, into which I can not enter in any detail, and several different ways of defining a fermionic lattice action. For our present purposes it suffices to state that the matter field action will be an expression bilinear in the fermionic fields, $\bar{\psi}^i_x$ and ψ^i_x, which we denote by

$$S_M = \Sigma_i \, \Sigma_{xy} \, \bar{\psi}^i_x \left\{ D(U)_{xy} + m\delta_{xy} \right\} \psi^i_x \, . \tag{4.1}$$

Spin and color indices have been left implicit, a common mass has been assumed for all flavors and x, y denote coordinates of lattice sites. D(U) stands for the lattice equivalent of the continuum operator $\gamma^\mu(\partial_\mu + i\ g\ A_\mu^{\ \alpha}\ \lambda_\alpha)$. In most numerical applications D(U) has non-vanishing matrix elements only between fields at neighboring sites x and $y = x + \hat{\mu}$ and depends on the gauge field variables through the transport operators needed to make the couplings gauge invariant (see Eq. (2.3)).

To obtain information on masses e.g. in the meson sector, one would consider the quantum expectation value

$$G^{ij}_{xy} = \langle \, \bar{\psi}_x^{\ i} \ \Gamma \ \psi^j_x \ \bar{\psi}^j_y \ \Gamma \ \psi^i_y \rangle , \tag{4.2}$$

where Γ is a suitable matrix in spin space. $\bar{\psi}^j_y \ \Gamma \ \psi^i_y$ acts as a source for mesonic states with definite quantum numbers, which propagate from y to x and are then annihilated. The propagation occurs in Euclidean time and therefore involves exponentially decreasing rather than oscillating factors. The asymptotic rate of decay is determined by the lowest energy eigenstate and thus the lowest mass in the spectrum can be found examining the behavior of G^{ij}_{xy} for large separation. Actually, it is more convenient to examine the rate of decay of the quantity

$$\tilde{G}^{ij}_t = \Sigma_{\vec{x}} \ G^{ij}_{\vec{x},t;0} \tag{4.3}$$

where y has been placed at the origin and one sums over all space positions $\vec{x}$ at fixed time coordinate t. By expanding G_t^{ij} through the insertion of a complete set of physical states $|n\rangle$ one finds

$$G_t^{ij} = \sum_{n,\vec{p}_n=0} \left| \langle n,\vec{p}_n=0 | \bar{\psi}_o^j \, \Gamma \, \psi_o^i | \emptyset \rangle \right|^2 e^{-m_n t} \tag{4.4}$$

where $|\emptyset\rangle$ stands for the vacuum. The sum over spatial positions x projects over states with vanishing lattice space momentum $\vec{p}_n$; the energy eigenvalue then reduces to the mass and the determination of masses can be done with more accuracy. We shall consider $i \neq j$, which implies that the states $\langle n|$ can not have vacuum quantum numbers, in order to achieve some simplifications in the following equations. (This does not represent, however, any fundamental restriction.)

The expectation value in Eq. (4.3) can be expressed as an integral over all dynamical variables. The fields ψ and $\bar{\psi}$ are to be considered as elements of a Grassmann algebra and the integration over them defined accordingly. However, for the sake of later considerations, let us not be specific on the nature of the ψ and $\bar{\psi}$ fields and allow these variables to stand for <u>either</u> fermionic <u>or</u> bosonic fields. In any case G_{xy}^{ij} is given by

$$G_{xy}^{ij} = Z^{-1} \int \prod_{x,\mu} d\, U_x^\mu \prod_{x,i} (d\bar{\psi}_x^i \, d\psi_x^i) < \bar{\psi}_x^i \, \Gamma \, \psi_x^j \, \bar{\psi}_y^j \, \Gamma \, \psi_y^i >$$

$$\times \; e^{-S_G - \Sigma_i \bar{\psi}^i (D+m) \psi^i} \; , \tag{4.5}$$

where Z is given by an analogous integral over the measure alone.

A crucial point is that, because of the Gaussian nature of the integration over ψ and $\bar{\psi}$, the whole expression for G_{xy}^{ij} can be cast in two further, equivalent forms.

$$G^{ij}_{xy} = Z^{-1} \int \prod_{x,\mu} d\,U^{\mu}_{x} \prod_{x,i} (d\bar{\psi}^{i}_{x}\, d\psi^{i}_{x})$$

$$\times \operatorname{Tr}\left[\Gamma(D+m)^{-1}_{xy}\ \Gamma(D+m)^{-1}_{yx}\right] e^{-S_G + \Sigma_i \bar{\psi}^i (D+m)\psi^i} \tag{4.6}$$

where the trace is over spin and color degrees of freedom, or

$$G^{ij}_{xy} = Z^{-1} \int \prod_{x,\mu} d\,U^{\mu}_{x} \operatorname{Tr}\left[\Gamma(D+m)^{-1}_{xy}\ \Gamma(D+m)^{-1}_{yx}\right] e^{-S_G + n_f \ln \operatorname{Det}(D+m)} \tag{4.7}$$

where $|n_f|$ is the number of flavors and n_f should be set equal to $|n_f|$ for a fermionic theory, to $-|n_f|$ for a bosonic theory. Equation (4.6) follows immediately from the bilinearity of the matter field action; Eq. (4.7) is obtained using

$$\int \Pi (d\bar{\psi} d\psi)\ e^{-\psi(D+m)\psi} = \begin{cases} \operatorname{Det}(D+m) & \text{for fermionic fields} \\ \operatorname{Det}(D+m)^{-1} & \text{for bosonic fields.} \end{cases}$$

We shall denote by $\langle \psi_x \bar{\psi}_y \rangle_U$ the matrix elements $(D(U)+m)^{-1}{}_{xy}$ of the inverse Dirac operator, since they also are expectation values of product of fermi fields, or, equivalently, fermionic propagators, in the background of a given gauge field configuration $\{U^{\mu}{}_{x}\}$.

Equations (4.5), (4.6) and (4.7), although they represent identical mathematical expression, imply rather different calculational schemes when they are used in the context of a numerical approximation of the Monte Carlo type. Let us begin our discussion by assuming that ψ and $\bar{\psi}$ are bosonic fields (and, incidentally, also that S_M is positive definite). We could then proceed in three different ways for an approximate determination of G^{ij}_{xy}.

i) We could perform a Monte Carlo simulation on a combined system of U_x^μ, ψ_x^i and $\bar{\psi}_x^i$ dynamical variables and evaluate G_{xy}^{ij} by averaging $\bar{\psi}^i{}_x \Gamma \psi^j{}_x \bar{\psi}_y{}^j \Gamma \psi_y{}^i$ in the course of the simulation (this reflects Eq. (4.5)).

ii) We could still perform a simulation on the U, $\bar{\psi}$, ψ system, but evaluate, every so many iterations, the propagators $\langle\psi_x \bar{\psi}_y\rangle_U$ in the background of the current configuration, and then average the trace of the products of the propagators (this reflects Eq. (4.6)).

iii) We could dispense with the variables $\psi\bar{\psi}$ altogether and run a simulation over the gauge variables alone, but with a new action

$$S_{eff} = - S_G + n_f \ln \operatorname{Det}(D+m) \ . \tag{4.8}$$

We would then evaluate the propagators as before and average their products.

What are the advantages and disadvantages of the three methods of proceeding? Method i) is computationally faster; however, the value of G_{xy}^{ij}, as the distance between x and y increases to several lattice spacings, in general will fall off by orders of magnitude. (And it is precisely these small values at large separations which one seeks to determine, to reliably estimate the lowest masses in the spectrum.) In the averaging procedure such small numbers ought to result from cancellations among very many values for $\bar{\psi}^i{}_x \Gamma \psi^j{}_x\bar{\psi}^j{}_y \Gamma \psi^i{}_y$ which are all O(1). In practice, statistical fluctuations make the determination of G_{xy}^{ij} totally unreliable as soon as the separation goes beyond just a few lattice spacing. [As a side remark, here lies the difficulty to obtain good MC information for the spectrum of a pure gauge quantized spectrum[18], where, lacking a separation between gauge and matter field variables, options ii) and iii) are not available.]

If we follow the second alternative, the values for $\langle\psi_x\bar{\psi}_y\rangle_U$ can be accurately determined, on any given configuration, at almost arbitrary separations. One needs to solve a system of linear equations in many variables but with a very sparse matrix, which can be done efficiently by a variety of techniques. However, finding the above propagators is still a time consuming enterprise. Typically it takes more time (by a factor, not orders of magnitude) to calculate $[D(U)+m]^{-1}{}_{xy}$ for arbitrary x but fixed y, than to upgrade all $U_x{}^\mu$ variables in the lattice. Thus, although in principle one would like to average the propagators over many configurations, in practice the averaging is rather limited and may thus give only a poor account of some of the dynamical effects due to the quantized gauge field.

The third way of proceeding presents the only advantage of not needing the variables $\bar{\psi}$ and ψ. These are however eliminated at the expense of making the effective action for the gauge variables badly nonlocal (such is Det(D(U)+m)). The locality of S was crucial in allowing a fast implementation of the MC algorithm. The calculation of the variation of Det(D(U)+m) induced by the variation in any of the $U_x{}^\mu$ is so time consuming to make a direct implementation of this last option, with no approximations involved, impossible for any system of meaningful size.

If we consider now the case where the matter fields are fermionic, we face the difficulty that options i) and ii) can not be implemented on the computer, which deals with real or complex numbers and not elements of a Grassmann algebra. It could be argued that the Grassmann integration is nothing more than a bookkeeping for binary sums over fermionic occupation numbers, which can be 0 or 1. Indeed, for two dimensional systems it has been possible to reformulate the integral over fermionic fields precisely in this way, obtaining very efficient algorithms for the simulation of models with bosons and fermions[19]. But when one tries to extend the procedure to more than one-space one time dimensions, one runs into insurmountable problems determined by the fact that the measure loses its character of positive definiteness. Thus, in the case of fermions, one is left with the third (and less appealing) option.

Since an exact implementation of the algorithm, as mentioned before, can be done only on very small lattices, one must resort to some form of approximation. A procedure which has been followed in several investigations consists in neglecting the $\ell n\ \mathrm{Det}(D+m)$ term altogether in Eqs. (4.7) and (4.8) and performing the quantum averages with the pure gauge measure $\exp\{-S_G\}$. This approximation, which is commonly called the quenched[20] or valence[21] approximation, looks very drastic, but may actually produce reasonable results for the hadronic spectrum. Its significance can be understood noting that, in a perturbative expansion, it would correspond to neglecting all diagrams which contain internal fermionic loops, but otherwise summing over all orders of gluonic (i.e. gauge field) interactions. Many arguments, of phenomenological nature or derived from large N expansions in an SU(N) gauge theory, have been presented in support of the validity of the above approximation.

Earlier studies of the hadronic spectrum[21-27], all using the quenched approximation, were done on rather small lattices, with typical sizes ranging from $5^3{\times}8$ to $6^3{\times}14$ sites (the longer extent being in the time direction). Inputs were the independently measured string tension, to fix the lattice spacing corresponding to the value selected for the coupling parameter, and the quark mass adjusted so as to fit the mass of the lightest psuedoscalar. Mass values for the lowest mesons and baryons were then derived. Otherwise, in some of the calculations the lattice spacing was fixed by requiring that the mass of the ρ-meson takes its correct value. The physical value of the string tension is then an output, which can be tested against its phenomenological value. Summarizing in very few words the results of all these investigations, two major features appeared to emerge:

i) definite evidence for a dynamical realization of chiral symmetry, with the mass squared of the pion approaching zero as the quark mass m_q is let to zero, and with a nonvanishing condensate $\langle\bar{\psi}\psi\rangle \neq 0$ for $m_q = 0$.

ii) A general consistency between the independent Monte Carlo evaluations of the string tension and the hadronic masses, and a broad agreement (to the level of 10 to 20%) between the calculated and experimental values.

Discrepancies between the outcomes of separate investigations could be attributed to rather strong finite size effects (at $\beta = 6$, where most of the calculations were done, the lattice spacing does not exceed .17 fm), ambiguities in extrapolations of the quark mass (the numerical determination of the propagators becomes very time consuming or may even fail to converge on a small lattice if m_q is taken too small), and limited statistics in the Monte Carlo simulation.

More recently, calculations of hadronic masses in the quenched approximation have been repeated on larger lattices[28-29]. The evidence for spontaneous breaking of chiral symmetry remains strong, however it is found that lower values of the lattice spacing a, at a given β, are necessary to fit the experimental masses. In Ref. (28), for instance, using as inputs the masses of the pion and ρ-meson to fix the quark mass parameter and the lattice spacing, the value $a = (2.1\ \mathrm{GeV})^{-1} \approx 0.1$ fm is found at $\beta = 6$ (with a possible error from statistical fluctuations of $\approx 7\%$). This gives in turn values of 1.20 ± 0.15 GeV and 1.30 ± 0.20 GeV for the proton and Δ masses.

The discrepancy between the values of the lattice spacing, or equivalently of the scale parameter Λ in Eq. (2.7), obtained from the numerical evaluation of the string tension and of hadronic masses on large lattices represents an inconsistency which must be resolved. It can be argued that the determination of the string tension, being based on an estimate of the force at a separation of 6 lattice spacings at most, does not reflect the true asymptotic value of the force itself. The string tension may thus have been overestimated for σ leading to a corresponding understimate of a^{-1}. However, the string tension is not the only observable of a pure gauge theory which has been determined by Monte Carlo simulations. A variation of a^{-1} by a factor ≈ 2 would bring with itself the rather drastic consequence that the current estimates for the deconfining temperature ($\approx$ 200 MeV) and mass-gap ($\approx$ 800 MeV) should be re-scaled upward of the same factor. Another possibility to keep in mind is that the hadron masses are underestimated by the present computational procedures. The expectation values for the propagators of the composite states are obtained by averages extended to a very limited sample of configurations. It could be that, as the time of propagation increases, many more terms in the averages become necessary to reproduce the cancelation of phases which are ultimately responsible for the confinement of quarks into bound states. Thus the current calculations might account for confining effects only partially and lead to underestimates of mass values.

Although, as mentioned before, it is believed that the quenched approximation would not affect too much the spectrum of non-singlet states, this should be eventually checked by incorporating into the simulations the effects of the fermionic determinant in the measure. Also, simulations with fully dynamical fermions would be useful to evaluate properties of the theory, such as the value of the condensate $\langle\bar{\psi}\psi\rangle$ for $m_q = 0$ or the characteristics of the deconfining transition, which are particularly sensitive to the inclusion of inner fermionic loops. Among the algorithms which have been proposed to account for the fermionic determinant, let me briefly describe the so-called method of pseudofermions[30]. The basic idea is to approximately implement with fermions the calculational procedure exemplified by Eq. (4.6) and described as option ii) above.

The crucial step in the Monte Carlo algorithm for the exact simulation of a fermionic system (see Eq. (4.7) and option iii) above) consists in calculating the variation

$$\Delta S_{eff} = \Delta S_G - n_f \ \Delta \ \ell n \ \mathrm{Det}(D(U)+m), \tag{4.9}$$

induced by the temptative upgrade

$$U_x^\mu \rightarrow U_x^{\ \mu} + \Delta U_x^{\ \mu}. \tag{4.10}$$

It is convenient to rewrite the r.h.s. of Eq. (4.9) in the form

$$\Delta S_{eff} = \Delta S_G - n_f \Delta \Sigma_x \mathrm{Tr}\left[\ell n(D(U)+m)\right]_{xx}, \tag{4.11}$$

where, in conformity with previous notation, Tr stands for the sum over implicit spin and color indices and the trace over positional indices has been explicitly written. Let us assume now that the variation $\Delta U_x{}^\mu$ in Eq. (4.10) is small, so that quantities which are $O((\Delta U)^2)$ can be neglected. In principle, the magnitude of such a variation is a free parameter and can always taken to be small, although this may reduce the efficiency of the Monte Carlo algorithm. We then expand

$$\Delta S_{eff} \approx \Delta S_G - n_f \Sigma_{xy} \mathrm{Tr}\left[(D(U)+m)^{-1}_{xy} \ \Delta(D(U)+m)_{yx}\right], \tag{4.12}$$

(where terms $O((\Delta U)^2)$ have been neglected) and notice that all quantities in the r.h.s. are of local nature, and straightforward to compute, with the exception of the propagator

$$(D(U) + m)^{-1}_{xy} \equiv \langle \psi_x \bar{\psi}_y \rangle_U \quad . \tag{4.13}$$

However only matrix elements of the propagator where x and y are neighboring sites are needed, because $(D(U)+m)_{yx}$ and, a fortiori, $\Delta(D(U)+m)_{yx}$ are non-vanishing only when x and y differ by one lattice spacing at most. This suggests that $\langle\psi_x\bar{\psi}_y\rangle_U$ can be efficiently calculated by running a Monte Carlo simulation over a set of parallel C-number variables ϕ_x and $\bar{\phi}_x$, the pseudofermions, distributed according to the quadratic measure

$$\exp\{-\Sigma_{xy}\ \bar{\phi}_x\ (D(U)+m)_{xy}\ \phi_y\}. \tag{4.14}$$

As a matter of fact, the matrix $(D(U)+m)$ is not positive definite (not even Hermitian), and the algorithm can not be implemented in such a straightforward manner. But this is not a major obstacle. One can use for measure

$$\exp\{-\Sigma_{xy}\ \bar{\phi}_x[(D^\dagger(U)+m)(D(U)+m)]_{xy}\ \phi_y\} \tag{4.15}$$

and express the required matrix elements of $(D(U)+m)^{-1}$ as

$$(D(U)+m)^{-1}_{xy} =$$

$$= \Sigma_z[(D^\dagger(U)+m)(D(U)+m)]^{-1}_{xz}\ (D^\dagger(U)+m)_{zy}$$

$$= \Sigma_z \ll \phi_x\ \bar{\phi}_z \gg (D^\dagger(U)+m)_{zy}, \tag{4.16}$$

where $\ll\gg$ denotes averages taken with respect of the measure (4.15). Making the variation in Eq. (4.12) explicit, one eventually finds an expression of the form

$$\Delta S_{eff} \approx \Delta S_G - n_f\,\frac{1}{2}\,\{\Sigma_{x\mu}\ \mathrm{Tr}\ J_x^{\mu\dagger}\ \Delta U_x^{\mu} + \mathrm{h.c.}\}, \tag{4.17}$$

where

$$J^\mu{}_x = \frac{1}{2}\,\Gamma^\mu{}_x[(D(U)+m)^{-1}_{x+\hat{\mu}\ x} - (D^\dagger(U)+m)^{-1}_{x+\hat{\mu}\ x}] \tag{4.18}$$

and $\Gamma_x{}^\mu$ stand for the lattice equivalent of Dirac's γ

matrices. The physical significance of the method of pseudofermions is now apparent. For small steps $U \to U + \Delta U$ the effective action which determines the distribution of gauge field configurations varies by a term ΔS_G plus a term where the variation ΔU is coupled to the current induced by the gauge field U itself. This current is formally the same, whether the system is fermionic or bosonic, only the sign by which it enters into ΔS_{eff} changes. Thus one may estimate this current by performing Monte Carlo averages over a C-number system, i.e., the pseudofermions.

The algorithm described above becomes exact if one is simulating a true bosonic system, irrespective of the magnitude of the step ΔU or of how many MC iterations are used to estimate J_x^{μ} (through the estimates of $\langle\langle \phi_x \bar{\phi}_y \rangle\rangle$); it reduces then, indeed, to the ii) algorithm previously considered for bosonic systems. For fermions it is only approximate, becoming exact only in the limits where $\Delta U \to 0$ and an infinite number of MC iterations are used to calculate J_x^{μ}. The crucial question is whether it can be used in practical simulations, without having to take an excessively small upgrading step or too many iterations over the pseudofermions. The method has been tested by applying it to simulations of the 2-dimensional[20,31,32] Schwinger model and 4-dimensional QCD[33] and the indications are that it may give a reasonable approximation to the effects of dynamical fermions without increasing the CP time needed for a simulation by more than one order of magnitude.

Let me conclude by recapitulating the most important points. The lattice formulation of quantum gauge theories has demonstrated itself as a viable technique for quantitative studies of non-perturbative effects in QCD. The lattice action for the gauge field is not unique, but there is evidence that whole classes of actions produce a universal continuum limit. The situation is more problematic for the fermionic sector, where different actions lead to widely different formal properties and an optimal lattice action for the fermionic fields may not yet have been found.

There are some discrepancies between numerical results from Monte Carlo simulations for the pure gauge system and for the system with gauge and quark fields (treated in the quenched approximation), which ought to be resolved.

Techniques for incorporating dynamical effects from fermions in Monte Carlo simulations are available. Eventually these should be used to test the validity of the quenched approximation, but such checks do not make much sense until the discrepancies just mentioned above are eliminated. On the other hand calculations with dynamical fermions are very relevant for studies of properties such as the formation of a quark condensate or the deconfining transition, which are likely to be most affected by fermionic degrees of freedom.

Numerical calculations for Quantum Chromo Dynamics require very substantial computational resources. The systems considered up to now have been, in general, of a size barely adequate to produce meaningful results. Progress is being made however, through the use of powerful vector processors or special purpose machines, in extending the scope and magnitude of the calculations, and one may reasonable expect that in the near future good quantitative predictions will be obtained for QCD.

REFERENCES

1. K. Wilson, in: New Phenomena in Subnuclear Physics, ed. A. Zichichi (Plenum Press, N.Y. 1977).
2. K. Wilson, Phys. Rev. D10 (1974) 2445.
3. M. Creutz, L. Jacobs and C. Rebbi, Phys. Reports 23 (1975) 1331.
4. Lattice Gauge Theories and Monte Carlo Simulations, ed. C. Rebbi, World Scientific, Singapore (1983).

5. E. Pietarinen, Nucl. Phys. B190[FS3] (1981) 349;
M. Creutz and K.J.M. Moriarty, Phys. Rev. D26 (1982) 2166.
6. J.M. Drouffe and J.B. Zuber, Phys. Reports, to be published.
7. C. Rebbi, Phys. Rev. D21 (1980) 3350.
8. D. Petcher and D. Weingarten, Phys. Rev. D22 (1980) 2465.
9. G. Bhanot and C. Rebbi, Nucl. Phys. B180 (1981) 469.
10. G. Bhanot and C. Rebbi, Phys. Rev. D24 (1981) 3319.
11. J. Stack, Phys. Rev. D27 (1983) 412.
12. M. Creutz, Phys. Rev. Lett. 45 (1980) 313.
13. C.B. Lang and C. Rebbi, Phys. Lett. 115B (1982) 137.
14. C.B. Lang, C. Rebbi, P. Salomonson and B.S. Skagerstam, Phys. Lett. 101B (1981) 173; Phys. Rev. D26 (1982) 2028.
15. K.H. Mütter and K. Schilling, Phys. Lett. 121B (1983) 267.
16. R.V. Gavai, F. Karsch, H. Satz, Nucl. Phys. B220 (1983) 223.
17. D. Barkai, K. Moriarty, C. Rebbi, BNL preprint, (1983).
18. B. Berg, A. Billoire and C. Rebbi, Ann. of Phys. 142 (1982) 185;
M. Falcioni et al., Phys. Lett. 110B (1982) 295;
K. Ishikawa, G. Schierholz and M. Teper, Phys. Lett. 110B (1982) 399;
B. Berg and A. Billoire, Phys. Lett. 113B (1982) 65.
19. R. Blankenbeckler, J. Hirsch, D. Scalapino and R. Sugar, Phys. Rev. Lett. 47 (1982) 1628.
20. E. Marinari, G. Parisi and C. Rebbi, Nucl. Phys. B190 (1981) 266.
21. D. Weingarten, Phys. Lett. 109B (1982) 57 and Nucl. Phys. B215[FS7] (1983) 1;
22. H. Hamber and G. Parisi, Phys. Rev. Lett. 47 (1982) 1792 and Phys. Rev. D28 (1983) 247.
23. F. Fucito, G. Martinelli, C. Omero, G. Parisi, R. Petronzio and F. Rapuano, Nucl. Phys. B210[FS6] (1982) 407.
24. G. Martinelli, C. Omero, G. Parisi and R. Petronzio, Phys. Lett. 117B (1982) 434.
25. A. Hasenfratz, P. Hasenfratz, C.B. Lang and Z. Kunszt, Phys. Lett. 117B (1982) 81.
26. C. Bernard, T. Draper, K. Olynyk, Phys. Rev. D27 (1983) 227.
27. R. Gupta and A. Patel, Caltech preprint CALT-68-966 (1982).
28. H. Lipps, G. Martinelli, R. Petronzio and F. Rapuano, CERN preprint TH.3548 (1983).
29. K.C. Bowler, G.S. Pawley, D.J. Wallace, E. Marinari and F. Rapuano, Nucl. Phys. B220 (1983) 137.
30. F. Fucito, E. Marinari, G. Parisi and C. Rebbi, Nucl. Phys. B180 (1981) 369.
31. S. Otto and M. Randeria, Nucl. Phys. B220 (1983) 479.
32. A.N. Burkitt and R.D Kenway, Edinburgh Univ. preprint (1983).
33. H. Hamber, E. Marinari, G. Parisi and C. Rebbi, Phys. Lett. 124B (1983) 99.

DISCUSSIONS

CHAIRMAN : C. REBBI

Scientific Secretary: C.P. van den Doel

DISCUSSION 1

- *VAN DEN DOEL :*

At strong coupling the β-function defined by keeping the baryon mass or the chiral symmetry restoration temperature fixed is greater than zero. Since it is negative at weak coupling this implies a phase transition. Does this not make strong coupling expansions useless ?

- *REBBI :*

Some quantities may exhibit singularities in the coupling constant which prohibit a straightforward use of strong coupling expansions. In general, strong coupling expansions in lattice guage theories do not seem adequate to extrapolate to the scaling regime. This does not rule out the possibility of obtaining better results concerning strong coupling expansions with known analytical properties, such as the expected scaling behaviour, of physical variables.

- *VAN DEN DOEL :*

Could you say something about finite volume effects in Monte-Carlo calculations ?

- *REBBI :*

Finite volume effects can be very relevant because in four dimensions we are forced to consider lattices of rather small extent. It is very important to keep them under control and efforts to work with larger lattices are motivated precisely by this.

- *AMI* :

Do you think other lattice structures than cubical ones may be advantageous ?

- *REBBI* :

Taking other structures which are more isotropic may be not so convenient because there are more nearest neighbours involving more dynamical variables to be controlled.

- *BALLOCCHI* :

Do you plan to use concurrent processors to do lattice computations ?

- *REBBI* :

I think it is a good approach but, personally, I am currently only involved in investigations using conventional computers.

- *LEMARCHE* :

Have you thought of using an analog computer ?

- *REBBI* :

There is one in Denmark on which preliminary calculations on a simple problem have been done. The generation of random numbers with good uniformity properties may be problematic with analog computers.

- *BERNSTEIN* :

What is the physical significance of the lattice size ?

- *REBBI* :

Originally, the lattice spacing a enters trivially in the expressions for the observables, just to provide a dimensional factor, while the physics is embodied in dimensionless functions of the (bare) coupling constant. A magnitude (function of the bare coupling constant) can be assigned to the lattice spacing the moment when, in the process of renormalization, one demands that some observable takes a definite value.

- *CAMPOSTRINI* :

Weak coupling expansions are difficult on the lattice. Why is it useful ?

- *REBBI* :

It is important in relating the scale of the theory with the lattice regularization to the scale in the continuum regularization.

- *BALL* :

What is the present state of the lattice argument for confinement? I am thinking of Creutz' calculation in the mixed SU(2)-SO(3) system.

- *REBBI* :

From the work of Tomboulis there is now a lower bound on the string tension for all values of the coupling constant. The bound is below the expected renormalization group behaviour so it does not provide as yet a rigorous proof of confinement.

Regarding the mixed action calculations, a whole structure of phase transitions was discovered. However, there are paths to the continuum limit which avoid the critical points and the existence of these phase transitions does not disprove confinement.

- *HWANG* :

How do you handle magnetic monopoles and other solitons on the lattice ?

- *REBBI* :

There are topological excitations on the lattice (for instance, Abelian monopoles can be defined without any singularities), but in general the topological properties of continuum and lattice systems are different. Excitations with topological stability in the continuum would typically not be stable on the lattice. Lattice equivalents of continuum topological excitations can however be defined assuming that the configuration of fields on the lattice becomes regular enough as one approaches the continuum limit.

- *RIBARICS* :

Can the lattice regularization be applied to supersymmetric theories ?

- *REBBI :*

There are a few lattice models which embody some degree of supersymmetry, but there is no lattice analog of the full supersymmetry of the continuum system.

- *OLEJNIK :*

Is there a proof that the continuum theory is recovered in the limit of zero lattice spacing ?

- *REBBI :*

There are perturbative arguments. As far as I know there is no rigorous proof excluding possible (unlikely) non-perturbative effects, which might spoil the continuum limit.

DISCUSSION 2   Scientific Secretaries: G. Ballocchi and M. Ogilvie

- *HOU :*

Do present-day calculations of glueball masses include mixing with quarks ?

- *REBBI :*

No.

- *HOU :*

When you see a state, how do you see that it is a bound state ?

- *REBBI :*

With present lattice sizes it can be very difficult to tell whether a singularity represents a resonant bound state or a cut coming from two particles propagating in a closed volume. This is my main criticism of attempts to calculate higher mass glueballs. You can give examples of very simple models where you cannot differentiate a bound state from a superposition of two free particles. Of course, if you consider a state with definite quantum numbers, like those of the ρ, and you find a low mass excitation, you assume it does correspond to the expected state.

- *HOU* :

Because the temporal lattice size is finite, you do not work at zero temperature. Are finite-temperature effects important ?

- *REBBI* :

You can tell whether you are in a low or high temperature regime by looking at the expectation value of certain operators.

- *VAN DEN DOEL* :

I computed $<\bar{\psi}\psi>$ in the strong coupling limit both in the full theory and in the quenched approximation. The difference was about 0.5%. Does this result hold at weak coupling ?

- *REBBI* :

$<\bar{\psi}\psi>$ is sensitive to quenching and the sensitivity depends on the number of flavours and quark masses too. With two flavours in the Susskind formulation we find about a 5% effect.

- *VAN DEN DOEL* :

Can $\eta - \eta'$ splitting be computed in the quenched approximation?

- *REBBI* :

No, but it can perhaps be done with a more sophisticated approximation.

- *BALLOCCHI* :

Which are the most interesting results from Monte Carlo calculations ?

- *REBBI* :

The most interesting results that have been obtained are for the pure gauge theory, without quarks, the ratios between Λ_{QCD}, the string tension, and the deconfinement temperature.

In the quenched approximation, ratios between these quantities, $<\bar{\psi}\psi>$, and various hadron masses have been obtained.

- *KLEVANSKY* :

The quenched approximation neglects virtual fermion pairs. How good is this ?

- *REBBI* :

This would be a good approximation for positronium, and we believe it is good for many hadrons. It includes the valence quarks, in the spirit of the naive quark model. After all, we do think of a meson as a quark and an antiquark held together by glue. Also, it incorporates and goes beyond the planar approximation, whose validity finds support in theoretical and phenomenological arguments.

- *BERNSTEIN* :

What is the 'best prediction' for the meson mass spectrum ?

- *REBBI* :

Most calculations quote errors of 20-30%. It happened that the first calculations gave central values quite close to the actual values.

- *BERNSTEIN* :

What is the best value for a glueball mass ?

- *REBBI* :

There have been several calculations using several different methods. All of the results are consistent with a lowest mass glueball of mass 1 GeV for SU(2) and 800 MeV for SU(3). This assumes that the calculation of the string tension is correct.

- *D'AMBROSIO* :

How good is the quenched approximation ? Can it be justified by the $^1/_N$ expansion ?

- *REBBI* :

The quenched approximation has been tested, for example, in the Schwinger model (QED_2). In that case, the expectation value $<\bar{\psi}\psi>$ is 20-30% higher in the quenched approximation. We expect better results in four dimensions, and this seems to be the case. Further tests should be made.

The $^1/_N$ expansion can justify the quenched approximation. However, the quenched approximation includes non-planar graphs in the gluonic sector, which are non-leading in the $^1/_N$ expansion.

- *MANA* :

How do you estimate non-statistical errors in Monte Carlo calculations ?

- *REBBI* :

You may consider different size lattices, different couplings, and different actions, which give clues to systematic errors, due, for example, to finite size effects.

- *HWANG* :

How do you fix the gauge on a lattice ?

- *REBBI* :

Gauge fixing is not necessary on the lattice, because the gauge volume is finite, and drops out when you consider expectation values. If you need to fix the gauge, you can in the standard way.

- *NEMESCHANSKY* :

The Wilson formulation of fermions on a lattice breaks chiral symmetry explicitly. Can one restore this symmetry by setting $m_\pi = 0$?

- *REBBI* :

There is another way of placing fermions on a lattice, the generalized Susskind formulation, which has a chiral symmetry, but also has other problems.

- *LAMARCHE* :

How is special relativity recovered in lattice calculations ?

- *REBBI* :

In Euclidean spacetime, Lorentz transformations become rotations. In the continuum limit, this O(4) symmetry reappears.

- *OLEJNIK* :

Could you explain the physical meaning of the interquark potential extracted from lattice calculations ?

- *REBBI* :

You can take the pure gauge theory and introduce two static sources. The potential is the difference between the ground state energy with and without sources. In this sense, it is the potential between infinitely massive quarks.

- *BURGESS* :

In chirally invariant formulations of the lattice, how do the anomalies arise ? Is the classical action not $U_A(1)$ invariant or can you see the anomaly appear in the calculation ?

- *REBBI* :

The fermionic action which has some chiral invariance is not $U_A(1)$ invariant. The chiral symmetries present are of the non-singlet type.

- *ZEPPENFELD* :

When taking quark masses equal to zero the mass of the lightest mesons should vanish in those lattice theories with chiral invariance. Does this happen ?

- *REBBI* :

All numerical evidence indicates that this is what happens.

THE U(1) PROBLEM: INSTANTONS, AXIONS, AND FAMILONS

Frank Wilczek

Institute for Theoretical Physics
University of California
Santa Barbara, CA 93106

The U(1) problem began as a rather obscure difficulty in current algebra, but over time it has stimulated some remarkable theoretical ideas which may even in a quite literal sense describe the bulk of the Universe!

In these lectures this whole complex of ideas is discussed in as self-contained a fashion as possible. I'll start by reviewing the original U(1) problem in the strong interactions and its resolution in QCD. The solution implies a new problem, that it becomes difficult to understand why the strong interactions respect CP to a good approximation. The most convincing solution to this problem requires the existence of a new particle, the axion, with remarkable properties. I'll spend a while discussing the motivation for introducing the axion and its mass and couplings to matter. Axions may play an important role in cosmology -- they are a serious candidate to provide the nonluminous mass for which astronomical evidence is accumulating. Conversely cosmological considerations place significant constraints on the properties of axions. I'll discuss all these things, and then wind up by discussing the framework in which axions most naturally appear, the idea that family symmetries are broken only spontaneously and not intrinsically.

It may strike the reader (and has struck the author) that enormous theoretical superstructures are here being erected upon a very narrow foundation. The whole superstructure of axions could be made obsolete if a good alternative approach to the problem of strong CP invariance were found. Even if this does happen, I am confident that techniques for dealing with approximate Nambu-Goldstone bosons and their phenomenological (including cosmological) implications will be of enduring interest -- so I won't be completely wasting your time.

THE STRONG U(1) PROBLEM

Qualitative Discussion

In the late 1960s and early 1970s an enormously successful phenomenology was built up based on the idea that the strong interactions have an approximate SU(3) × SU(3) chiral flavor symmetry which is spontaneously broken to ordinary flavor SU(3). When combined with the hypotheses of current algebra, this idea leads to such successful results as the Goldberger-Treiman relation, the Adler-Weisberger sum rule, the Callan-Treiman formulas, and many others.

An important feature of the hadron spectrum "visible to the naked eye" -- the existence of an octet of pseudoscalar mesons much lighter than the other hadrons -- is convincingly explained in this framework. The pseudoscalar mesons are the Nambu-Goldstone bosons of the approximate chiral SU(3) symmetry which is spontaneously broken.[2] They would be precisely massless (Goldstone's theorem) if the original chiral symmetry were exact, and the smallness of their mass is a measure of how close we are to this situation, as will be quantified shortly.

The strong U(1) problem arises when one tries to join the successful phenomenology of chiral symmetry breaking to a microscopic theory of the strong interactions where quarks are the fundamental degrees of freedom.[3] In terms of quarks the hypothesis of approximate chiral symmetry is that the Lagrangian for the strong interactions is almost invariant under independent SU(3) rotations involving the left-handed quarks u_L, d_L, s_L or the right-handed quarks u_R, d_R, s_R. This entails in particular that quark bare mass terms

$$- L_{mass} = m_u \bar{u}_L u_R + m_d \bar{d}_L d_R + m_s \bar{s}_L s_R + h.c. \qquad (1)$$

are in some sense small, since such terms are invariant under chiral $SU(3) \times SU(3)$ only if $m_u = m_d = m_s = 0$ (whereas ordinary flavor SU(3) requires only $m_u = m_d = m_s$). The spontaneous breaking of chiral $SU(3) \times SU(3)$ down to ordinary flavor SU(3) occurs very simply if in the vacuum one finds nonvanishing expectation values for quark bilinears:

$$\langle \bar{u}u \rangle = \langle \bar{d}d \rangle = \langle \bar{s}s \rangle \neq 0 \; . \qquad (2)$$

The apparent difficulty with this picture is that as the masses of all the quarks vanish one finds not only $SU(3) \times SU(3)$ but $U(3) \times U(3)$ symmetry -- multiplication of all the left- or all the right-handed quarks by a common phase tends to leave the Lagrangian invariant.

(One can imagine complicated interactions which are invariant under $SU(3) \times SU(3)$ but not the extra U(1) -- of the general form $\bar{u}_L u_R \bar{d}_L d_R \bar{s}_L s_R$ -- but one would not like to have them in as fundamental interactions since they are <u>ad hoc</u>, complicated, and unrenormalizable. As we shall see below, just such an interaction, but nonlocal and soft, is generated in QCD.) The extra chiral U(1) is of course also broken by the vacuum expectation values of quark bilinears (2). One would therefore expect to find an extra light pseudoscalar meson which is the approximate Nambu-Goldstone boson of axial U(1) symmetry. The η' meson has the correct quantum numbers, but it is much too heavy.

Good and Bad Gell-Mann-Okubo Formulas

To quantify the difficulty with the η' meson, we'll recall how the Gell-Mann-Okubo formula for $(\text{masses})^2$ of pseudoscalar mesons is derived, and show (following Weinberg[3]) how the same manipulation leads to a disastrous result for the η' mass, under the assumption that axial U(1) is mainly broken spontaneously by the nonzero vacuum expectation value of quark bilinears.

For expository purposes it is actually convenient to start with the bad Gell-Mann-Okubo formula. We are supposing that the Lagrangian of the strong interactions is invariant under chiral U(3) × U(3) rotations except for the small term L_{mass} [Eq. (1)] which will be treated as a perturbation. We are also supposing that the quark bilinears acquire vacuum expectation values [Eq. (2)]. Now if the vacuum is a state where

$$\langle \bar{u}_L u_R \rangle = \langle \bar{d}_L d_R \rangle = \langle \bar{s}_L s_R \rangle = \frac{1}{2} v \tag{2}$$

(note $v < 0$) then there are states related by axial baryon number, hyperchange, and I_3 transformations for which

$$\langle \bar{u}_L u_R \rangle = v e^{i\left(\tilde{a} + \frac{\tilde{\beta}}{\sqrt{3}} + \sqrt{\frac{2}{3}}\, \tilde{\gamma}\right)}$$

$$<\bar{d}_L d_R> = ve^{i\left(-\tilde{\alpha} + \frac{\tilde{\beta}}{\sqrt{3}} + \sqrt{\frac{2}{3}}\,\tilde{\gamma}\right)} \tag{4}$$

$$<\bar{s}_L s_R> = ve^{i\left(-\frac{2\tilde{\beta}}{\sqrt{3}} + \sqrt{\frac{2}{3}}\,\tilde{\gamma}\right)} \,.$$

In the limit m_u, m_d, m_s 0 where chiral symmetry is good these states are all degenerate in energy with the vacuum (3). There are also states where $\tilde{\alpha}$, $\tilde{\beta}$, and $\tilde{\gamma}$ vary in space; these do cost energy which we can parametrize (to lowest order in gradients, i.e. for long wavelengths) as

$$L_{Kin} = \frac{f_\pi^2}{2}\left((\partial_\mu \tilde{\alpha})^2 + (\partial_\mu \tilde{\beta})^2\right) + \frac{f_s^2}{2}\,(\partial_\mu \tilde{\gamma})^2 \,. \tag{5}$$

The same coefficient applies to $\tilde{\alpha}$ and $\tilde{\beta}$ since they are in the same SU(3) octet. In the quantum field theory, the $\tilde{\alpha}$, $\tilde{\beta}$, and $\tilde{\gamma}$ fields create massless particles, the Nambu-Goldstone bosons.

When we allow the quark masses to vary a little away from zero, we find there is energy associated even with very long-wavelength excitations. Indeed, expanding to second-order in α, β, and γ we find mass terms

$$H_{mass} = -L_{mass} = -m_u \frac{<\bar{u}u>}{2}\left(\tilde{\alpha} + \frac{\tilde{\beta}}{\sqrt{3}} + \sqrt{\frac{2}{3}}\,\tilde{\gamma}\right)^2$$

$$-m_d \frac{<\bar{d}d>}{2}\left(-\tilde{\alpha} + \frac{\tilde{\beta}}{\sqrt{3}} + \sqrt{\frac{2}{3}}\,\tilde{\gamma}\right)^2 - m_s \frac{<\bar{s}s>}{2}\left(-\frac{2\tilde{\beta}}{\sqrt{3}} + \sqrt{\frac{2}{3}}\,\tilde{\gamma}\right)^2$$

$$+ \text{const.} \tag{6}$$

(By the way, notice that $\alpha = \beta = \gamma = 0$ does minimize the energy, so it represents the true vacuum). For the physical interpretation of this we must pass to the fields with properly normalized kinetic energy

$$\alpha = f_\pi \tilde{\alpha}$$
$$\beta = f_\pi \tilde{\beta} \qquad (7)$$
$$\gamma = f_s \tilde{\gamma}$$

and diagonalize the mass matrix from (6):

$$H_{mass} = \frac{1}{2} (\alpha,\beta,\gamma)\, M^2 \begin{pmatrix} \alpha \\ \beta \\ \gamma \end{pmatrix} \qquad (8)$$

where

$$M^2 = -v \begin{pmatrix} \dfrac{m_u+m_d}{f_\pi^2} & \dfrac{m_u-m_d}{\sqrt{3}\, f_\pi^2} & \dfrac{1}{f_s f_\pi} \dfrac{\sqrt{2}}{3} (m_u-m_d) \\ \dfrac{m_u-m_d}{\sqrt{3}\, f_\pi^2} & \dfrac{m_u+m_d+4m_s}{3f_\pi^2} & \dfrac{1}{f_s f_\pi} \dfrac{\sqrt{2}}{3} (m_u+m_d-2ms) \\ \dfrac{1}{f_s f_\pi} \dfrac{\sqrt{2}}{3} (m_u-m_d) & \dfrac{1}{f_s f_\pi} \dfrac{\sqrt{2}}{3} (m_u+m_d-2m_s) & \dfrac{2}{3f_s^2} (m_u+m_d+m_s) \end{pmatrix}$$

For convenience let us now suppose $m_u = m_d$, the essential issue doesn't depend upon this assumption. If $m_u = m_d$ the M^2 matrix simplifies to

$$M^2 = -v \begin{pmatrix} \frac{m_u+m_d}{f_\pi^2} & 0 & 0 \\ 0 & \frac{m_u+m_d+4m_s}{f_\pi^2} & \frac{1}{f_s f_\pi}\frac{\sqrt{2}}{3}(m_u+m_d-m_s) \\ 0 & \frac{1}{f_s f_\pi}\frac{\sqrt{2}}{3}(m_u+m_d-m_s) & \frac{1}{f_s^2}\frac{2}{3}(m_u+m_d+m_s) \end{pmatrix}$$

Clearly we have one particle, to be identified with the physical π^o, with $(\text{mass})^2$

$$m^2 = \frac{-1}{f_\pi^2} v(m_u + m_d) \quad . \tag{10}$$

The π^o corresponds to the phase $\tilde{\alpha}$. The other eigenstates of mass are mixtures of $\tilde{\beta}$ and $\tilde{\gamma}$. One of these is very light. To see this, consider the vector

$$\phi = \frac{1}{\sqrt{f_\pi^2+2f_s^2}} \begin{pmatrix} 0 \\ f_\pi \\ \sqrt{2f_s} \end{pmatrix} \tag{11}$$

This is normalized and orthogonal to the π^o, to according to general principles one has an upper bound on the second eigenvalue

$$m^2 \le \phi^\dagger M^2 \phi = \frac{-3(m_u+m_d)v}{f_\pi^2+2f_s^2} \le 3m_\pi^2 \; . \tag{12}$$

This is apparently a disaster, since there is no meson of the appropriate mass.

The good Gell-Mann-Okubo formula is arrived at if the extra U(1) symmetry is ignored. We can think about it this way: The extra light meson found above is associated with the mixture of hypercharge and singlet chiral rotations which does not affect the strange quark, so it has a $(\text{mass})^2$ proportional to the light quark masses and therefore comparable to the π^o $(\text{mass})^2$. If the extra U(1) symmetry can be eliminated, there will be no singlet for the η to mix with and it will have a $(\text{mass})^2$ proportional to m_s.

To get the good Gell-Mann-Okubo formula we must widen our notation slightly to accommodate more general transformations of the quarks than the phase multiplications in (4). In fact if we parametrize the vacuum expectation values as a matrix

$$U^i_j \equiv \langle \bar{q}_{iL} q^j_R \rangle \tag{13}$$

then chiral SU(3) × SU(3) transformations will give us degenerate vacuua parametrized by

$$\begin{aligned} U' &= V^{-1}\, U\, W \\ V,W &\in SU(3)\ . \end{aligned} \tag{14}$$

Suppose that in one of the vacuua $\langle U\rangle = 1$ [as in Eq. (3)]. Then the transformations with V = W leave the vacuum invariant -- this is ordinary flavor SU(3) -- whereas say with V = 1 the different W produce different states with the same energy as the vacuum, for which U' is an arbitrary unitary matrix with determinant one.

As before, we parametrize the energy of long-wavelength variations in U by the simplest term of appropriate symmetry:

$$L_{kin} = \frac{f_\pi^2}{2}\, \mathrm{Tr}(U^{-1}\partial_\mu U)^2 \tag{15}$$

and take into account the quark mass term contribution to the energy density

$$- L_{mass} = \mathrm{Tr}\, MU + h.c. \tag{16}$$

$$M = \begin{pmatrix} m_u & 0 & 0 \\ 0 & m_d & 0 \\ 0 & 0 & m_s \end{pmatrix} \tag{17}$$

Finally, expand U around the identity

$$U = 1 + i \frac{\lambda^a}{2} \tilde{p}^a + \ldots \tag{18}$$

where the λ^a are the usual Gell-Mann matrices and the $\tilde{p}^a$ are pseudo-scalar meson fields. Expanding out the fields and normalizing, as before, one finds the meson mass formulas (assuming $m_u = m_d$)

$$\begin{aligned} m_\pi^2 &= \frac{-1}{f_\pi^2} (m_u + m_d) v \\ m_k^2 &= \frac{-1}{f_\pi^2} (m_u + m_s) v \\ m_\eta^2 &= \frac{-1}{f_\pi^2} \frac{(m_u + m_d + 4m_s)}{3} v \quad . \end{aligned} \tag{19}$$

The π and η masses are what we had before, in the limit $f_s \to \infty$ where the singlet can't vary and decouples.

These formulas are of the same type: they state that the $(\mathrm{mass})^2$ of the meson is proportional to the linear mass of its quark constituents, weighted by their amplitudes in the meson.

From the formulae (19) one gets the famous Gell-Mann-Okubo formula

$$3m_\eta^2 + m_\pi^2 = 4m_k^2 \tag{20}$$

which agrees very well with experiment. [With $m_\eta = 550$ MeV, $m_k = 500$ MeV, $m_\pi = 140$ MeV, the left-hand side is .93 $(\mathrm{GeV})^2$ and the right-hand side is 1.0 $(\mathrm{GeV})^2$.]

Another important success of this framework is to explain why the pseudoscalar mesons are so different from the vector mesons, where one has essentially "magic mixing" -- the ϕ meson is essentially an $\bar{s}s$ bound state and far from an SU(3) eigenstate. Also, we gain an understanding of why the pseudoscalar masses break SU(3) by a large amount: indeed the pseudoscalar $(\mathrm{mass})^2$ ratios are determined by quark mass ratios and could be very different from unity even as m_u, m_d, $m_s \to 0$.

It should be mentioned that the Gell-Mann-Okubo formula is one of the few unambiguous successes of chiral SU(3) × SU(3) [as opposed to SU(2) × SU(2)] symmetry. In this sense, it is evidence that the strange quark mass is small compared to other strong interaction scales.

To summarize, we have a very successful picture of pseudoscalar mesons without the extra axial U(1), and an apparent disaster with the extra U(1). Since the hypotheses used in this analysis (existence of chiral symmetries broken only by quark mass terms) are apparently valid in QCD, and QCD is presumably the microscopic theory of the strong interaction, one must understand why the apparent extra U(1) chiral symmetry is broken in QCD.

Anomaly and 't Hooft Vertex

This problem has in fact been solved. You will find the solution beautifully exposed in Coleman's Erice lectures "The Uses of

Instantons."[4] I will try to recall just enough of this material for our further purposes. Fortunately, only some qualitative and group-theoretical features of the results will required -- the rather formidable (and physically incomplete) derivations won't be needed.

It is a very general phenomenon that a classical Lagrangian may have symmetries which are inevitably destroyed by quantization. The most famous and important example of this is probably scale invariance in QCD with massless quarks. In the present context, the relevant remark is that axial baryon number is <u>not</u> strictly conserved in QCD with massless quarks due to an anomaly in the triangle graph where the current couples to two gluons (Fig. 1). One finds that any regulator procedure which respects such general principles as Lorentz invariance and gauge invariance introduces a nonzero contribution to the divergence of the axial baryon number current:[5]

$$\partial_\mu(\bar{u}\gamma_\mu\gamma_5 u + \bar{d}\gamma_\mu\gamma_5 d + \bar{s}\gamma_\mu\gamma_5 s) = 2(m_u\bar{u}\gamma_5 u + m_d\bar{d}\gamma_5 d + m_s\bar{s}\gamma_5 s)$$
$$+ 3\,\frac{g^2}{8\pi^2}\,\mathrm{Tr}G_{\mu\nu}\tilde{G}_{\mu\nu}\,. \qquad (21)$$

Here $G_{\mu\nu}$ is the gluon field strength and $\tilde{G}_{\mu\nu} \equiv \frac{1}{2}\,\varepsilon_{\mu\nu\rho\sigma}G_{\rho\sigma}$, so that the anomalous term is basically the color version of $\vec{E}\cdot\vec{B}$. Given the anomaly, axial baryon number is not conserved even in the limit $m_u, m_d, m_s \to 0$. Turning this around, a chiral rotation on the quark fields i.e., in the infinitesimal form, adding $m_u\bar{u}\gamma_5 u + m_d\bar{d}\gamma_5 d + m_s\bar{s}\gamma_5 s$ to the Lagrangian, will not leave the action invariant; indeed according to (21) it will add a term proportional to $\mathrm{Tr}G_{\mu\nu}\tilde{G}_{\mu\nu}$ as well as a harmless total divergence.

(You may be puzzled about where this extra term comes from, or indeed about how a classical symmetry can ever be broken by

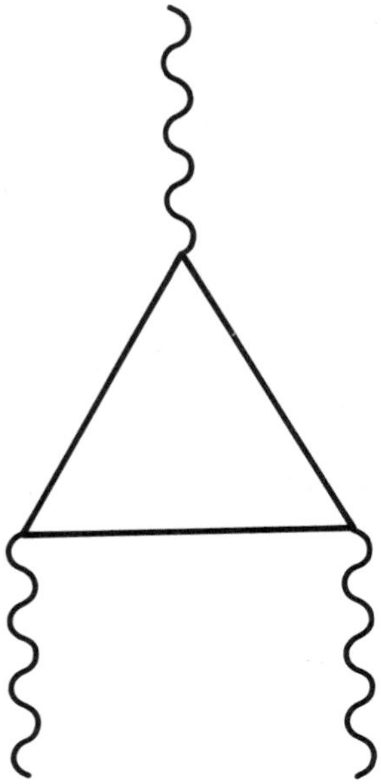

Fig. 1. The triangle graph coupling a current to two gluons. The quantum effect represented by this graph can break symmetries present in the classical Lagrangian.

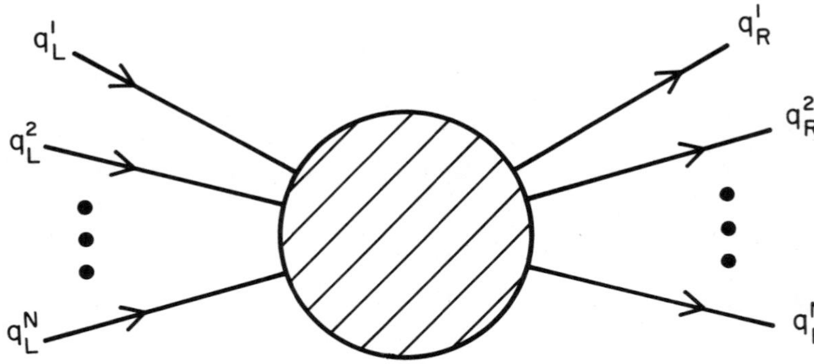

Fig. 2. Propagation in a background gauge field with $\int G_{\mu\nu}\tilde{G}_{\mu\nu} > 0$ causes left-handed quarks to be absorbed and right-handed quarks to be emitted. This effect can be represented as as effective interation, the 't Hooft vertex.

quantization, if you think in terms of path integrals. After all, the weight of different paths is determined by the classical action. The point is that the measure in the path integral must be regulated and defined by a limiting procedure, and the regularized measure does not in general respect the classical symmetries. This point of view toward anomalies has been stressed by Fujikawa[6] in some nice papers.)

We cannot long remain satisfied with this "solution" of the U(1) problem, however, because $\mathrm{Tr}G_{\mu\nu}\tilde{G}_{\mu\nu}$ itself is a total divergence

$$\mathrm{Tr}G_{\mu\nu}\tilde{G}_{\mu\nu} = \partial_\mu K_\mu$$
$$K_\mu = \epsilon_{\mu\nu\alpha\beta}\mathrm{Tr}\; A_\nu(\partial_\alpha A_\beta + \frac{2}{3}\, gA_\alpha A_\beta) \tag{22}$$

so it looks like we've gotten nowhere, that we still have a chiral U(1) invariance.

In reality, however, although $\mathrm{Tr}G_{\mu\nu}\tilde{G}_{\mu\nu}$ is formally a total divergence, it does change the physics when it is added to the Lagrangian. The point is that K_μ, the thing of which it is a divergence, is potentially a singular object. In fact, K_μ is gauge-dependent so it may have singularities or bad behavior at infinity which do not represent bad behavior of any physical, gauge-independent quantity (compare coordinate singularities in general relativity). To see whether such singularities do arise requires a dynamical calculation. Any relevant dynamical calculation must go beyond ordinary perturbation theory, since $\mathrm{Tr}G_{\mu\nu}\tilde{G}_{\mu\nu}$ like any total divergence -- formal or not -- gives no contribution to the Feynman graph rules.

The most important relevant calculation was done by 't Hooft in 1976.[7] He used the semiclassical approximation, where one expands the functional integral around solutions of the classical field equations. There are solutions of the source-free gauge field equations in Euclidean space, instantons, which have the required feature that $\int \mathrm{Tr} G_{\mu\nu}\tilde{G}_{\mu\nu} d^4x \neq 0$ although $\int \mathrm{Tr} G_{\mu\nu}\tilde{G}_{\mu\nu}$ is finite. The physical meaning of this is that in calculating the partition function of QCD, shich is done using Euclidean path integrals (temperature $\sim$ imaginary time) we will encounter amplitudes which violate axial baryon number, due to the anomaly. This holds in particular for the 0-temperature limit of the partition function -- i.e., the ground state will be a coherent mixture of states with different axial baryon number.[8]

To fill in this sketch a bit: the instanton solution[9] takes the form for an SU(2) gauge group or subgroup of color SU(3)

$$A_\mu(x) = \frac{1}{g} f(\tau)\eta^{-1}\partial_\mu \eta$$

$$\eta \equiv (x_0 + i\vec{x} \cdot \vec{\sigma})/\tau \tag{23}$$

$$\tau^2 \equiv x_0^2 + \vec{x}^2$$

where

$$f(\tau) = \frac{\tau^2}{\tau^2+\lambda^2}\,, \qquad \lambda \text{ arbitrary .} \tag{24}$$

Some qualitative features of this solution are notable:

a) The field strength is proportional to $\frac{x^2}{(\lambda^2+x^2)^2}$ and is localized in space and time, hence the term instanton.

b) The potential falls only as $1/\tau$ for large τ; K_μ falls as $1/\tau^3$ and gives a nonvanishing surface term as anticipated above. The field strength falls faster than $1/\tau^2$ because the derivatives cancel against commutator terms. Notice this can only happen is $A \sim 1/g$.

c) The action integral is

$$S = \frac{1}{2} \int TrG_{\mu\nu} G_{\mu\nu} = \frac{8\pi^2}{g^2} . \tag{25}$$

In the semiclassical approximation a factor e^{-S} accompanies the amplitude, so the amplitude has a factor $e^{-8\pi^2/g^2}$. The perturbative expansion of this for small g vanishes, as we know it must.

In a more exact treatment, we would find the factor $e^{-8\pi^2/g^2(\lambda)}$ involving the effective coupling constant of QCD at the appropriate scale. This brings up the fundamental difficulty in trying to use instantons as a quantitative tool in QCD: at large spatial scales $g(\lambda)$ grows and the semiclassical expansion fails just when the instantons are becoming important, i.e., just when $e^{-8\pi^2/g^2(\lambda)}$ is not $\ll 1$.

d) $G_{\mu\nu} = \tilde{G}_{\mu\nu}$, so according to (21) the integrated divergence of the axial baryon number is 6. (For N quarks, it would be 2N. For N purely right-handed quarks, it would be N.)

The connection between instantons and fermions suggested by d) and confirmed by the detailed semiclassical calculationis the following: an instanton fluctuation in the gauge field is inevitably accompanied by emission and absorption of quarks such that the axial baryon number changes by 2N. Of course, in this process the chiral SU(3) × SU(3) symmetries must be respected, these are not spoiled by anomalies. We can picture this result as an effective vertex shown in Fig. 2, where an instanton fluctuation in the gauge fields sprouts fermion legs. I'm calling this the 't Hooft vertex.

The physical origin of these fermion legs can be understood as follows. At large or small times $G_{\mu\nu}$ for an instanton vanishes, so it is a pure gauge, i.e. equivalent to zero by a gauge transfor-

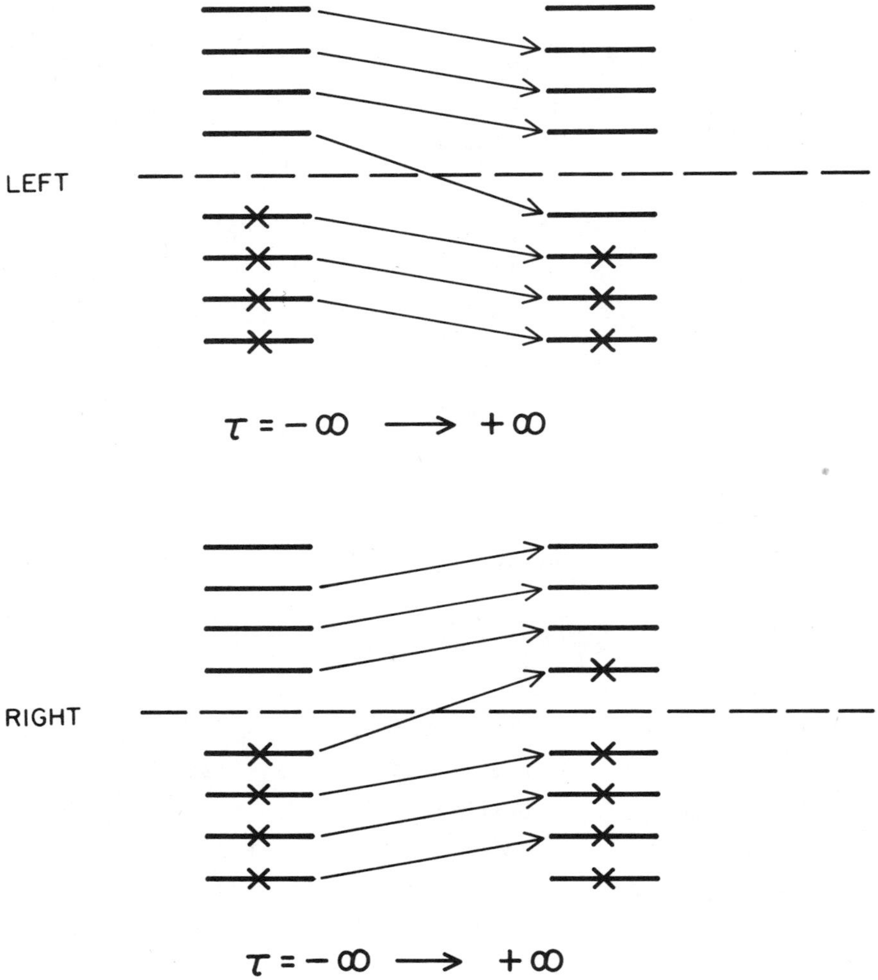

Fig. 3. Energy levels of the Dirac equation for mass quarks of different chirality in the presence of an instanton field. As (imaginary) time progresses these levels are rearranged. The consequence of this is that the fully occupied Dirac sea at $\tau = -\infty$ develops a hole or is promoted to a state containing a particle, at $\tau = +\infty$.

mation. The energy levels of the Dirac equation are then of course those of a free particle. As (imaginary) time unfolds forward, the levels of the Dirac equation move around and get reshuffled. In our example a right-handed hole state originally in the Dirac sea is excited to become a particle, and a hole state is created in the left-handed Dirac sea (see Fig. 3).

The 't Hooft vertex comes multiplied by a phase factor $e^{i\Theta}$, if the Lagrangian contains the term

$$L_\Theta = \frac{g^2\Theta}{16\pi^2} \, \mathrm{tr} G_{\mu\nu} \tilde{G}_{\mu\nu} d^4x \ . \tag{26}$$

This factor fits in nicely with the previous discussion of the anomaly equation and axial baryon number. If we make an axial baryon number phase rotation, say multiplying the right-handed quark fields by $e^{i\alpha}$ and the left-handed ones by $e^{-i\alpha}$, then the 't Hooft vertex is multiplied by $e^{6i\alpha}$. In other words, we have a new Θ, $\Theta_{new} = \Theta_{old} + 6\alpha$. As promised by the anomaly equation, even for massless quarks an axial baryon number phase rotation changes the Lagrangian, requiring us to add an extra $G_{\mu\nu}\tilde{G}_{\mu\nu}$ piece.

To summarize: the naive expectation that the gluon anomaly spoils axial baryon number as an approximate symmetry turns out to be correct, although a very deep analysis is necessary to demonstrate this. The result of this analysis is that although $\int G_{\mu\nu}\tilde{G}_{\mu\nu}$ is formally a total divergence it can appear in the action and then has nontrivial physical effects. For the tutored eye this is encapsulated in the 't Hooft vertex, surely "a picture worth a thousand words."

Quark Mass Ratios

In our further considerations the ratios of light quark masses will be very important, so I want to digress a moment to discuss them. A good up-to-date review of this is the Physics Reports article by Gasser and Leutwyler.[10]

The Gell-Mann-Okubo formula gives

$$\frac{m_s}{m_u+m_d} \cong \frac{m_K^2}{m_\pi^2} \cong 12 \pm 1 \ . \tag{27}$$

The discussion of $m_u - m_d$ is more delicate because one must be careful to separate out electromagnetic isospin-breaking. The best estimates come from analyses of meson and baryon masses. A very nice agreement is found with

$$\frac{m_s}{m_d-m_u} \text{ "="} \frac{m_k^2-m_\pi^2}{m_{k^o}^2-m_{k^+}^2}$$

$$\text{"="} \frac{m_\Xi-m_N}{m_{\Sigma^-}-m_{\Sigma^+}}$$

$$\text{"="} \frac{\frac{1}{2}(m_\Xi-m_N) - \frac{3}{4}(m_\Sigma-m_\Lambda)}{m_N-m_p}$$

$$\text{"="} \frac{\frac{1}{2}(m_\Xi-m_n) + \frac{3}{4}(m_\Sigma-m_\Lambda)}{m_{\Xi^-}-m_{\Xi^o}}$$

$$\text{"="} \ 45 \pm 3 \ . \tag{28}$$

Here the quotation marks around the equal signs indicate that electromagnetic and certain kinematic effects have been eliminated.

Putting all this together, we have

$$m_u:m_d = 1:2 \tag{29}$$

$$m_d : m_s = 1:20 \ . \tag{30}$$

THE STRONG CP PROBLEM AND AXIONS

Qualitative Discussion

As we have seen, the amelioration of the U(1) problem in QCD involves two essential observations. The first observation is that the axial baryon number current has an anomaly; its divergence (in the limit of massless quarks) is proportional to $g^2 \mathrm{tr} G_{\mu\nu} \tilde{G}_{\mu\nu}$. The second observation is that such a term, although it is formally a total divergence, does change the physics when it is added to the Lagrangian. Putting these two observations together, an axial baryon number transformation, which multiplies the right-handed quarks by $e^{i\eta}$ and the left-handed quarks by $e^{-i\eta}$, effectively adds a term proportional to $\eta g^2 \mathrm{tr} G_{\mu\nu} \tilde{G}_{\mu\nu}$ to the Lagrangian and does not represent a symmetry.

The possibility of having a term like $g^2 \mathrm{tr} G_{\mu\nu} \tilde{G}_{\mu\nu}$ in the QCD Lagrangian is very disturbing, since this term violates the space-time symmetries P and T. In the real world, as we shall quantify more precisely below, its conventionally normalized coefficient

$$L_\Theta = \frac{\Theta}{16\pi^2} g^2 \int \mathrm{tr} G_{\mu\nu} \tilde{G}_{\mu\nu} d^4x \tag{31}$$

must be very small, $|\Theta| < 10^{-8}$. This limit follows from the observed smallness of the neutron electric dipole moment.

The observed smallness of Θ presents a puzzle. If QCD were a closed theory, we might be content to take $\Theta = 0$ as a manifestation of P and T invariance. However, QCD is not a closed theory nor are P and T exact symmetries of the world. The weak interactions which break P and T will in general induce, by radiative corrections, infinite renormalization of the effective Θ parameter.

Let me be more precise about this. The weak interactions will in general present us with a mass matrix for the quarks:

$$L_{mass} = \bar{Q}_L \, M \, Q_R \tag{32}$$

$$Q = \begin{pmatrix} U \\ D \\ S \\ \cdot \\ \cdot \\ \cdot \end{pmatrix} \tag{33}$$

which, by redefinition of the quark fields (chiral SU(3) × SU(3) transformations) we can make diagonal and real <u>up to an overall phase</u>:

$$M = e^{i\Theta'} D$$

$$D = \begin{pmatrix} m_u & & & 0 \\ & m_d & & \\ & & m_s & \\ 0 & & & \cdots \end{pmatrix} . \tag{34}$$

without changing anything else in the QCD Lagrangian. We must however be careful about the overall phase. It would appear that this too could be removed by redefining the quark fields -- i.e. taking as new right-handed fields $Q_{R\ new} = e^{i\Theta'} Q_{R\ old}$. However, here we encounter the anomaly, and such a transformation must be accompanied by a change in the coefficient of $g^2 \mathrm{tr} G_{\mu\nu} \tilde{G}_{\mu\nu}$ in the Lagrangian. In a path integral formulation, we would find that the Jacobian of this redefinition of the fermion fields gave us the extra gluon term. Since the quark mass matrix, including its phase, is in general infinitely renormalized by the weak interactions, it is very puzzling why in nature the <u>net effective</u> Θ parameter is so small.

To summarize, the amelioration of the original strong U(1) problem brings us a new puzzle: why is the effective Θ parameter so small?

I think puzzles like this should be thought of as opportunities. If some physical parameter turns out to be very different from what we might "naturally" expect, we get very suspicious that there should be a simple qualitative reason for it. Searching for the reason might help to uncover new physical ideas.

As we shall elaborate at length below, there is a simple mechanism which will explain why the effective Θ-parameter is so small in nature. The basic ideas is to embed QCD into a larger theory such that the only explicit axial baryon number symmetry breaking is by the anomaly, not as in QCD by a combination of the anomaly and quark masses. (More precisely, the requirement is that one be able to change the Θ parameter multiplying the 't Hooft vertex, and no other parameter in the Lagrangian, by a transformation of fields.) The quasi-symmetry involved was first discussed by Peccei and Quinn[11] and from now on I will refer to PQ quasi-symmetry. In a theory with PQ quasi-symmetry, the effective Θ-parameter is chosen by the spontaneous breakdown of PQ symmetry, i.e. by minimization of an effective potential. Perhaps the clearest way to state the idea is that the effective Θ-parameter becomes a dynamical variable, rather than a fixed coupling constant. In many models the potential will have a minimum where the effective Θ parameter is very small.

(This is the usual case, because the lowest-order term in the effective Lagrangian is generally CP invariant and therefore symmetric in $\Theta \to -\Theta$ and stationary at $\Theta = 0$. Weak corrections here are really small, not infinite as for quark masses, because the instanton vertex is soft.)

This mechanism for explaining why Θ is so small has an important additional physical consequence.[12] If the PQ quasi-symmetry rigorously were broken only spontaneously, there would be an associated massless Nambu-Goldstone boson. Explicit breaking of PQ quasi-symmetry by the gluon anomaly generates a tiny mass for this particle, called the axion.

In terms of our previous discussion of the low-energy effective theory, the effect of the PQ quasi-symmetry is to reinstate the "bad" Gell-Mann-Okubo formula, with the singlet piece basically representing the axion. The effect of the anomaly is mainly to raise the mass of the η', which then can be regarded as heavy and essentially decouples from the axion-π^o-η system. The new element is that f_s is no longer associated with any strong-interaction scale, but rather is the scale at which PQ quasi-symmetry spontaneously breaks. If f_s is large, the axion mass can be exceedingly small, as follows from our previous analysis:

$$M_a^2 \cong \frac{f_\pi^2}{f_s^2} \; m_\pi^2 \quad . \tag{35}$$

The other side of the coin is that the couplings of this particle are also inversely proportional to f_s. It may not be absurd to think that our world is full of ultralight but very weakly interacting particles yet to be observed. The phenomenology and cosmology associated with such very light, very weakly coupled particles will be analyzed considerably below.

Are there alternative mechanisms for understanding why the Θ parameter is small in nature? Some ideas which have been mentioned in the literature are these:

a) Vanishing u-quark mass: If $m_u = 0$, the phase angle Θ in the 't Hooft vertex may be altered by redefining the right-handed u quark field, without changing anything else in the QCD Lagrangian. This possibility persists at the SU(2) × U(1) level, but apparently not in simple versions of SU(5).[13] Since moreover the successful current algebra treatment of meson and baryon masses depends on $m_u \neq 0$, the hypothesis $m_u = 0$ does not seem to be an attractive idea to eliminate the strong CP problem.

b) Spontaneous CP violation: If CP is broken only spontaneously (or softly) one can argue that radiative corrections to the Θ angle are finite and perhaps they may turn out small. Models implementing this idea are necessarily complicated; moreover it appears that it is difficult to make the effective Θ angle sufficiently small in a natural way without axions. Indeed, without a dynamical degree of freedom (the axion) which responds directly to the effective Θ angle, it requires something of a dynamical miracle for the energy functional of the other variables to find its minimum near enough to $\Theta_{eff} = 0$.

A graphic way of saying this: what we have learned from the solution of the strong U(1) problem is that with three families the quark mass matrix contains two CP violating angles, the usual one and the overall phase. The usual one as perceived through K-decays is probably not particularly small ($10^{-1} - 10^{-2}$) compared to Θ, which must be quite small ($< 10^{-7}$). Any serious proposal for solving the strong CP problem must address this disparity!

c) There are other frameworks involving supersymmetry or Kaluza-Klein theory, which might make the hypothesis that the coefficient of $\int \mathrm{tr} G_{\mu\nu} \tilde{G}_{\mu\nu}$ is zero seem natural. However, as far as I am aware neither of these address the problem of why the effective Θ parameter, which is a combination of this coefficient and the determinant of the quark matrix, is small. (This is basically the same question as arose for spontaneous

CP violation.) Similar remarks apply to technicolor theories, for which indeed it is a terrible problem to generate fermion masses in general.

To summarize, there is no proposal for solving the strong CP problem on the market which is sufficiently convincing that we should stop thinking about axions.

<u>Strong CP Problem in Minimal SU(2) × U(1)</u>

To make the discussion more definite and as a warmup to more complicated things, I'll briefly discuss how the strong CP problem looks in a minimal SU(2) × U(1) model of electroweak interactions.

In this model the quark masses are generated when a doublet Higgs field acquires a vacuum expectation, starting from interactions like:

$$- L_{mass} = \bar{\psi}^i_{\alpha L} U^j_R g_{ij} \phi^\alpha + \bar{\psi}^i_{\alpha L} D^j_R h_{ij} \varepsilon^{\alpha\beta} \phi^*_\beta \quad . \tag{36}$$

In this expression the Greek indices α,β are gauge SU(2) indices, the Latin indices i,j are family indices, U is the right-handed charge 2/3 quark field, etc. Spontaneous symmetry breaking occurs as ϕ_1 develops a nonvanishing vacuum expectation value, leading the mass matrix

$$(\bar{U}_L \bar{D}_L) \begin{pmatrix} \langle\phi_1\rangle g_{ij} & 0 \\ 0 & \langle\phi_1^*\rangle h_{ij} \end{pmatrix} \begin{pmatrix} U_R \\ D_R \end{pmatrix} + \text{h.c.} \tag{37}$$

The quantity relevant to strong CP violation is the phase of this determinant:

$$\det M = |\phi|^{2f} \det g \det h$$

$$\arg \det M = \arg \det g \det h \qquad (38)$$

(f = number of families). In this model, this phase is just a jumble of renormalized coupling constants, some of which are guaranteed to be complex so that we get CP violation in K-decay. It is very puzzling in this model why the effective should come out small.

Minimal Model with PQ Quasi-Symmetry

It is a fairly simple matter, and instructive, to expand the standard SU(2) × U(1) model slightly so that it has a PQ quasi-symmetry. In fact, the fiasco that occurred in Eq. (38), that the determinant of the quark mass matrix is fixed in terms of renormalized couplings, would not have occurred if the phase of the vacuum expectation value of the Higgs field hadn't cancelled. This suggests that if we have more than one Higgs field giving mass to the quarks, the relative pahse of the vacuum expectation values will determine the overall phase of the quark mass matrix. Since the relative phase of the vacuum expectation values is dynamically determined, the theory has a chance of choosing $\Theta_{eff} = 0$ dynamically.

So let us replace Eq. (36) with

$$- L_{mass} = \overline{\psi}^{i}_{\alpha L} U^{j}_{R} g_{ij} \phi^{\alpha}_{I} + \overline{\psi}^{i}_{\alpha L} D^{j}_{L} h_{ij} \varepsilon^{\alpha\beta} \phi^{*}_{II\beta} \qquad (39)$$

where ϕ_I, ϕ_{II} are <u>different</u> Higgs fields. The determinant of the quark mass matrix is now

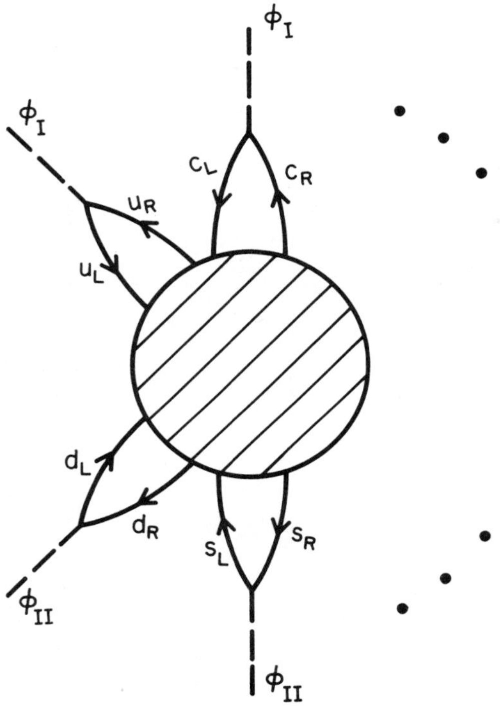

Fig. 4. The 't Hooft vertex generates an interaction which is sensitive to the relative phase of ϕ_I and ϕ_{II}. In theories with PQ quasi-symmetry, this is the only such interaction.

$$\det M = (v_I v_{II}^*)^f \det g \det h$$

$$v_I \equiv \langle\phi_I\rangle \qquad v_{II} \equiv \langle\phi_{II}\rangle \tag{40}$$

and can assume any phase, depending on the relative phase of v_I and v_{II}.

Is it plausible that these fields will choose the <u>right</u> phase, in other words that the energy is minimized when $\Theta_{eff} \approx 0$? Let us suppose for a moment that the usual potential for ϕ_1, ϕ_2 does not depend on their relative phase, that all dependence of the potential on this phase comes through the 't Hooft vertex (Fig. 4). Symmetry arguments, or explicit calculations, show that this lowest-order contribution is proportional to

$$e^{i\Theta}(v_I v_{II}^*)^f \det g \det h + \text{h.c.} \sim \cos \Theta_{eff} \, . \tag{41}$$

This is minimized precisely where we want it, at $\Theta_{eff} = 0$. v_I and v_{II} adjust themselves so that the overall phase of the quark mass matrix just cancels the initial Θ parameter.

Higher-order diagrams will alter this result by a small amount, as we shall discuss more quantitatively later. For the present, let me just remark that the corrections are finite in the ultraviolet, because the instanton vertex itself dies rapidly in the ultraviolet.

To summarize: the effective Θ parameter will very plausibly turn out small, if the only dependence of the energy functional on the relative phase of v_I and v_{II} is through instanton effects.

The next, crucial, question is why the relative phase enters the energy only in this particular way. Concretely, we must address why terms like

$$L_{bad} = \phi_I \phi_{II}^+ \quad \text{or} \quad (\phi_I \phi_{II}^+)^2 \tag{42}$$

or couplings of a Higgs field to the "wrong" kind of quarks do not occur.

This is where the Peccei-Quinn quasi-symmetry comes in. If we assume our Lagrangian is invariant under the PQ transformation

$$\begin{aligned} U_R &\to e^{i\eta} U_R \\ D_R &\to e^{i\eta} D_R \\ \phi_I &\to e^{-i\eta} \phi_I \\ \phi_{II} &\to e^{i\eta} \phi_{II} \end{aligned} \tag{43}$$

then all the "bad" terms are in fact forbidden. Of course this quari-symmetry has a rather peculiar status, since it is intrinsically broken by instanton effects. This represents a very special, physically distinct kind of soft breaking (i.e., true PQ symmetry is rapidly restored at large momenta). Perhaps a more palatable phrasing of the assumption, then, is that the PQ transformation is an asymptotic symmetry approached rapidly at large momentum.

Another important point is that a discrete subgroup of the PQ transformations, namely when $\eta = \frac{\pi}{f} \times$ integer, is exact. For $f = 3$, this exact discrete symmetry forbids all the "bad" terms. So we could postulate this discrete exact symmetry; then the full PQ quasi-symmetry arises as an accidental consequence!

In my view neither of these postulates, of an asymptotic symmetry or of a discrete symmetry, is fully satisfactory. I will return to this point in Lecture IV.

It is easy to see where the axion comes from and its qualitative properties in this model. I will analyze it only very crudely now, because more general examples will be analyzed below (and because the precise analysis is a little complicated to take in all at one blow).

The crucial point is that the energy functional depends on the relative phase between ϕ_I and ϕ_{II} only through terms of typical strong interaction size, since the Higgs couplings turn into quark mass terms. Thus after the magnitudes of ϕ_I, ϕ_{II} are fixed the Lagrangian for the axion kinetic energy and mass will look like

$$L_a = (\partial_\mu\phi_I)^2 + (\partial_\mu\phi_{II})^2 + (\text{'t Hooft vertex})$$

$$\tilde{\to} (v_I^2 + v_{II}^2)(\partial_\mu\alpha)^2 - \varepsilon\cos\alpha \qquad (44)$$

where $\varepsilon \approx (200\text{MeV.})^4$ is a typical strong interaction energy density and the relative phase of the Higgs fields. Normalizing the kinetic energy correctly and expanding for small $a = \frac{\alpha}{\sqrt{v_1^2+v_2^2}}$, we find the mass of a to be of order

$$m_a^2 \sim \varepsilon/(v_I^2 + v_{II}^2) \ . \qquad (45)$$

v_I and v_{II} are presumably of order 100 GeV., so $M_a \sim 10^5$ eV., very light. The couplings to quarks is calculated like this

$$g\phi\bar{q}_L q_R \overset{1}{\to} m_q e^{i\alpha}\bar{q}_L q_R \overset{2}{\to} \alpha m_q \bar{q}\gamma_5 q \overset{3}{\to} a \frac{m_q}{\sqrt{v_1^2+v_2^2}} \bar{q}\gamma_5 q \qquad (46)$$

as: 1) ϕ acquires a vacuum expectation value, 2) α is expanded to first order, and 3) α is replaced by the axion field with properly normalized kinetic energy. Thus we see that a couples quite weakly.

The axion model described in this section was the original version; it is pretty clearly ruled out by experiments as we shall discuss in Lecture III. Before confronting the phenomenology, however, I'd like to describe a more general class of models.

Generalized Models - Invisible Axions

The characteristic properties of the axion, its mass and couplings, are essentially determined by the mass scale at which the PQ quasi-symmetry spontaneously breaks, i.e. by the largest vacuum expectation value of a Higgs field carrying PQ charge (that is, a Higgs field which changes under a PQ quasi-symmetry transformation). Both the mass and the couplings of the axion are inversely proportional to the PQ breaking scale. In the previous section, the PQ breaking scale was essentially equal to the Weinberg-Salam breaking scale, because the Higgs fields carrying PQ charge also carried $SU(2) \times U(1)$ charge. It is quite possible that there are Higgs fields which are $SU(2) \times U(1)$ singlets yet carry PQ charge, and these could have much larger vacuum expectation values.

Let me first discuss the simplest prototype of such a model. To the model of the previous section let us add a complex scalar field ϕ, an $SU(2) \times U(1)$ singlet, which under the PQ transformation (43) transforms as

$$\phi \rightarrow e^{i\eta}\phi \quad . \tag{43'}$$

The coupling

$$- L_{new} = g^2 \phi^2 \phi_{II} \phi^+_{II} + h.c. \tag{47}$$

is now allowed by PQ quasi-symmetry. If the potential for ϕ is such that it acquires a large vacuum expectation value of magnitude F, the PQ quasi-symmetry will be broken at the scale F. The interaction (47) supplies a mass cross-term for ϕ_I and ϕ_{II}. Suppose that the bare $(mass)^2$ terms for ϕ_I and ϕ_{II} are μ_1^2, μ_2^2; then the $(mass)^2$ matrix in that sector looks like

$$\begin{pmatrix} \mu_1^2 & g^2F^2 \\ g^2F^2 & \mu_2^2 \end{pmatrix} . \tag{48}$$

We presumably want some linear combination of ϕ_I, ϕ_{II} to be the usual Weinberg-Salam doublet, with a negative $(mass)^2$ much less than F^2 in magnitude. Also, it is presumably desirable that the linear combination have roughly equal parts of ϕ_I, ϕ_{II}, since the masses of quarks with different charge are not drastically different. For all this to occur clearly requires some conspiracy among μ_I, μ_{II}, and F. I shall not attempt to explain this conspiracy, which is closely related to the hierarchy problem, because I don't know a good explanation yet.

Assuming the desired behavior occurs, the effective theory at low energies will be simply the usual SU(2) × U(1) model with a single Higgs doublet. Of course, more than one doublet is possible too; the point I am making is that one doublet is sufficient. In addition, there will be a very light axion field, essentially the Nambu-Goldstone boson of spontaneously broken PQ quasi-symmetry. To isolate it, we follow the now familiar procedure of writing the scalar fields as

$$\langle\phi_I\rangle = e^{-i\alpha} v_I$$

$$\langle\phi_{II}\rangle = e^{i\alpha} v_{II} \qquad (49)$$

$$\langle\phi\rangle = e^{2i\alpha} F$$

so as to isolate η, which according to PQ quasi-symmetry appears in the energy functional only in terms involving instantons.

(In these expressions, SU(2) indices of the ϕ_I and ϕ_{II} are suppressed; also both v_I and v_{II} are supposed real, which can be arranged by a hypercharge U(1) gauge transformation. It is worth remarking that the ratio v_I/v_{II} is determined at high scales according to what linear combination of ϕ_I and ϕ_{II} forms the Weinberg-Salam doublet, as discussed above. F may be complex; its phase is such as to make $\Theta_{eff} \approx 0$ similarly to what we discussed in the previous section.)

The properly normalized axion field comes from fixing the kinetic energy terms

$$|\partial_\mu\phi_I|^2 + |\partial_\mu\phi_{II}|^2 + |\partial_\mu\phi|^2 \rightarrow (4F^2 + v_I^2 + v_{II}^2)(\partial_\mu\alpha)^2 \qquad (50)$$

and is, for $F << v_I, v_{II}$,

$$a \cong 2F\alpha \quad . \qquad (51)$$

The energy density controlled by α is the coefficient of the 't Hooft vertex, i.e.

$$\cos(2N\alpha)\varepsilon$$

where $\varepsilon = (200 \text{ MeV.})^4$ is a typical strong interaction energy density. Expanding to second order in $a = \alpha/2F$, we find the mass of a to be

$$m_a \cong \frac{N}{F} \varepsilon^{1/2} \quad . \tag{52}$$

The coupling to fermions comes from expanding the mass terms (36) to first order in a, which boils down to

$$- L = \frac{a}{F} (m_u \bar{u}\gamma_5 u + m_d \bar{d}\gamma_5 d + \ldots) \quad . \tag{53}$$

Although the mass and interactions for axions deduced above are qualitatively correct, spontaneous breaking of chiral symmetry modifies them significantly. This will be analyzed in the next lecture, and used to assess axion phenomenology.

AXION PHENOMENOLOGY

The Effective Lagrangian

For an accurate discussion of axions, it is necessary to take into account their mixing with πs and ηs. Even though the mixing is small, the πs and ηs are so much more strongly coupled than the bare (Higgs field) axion that the small mixing significantly alters the properties of the axion.

(In preparing this part, I have benefitted greatly from unpublished notes by and conversations with John Preskill.)

To discuss this problem, we consider the dependence of the energy function on chiral phase rotations, to lowest order in quark masses. With

$$H = m_u \bar{u}_L u_R + m_d \bar{d}_L d_R + m_s \bar{s}_L s_R + \text{h.c.} \tag{54}$$

and

$$\begin{aligned}
\langle \bar{u}_L u_R \rangle &= v e^{i\left(\tilde{\alpha} + \frac{\tilde{\beta}}{\sqrt{3}} + \sqrt{\frac{2}{3}}\,\tilde{\gamma} + \frac{\theta}{3}\right)} \\
\langle \bar{d}_L d_R \rangle &= v e^{i\left(-\tilde{\alpha} + \frac{\tilde{\beta}}{\sqrt{3}} + \sqrt{\frac{2}{3}}\,\tilde{\gamma} + \frac{\theta}{3}\right)} \\
\langle \bar{s}_L s_R \rangle &= v e^{i\left(-\frac{2\tilde{\beta}}{\sqrt{3}} + \sqrt{\frac{2}{3}}\,\tilde{\gamma} + \frac{\theta}{3}\right)}
\end{aligned} \tag{55}$$

the energy functional is

$$\langle H \rangle = v \left\{ m_u \cos\left(\tilde{\alpha} + \frac{\tilde{\beta}}{\sqrt{3}} + \sqrt{\frac{2}{3}}\,\tilde{\gamma} + \frac{\theta}{3}\right) + m_d \cos\left(-\tilde{\alpha} + \frac{\tilde{\beta}}{\sqrt{3}} + \sqrt{\frac{2}{3}}\,\tilde{\gamma} + \frac{\theta}{3}\right) + m_s \cos\left(-\frac{2\tilde{\beta}}{\sqrt{3}} + \sqrt{\frac{2}{3}}\,\tilde{\gamma} + \frac{\theta}{3}\right) \right\} . \tag{56}$$

The physics behind this expression is the following. $\tilde{\alpha}$ and $\tilde{\beta}$ represent the ordinary chiral rotations. The associated gradient energy, if $\tilde{\alpha}$ and $\tilde{\beta}$ vary in space and time, is $(f_\pi^2/2)\left[(\partial_\mu \tilde{\alpha})^2 + (\partial_\mu \tilde{\beta})^2\right]$. $\tilde{\gamma}$ represents the effect in the quark sector of the PQ quasi-symmetry. It is associated with the large gradient energy $\frac{1}{2} f_s^2 (\partial_\mu \tilde{\gamma})^2$, because it involves rotating the phase of a Higgs field with a large vacuum expectation value.

We will want to be discussing CP-violating effects, so I have kept the effective parameter θ. As written in (56) this parameter can, of course, be eliminated by redefining $\tilde{\gamma}$. However (56) is not the complete energy expression. As discussed in Appendices II and especially III, the complete energy expression will be minimized with a mismatch between the phases of quark masses and the vacuum expectation values (55). To take this into account, I am not going to minimize the partial energy (56) with respect to $\tilde{\gamma}$; rather I will expand around $\tilde{\gamma} = 0$ which is by definition the minimum of the full energy.

In the realistic case $m_u, m_d \ll m_s$, $\theta \ll 1$ one easily finds the minimum of (56) with respect to $\tilde{\alpha}, \tilde{\beta}$ to occur at

$$\tilde{\alpha}_m = -\frac{\theta}{2}\,\frac{(m_u - m_d)}{(m_u + m_d)}$$
$$\tilde{\beta}_m = \frac{\theta}{2\sqrt{3}}\quad . \tag{57}$$

It is convenient to translate $\tilde{\alpha}$ and $\tilde{\beta}$ so that their minima occur at zero; implementing this (56) becomes

$$\langle H\rangle = v\left\{m_u \cos\left(\tilde{\alpha} + \frac{\tilde{\beta}}{\sqrt{3}} + \sqrt{\frac{2}{3}}\,\tilde{\gamma} + \frac{m_d}{m_u + m_d}\right)\right.$$
$$+ m_d \cos\left(-\tilde{\alpha} + \frac{\tilde{\beta}}{\sqrt{3}} + \sqrt{\frac{2}{3}}\,\tilde{\gamma} + \frac{m_u}{m_u + m_d}\right)$$
$$\left. + m_s \cos\left(-\frac{2\tilde{\beta}}{\sqrt{3}} + \sqrt{\frac{2}{3}}\,\tilde{\gamma}\right)\right\} \tag{58}$$

Our task now is to identify the eigenstates of mass; then we can read off the linear couplings of these mesons by expanding (54) and (55) (with the vacuum expectation values removed) to first order around the minimum.

The necessary calculations are very similar in spirit to ones we did in Lectures I and II, but even more laborious. I will not record all the intermediate steps here but rather proceed directly to describe the main results.

i) The axion $(\text{mass})^2$ is

$$m_a^2 = \frac{f_\pi^2}{f_s^2} m_\pi^2 \frac{m_u m_d}{(m_u+m_d)^2} \quad . \tag{59}$$

Note that the mass vanishes as m_u or $m_d \to 0$: we know it must, because in this limit there is a true symmetry associated with a phase rotation of the appropriate right-handed quark.

ii) The pseudoscalar (CP conserving) couplings of the axion are

$$- L_{eff} = \frac{a}{f_s} \frac{m_u m_d}{m_u+m_d} i \, (\bar{u}\gamma_5 u + \bar{d}\gamma_5 d) \; . \tag{60}$$

Despite appearances this coupling is <u>not</u> isoscalar![15] The point is that it is really the current $\bar{u}\gamma_\mu\gamma_5 u + \bar{d}\gamma_\mu\gamma_5 d$ and its divergences $m_u\bar{u}\gamma_5 u + m_d\bar{d}\gamma_5 d$ which commute with the isospin current. Insofar as $m_u \neq m_d$, this is not the same as occurs in (60).

iii) The scalar (CP violating) couplings of the axion are

$$- L_{eff} = \frac{a}{f_s} \theta_{eff} \frac{m_u m_d}{m_u+m_d} (\bar{u}u + \bar{d}d) \; . \tag{61}$$

iv) The scalar (CP violating) couplings of the pion are

$$- L_{eff} = \frac{\pi^0}{f_\pi} \theta_{eff} \frac{m_u m_d}{m_u+m_d} (\bar{u}u - \bar{d}d) \quad . \tag{62}$$

v) The dominant coupling of axions to electrons is

$$- L_{eff} = \frac{a}{f_s} m_e \, i\bar{e}\gamma_5 e \; . \tag{63}$$

Laboratory and Astrophysical Constraints

Several types of laboratory experiments put significant lower limits on f_s.[16] Among them the following are particularly significant:

i) Search for $K^+ \to \pi^+ a$: The axion is not directly detected, but one can look for a peak in the π^+ spectrum at the endpoint. There are serious backgrounds from $K^+ \to \pi^+\pi^0$ and from $K^+ \Xi \mu^+\bar{\nu}$. The limit from observations of this type is $f_s > 10^5$ GeV.

ii) "Beam dump" experiments: If $f_s < 10^4$GeV., the axions are produced with roughly semiweak strength and likewise interact with semiweak strength, so they are comparable to neutrinos which are produced "strongly" (as π decay products) and interact weakly, as sources of events in beam dump experiments. Although it is difficult to quantify precisely, the results of such experiments probably require $f_s > 10^4$ GeV.

iii) Reactor experiments: This is similar to beam dumps, and requires $f_s > 10^4$ GeV.

An extremely powerful constraint on axion properties derives from astrophysics.[17] Axions would be photoproduced in stars, and if they are as weakly coupled as required by the laboratory experiments, they will escape from the star and cool it. Comparison with the observed energy loss of stars constrains $f_s > 10^8$ GeV.

Axion Cosmology

If axions exist, they have very significant cosmological implications. Indeed, depending on the value of f_a, a significant or even dominant contribution to the mass density of the Universe can be made by axions.[18] For other values of f_a too much mass will be in axions, inconsistent with observations.

It should be noted that there is increasing evidence for the existence of large amounts of dark matter in the Universe, and that axions are a serious candidate to provide this matter.

I will first describe the cosmological mechanism for axion production in words, then in equations. An important qualitative feature of the axion field is that its potential is strongly temperature dependent. Indeed, the only effects which explicitly break PQ quasi-symmetry and keep the associated axion potential from being completely flat are instanton effects, and instanton effects are heavily suppressed at high temperature. Since the effective potential at $T \sim f_s$ where PQ quasi-symmetry breaks is issentially independent of the axion field, this field will align itself at random, and will generally <u>not</u> sit at the eventual minimum. These random values will be frozen in, there being essentially no energy gradients driving them, until at $T \sim 1$ GeV. instanton effects turn on. The axion field then "learns" that it has chosen a value away from the minimum, and moves toward the minimum. However, since the axion field is exceedingly weakly coupled it cannot readily lose energy, and instead of settling at the minimum it oscillates. Cosmologically significant amounts of energy can be stored in these oscillations.

The preceding field picture can also be interpreted in particle language, as the non-thermal production of axions at $T \sim 1$ GeV.

To make the discussion more quantitative, consider the action for the axion field a in a curved space

$$S = \int \sqrt{g}\, \frac{1}{2} \left\{ (\partial_\mu a)^2 - f^2 m_a^2 h\left(\frac{a}{f}\right) \right\} d^4x \, . \tag{64}$$

Here m_a is the mass of the axion, f its scale parameter, and h the normalized potential. To lowest order, $h(a/f) = a^2/f^2$. For the Friedmann-Robertson-Walker metric $ds^2 = dt^2 - R^2(t)d\vec{x}^2$ and a spatially constant axion field we have the action

$$S = \int R^3(t)\, \frac{1}{2} \left\{ \dot{a}^2 - m_a^2 a^2 \right\} d^4x \tag{65}$$

and the equation of motion, for small a

$$\ddot{a} + 3\,\frac{\dot{R}}{R}\,\dot{a} + m^2(t)a = 0 \, . \tag{66}$$

Note that m^2 depends on the temperature and therefore on t.

The restriction to spatially constant a is appropriate for the following reason. We expect the characteristic scale for spatial variations in the phase angle a/f to be the horizon size. Variations on shorter scales will be damped as they are energetically unfavorable, but on the horizon scale or larger the fields cannot be correlated -- they are causally disconnected. The horizon scale is essentially the distance that light can have travelled in the age of the Universe, $l \sim M_{pl.}/T^2$. At $T \sim 1$ GeV. or less the characteristic gradient energy density of order a^2/l^2 is substantially less than the potential energy m^2a^2 and can be ignored in the local dynamics. In particle language, this means that axions are produced essentially at rest. Also it is important to note that (rephrasing the previous discussion), although the horizon scale at these temperatures is much larger than the Compton wavelength of the axion, very little baryon number, in cosmological terms, is contained in a horizon volume: $\eta_B \sim 10^{-10}T^3(M_{pl.}/T^2)^3 \sim 10^{47}$ for $T = 1$ GeV., not enough for even one star! So, although variations in a/f are negligible for the local dynamics, on cosmologically interesting scales we must average over this quantity.

Now let's return to the analysis of the spatially constant case. The equation (66) is readily recognized as the equation for motion of an harmonic oscillator with a friction term and variable mass. When the mass is negligible a does not move, as we said in the qualitative discussion. To analyze the behavior it is reasonable to make the adiabatic assumptions $\dot{m}/m$, $\dot{R}/R \ll m$. Since m increases rapidly as the temperature decreases, this approximation becomes good very quickly soon after m becomes of order $\dot{R}/R$ -- and before that, essentially nothing happens.

In the adiabatic regime we expect an approximate solution of the form

$$a(t) = A(t) \cos \int^{t} m(t')dt' \tag{67}$$

with $\dot{A}/A \ll A$. The equation for the slow variation in A is then found to be

$$2 \frac{\dot{A}}{A} + \frac{\dot{m}}{m} + 3 \frac{\dot{R}}{R} = 0 \tag{68}$$

from which

$$\frac{m_f^2 A_f^2 R_f^3}{m_i^2 A_i^2 R_i^3} = \frac{m_f}{m_i} \quad . \tag{69}$$

The quantity on the left-hand side of this equation is the ratio of final to initial energies in a comoving volume. From this equation we see that apart from the factor m_f/m_i the energy in axions is conserved, as for nonrelativistic matter and different from nonrelativistic matter where the energy in a comoving volume goes as 1/R due to the redshift. The extra factor is also easy to interpret in particle language -- the axion number is conserved.

Putting it all together, the average energy density in the axion field at present is

$$\bar{\rho}_a \simeq f_a^2 m_i^2 \bar{h} \left(\frac{T_f}{T_i}\right)^3 \frac{m_f}{m_i} \simeq f_\pi^2 m_\pi^2 \left(\frac{T_f}{T_i}\right)^3 \frac{m_i}{m_f} \tag{70}$$

where T_f, m_f are the present temperature and axion mass (actually, the temperature has to be slightly corrected from the 3° K background because the photon gas has been heated by particle-antiparticle annihilations) and T_i, m_i are the corresponding quantities when the axion field started oscillating -- i.e. when $m_i(T_i) \sim m_{pl.}/T_i^2$.

Now because the axion mass depends very steeply on the temperature in the relevant regime ($T \sim 800$ MeV.) where the oscillations set in, changes in f will change m_i and T_i only by small amounts. Also, the temperatures are sufficiently high that an accurate evaluation of instanton effects can be done, enabling us to do a good estimate of m_i and T_i. The detailed calculation of instanton effects is quite involved, but now you can appreciate the qualitative form of the answer which is

$$\Omega_a \equiv \frac{\rho_a}{\rho_{crit}} \simeq 10^7 \left(\frac{f_a}{M_{pl.}}\right)\left(\frac{800 \text{ MeV.}}{T_i}\right) \tag{71}$$

for the ratio of axion energy density to the critical density. According to observations, the Universe cannot be overclosed by much, so $f_a < 10^{12}$ GeV. If $f_a \gtrsim 10^{12}$ GeV., axions could provide the dark matter in the Universe.

The foregoing analysis was all done assuming the simplest possible extrapolation of Big Bang cosmology to higher temperatures, and variations are certainly possible. Perhaps the best motivated would be to assume that inflation takes place after the PQ phase transition. Then the initial value of the axion phase a/f_a is really a universal constant, i.e. the horizon scale includes the whole visible Universe. If this constant $(a/f_s)^2 \approx 10^{-3}$ then $f_a \sim 10^{15}$ GeV. would be preferred, so that the PQ symmetry breaking scale could be the same as the unification scale.

Very different pictures of growth of fluctuations in the Universe are implied if the dark matter is produced cold (as for axions) or relativistic (as for 30 eV. neutrinos). I will discuss this in detail elsewhere.

Macroscopic Forces Mediated by Axions

The axion mass is so small that its Compton wavelength may be a macroscopic distance. In fact we have

$$\lambda \cong 2 \text{ cm.} \left(\frac{f_a}{10^{12} \text{ GeV.}} \right) \quad . \tag{72}$$

Accordingly axions mediate forces with a range that might be accessible in experiments with macroscopic bodies. The intrinsically incredibly weak force between individual particle may be substantially enhanced when we have coherent contributions from $\sim 10^{23}$ particles.

As discussed above, the axion has a scalar type coupling proportional to the CP violating parameter Θ, and a pseudoscalar type coupling. The pseudoscalar type coupling generates a spin-dependent coupling. We can discuss three types of experiments (This summarizes work done with J. Moody):

i) monopole-monopole: Here we are looking for a potential of the form

$$U(r) \propto \frac{-1}{r} e^{-m_a r} \tag{73}$$

due to two scalar vertices. This represents an attractive addition to the normal gravitational force at short distances. The ratio of axion to gravitational forces is roughly

$$\frac{\theta^2 m^2_{pl.}}{f_a^2} \approx \theta^2 10^{14} \left(\frac{10^{12}\ \text{GeV.}}{f_a}\right)^2 . \tag{74}$$

Torsion balance experiments sensitive to forces 10^{-4} of gravity at $\sim$ 1 cm. have been carried out. Our best estimates on limits to axion parameters from existing and reasonably foreseeable experiments are shown in Fig. 5. From our point of view, the most interesting direction to pursue further is experiments at shorter distances.

ii) monopole-dipole: Probably the most promising and novel kind of experiment would be to look for a potential between a body with aligned spins and an ordinary body of the form

$$U(r) \propto \frac{1}{m_a r^2} (\hat{r} \cdot \hat{s})(1 + m_a r)\, e^{-m_a r}$$

This represents a T-violating macroscopic force. It would be generated by axion exchange where the coupling is scalar at one vertex and pseudoscalar at the other. This type of interaction apparently has an important experimental advantage; that is, one can periodically modulate the alignments of the spins and use resonance techniques of remarkable sensitivity. Our estimates of what can be achieved with present or readily foreseeable technology is shown in Fig. 6. It is very possible that we have not found the best experimental

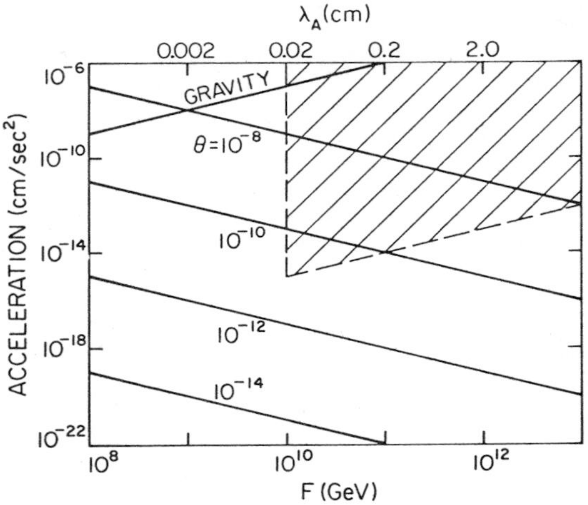

Fig. 5. The acceleration due to axion-mediated monopole-monopole forces, for macroscopic samples of iron, for different values of θ and F. The most important limitation on experiments to detect such forces are metrological. We estimate (Ref. 9) that the upper right-hand corner of parameter space gives a measurably large force.

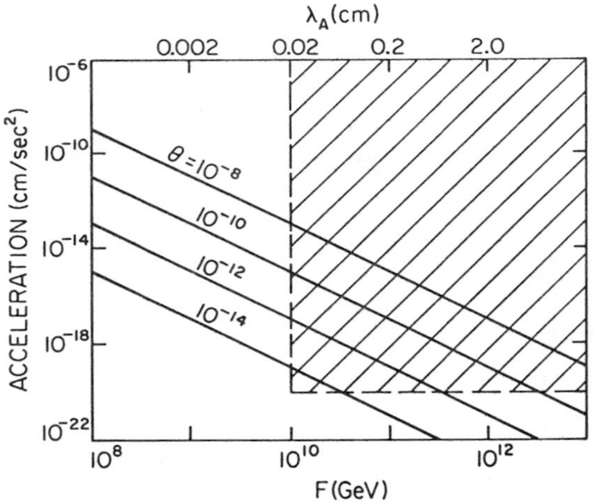

Fig. 6. Similar to Fig. 5, for the monopole-dipole force. This type of force seems particularly attractive from an experimental point of view, since it is T-violating (thus lending itself to null experiments) and can be modulated by varying the spin source (thus lending itself to resonance techniques).

design. Also, the torsion balance technique is seismic noise limited and could give much better results if suitable seismic shielding could be developed.

iii) dipole-dipole: Axion exchange with two pseudoscalar vertices gives a potential of the form

$$U(r) \propto \frac{e^{-m_a r}}{m_a^2 r^3} \left\{ \hat{s}_1 \cdot \hat{s}_2 (1 + m_a r) - \hat{s}_1 \cdot \hat{r} \hat{s}_2 \cdot \hat{r} (3 + 3m_a r + m_a^2) \right\} . \quad (75)$$

Unfortunately, this type of spin-spin force is also generated by magnetic interactions, which leads to terrible problems in attempting to observe the tiny corrections we are considering. Our best, rather discouraging, estimate of what can be achieved in this kind of experiment is shown in Fig. 7.

In this context it is also appropriate to mention the macroscopic forces mediated by true Nambu-Goldstone bosons, familons (see Lecture 4). These are rigorously massless particles, with derivative couplings to matter. Their couplings are basically of the pseudoscalar type, so unfortunately they could only be seen in dipole-dipole experiments. It should also be mentioned that the pseudoscalar vertex really gives a coupling to spin, not to angular momentum. Partly as a result, it is safe to ignore familons compared to gravity in any astrophysical context, or for the Stanford gyroscope experiment. As a modest example, the angular momentum of the earth is about 10^{15} times the spin angular momentum you would obtain by aligning every atom in the earth. The intrinsic strength of familon couplings is limited by the same considerations on stellar cooling as for axions, and cannot be nearly enough to make up for this enormous suppression.

Quite different possibilities for experiments to find axions have been proposed by Sikivie.[19]

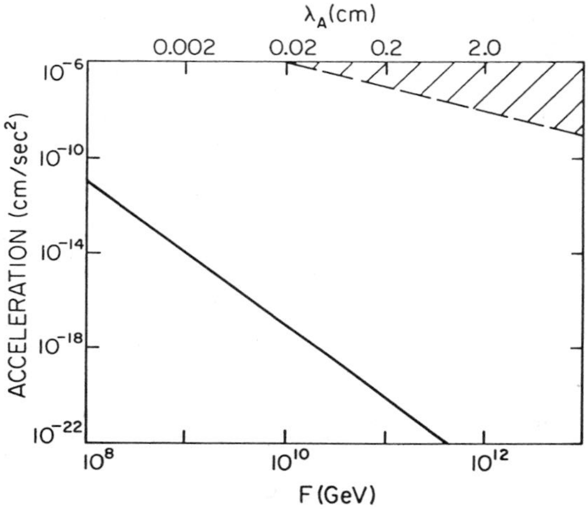

Fig. 7. Similar to Fig. 5, for the dipole-dipole force. The most important limitation here is background magnetic fields, which give a force of the same form.

AXIONS IN CONTEXT: FAMILONS AND COSMIONS

Critique of PQ Quasi-Symmetry

Although PQ quasi-symmetry seems to be the only available idea to understand why the strong interactions conserve CP for high accuracy, it is rather unsatisfying in several regards. In particular,

i) Axial baryon number is only one small part of a very large family symmetry group that holds in the limit of massless quarks. Why should it be treated on a special footing?

ii) The requirement that the theory should have a symmetry, like the PQ quasi-symmetry, "except for instanton effects" seems rather artificial. The PQ quasi-symmetry can be consistently postulated as an asymptotic (large momentum) symmetry or as an accidental consequence of a discrete symmetry, as was mentioned before. (In fact, the discrete symmetry idea may have severe problems, as we shall discuss shortly.) Nevertheless, it would certainly seem more satisfactory to have the possible interactions which were banished by appeal to PQ quasi-symmetry forbidden instead for honest symmetry reasons.

iii) The PQ U(1) quasi-symmetry typically has a discrete subgroup which is <u>not</u> intrinsically broken but which is spontaneously broken at the large PQ-breaking scale ($F \gtrsim 10^{12}$ GeV.) discussed above. In this circumstance, there are topologically stable domain walls separating regions where the vacuum changes from one of the discrete set of degenerate possibilities to another.[21] These domain walls would be produced in the early Universe. One can show that they lead to severe difficulties for cosmology, closely related to the cosmological phenomena discussed in Lecture 3. In Lecture 3 we found that it was difficult to get rid of the energy in the axion field locally; with domain walls we get a similar global problem as the energy piles up at the domain walls.

All these weaknesses can be relieved if we suppose that there is a real family symmetry, unbroken by anomalies.[22] An appealing way to say this is the following. The PQ mechnaism is a method for making the phase of the determinant of the quark mass matrix into a dynamical variable. If this variable is coupled in a symmetric way, we have the picture that family symmetry breaking -- such as $m_e \neq m_\mu$ and $m_d \neq m_s$ -- is a dynamical effect.

Let's see how this idea relieves the three weaknesses:

i) The axion is naturally put into a larger context, where the whole mass matrix is dynamical. This means that a large set of Higgs fields is introduced, in the most extreme form, one for each element of the mass matrix. Most of the Higgs particles of course are supposed to acquire very large masses, $\sim 10^{12}$ GeV. or greater.

ii) As a toy example of how PQ quasi-symmetry might be an accidental consequence of a true higher family symmetry, consider a family symmetry SU(6) × SU(6) under which the mass matrix transforms as

$$M \to U^{-1}MV$$
$$U,V \in SU(6) \quad . \tag{76}$$

Consistent with this symmetry, the potential can contain terms like

$$\Delta L_{invariant} \sim trMM^{+}, \quad tr(MM^{+})^2 , \tag{77}$$

but not for instance

$$\Delta L_{non-inv.} \sim trM^2, \quad trM^4 . \tag{78}$$

The important point is that all the allowed terms do not depend on the phase of M (detM is forbidden by renormalizability), so that there is an accidental PQ quasi-symmetry.

iii) The discrete subgroup of PQ quasi-symmetry which is not intrinsically broken is now embedded in a continuous group. Upon spontaneous symmetry breaking no stable domain walls can form; the configurations which were previously stable can now relax away by emitting massless Nambu-Goldstone bosons of broken family symmetry!

Spontaneously Broken Family Symmetry

As just discussed, there are good reasons to suspect that family symmetries might be spontaneously rather than intrinsically broken. If, furthermore, these symmetries are global rather than local symmetries then there will be characteristic (strictly massless, spin zero, neutral) Nambu-Goldstone bosons.[22-24] I call these objects familons.

Massless particles are in principle accessible to low-energy experiments; it is a practical question of how weakly they couple. As is by now agonizingly familiar, the coupling of a Nambu-Goldstone boson f is generically of the form

$$- L_{int.} = \frac{1}{F} \partial_\mu f \; j_\mu \tag{79}$$

where F is the symmetry-breaking scale and j_μ the current which generates the symmetry.

The case of most experimental interest seems to be the spontaneous breakdown of strangeness conservation, where $j_\mu = \bar{s}\gamma_\mu d$. We can have K^+ decaying into π^+ and a familon, with the branching ratio[22,24]

$$\frac{\Gamma(K^+ \to \pi^+ f)}{\Gamma(K^+ \to \pi^+ \pi^0)} = \frac{1.3\times10^{14}}{F^2} \text{ GeV.}^2 \tag{80}$$

The existing limit corresponds to $F > 10^{10}$ GeV. We see that this decay allows us to probe enormous scales. $F > 10^{12}$ GeV., which as we have seen is an extremely interesting scale, does not appear to be beyond experimental reach.

Another exotic process that could be induced by familons is neutrino decay. The relevant interaction here is

$$- L_{int.} = \frac{1}{F'} \bar{\nu}' \gamma_\mu \nu \, \partial_\mu f' \tag{81}$$

where ν' is a heavy and ν a light neutrino, and F', f' are respectively scales and familon fields, in principle different from those appearing in (79). The interesting quantity here is the neutrino lifetime

$$\tau(\nu' \to \nu f) = 3 \times 10^9 \left(\frac{F}{10^{10} \text{ GeV.}}\right)^2 \left(\frac{100 \text{ keV}}{m_{\nu'}}\right)^3 \text{ sec.} \tag{82}$$

An interesting consequence is that familon decay can destabilize neutrinos on cosmological time scales. As a consequence, they might make it possible to have fairly massive neutrinos which do not contribute to the present mass density of the Universe. For <u>stable</u> neutrinos, masses in the range $\sim$ 100 eV. $\to \sim$ 2 GeV. are forbidden according to this consideration.

We can attempt to use broken family symmetry to relate fermion masses and mixing angles. As a simple example, let us suppose that left- and right-handed charge -1/3 quarks transform in a 3 and $\overline{3}$ representation of family SU(3). (This is a natural assignment in the context of an SU(5) unified model.) Let us further suppose that the Higgs field which survives down to low energies and breaks SU(2) × U(1) electroweak symmetry is a linear combination of a large flavor $\underline{6}$ (symmetric tensor) piece, the 33 component, and a smaller $\underline{3}$ (antisymmetric tensor) piece mostly but not entirely aligned in the orthogonal direction. Then the mass matrix will look like

$$M \sim \begin{pmatrix} 0 & A & 0 \\ -A & 0 & B \\ 0 & -B & C \end{pmatrix} . \tag{83}$$

with C >> B >> A. The physical masses are then $m_b \cong C$, $m_s \cong B^2/C$, $m_d \cong A^2C/B^2$. Assuming that the charge 2/3 quarks have a diagonal mass matrix in the same basis, we find the mixing angles

$$\tan^2\theta_{s\to u} = \tan^2\theta_{cabbibo} \cong m_d/m_s \tag{84}$$

$$\tan^2\theta_{b\to c} \cong m_s/m_b \quad . \tag{85}$$

The first of these is known to be good; the second angle is becoming accessible through measurements of the lifetime of mesons containing b-quarks -- it appears to be very small, consistent with (85).

One can also relate quark to lepton masses in this framework. At present there are too many arbitrary assumptions in these schemes for us to take their detailed results very seriously; nevertheless I think it's intriguing that fairly simple symmetry-breaking patterns are not manifestly at variance with experiment.

Cosmions

Probably everyone who's thought hard about axions has toyed with the idea that a similar mechanism might serve to cancel off another embarrassing coupling, the cosmological constant.

Recall that that the problem here is that modern gauge theories lead us to believe that the physical vacuum is a very complicated state picked out by spontaneous symmetry breaking at enormous mass scales. The typical energy densities one encounters in comparing the broken-symmetry vacuum to the symmetric vacuum are of order $(10^{15}\ \text{GeV.})^4$ for typical (non-supersymmetric) grand unified theories. On the other hand gravity does not seem to notice all this energy density; there are limits of order $|\Lambda| < (10^{-12}\ \text{GeV.})^4$ for a contribution of the form $\langle T_{\mu\nu}\rangle = \Lambda g_{\mu\nu}$ to the energy density of the Universe on large scales.

There are similar problems, perhaps less than a factor 10^{128} but still very impressive, connected with reasonably well-understood broken symmetries such as chiral symmetry in the strong interactions.

It is certainly very reasonable, in this circumstance, to imagine that there is some universal mechanism which will cancel off any Λ supplied by the behavior of matter. In other words, if there were further phase transitions, another change in the vacuum energy, it seems very reasonable to suppose that the effective Λ as measured by gravity would still be very nearly zero -- perhaps settling down to zero after a period of adjustment. You can think of this as a generalized Copernican principle.

If this generalized Copernican principle is true, the cosmological constant problem appears very similar to the strong CP problem we have been concerned with in the preceding. In that problem, the weak interactions feed us an arbitrary θ but the dynamical axion field always adjusts to make the effective θ parameter very small. Could there be a similar "cosmion" field ϕ which does this job for the cosmological constant?

Putting off for the present the question of its origin in a microscopic theory, let's consider what the properties of such a ϕ field would have to be like.

The first requirement of the theory, that ϕ be capable of cancelling off the cosmological term, is trivial to satisfy. It only requires that ϕ have an unbounded, or at least very large, potential $V(\phi)$. We should be satisfied to cancel off only negative vacuum energy, so $V(\phi)$ could be bounded below. So in this sense it is easy (all too easy!) to make Λ a dynamical variable. The second, crucial requirement is that ϕ dynamically adjusts itself to give nearly vanishing effective cosmological term.

At first glance this requirement seems impossible to satisfy -- how is the total potential clever enough so that all minima, with very different values for expectation values of fields other than ϕ, have the same value of energy density once we minimize with respect to ϕ? Very little experimentation is necessary to convince yourself that this never happens.

More promising is to suppose that ϕ favors $\Lambda_{eff} = 0$ for the indirect reason that scalar curvature is energetically unfavorable. In other words, there are couplings like $L = -\phi^2R^2$ (ϕ^2R might also work) which make it energetically unfavorable to have nonzero R. Very small R of course requires $\Lambda_{eff} \approx 0$. Since ϕ has the capability of cancelling off Λ, and it is energetically favorable to do so, it is very plausible that it will.

The basic difficulty of this idea is that we are asking a very small tail to wag a very large dog. We want $V(\phi)$ to be able to take very large values; on the other hand we want the minimum of something like $V(\phi) + \phi^2 R$ to be determined mainly by the second term. Since R is small in any reasonable units (and this is exactly what we are trying to explain!), clearly $V(\phi)$ will have to be a very flat potential.

Let me quantify this a little. (The following summarizes some work with S. Deser.) Take as the action

$$\int \sqrt{g} \left\{ (\partial_\mu \phi)^2 + K^2 R - \eta\phi^2 R - V(\phi) + \Lambda) \right\} d^4x \ . \tag{86}$$

We want to see that there is a solution with ϕ = const., and R small with little dependence on Λ. Two relevant equations are the trace of the gravity equation and the ϕ field equation (for ϕ = const.):

$$R = \frac{V'(\phi)}{2\eta\phi} \tag{87}$$

$$\frac{V'(\phi)}{\eta\,\phi} K^2 + \eta\phi^2 = V(\phi) - \Lambda \ . \tag{88}$$

Now let's see if we can find a solution for the kind we want. I am going to analyze a couple of special choices for $V(\phi)$. First take $V(\phi) = B/\ln(\phi/\mu)$, where B is a very large quantity (say $\sim M^4_{pl.}$) -- this is a not unsuggestive form for the vacuum energy in gravity. Assuming $\eta\phi^2 < K^2$, the equation for ϕ is

$$2RK^2 \approx V(\phi) - \Lambda \ .$$

The left-hand side is the effective Λ term, which we suppose small so $B/\ln(\phi/\mu) = V(\phi) \approx \Lambda$. The crucial consistency check is then to evaluate (87) under this hypothesis; the result is

$$R = -\frac{B}{2\eta\phi^2(\ln\phi/\mu)^2} \approx -\frac{\Lambda^2}{BK^2} \, . \tag{89}$$

This result indicates that the effective cosmological term is changed from Λ to Λ^2/B. An even more effective suppression is found for fractional powers of the logarithm; finally, if $V(\phi) \approx B/\ln\ln(\phi/\mu)$ the suppression is exponential: $\Lambda_{eff} \sim \Lambda e^{-B/\Lambda}$.

Many difficult questions remain to be addressed even at the level of these effective Lagrangians: how large is the domain of attraction of the solution we have found; in particular how does the system respond to phase transition in the matter sector; the ϕ field is very reminiscent of Brans-Dicke theory -- is it consistent with tests of general relativity? All these questions must be addressed; meanwhile there is the unavoidable suggestion in this framework of a very light scalar -- the cosmion -- coupling exclusively to spatial curvature, and to itself.

The next, all important, question is whether and how couplings such as that we postulated for our ϕ field might arise in a microscopic theory. I have no real answer to this at present; let me just say that inverse fractional powers of logarithms in effective potentials do arise as coupling constant renormalizations in an asymptotically free theory.

APPENDIX A. An Important Footnote to Lecture I: the "Bad" Gell-Mann-Okubo Formula Can Become "Good"

In the lecture we discussed how the meson spectrum would look quite different is the axial baryon number symmetry were good. This symmetry, however, is broken by instanton effects. It is well worth remarking that under certain physical circumstances we can expect that the instanton effects are suppressed.[25]

Specifically, at finite temperatures or finite baryon number density, we expect instanton effects to be suppressed. This is because a background plasma of charged particles, such as we have in either of these circumstances, produces screening coherent fluctuations of the electric field at wavelengths greater than the screening length are heavily suppressed. Large instantons do involve coherence over large distances or times, and are suppressed by the screening. Small instantons are always suppressed, because their amplitude goes as $e^{-8\pi^2/\bar{g}^2}$ and the effective coupling $\bar{g}$ is small at small distances. So at finite temperatures or densities the instantons may be suppressed altogether, and axial baryon number symmetry restored.

The consequences of symmetry restoration for the pseudoscalar mesons are quite dramatic. The η' mass will drop precipitously, as will the η mass. Also one may have isospin violating effects of order $(m_u-m_d)/(m_u+m_d)$ [instead of the usual $(m_u-m_d)/m_s$)]. The basic phenomenon here is that the mass eigenstates become the pseudoscalar mesons with quark constant $\bar{u}u$, $\bar{d}d$, $\bar{s}s$, and have $(\text{masses})^2$ in the ratio $m_u:m_d:m_s$.

APPENDIX B. Mass Scale of θ-Dependence

In estimating θ_{eff}, it will be important to understand how the dependence of the phase of the instanton vertex on the phase of heavy quark fields arises. I address this preliminary question first.

According to our discussion of the anomaly equation, a chiral rotation of the quark through an angle α should produce a change in the phase of the whole functional integral by $e^{i\alpha}$, in the 1-instanton sector. In an equation:

$$\frac{\det(iD + me^{i\alpha\gamma_5})}{\det(iD + m)} = e^{i\alpha} \quad . \tag{B.1}$$

Here D is the Dirac operator in the presence of a background instanton, and the determinant is the factor one gets on integrating out the quarks.

Let the eigenfunctions f(n) of the Dirac equation with m = 0 be decomposed into left- and right-handed components $f(n) = u_L^{(n)} + v_R^{(n)}$, so that

$$\begin{aligned} iD\, u_L^{(n)} &= i\lambda^{(n)}\, v_R^{(n)} \\ iD\, v_R^{(n)} &= i\lambda^{(n)}\, u_L^{(n)} \quad . \end{aligned} \tag{B.2}$$

Then different linear combinations of $u_L^{(n)}$ and $v_R^{(n)}$ also give the eigenvalues of the Dirac equation with mass. In fact, the eigenvalue equation becomes

$$\begin{aligned} (iD + me^{i\alpha\gamma_5})(au_L^{(n)} + bv_R^{(n)}) &= (b\lambda^{(n)} + ame^{-i\alpha})u_L^{(n)} \\ &\quad + (a\lambda^{(n)} + bme^{i\alpha})v_R^{(n)} \\ &= \mu^{(n)}(au_L^{(n)} + bv_R^{(n)}) \end{aligned} \tag{B.3}$$

which is solved for a/b, $\mu^{(n)}$ to give a pair of eigenvalues

$$\begin{aligned} a/b &= \frac{-m \sin\alpha \pm \sqrt{\lambda^{(n)2} + m^2\sin^2\alpha}}{\lambda^{(n)}} \\ \mu^{(n)} &= m\cos\alpha \pm i\sqrt{\lambda^{(n)2} + m^2\sin^2\alpha} \quad . \end{aligned} \tag{B.4}$$

As long as $\lambda^{(n)} \neq 0$ the eigenvalues are paired, and the product of those with the upper and lower signs in (B.4) is $m^2 + \lambda^{(n)2}$, independent of α. The only way to satisfy (B.4) is to have some zero eigenvalues of the massless Dirac equation. Each left-handed solution will give a factor $me^{-i\Theta}$ and each right-handed solution a factor $me^{i\Theta}$ in (B.1), so the conclusion is that there must be one more right- than left-handed solution. In fact, there is exactly one right- and no left-handed solution in the case at hand; its explicit form is known and was alluded to in Lecture I.

For the present purpose, the important conclusion is that the zero-mode contains all the dependence of the determinant in α. The zero-mode contains momenta of order $1/\rho$, where ρ is the instanton size.

APPENDIX C. Approximate Magnitude of θ in Minimal Model

I will now outline the principles of how the effective value of θ is determined in a theory with axions and estimate it roughly in the minimal Kobayashi-Maskawa model of CP violation.

Let us work in a basis with all the quark masses real. The relevant phases for us are then the phase accompanying the 't Hooft vertex and the phase of the quark condensates $\langle \bar{q}_L q_R \rangle$. The phase associated with the 't Hooft vertex is a separate dynamical parameter in a theory with PQ quasi-symmetry. To minimize the energy, it will adjust itself so that the contribution to the vacuum energy arising from saturating the 't Hooft vertex with chiral condensates or quark masses is extremized. This will occur when the instanton and anti-instanton contributions have the same phase, so their magnitudes add.

The phase of the condensates are to a first approximation real, since we want to minimize the vacuum energy density

$$\langle H \rangle = m_u \langle \bar{u}_L u_R \rangle + m_d \langle \bar{d}_L d_R \rangle + m_s \langle \bar{s}_L s_R \rangle + \text{h.c.}$$

in the manifold of possible vacuua picked out by the strong breaking $SU(3)_L \times SU(3)_R \rightarrow SU(3)_{L+R}$. For present purposes this amounts to the magnitudes of condensates for different quarks being fixed but their relative phases variable. The overall phase, i.e. $\arg\{\langle \bar{u}_L u_R \rangle \langle \bar{d}_L d_R \rangle \langle \bar{s}_L s_R \rangle\}$, is also to be determined by minimizing the energy in the no-instanton sector, in a theory with PQ quasi-symmetry. The full Hamiltonian includes not only the mass term, but also for instance four-quark operators of the general type $\bar{s}_L \gamma_\mu s_L \bar{d}_L \gamma_\mu d_L$ generated by the weak interactions. The coefficient of this operator is complex, relfecting CP violation in the weak interaction. The operator $\bar{s}_L \gamma_\mu s_L \bar{d}_L \gamma_\mu d_L$ can be connected to the vacuum by a mass insertion together with a chiral pairing of quarks, so its contribution to the energy (along with that of its Hermitian conjugate) likewise depends on the relative phase. The complex co-efficient of the operator implies that its contribution, unlike that from the mass term proper, will be minimized with a nontrivial relative phase between the masses and the condensates.

The location of the true minimum will be governed by the competition between these two effects. In the minimal Kobayashi-Maskawa model, one can estimate the resulting relative phase, which is essentially the effective -parameter of the text, to be

$$\theta_{eff} \sim G_F^2 m_c^2 \varepsilon \rho^2 \sim 10^{-14}$$

where $G_F^2 m_c^2 \varepsilon$ is a measure of CP violation in the coefficient of the operator (m_c = charmed quark mass, $\varepsilon \sim 10^{-3}$ from K-decays, and $\rho \sim 300$ MeV = typical strong interaction scale). This estimate is very crude and an order of magnitude deviation either way would not be shocking.

In models of CP violation other than the KM model the GIM suppression implicit in this estimate will not occur and substantially larger values of θ_{eff} will result.

REFERENCES

1. Many of the original papers appear in S. Adler and R. Dashen, "Current Algebras," Benjamin, New York (1969). A good pedagogical review is S. Treiman in "Lectures on Current Algebra and Its Applications," Princeton University Press (1972).
2. Important treatments of current algebra emphasizing this point of view are S. Weinberg, 1970 Brandeis lectures, MIT Press (1971); R. Dashen, Phys. Rev. 183:1245 (1969).
3. The first mention of the strong U(1) problem seems to have been made by Glashow here at Erice (1963). An influential review is S. Weinberg, Phys. Rev. D 11:3583 (1975).
4. S. Coleman, Erice lectures (1977).
5. For reviews of the triangle anomaly, see especially S. Adler, 1970 Brandeis lectures, MIT Press (1971); R. Jackiw in "Current Algebra and Its Applications," Princeton University Press (1972).
6. K. Fujikawa, Phys. Rev. Lett. 42:1195 (1979).
7. G. 't Hooft, Phys. Rev. Lett. 37:8 (1976); Phys. Rev. D 14:3432 (1976).
8. C. Callan, R. Dashen, and D. Gross, Phys. Lett. 363:334 (1976); R. Jackiw and C. Rebbi, Phys. Rev. Lett. 37:172 (1976).
9. A. Belavin, A. Polyakov, A. Schwartz, and Y. Tyupkin, Phys. Lett. B 59:85 (1975).
10. J. Gasser and H. Lentwyler, Phys. Rep. 87:78 (1982).
11. R. Peccei and H. Quinn, Phys. Rev. D 16:1791 (1977).
12. S. Weinberg, Phys. Rev. Lett. 40:223 (1978); F. Wilczek, Phys. Rev. Lett. 40:279 (1978).
13. The observable residue of this, viz. large T-violation in nucleon decay, is of course not yet ruled out. See A. Hurlbert and F. Wilczek, Phys. Lett. B 93:274 (1980).
14. J. Kim, Phys. Rev. Lett. 43:103 (1979); M. Dine, W. Fischler, and M. Srednicki, Phys. Lett. B 104:199 (1981); M. Wise, H. Georgi, and S. Glashow, Phys. Rev. Lett. 47:402 (1981).
15. D. Gross, S. Treiman, and F. Wilczek, Phys. Rev. D 19:2188 (1979).
16. T. Donnelly, S. Freedman, R. Lytel, R. Peccei, and M. Schwartz, Phys. Rev. D 18:1607 (1978).
17. D. Dicus, E. Kolb, V. Teplitz, and R. Wagoner, Phys. Rev. D 18:1829 (1978).
18. J. Preskill, M. Wise, and F. Wilczek, Phys. Lett. B 120:127 (1983); L. Abbot and P. Sikivie, Phys. Lett. B 120:133 (1983); M. Dine and W. Fischler, Phys. Lett. B 120:137 (1983).
19. J. Moody and F. Wilczek, "New Macroscopic Forces?" ITP preprint (1983).
20. P. Sikivie, Phys. Rev. Lett. 51:1415 (1983).
21. P. Sikivie, Phys. Rev. Lett. 48:1156 (1982).
22. F. Wilczek, Phys. Rev. Lett. 49:1549 (1982).

23. D. Reiss, Phys. Lett. B 115:217 (1982).
24. G. Gelmini, S. Nussinov, and T. Yanagida, Nucl. Phys. B 219:31 (1983).
25. R. Pisarski and F. Wilczek, "Remarks on the Chiral Phase Transition in Quantum Chromodynamics," ITP preprint, to appear in Phys. Rev. D (1984).

DISCUSSION

CHAIRMAN: Prof. F. Wilczek

Scientific Secretaries: C. Burgess and P. Simic

DISCUSSION SECTION 1

BERNSTEIN: What is the diagrammatic significance of the $tr(G_{\mu\nu}\tilde{G}_{\mu\nu})$ term in the action?

WILCZEK: Because it is a total divergence, $tr(G_{\mu\nu}\tilde{G}_{\mu\nu})$ does not contribute to the Feynman rules. In a semiclassical approximation, however, you expand about a solution of the classical field equations. Then the classical potentials may fall off at infinity slowly enough to give a nonzero contribution to $\int tr(G_{\mu\nu}\tilde{G}_{\mu\nu})dV$. This can be done while keeping $\int tr(G_{\mu\nu}\tilde{G}_{\mu\nu})dV$ finite provided the potential, A_μ, is of order $1/g$. Such a field configuration contributes to amplitudes with a factor $\sim \exp(-8\pi^2/g^2)$ which would not appear in an expansion about $g = 0$.

VAN DEN DOEL: You said that the $U_A(1)$ problem was solved when 't Hooft found that instantons contribute to the axial anomaly. However, in two dimensions there are no instantons but the $U_A(1)$ symmetry is still anomalous. Why then do you need instantons to solve the $U_A(1)$ problem?

WILCZEK: There may be other mechanisms besides instantons which remove the $U_A(1)$ from the theory. The only one known to operate in QCD in a controllable approximation is instantons. Although one cannot do quantitative calculations with instantons at zero temperature one can draw qualitative conclusions, such as that the classical $U_A(1)$ symmetry does not survive to the quantum theory. The important result to be used in what follows is the anomaly equation, which is an exact relation, not relying on the semiclassical approximation.

VAN DEN DOEL: Does the restoration of the $U_A(1)$ symmetry at high temperatures have any cosmological implications?

WILCZEK: I don't think there are any interesting cosmological implications. There is no danger of being stuck in a metastable "vacuum" because the restoration of the $U_A(1)$ symmetry is not a real phase transition and so proceeds continuously. It is more like the ionization of a gas; it is never an exact symmetry but it gets better and better at higher temperatures.

BAGGER: Usually we think of the pions being light as a consequence of the fact that the "up" and "down" quarks are light on a hadronic scale. Is there any way to understand the magnitude of η' mass? Is there any parameter, analogous to the quark masses, that can be dialed to change the η' mass?

WILCZEK: There has been no convincing calculation of the η' mass in QCD. All one can say is that there is no reason for the η' to be viewed as a light Goldstone boson. The $U_A(1)$ is broken by quantum corrections. One parameter to dial is the temperature. At high temperatures the $U_A(1)$ becomes a better symmetry and the η' becomes more like a Goldstone boson.

BAGGER: How should the process of thermalization in heavy ion collisions be understood?

WILCZEK: The modern picture of heavy ion collisions has the valence quarks passing through each other, with the sea quarks (or "wee partons") thermalizing in the central region.

SALATI: Could you explain the difference between the constituent masses and current masses of the quarks and the role chiral symmetry breaking plays in their difference?

WILCZEK: Because the quarks aren't observed as free particles their masses can't be defined as poles in a propagator. The current algebra masses are the masses in the Lagrangian which parameterize the amount by which chiral symmetry is broken. They appear there in the combination $m_q\bar{q}q$. Both m_q and $\bar{q}q$ are renormalized and so have an implicit scale dependence. However, their product is scale invariant and so can appear in physically significant relations, such as those derived in the lecture. Likewise the ratio of light quark masses has scale independent meaning.

Constituent masses are a more phenomenological concept. Because of its confinement in a hadron, a quark always has a minimum kinetic energy by the uncertainty principle. This kinetic energy is of the order of the confinement radius, R, inverse. It provides a contribution to the hadron mass which is independent of the "bare" quark mass for light quarks and is called the constituent mass.

OGILVIE: Do you think that instantons are useful for most purposes?

WILCZEK: My personal opinion is that instantons are not a useful calculational tool in zero temperature QCD. They do however suggest qualitative behaviors. They can also be useful quantitatively at finite temperature.

HOU: How is your parameter, K, related to, say, the QCD scale?

WILCZEK: Although it is in principle related to Λ_{QCD} it is very difficult to calculate.

CAPSTICK: You mentioned that the η' should have glue in it. Do you know of any spectroscopic calculations of its mass which include excitations of the glue?

WILCZEK: No, I don't know how that would be done.

OHTA: Although the $U_A(1)$ current is anomalous there is still a conserved current because the anomaly is a total divergence. How is the Nambu-Goldstone theorem evaded for this current?

WILCZEK: This conserved current is gauge dependent. This allows the non-appearance of Nambu-Goldstone poles in physical quantities; the zero-energy excitations are merely gauge transformations.

DISCUSSION SECTION 2 (Scientific Secretaries: C. Burgess, E. Milotti)

KLEVANSKY: Can you explain the topological significance of instantons?

WILCZEK: Instantons are field configurations which are localized in Euclidean space-time, in that $G_{\mu\nu} \to 0$ at infinity faster than $1/r^2$. Since $G_{\mu\nu} \to 0$ the potential becomes pure gauge at infinity, that is:

$$A_\mu \underset{r\to\infty}{} \Omega^{-1}\partial_\mu \Omega$$

for some function Ω taking values in the gauge group. In particular, this must be true at $t \to \pm \infty$, so for these times Ω is a function from space ($\mathbb{R}^3$) to the gauge group, such as SU(2). Since the instantons are localized in space, we may just as well take space as being a three-sphere, S_3. The group SU(2) is also equivalent to S_3 so the instanton corresponds to a function from S_3 to itself at $t = \pm \infty$.

Maps from S_3 to S_3 can be topologically nontrivial. They fall into classes labelled by an integer, the winding number, which counts how many times the function wraps the original S_3 around the image S_3.

The difference between the winding number of the instanton at $t = \infty$ and that at $t = -\infty$ is given by:

$$\Delta n = \frac{g^2}{32\pi^2} \int \mathrm{Tr}(G_{\mu\nu}\tilde{G}_{\mu\nu})dV .$$

The instanton interpolates between pure gauge configurations of different winding number.

NEMESCHANSKY: You have shown that the Goldstone boson has a mass squared which is proportional to the quark mass. What is the situation in supersymmetric theories?

WILCZEK: For Goldstone bosons the mass formula is independent of whether or not the theory is supersymmetric. If supersymmetry itself were explicitly broken by a small piece of the action, as well as being spontaneously broken, an analogous formula would hold for the goldstino mass.

VAN DEN DOEL: At infinite temperatures QCD becomes effectively a three dimensional theory. You say that the axial $U_A(1)$ symmetry gets restored at infinite temperature, so doesn't this imply that $U_A(1)$ symmetry is unbroken in three dimensions? This doesn't seem plausible.

WILCZEK: I don't see why it doesn't seem plausible. You must bear in mind that there is no analogue of the matrix in three dimensions, so "axial" symmetries in four dimensions don't appear as space-time related symmetries in three dimensions.

Nevertheless it would be interesting to think more about the three dimensional theory.

OGILVIE: How is the current associated with the η' related to that of the axion? Isn't there any mixing between the η' and the axion? Why isn't the axion heavy like the η'?

WILCZEK: The PQ symmetry associated with the axion also involves transformations of the Higgs fields and so is much closer to being a symmetry than the old axial $U_A(1)$ which just rotates quarks. Consequently the PQ pseudo-Goldstone boson, the axion, is much lighter than that of the $U_A(1)$.

More formally, this shows up in the formalism in the axion kinetic term: $\frac{1}{2} F^2(\partial_\mu \alpha)^2$.

Because the vacuum expectation value which breaks the PQ symmetry is large, F is large and the axion mass, being inversely proportional to F, is small. The analogous quantity for the η' comes from the quark sector and so is not so large.

BURGESS: You've used several times the anomaly equation

$$\partial \cdot j_A \propto Tr(G_{\mu\nu}\tilde{G}^{\mu\nu})$$

Is there an analogue of the Adler-Bardeen theorem in QED for non-abelian theories which ensures this is an exact equation?

WILCZEK: Yes, I believe there is.

VAN DEN DOEL: You said that CP is not violated for $\theta = 0$ or $\theta = \pi$. Why then can θ not be near π instead of being close to zero?

WILCZEK: $\theta = \pi$ can be changed to $\theta = 0$ by making a suitable chiral transformation at the expense of making the up quark mass, say, negative. Once we take all the quark masses positive, CP preservation implies $\theta = 0$ (mod 2π). The quark masses as determined phenomenologically are all positive.

VAN DEN DOEL: You said that terms like $\mu^2 \phi_I \phi_{II}^*$ were forbidden by PQ quasi-symmetry. Since this is only a quasi-symmetry won't such terms be generated radiatively?

WILCZEK: Yes, they are generated radiatively; however, they don't contribute independently to the potential which fixes θ_{eff}. This is because they themselves are generated only through the instanton vertex and don't represent an independent source of CP violation.

BERNSTEIN: I don't understand the connection between θ_{eff} and the CP violating phase in the Kobayashi-Maskawa matrix. Could you explain this?

WILCZEK: In a theory without axions the θ-parameter and the Kobayashi-Maskawa phases are completely independent. In a theory with axions, if you know all of the CP-violating phases in the Lagrangian, then the effective θ-parameter is completely determined.

SIMIC: Suppose θ is treated as an additional coupling constant and so is renormalized. Could you comment on the possibility that the β function for θ may be positive near $\theta = 0$ and that perhaps θ is driven to zero in the limit of infinite cutoff.

Alternatively, perhaps there is a fixed point near the origin. In either case θ would remain small.

WILCZEK: What you say is certainly conceivable. The most definite thing along these lines which comes to mind is that θ_{eff} is related to the quark masses. When the masses are renormalized θ_{eff} will change. I don't think there is any indication of a fixed point near the origin.

All of the parameters are also probably small, and likely don't get renormalized much.

Finally, if the small size of θ_{eff} were to be explained in this way you'd have then to explain why the Kobayashi-Maskawa phase is so large. However, I can't exclude what you are suggesting.

DISCUSSION SECTION 3 (Scientific Secretaries: J. Bagger, D. Nemeschansky)

OLEJNIK: What is the pseudoscalar coupling of the axion to the electron? What would be the coupling of a massive (Dirac) neutrino?

WILCZEK: Ignoring chiral symmetry breaking, which complicates things for quarks, the universal coupling of axions to quarks or leptons is proportional to

$$\frac{a}{F} m_f \bar{f}\gamma_5 f ,$$

where a is the axion and f is a quark or a lepton. For quarks, the axion mixes with the π^0 and the η. This changes the situation.

MOUNT: Can you describe more fully the experimental signs to be expected from the monopole-dipole and dipole-dipole interactions?

WILCZEK: I don't want to say too much because I'm not yet sure what the optimal configuration will be. The basic phenomenon is a weak but coherent macroscopic force on scales less than 2 cm. This generates potentials for the monopole-dipole force of the form

$$U(r) \sim \frac{1}{F^2} \frac{\hat{s}\cdot\hat{r}}{r^2} e^{-mr} .$$

ZICHICHI: Where does the distance 2 cm. come from?

WILCZEK: It is the Compton wavelength of the axion if F is 10^{12} GeV., the cosmological upper bound. To measure the monopole-dipole force one would look for an attraction between one body with a large amount of spin and, say, a torsion balance of resonant frequency ω_0:

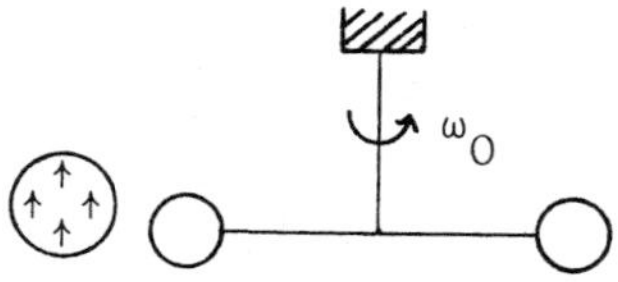

(Note that the arrows represent <u>spin</u> and not orbital angular momentum.) One would switch the spin of the large body at frequency ω_0, driving the oscillator at its resonant frequency, and then look for very small displacements. Current technology allows one to measure displacements of 10^{-17} cm.

The spin-spin force is more or less the same, except that one has to worry about background magnetic fields. This is a severe technical problem.

<u>HOU</u>: Can one decide whether axions are fundamental or composite?

<u>WILCZEK</u>: Axions could be composite, but only on a scale of order 10^{12} GeV. We need not worry about this for low energy phenomena.

<u>TAVANI</u>: Can one solve the hierarchy problem associated with strong CP by introducing gravity?

<u>WILCZEK</u>: I don't see how it helps.

<u>VAN DEN DOEL</u>: Are there any ultralight particles that could have a macroscopic force associated with them?

<u>WILCZEK</u>: Over the years, there have been many suggestions in the literature: gravitons, photinos, spin-1 antigravitons, cosmions,... Even if a new macroscopic force were found, one would have to check whether it was consistent with axions. This could be done by using different materials (with different neutron/proton abundances) as sources.

<u>MARUYAMA</u>: Axions are produced almost at rest, they are bosons, and, furthermore, they attract each other. Can they collapse into black holes?

<u>WILCZEK</u>: Yes, they can. However, it's not as bad as it seems, because axions have a very hard time losing energy (The jargon for that is dissipationless). As the axions collapse they move right through each other, and rarely form black holes. The same is true for any cosmological "dust," (that is, dissipationless and non-relativistic matter), regardless of whether it is fermionic or bosonic.

<u>SIMIC</u>: There is a relationship between the electric charge on a

magnetic monopole and the θ-parameter, given by Witten's formula

$$Q = \frac{eg}{2\pi} \theta .$$

In a theory of axions, with dynamical θ, one would expect interactions between monopoles and axions. What are their consequences?

WILCZEK: There is the interesting effect that if θ_{eff} varies over space, the electric charge on the monopole also varies. That means the monopole has a very extended form factor.

DISCUSSION SECTION 4 (Scientific Secretaries: J. Bagger, D. Nemeschansky)

HOU: How can one hope to exploit the idea of spontaneously broken family symmetry to explain the quark and lepton masses and mixing angles?

WILCZEK: Let me give you an example. Suppose there are three families in the standard SU(5) gauge theory, that is, three 5's and three 10's. Let us also suppose there is an SU(3) global family symmetry. There are an enormous number of Higgs fields which generate the quark and lepton masses. They must be 5's under the usual SU(5), and they also have to transform under the global SU(3). Let me assume that they are in the 6-dimensional representation of SU(3). That corresponds to a 3 × 3 symmetric tensor under SU(3). Now let's suppose the (3,3) component gets a large vacuum expectation value, breaking the family symmetry. This gives the following mass matrix

$$\begin{pmatrix} 0 & 0 & 0 \\ 0 & 0 & 0 \\ 0 & 0 & C \end{pmatrix}$$

for the Q = − 1/3 quarks. This is the correct mass matrix (to zeroth order). One quark is heavy and the others are massless. Moreover, with this argument, we get the successful relation

$$m_b/m_\tau = 3$$

after renormalization. To give the light quarks and leptons their masses, it turns out that one should add an SU(3) triplet Higgs field. I assume the triplet aligns mostly in the same direction, but not entirely. The 3 represents an antisymmetric matrix, so the full mass matrix is

$$\begin{pmatrix} 0 & A & 0 \\ -A & 0 & B \\ 0 & -B & C \end{pmatrix}$$

where C >> B >> A. If we assume the charge 2/3 quark mass matrix is approximately diagonal in the same basis, then we can read off the mixing angles and masses, and derive the following relations:

1) $\tan^2\theta_c = m_d/m_s$

2) $\dfrac{m_e}{m_\mu} = \dfrac{1}{g}\ \dfrac{m_d}{m_s}$

3) $\tan^2\theta_{b\to c} = \dfrac{m_s}{m_b} = \dfrac{1}{g}\ \dfrac{m_\mu}{m_\tau}$.

Relations 1) and 2) are successful. Relation 3) gives $\theta_{b\to c}$ perhaps a bit too large, but we must learn more about b decay.

ZICHICHI: Apart from the smallness of the pion mass, do we have any evidence in the real world of the spontaneous breaking of a global symmetry?

WILCZEK: Yes; the Goldberger-Treiman relation, the Callan-Treiman relation, the Adler-Weisberg relation, π-N scattering lengths, the quadratic Gell-Mann-Okubo formula, and many others...

ZICHICHI: I am speaking about the existence of Goldstone bosons.

WILCZEK: Examples from other branches of physics include magnons and phonons. There are many other examples.

WIGNER: Does it bother you that SU(3) is not exact?

WILCZEK: Yes, it does, and I would like to have it be an exact symmetry, spontaneously broken.

WIGNER: Why is SU(3) not exact?

WILCZEK: We know experimentally that it is not exact. Within the framework of QCD, it is a consequence of the fact that the quarks have different masses.

WIGNER: Why do they have different masses?

WILCZEK: That's the deep question. I'd like to think they are not equal because of a spontaneous symmetry breakdown.

HOU: Can the family symmetry be gauged?

WILCZEK: This is an important issue, because if the family symmetries were gauged, the Nambu-Goldstone bosons would be absorbed into heavy vector mesons, and would not be physical particles. However, there are some reasons why one might prefer to have the symmetry global rather than local:

1) If one tries to gauge a large family symmetry, one often runs into trouble with anomalies.

2) I suspect that one cannot get rid of domain walls if one has a non-abelian symmetry, but I'm not sure about this. The reason I suspect this is that the physical mechanism which allows one to get rid of domain walls is the continuous emission of Goldstone bosons. This permits the domain walls to relax away. If these Goldstone bosons were the longitudinal modes of heavy vector mesons, this mechanism would not be available. It would then be very rare for the domain walls to relax away.

GLASHOW: I'd like to add another reason against gauging family symmetries. If you gauge the family symmetry, you get a new type of interaction. There is no evidence for such an interaction. The best limit comes from the agreement of the K_1, K_2 mass difference with ordinary theory. This tells you that the horizontal gauge bosons must have masses of more than 10^6 GeV.

VAN DEN DOEL: Instantons leave a discrete symmetry intact. Could other nonperturbative effects break this discrete symmetry?

WILCZEK: Of course it's possible, but I would very surprised. The discrete symmetry is related to the anomaly equation, which I believe to be exact.

VAN DEN DOEL: How can one detect the long-range forces of the familons? Are they similar to axions?

WILCZEK: The familons are different: they are exact Nambu-Goldstone bosons. The family symmetry is not anomalous, so the familon couplings are purely derivative. Therefore, the potential goes as $1/r^3$, and in the non-relativistic limit, there is only a spin-dependent force. Familons give dipole-dipole forces, so they are hard to discover.

TAVANI: Are there problems with the inflationary universe?

WILCZEK: Yes, there are a few. A severe conceptual problem is that the inflationary universe scenario leans heavily on a nonzero cosmological constant in the early universe. The cosmological constant is the weakest point in our understanding of the physical world. The detailed models of inflation also tend to predict too large a value for $\delta\rho/\rho$. However, the inflationary universe picture explains the homogeneity and isotropy of the universe, predicts the correct spectrum of $\delta\rho/\rho$, gets rid of monopoles, and explains why the universe is so flat. There is impressive phenomenological evidence that the basic idea of inflation is correct.

LAMARCHE: Is there a principle which tells when a parameter can be made into a dynamical variable?

WILCZEK: I think that with a little imagination, one can turn any parameter into a dynamical variable, subject to restrictions such as renormalizability and gauge invariance.

SIMIC: Does the sign of θ have any physical significance?

WILCZEK: Yes. For example, it determines the sign of the $\hat{s}\cdot\hat{r}$ force, and the sign of the neutron electric dipole moment.

SIMIC: Does the θ-parameter get renormalized?

WILCZEK: Yes, the effective θ-parameter gets small renormalization corrections from the renormalizaton of the quark mass matrix.

SPONTANEOUS SUPERSYMMETRY BREAKING IN N=1 AND N=2 SUPERGRAVITY THEORIES COUPLED TO MATTER SYSTEMS

S. Ferrara

CERN
Geneva
Switzerland

In these lectures we will discuss some aspects of spontaneously broken supersymmetric gauge theories with N=1 and N=2 local supersymmetry. Examples of the super-Higgs effect with vanishing cosmological constant are exhibited and N=2 broken supergravity models with flat potentials are explicitly constructed. A subclass of the latter is connected to gauged supergravity models in five-dimensional space-time.

Supersymmetric models of fundamental particle interactions offers nowadays an interesting extension of the present Standard Model based on the minimal low-energy gauge symmetry $SU_c(3) \times SU_L(2) \times U(1)$[1].

In particular, if broken supersymmetry is related to the so-called hierarchy problem[2] of different gauge symmetry-breaking patterns, then new physics, with a plethora of new particles, is soon to be expected in the region of the Fermi scale $G_F^{-\frac{1}{2}}$ (G_F being the Fermi constant), and its consequences could be tested with future (LEP) or even present accelerators ($p\bar{p}$ Collider?)[3].

In spite of the fact that supersymmetry seems nowadays more a general framework than a specific theory, the possibility that it may show up at moderate energies is fascinating.

Surprisingly enough, supersymmetric models for particle physics offer a unique interplay between gauge symmetries of usual type, such as color and charge and space-time symmetries which enter in the dynamics through gravity.

In fact, it has been recognized[1,3], on general grounds that supergravity corrections to globally supersymmetric theories become important at "low energies" ($E \sim 100$-1000 GeV) if the supersymmetry-breaking scale M_S is $O[(M_W M_P)^{\frac{1}{2}}]$, i.e. intermediate between the weak scale and the Planck scale.

On the other hand, such a large value of M_S seems to be compelling for constructing a realistic model exhibiting an approximate supersymmetric structure. For example, unwanted mass relations between quarks, leptons, and their scalar superpartners, which are typical of spontaneously broken globally supersymmetric models, can be avoided since scalar masses get supergravity corrections typically of the form

$$\delta mi \cong O(\kappa M_s^2)$$

in which κ is the gravitational coupling constant ($\kappa = \sqrt{8\pi}/M_P$).

The phenomenon yielding the non-trivial interplay between gravity and "low-energy" physics, is called the super-Higgs effect[4].

It is known, from the supersymmetry algebra, that gravity coupled to global supersymmetry implies local supersymmetry and vice versa. The gauge field of local supersymmetry is a spin $^3/_2$ field $\psi_{\mu\alpha}$, which, in the absence of supersymmetry breaking, is supposed to describe a massless spin $^3/_2$ particle, the gravitino. If supersymmetry is spontaneously broken, the super-Higgs effect occurs, namely the Goldstone mode is eaten up by the gravitino which becomes massive, with Mass[4].

$$m_{3/2} = \sqrt{\frac{8\pi}{3}}\ \frac{M_s^2}{M_P} .$$

The gravitino mass depends on M_S quadratically. If $M_S = O((M_W M_P)^{\frac{1}{2}})$, then $m_{3/2} = O(M_W)$, and supergravity effects can become important in the region of the Fermi scale.

The relation between $M_{3/2}$ and M_S is valid in absence of a cosmological constant. This is possible in local supersymmetry, since the scalar potential is no longer positive definite, as it is in the case in global supersymmetry.

The aim of this section is to discuss in some detail this phenomenon and its implications for the low-energy effective Lagrangian at energies comparable to the weak vector boson masses. It will appear that at these energies one can still neglect gravitational corrections, but not supergravity corrections which will manifest as a soft explicit breaking of global supersymmetry.

The general coupling of chiral and vector multiplets to N=1 super-gravity is specified by two functions of the complex scalar fields z^i contained in chiral multiplets[5,6]. An analytic function $f_{ab}(z) = f_{ba}(z)$, related to the Yang-Mills part of the Lagrangian, gives for the kinetic terms of the gauge fields,

$$-\frac{1}{4}(\mathrm{Re}\, f_{ab})F^a_{\mu\nu}\, F^{b\mu\nu} + \frac{i}{4}(\mathrm{Im}\, f_{ab})\, F^a_{\mu\nu}\, \tilde{F}^{b\mu\nu} \; . \tag{1.1}$$

(a,b are indices of the adjoint representation of the gauge group G). Then, a real function $G(z,z^*)$, the Kähler potential, defines the scalar kinetic terms, given by*

$$-\, G^i_{\;j}\, (\partial_\mu z^j)(\partial^\mu z^*_i). \tag{1.2}$$

This function is gauge invariant:

$$G_i (T^a)^i_{\;j} z^j = G^i (T^a)^j_{\;i} z^*_j \; . \tag{1.3}$$

The kinetic terms have a form characteristic of supersymmetric non-linear σ-models. They are invariant under Kähler transformations

$$G \rightarrow G + F(z) + F^*(z^*) \; , \tag{1.4}$$

where $F(z)$ is a gauge invariant analytic function. The scalar fields in N=1 supergravity span a Kähler manifold. Coset spaces corresponding to Kähler manifolds are of the form $H/K \otimes U(1)$, where $K \otimes U(1)$ is a maximal (compact) subgroup of H. It is convenient to split the Kähler potential in two parts:

$$G(z,z^*) = -\, J(z,z^*) + \ln|f(z)|^2 \; . \tag{1.5}$$

The gauge invariant superpotential $f(z)$ appears in the part of the Lagrangian which is not invariant under (1.4), induced by the coupling to gravitation. G is, of course, invariant under the transformations

$$J \rightarrow J + F'(z) + F'^*(z^*) \; ,$$

$$f \rightarrow \exp(F'(z)) f \; .$$

The discuss symmetry and supersymmetry breaking, we need the scalar potential V. The potential has two terms:

$$V = V_c + V_g \; . \tag{1.6}$$

---

*We use the notation

$$G_i = \frac{\partial G}{\partial z^i} \; , \quad G^i = \frac{\partial G}{\partial z^*_i} \; , \quad G^i_j = \frac{\partial^2 G}{\partial z^*_i \partial z^j} \quad , \; \ldots$$

The gauge potential V_g reads

$$V_g = \frac{1}{2}(\text{Re } f_{ab}{}^{-1}) D^a D^b , \tag{1.7}$$

where the real [see equation (1.3)] functions D^a are

$$D^a = g_a G_j (T^a)^j_i z^i \tag{1.8}$$

(g_a is the gauge coupling constant associated to the normalized generator T^a). V_g is invariant under Kähler transformations (1.4), and thus does not depend on the superpotential. The "chiral" potential[5,6] is

$$\begin{aligned} V_c &= \exp G(G_i G^j (G_i{}^j)^{-1} - 3) \\ &= -\exp(-J)\left((fJ_i - f_i)(f^* J^j - f^{*j})(J^j_i)^{-1} + 3|f|^2\right) \end{aligned} \tag{1.9}$$

The Lagrangian contains also an interaction between the gravitino ψ_μ and a particular spin ½ state ψ_L of the form

$$- \bar{\psi}_{\mu r} \gamma^\mu \psi_L + \text{c.c.} \tag{1.10}$$

ψ_L is given by:

$$\psi_L = - \exp(G/2) G_i \chi^i_L + \frac{i}{2} D^a \lambda^a_L , \tag{1.11}$$

where χ^i_L and λ^a_L are the spin ½ states of chiral and vector multiplets, respectively. When the vacuum expectation value of the scalar fields (denoted by <...>) is such that

$$\langle \exp(G/2) G_i \rangle \neq 0 , \tag{1.12}$$

or

$$\langle D^a \rangle \neq 0 , \tag{1.13}$$

the theory contains a Goldstone spinor η_L

$$\eta_L = - \langle \exp(G/2) G_i \rangle \chi^i_L + \frac{i}{2} \langle D^a \rangle \lambda^a_L ,$$

which can be rotated away using a local supersymmetry transformation. The super-Higgs effect is effective and supersymmetry is spontaneously broken.

It is apparent from the potential (1.9) that, unlike in the global case, spontaneously broken supersymmetry does not imply $\langle V \rangle > 0$. This is fortunate since one can now obtain the super-Higgs effect in Minkowski space ($\langle V \rangle = 0$) as well as in de Sitter ($\langle V \rangle > 0$) or anti-de Sitter ($\langle V \rangle < 0$) space. The theory contains also a gravitino mass term, $m_{3/2} \; \bar{\psi}_{\mu L} \sigma^{\mu\nu} \psi_{\nu L}$, with

$$m_{3/2} = \langle \exp(G/2) \rangle . \tag{1.14}$$

For unbroken supersymmetry, $m_{3/2}^2 = -\langle V\rangle/3$; when $\langle V\rangle \neq 0$, the parameter $m_{3/2}$ is not a physical mass; a massless gravitino in de Sitter or anti-de Sitter space does not mean $m_{3/2} = 0$.

A flat Kähler manifold, with

$$G^i{}_j = -J^i{}_j = \delta^i{}_j \qquad (1.15)$$

will lead to canonical scalar kinetic terms. Equation (1.15) implies

$$G = z^i z_i^* + \ln|f(z)|^2 . \qquad (1.16)$$

Victor kinetic terms can also be canonical, with the choice

$$f_{ab} = \delta_{ab} \qquad (1.17)$$

The minimal coupling, with canonical kinetic terms leads to the mass formula[5]:

$$\text{Supertrace}\, m^2 = \sum_{J=0}^{3/2} (-)^{2J}(2J+1)m_J^2$$
$$= (n-1)(2m_{3/2}^2 - \kappa^2 D^a D^a) + 2g_a D^a \mathrm{Tr}T^a . \qquad (1.18)$$

The first term is a supergravity correction to the mass formula of global N=1 supersymmetry, where only the last term is present. Equation (1.18) implies that the "mean squared mass" of the scalar fields is $m_{3/2}^2$. The phenomenological requirement of having scalars heavier than fermions is then naturally satisfied. In particular, in the attractive possibility where the gravitino scale is identified to the weak interaction scale, scalar partners of quarks and leptons will have masses of order M_W[1]. Notice, however, that the low-energy spectrum, when renormalization effects are taken into account, may be drastically different from the tree-level masses. One-loop corrections on scalar masses have been shown to be able to induce $SU(2)_L \times U(1)$ symmetry breaking at low energy, even if all scalars have positive tree-level "mean" squared masses of order $m_{3/2}^2 \sim M_W^2$ [3].

The simplest example of a Kähler potential (for one single neutral chiral field z) for which the super-Higgs mechanism occurs is the Polony model[7], for which the superpotential is

$$f(z) = \mu(z+\beta)/\kappa . \qquad (1.19)$$

The kinetic terms are canonical, i.e. G and f_{ab} are given by equations (1.16) and (1.17). The corresponding potential $V=V_c$ has a stationary point, which is a local minimum, with $\langle V\rangle = 0$ when β is tuned to take the value

$$\beta = (2-\sqrt{3})/\kappa . \qquad (1.20)$$

The gravitino mass is then related to the parameter μ by

$$m^2_{3/2} = \mu^2 \exp(\sqrt{3}-1)^2 \ , \tag{1.21}$$

and the two real scalars (z = a + ib) have masses

$$m^2_a = 2\sqrt{3}\, m^2_{3/2} \ , \tag{1.22}$$

$$m^2_b = 2(2-\sqrt{3}) m^2_{3/2} \ ,$$

which verify

$$m^2_a + m^2_b = 4m^2_{3/2} \ , \tag{1.23}$$

as derived from Equation (1.18), when n=1.

The simple form (1.19) can be easily generalized when several chiral multiplets are present. The superpotential reads[8].

$$f(z,y^i) = f(z)[1+f(z)^{q-2}/\langle f(z)\rangle^{q-1}]\ h(y) \quad q = 1,2,\ldots, \tag{1.24}$$

where z is the field which induces the super-Higgs effect and y^i the other scalars (which verify $\langle y \rangle \ll \langle z\rangle \sim M_P$), f(z) is the Polony potential [Equation (1.19)] and h(y) an arbitrary gauge invariant superpotential. One can now take the limit $\kappa \to 0$, with $m_{3/2}$ fixed, to get an effective Lagrangian valid at scales lower than M_P. With this class of superpotential, the field z consistently decouples in the effective theory: the interactions of the "hidden sector" z with the "visible sector" y^i are governed by κ. The effective potential has the form[8].

$$V_{eff}(y,y^*) = |\tilde{h}_i|^2 + (A-3)m_{3/2}\ (\tilde{h}+\tilde{h}^*) + \tfrac{1}{2}D^aD^a \ , \tag{1.25}$$

with $A - 3 = \sqrt{3}(q-2)$, $\tilde{h}(y) = \exp\,[(\sqrt{3}-1)^2/2]h(y)$ and

$$\tilde{h}_i = \frac{d\tilde{h}}{dy^i} + m_{3/2}\ y^*_i \ . \tag{1.26}$$

The effective theory is a supersymmetric gauge theory with chiral multiplets y^i, interacting through a superpotential $\tilde{h}$, softly broken by scalar interactions contained in V_{eff}. The effective theory is renormalizable as long as the superpotential $\tilde{h}(y)$ is a polynomial at most cubic in fields y^i. Notice that gaugino masses can be introduced simply, using non-minimal kinetic terms for the gauge fields; one could also obtain this way CP-violating $F_{\mu\nu}\tilde{F}^{\mu\nu}$ terms. Various SU(3) x SU(2) x U(1) or unified models coupled to N=1 supergravity have been constructed[1,3,8], resulting in an effective theory of this form at scales below Mp.

Up to now, we have discussed Kähler potentials for which the gravitino mass is essentially a free parameter μ. One has then to choose $\mu \sim O(M_W)$ to obtain a stable hierarchy of particle interactions. Moreover, the cosmological constant $\langle V \rangle = \Lambda$ is zero due to a fine tuning of the parameters of G. An additional parameter (β in the Polony case) is in general needed only to obtain a vanishing Λ.

There is, however, an elegant way to circumvent these two unsatisfactory points: there exist non-trivial Kähler potentials for which the chiral potential V_c is identically zero[9,10]. Supersymmetry is, however, broken. Vacuum expectation values are not determined by the classical theory; the same holds for the gravitino mass. The cosmological constant is naturally zero. One can then take the low-energy limit ($\kappa \to 0$ with $m_{3/2}$ undetermined but fixed) to get an effective theory with softly broken supersymmetry. This limit exists as long as the "hidden sector" (for which $V_c \equiv 0$) effectively decouples from the visible sector which does not participate in the super-Higgs effect. Radiative corrections are then applied in the effective theory to determine the various scales of gauge symmetry breaking[11], which in general will be closely related to the gravitino mass. These scales are in fact proportional to $m_{3/2}$, but the proportionality constants are strongly model-dependent and can be many orders of magnitude away from one. The consistency of this scheme is submitted to the very strong assumption that one can apply radiative corrections on an effective theory whose main properties (mainly the potential) are not affected by possible quantum gravitational effects.

Since it is, in principle, sufficient to require zero chiral potential only in the direction of the singlet scalar z, let us first discuss "flat potentials" in the case of a single chiral multiplet coupled to N=1 supergravity[9]. The chiral potential in terms of the Kähler potential can be written

$$V_c = \partial \exp(\tfrac{4}{3} G) G^{-1}_{z\bar{z}} \frac{d_2}{dzdz^*} \exp(-\tfrac{1}{3} G) \quad (G_{z\bar{z}} = \frac{d^2}{dzdz^*} G) \tag{1.27}$$

and $V_c \equiv 0$ implies

$$\frac{d^2}{dzdz^*} \exp(-\tfrac{1}{3} G) = 0 . \tag{1.28}$$

The solution is

$$G = -\ln(\phi(z) + \phi^*(z^*))^3 . \tag{1.29}$$

The scalar kinetic term $G_{z\bar{z}}(\partial_\mu z)(\partial_\mu z^*)$ is never canonical, and the gravitino mass is

$$m_{3/2} = \langle (\phi(z) + \phi^*(z^*))^{-3/2} \rangle . \tag{1.30}$$

$m_{3/2}$ is undetermined but non-zero since

$$\langle G_{z\bar{z}} \rangle = 3\langle |\phi_z|^2 \rangle (m_{3/2})^{4/3} \neq 0. \tag{1.31}$$

The geometry of the Kähler manifold when $V_c \equiv 0$ is particular. From Equation (1.28) one obtains

$$R_{z\bar{z}} = \frac{2}{3} G_{z\bar{z}} , \tag{1.32}$$

for the curvature, defined by

$$R_{z\bar{z}} = \frac{d^2}{dzdz^*} \ln G_{z\bar{z}} . \tag{1.33}$$

Equation (1.32) means that the Kähler manifold is an Einstein space (maximally symmetric space), i.e. that the scalar field z is a coordinate on the coset space SU(1,1)/U(1).* The non-compact SU(1,1) invariance can be checked explicitly in the whole Lagrangian apart from the gravitino mass term. A simpler way to understand its appearance is to consider the superfield formulation of the theory, in which SU(1,1) is a linear symmetry acting on the chiral superfield and on the compensating multiplet, when condition (1.28) is applied[12].

The condition (1.32) can indeed be derived from the mass sum rules which holds for any spontaneously broken N=1 supergravity theory with zero cosmological constant[13].

$$\begin{aligned} \text{Supertrace } M^2 &= 2m^2_{3/2} \ \langle G_z G_{\bar{z}} R_{z\bar{z}} / G^2_{z\bar{z}} \rangle \\ &= 2m^2_{3/2} \ \langle (\exp(-G)V+3) R_{z\bar{z}} / G_{z\bar{z}} \rangle , \end{aligned} \tag{1.34}$$

for one-chiral multiplet. When $V \equiv 0$ (and then $m_a = m_b = 0$), all values of the field z (n=1) are stationary points of the potential, so (1.34) holds for all values of z, with the corresponding gravitino mass $m^2_{3/2}(z) = \exp G(z)$. Equation (1.34) leads then immediately to the curvature constraint (1.32), since Supertrace $m^2 = 4m^2_{3/2}$.

It is interesting to notice[9] that the N=1 Lagrangian for one chiral multiplet with vanishing potential corresponds, up to the gravitino mass term, to a particular truncation of N=4 supergravity, which is known to possess an Su(1,1) non-compact global symmetry[14].

Vanishing chiral potentials for an arbitrary number n of chiral multiplets also exist. The generalization of the single field case is, however, not trivial due to the matrix structure of the Kähler metric G^j_i.

---

*Or equivalently U(1,1)/U(1) x U(1).

Let us first consider two cases[31] which will be useful also in the N=2 case we will discuss in the next section. It will be convenient to use the definition*

$$G = - \ln Y(z,z^*) \ , \tag{1.35}$$

for the Kähler function. Then, one can easily rewrite the chiral potential (1.9) in terms of $\tilde{Y}$

$$V_c = \frac{1}{\tilde{Y}} \left(\frac{W}{W-1} - 3\right) \ , \tag{1.36}$$

where

$$W = \frac{1}{\tilde{Y}} \frac{d\tilde{Y}}{dz^i} \left(\frac{d^2\tilde{Y}}{dz^i dz^*_j} \right)^{-1} \frac{d\tilde{Y}}{dz^*_j} \ . \tag{1.37}$$

The condition for vanishing potential is then

$$W = 3/2 \ . \tag{1.38}$$

The first class of solutions to this condition is obviously obtained when

$$\tilde{Y} = \sum_{i=1}^{n} \tilde{y}_{(i)}(z^i, z^*_i) \ , \tag{1.39}$$

which implies

$$W \tilde{Y} = \sum_{i=1}^{n} \frac{d\tilde{y}_{(i)}}{dz^i} \frac{d\tilde{y}_{(i)}}{dz^*_i} \left(\frac{d^2\tilde{y}_{(i)}}{dz^i dz^*_i} \right)^{-1}$$

$$\equiv \sum_{i=1}^{n} W_{(i)}\tilde{y}_{(i)} \ .$$

The potential will be zero if

$$\sum_{i=1}^{n} W_{(i)}\tilde{y}_{(i)} = \frac{3}{2} \sum_{i=1}^{n} \tilde{y}_{(i)} \ .$$

This is the case when each function of one field $\tilde{y}_i$ gives itself a vanishing potential in the single field case, i.e. when $\tilde{y}_{(i)}$ has the form [see Equation (1.29)]:

$$\tilde{y}_{(i)} (z^i, z^*_i) = (\phi_{(i)}(z^i) + \text{c.c.})^3 \ . \tag{1.40}$$

Another class of solutions to condition (1.38) is obtained in the following way. Assume that $\tilde{Y}$ is a homogeneous function of degree δ of the combinations $\alpha z_i + \beta z^*_i$. ($\tilde{Y}$ should in fact also be real, but

---

*$\tilde{Y}$ is related to the function Y, defined by Equation (2.10) of Section 2, by $\tilde{Y} = Y/|h(z)|^2$, h(z) being given in Equation (2.14).

the following derivation is independent of that constraint.) Homogeneity means that

$$\frac{\partial^2\tilde{Y}}{\partial z^i \partial z^*_j}(\alpha z^j + \beta z^*_j) = \beta(\delta-1)\frac{\partial\tilde{Y}}{\partial z^i} ,$$

$$\frac{\partial^2\tilde{Y}}{\partial z^i \partial z^*_j}(\alpha z^i + \beta z^*_i) = \alpha(\delta-1)\frac{\partial\tilde{Y}}{\partial z^*_j} ,$$

$$\frac{\partial^2\tilde{Y}}{\partial z^i \partial z^*_j}(\alpha z^i + \beta z^*_i)(\alpha z^j + \beta z^*_j) = \alpha\beta\delta(\delta-1)\tilde{Y} .$$

Then:

$$W = \frac{1}{\tilde{Y}}\frac{(\alpha z^i + \beta z^*_i)(\alpha z^j + \beta z^*_j)}{\alpha\beta(\delta-1)^2}\frac{\partial^2\tilde{Y}}{\partial z^i \partial z^*_j} = \frac{\delta}{\delta-1} , \qquad (1.41)$$

and the chiral potential is

$$V_c = \frac{\delta-3}{\tilde{Y}} \qquad (1.42)$$

V_c then vanishes when the function $\tilde{Y}$ is an arbitrary, homogeneous (real) function of degree 3, of a linear combination of z^i and z^*_i, like for instance $z^i + z^*_i$ or $i(z^i - z^*_i)$.

The last class of couplings of an arbitrary number of chiral multiplets with zero chiral potential we will mention here is in fact the most natural from a superfield point of view[12]. In this formalism, the condition $V_c \equiv 0$ can be solved using only one superfield, with scalar z, the coupling of the other fields, with scalars y^i, remaining unconstrained. This leads to the following class of Kähler potentials[11,12]:

$$G = -\ln(\phi(z) + \phi^*(z^*) + g(y^i, y^*_i))^3 . \qquad (1.43)$$

g is an arbitrary real function and ϕ is an analytic function of z only. One checks easily that this natural extension of Equation (1.29) gives also $V_c \equiv 0$.

The possibility of vanishing potentials in the hidden sector is attractive, since it allows naturally to break supergravity with zero cosmological constant. Subsequent scales of gauge symmetry breakings in the effective "low-energy" theory are then obtained through "dimensional transmutation", induced by the radiative corrections to the potential in the flat directions. The structure of the soft breaking terms, resulting from the choice of G, is then crucial to fix the scales. In particular, no scale larger than $m_{3/2}$ (like M_P, for instance) should appear in the effective theory, even when radiative corrections are included. A necessary condition for the consistency of the mechanism is that the soft breaking terms are such that

$$\text{Supertrace } m^2 = 0 \qquad (1.44)$$

in the effective theory. Such a condition is obviously not satisfied if one requires zero potential for the visible sector also. A class of models in which Equation (1.44) is fulfilled has been constructed[15]. The corresponding Kähler manifold is of the type $U(N,1)/U(N) \times U(1)$, but the potential is no more flat.

2. Super-Higgs Effect in N=2 Supergravity

Compared with the N=1 case, N=2 supergravity[16] coupled to N=2 matter multiplets (i.e. multiplets containing states of spin up to one) shows several new features, which appear in general in extended supergravities. Extended supersymmetry algebras possess an internal global SU(N) symmetry acting on the spinorial charges which is a symmetry of the supergravity theory. There is then the possibility to enlarge the local symmetry of the theory by using either the vector fields contained in the graviton supermultiplet, which are sufficient to gauge the SO(N) subgroup [SO(6) x U(1) for N=6], or additional vector multiplets or a combination of both options. The gauge group of a general N=2 theory will then have the form $G \times G_{int}$, the internal part G_{int} being $O(2) \sim U(1)$ [17], or eventually SU(2) [18] using additional vector multiplets. We will see that the gauging of this internal symmetry plays a crucial role when vector multiplets induce the super-Higgs effect.

Th N=2 supergravity and gravitino multiplets (with maximal helicity 2 and 3/2 respectively) do not contain scalar fields. N=2 offers then a convenient way to study the scalar sector and supersymmetry breaking pattern of larger N theories, which will correspond to some coupling (fixed by invariances) of N=2 matter multiplets. An important aspect of the N=2 theory, in particular if one is concerned with the σ-model structure[19] of the scalar sector of extended supergravities, is that it admits irreducible PCT self-conjugate multiplets. Scalars of such multiplets have an almost complex structure only, and the scalar manifold will not be Kählerian.

N=2 matter multiplets, like in the N=1 case, are of two kinds. Both, however, contain scalar fields and generate a characteristic scalar potential. Supersymmetry and gauge symmetry breaking can then be induced by using both sorts of matter multiplets.

To be more specific, vector multiplets[20] contain, as partners of each massless gauge field, two Majorana spinors (gauginos) and a complex scalar:

$$(A_\mu^a,\ \Omega_\pm^a,\ z^a),\ a = 1, \ldots, \dim G$$

All these fields belong to the adjoint representation of the gauge group G. Hypermultiplets[21] can be constructed from the action of supersymmetric charges on a Clifford vacuum having helicity $\pm\frac{1}{2}$ and belonging to some irreducible representation $\underset{\sim}{r}$ of G. Two cases arise[22]: if $\underset{\sim}{r}$ is pseudoreal, then PCT invariance is satisfied without doubling of states. The hypermultiplet contains two-component spinors in representation $\underset{\sim}{r}$ and real scalars transforming according to $(\underset{\sim}{r},\underset{\sim}{2})$ of G x SU(2) [SU(2) is the invariance of N=2 supersymmetry algebra]. Notice that the dimension of pseudoreal representations is always even. Pseudoreal hypermultiplets do not admit N=2 supersymmetric mass terms; this fact allows to construct finite Yang-Mills theories with some massless fermions[23]. On the contrary, if $\underset{\sim}{r}$ is real or complex, PCT invariance will require doubling the states. The hypermultiplet will then contain Dirac fermions in the representation $\underset{\sim}{r}$ and real scalars transforming according to $(\underset{\sim}{r}+\underset{\sim}{\bar{r}},2)$ of G x SU(2).

The scalar fields of the two sorts of matter multiplets will be coordinates on two classes of manifolds. Scalars from the vector multiplets, which are complex, will live on a Kähler manifold. Cosets corresponding to Kähler manifold have the form G/H ⊗ U(1), where H ⊗ U(1) is a maximal (compact, to avoid ghosts) subgroup of G. Scalars in hypermultiplets are in SU(2) doublets. They will span quaternionic manifolds[25,26]. The corresponding coset spaces have the structure G/H ⊗ SU(2); H ⊗ SU(2) is a maximal (compact) subgroup of G such that coset coordinates transform according to $(\underset{\sim}{R}_H,\underset{\sim}{2})$ of H ⊗ SU(2), which is precisely what is required for the scalar fields. The list of quaternionic coset spaces can be found in Reference 26. Notice that $\underset{\sim}{R}_H$ reads $\underset{\sim}{r} + \bar{\underset{\sim}{r}}$, with $\underset{\sim}{r}$ complex, only for Grassmannian manifolds $SU(n,2)/SU(n) \otimes U(1) \otimes SU(2)$. These manifolds are also Kählerian.

What are the implications of the multiplet structure and transformation properties of the theory for supersymmetry breaking pattern? Gravitinos are in an SU(2) doublet. That is also the case for the scalars of hypermultiplets. The complex scalars in vector multiplets are, however, SU(2) singlets. This fact implies that the number of unbroken supersymmetries allowed by vacuum expectation values of scalars belonging to vector multiplets is zero of two, since SU(2) remains unbroken. The massive gravitino multiplet of N=1 Poincaré supersymmetry contains two massive vector fields. Then, if we want to end up with unbroken N=1 supersymmetry we need to couple both hypermultiplets and vector multiplets to N=2 supergravity[27]. There is, however a general argument which forbids the breaking of N=2 supergravity into Poincaré N=1 supergravity (with zero cosmological constant)[28]. The only allowed breaking pattern is then into N=0. Notice, however, that the masses of the two gravitinos will be in general different in presence of hypermultiplets.

We now want to investigate some aspects of the super-Higgs effect for general couplings of n vector multiplets to N=2 supergravity.

The action for such couplings has been established by de Wit et al.[24,29], using N=2 tensor calculus. The scalar sector we are interested in can be obtained in the following way. The action is based on a gauge invariant function $F(X^I)$ of n+1 variables (I=0, 1, ..., n), required to be homogeneous of degree two. Conformal invariance is fixed by the gauge choice

$$X^I N_{IJ} X^{*J} = 1, \tag{2.1}$$

where

$$N_{IJ} = \frac{1}{4}(F_{IJ} + F^*_{IJ}) = \frac{1}{2} \mathrm{Re}\left(\frac{d^2F}{dX^I dX^J}\right) . \tag{2.2}$$

It is then convenient to define scalar fields z^I by

$$z^I = X^I/X^0 , \tag{2.3}$$

so that $z^0=1$. We further define

$$Y(z,z^*) = (X^0)^{-2} = z^I N_{IJ} z^{*J} . \tag{2.4}$$

Notice that Y and N_{IJ} remain invariant under the transformation

$$F \to F + iC_{IJ}X^I X^J \tag{2.5}$$

where C_{IJ} are real constants. This remains true for the whole theory. [X^0 can certainly be chosen real due to Equation (2.1).] Alternatively, one can define the coupling in terms of the fields $z^a=z^I$, a=1, ..., n, with an arbitrary function f(z) related to F by:

$$X^{0^{-2}} F(X^I) = F(z^I) = F(1,z^a) \equiv f(z^a) . \tag{2.6}$$

Then:

$$N_{oo} = \frac{1}{2}(f - f_a z^a + \frac{1}{2} f_{ab} z^a z^b) + \text{c.c.}, \tag{2.7}$$

$$N_{oa} = \frac{1}{4}(f_a - f_{ab} z^b) + \text{c.c.}, \tag{2.8}$$

$$N_{ab} = \frac{1}{4} f_{ab} + \text{c.c.}, \tag{2.9}$$

$$Y = \frac{1}{2}(f + f^* - \frac{1}{2}(f_a - f^*_a)(z^a - z^{*a})) . \tag{2.10}$$

Since the fields z^a are partners of the gauge bosons in the vector multiplets, they transform according to the adjoint representation of the gauge group.

As we have already mentioned, the SO(2) symmetry of the N=2 theory can be gauged.* The gauge field is in general an arbitrary linear combination of all abelian vector fields of the theory, including the one belonging to the gravity multiplet. The corresponding scalar combination is

$$g_I X^I = X^0(g_0 + g_a z^a) \ . \qquad (2.11)$$

The real coefficients g_I play the role of gauge coupling constants. The case

$$g_I = g' \ \delta_{I0} \qquad (2.12)$$

means that the SO(2) gauge field belongs to the supergravity multiplet.

Scalar fields z^a, like scalars of N=1 chiral multiplets, are coordinates on a Kähler manifold. The scalar kinetic Lagrangian and the potential will then be obtained from the Kähler function $G(z,z^*)$. In the N=2 case, we find[30]

$$G(z,z^*) = - \ln Y(z,z^*) + \ln |h(z)|^2 \qquad (2.13)$$

and the superpotential h(z) is

$$h(z) = 2\sqrt{2} \ g_I z^I = 2\sqrt{2} \ (g_o+g_a z^a) \ ; \qquad (2.14)$$

h(z) is then constant when the SO(2) gauge field is member of the gravity multiplet and vanishes when SO(2) is not gauged. The chiral part of the potential is then

$$\begin{aligned} V_c &= - 8 g_I g_J (N^{-1})^{IJ} - 16 |g_I X^I|^2 \\ &= - 8 g_I g_J (N^{-1})^{IJ} - 16 Y^{-1} |g_o + g_a z^a|^2 \ . \end{aligned} \qquad (2.15)$$

To obtain the gauge potential we first need the kinetic terms of the gauge fields. They are given by

$$- \frac{1}{4} \mathrm{Re} \ \tilde{f}_{IJ} \ F^I_{\mu\nu} \ F^{J\mu\nu} \ , \qquad (2.16)$$

with

$$\tilde{f}_{IJ} = - \frac{1}{4} F_{IJ} + \frac{\left(N_{IK} X^{*K}\right)\left(N_{JL} X^{*L}\right)}{X^{*K} N_{KL} X^{*L}} . \qquad (2.17)$$

*More general gaugings, in particular of the whole SU(2), can also be envisaged[18,29].

$F^0_{\mu\nu}$ is the field strength of the vector field in the gravity multiplet. Notice that the function $\tilde{f}_{IJ}$ is now dependent on z and z*. The z* dependence, which was not present in N=1, is related to the effect of the second N=1 gravitino multiplet necessary to form the N=2 theory. The second term, however, does not contribute to the gauge potential V_g, due to gauge invariance and transformation properties of the scalar fields. One finds

$$V_g = \frac{1}{2} \mathrm{Re}\, \tilde{f}^{-1}_{ab} D^a D^b , \tag{2.18}$$

with, like in the N=1 case,

$$D^a = g \frac{dG}{dz^c} C^a{}_{cd} z^d = g C^a{}_{cd} N_{cI} X^{*I} X^d . \tag{2.19}$$

Then:

$$V_g = g^2 N_{ab} (C^a{}_{cd} X^c X^{*d})(C^b{}_{ek} X^e X^{*k}) . \tag{2.20}$$

g is the coupling constant of the non-abelian part of the gauge group (assumed simple for simplicity) and $C^a{}_{bc}$ are its structure constants. The scalar fields which are partners of abelian gauge fields do not appear in V_g, due to the vanishing of the corresponding structure constants.

Notice that the N=1 formulation[30] of the N=2 theory is simply given by Equations (2.13), (2.14) and (2.17).

It will be useful to remark that one can reparametrize the theory in such a way that $g_I=0$ except for g_0 [31]. This is done with the help of real, linear transformations, acting on the scalar partners of abelian vector fields, according to

$$X^I \to U^I{}_J X^J ,$$

$$g_I \to g_J (U^{-1})^J{}_I , \tag{2.21}$$

$$N_{IJ} \to (U^{-1})^K_I N_{KL} (U^{T-1})^L{}_J .$$

Such a transformation does not act on the gauge potential since abelian multiplets do not appear in V_g. These transformations leave in fact F(x) and $g_I X^I$ form invariant.[8] Theories related by (2.21) are then equivalent, although expressed in terms of different parameters. This will allow us to consider in general the simpler case $g_I = g' \delta_{I0}$ for which the chiral potential is

$$V_c = - 8g'^2 ((N^{-1})^{00} + 2Y^{-1}) . \tag{2.22}$$

We still need to specify what are the conditions for unbroken supersymmetry[29] at a stationary point of the potential. It is convenient to split the potential into two parts

$$V = V_+ + V_- , \tag{2.23}$$

with

$$V_+ = V_g + g_I H^I , \tag{2.24}$$

and

$$V_- = - 24 |g_I X^I|^2 . \tag{2.25}$$

H^I is given by

$$H^I = 8(X^I(g_J X^{*J}) - (N^{-1})^{IJ} g_J) . \tag{2.26}$$

If negative-metric states are not present, one easily shows that

$$V_+ \geq 0.$$

Notice that V_-, in the N=1 formulation given by Equations (2.13) and (2.14), reads

$$V_- = - 3 \exp G \leq 0, \tag{2.27}$$

and V_+ is then the first term in the N=1 chiral potential given by Equation (1.9). Then, supersymmetry is preserved if

$$\langle H^I \rangle = 0, \tag{2.28}$$

which implies as usual $\langle V_+ \rangle = 0$. In terms of fields z^a, these conditions read:

$$\langle H^o \rangle = 0 : \langle g_o + g_a z^a \rangle = \langle Y(N^{-1})^{oI} g_I \rangle , \tag{2.29}$$

$$\langle H^a \rangle = 0 : \langle z^a \rangle = \frac{(N^{-1})^{aI} gI}{(N^{-1})^{oJ} gJ} . \tag{2.30}$$

The vacuum expectation values of z^a are real when supersymmetry is not broken. The second set of conditions,

$$\langle D^a \rangle = gC^a{}_{bc} \langle X^b \rangle \langle X^{*c} \rangle = 0, \tag{2.31}$$

are then automatically satisfied. The condition $\langle D^a \rangle = 0$ is, however, weaker than Equation (2.28). It only implies that $C^a{}_{bc}(\mathrm{Re}X^a)(\mathrm{Im}X^b) = 0$. In terms of the Kähler function G, one finds

$$H^I = - 8(g_L z^{*L})(N^{-1})^{IK} \frac{dG}{dz^{*K}} . \qquad (2.32)$$

Notice also that the cosmological constant Λ is given by

$$\Lambda = \langle V_- \rangle = - 24(g_o + g_a \langle z^a \rangle)^2 / \langle Y \rangle , \qquad (2.33)$$

at a supersymmetry preserving point. Λ is then negative for unbroken N=2 supersymmetry when the vector field of the gravity multiplet participates in SO(2) gauging.

N=2 supersymmetry will be broken (into N=0) when $\langle H^I \rangle \neq 0$ for some value of I. The corresponding gravitino mass term is, like in N=1 case:

$$- m_{3/2} \; \delta_{ij} \bar{\psi}_{\mu i} \sigma^{\mu\nu} \psi_{\nu j} \; , \; ij = 1, 2, \qquad (2.34)$$

with

$$m_{3/2} = \langle \exp(G/2) \rangle = |2\sqrt{2} \; g_I \langle X^I \rangle| . \qquad (2.35)$$

Our aim now is to analyze some of the aspects of the Higgs and super-Higgs effects in the N=2 theory we have just described.

As a first example, consider the so-called minimal coupling defined by

$$F(X) = (X^o)^2 - X^a X^a , \qquad (2.36)$$

for which

$$N_{IJ} = 2\delta_{Io}\delta_{Jo} - \delta_{IJ} , \qquad (2.37)$$

or equivalently by

$$f(z) = 1 - z^a z^a . \qquad (2.38)$$

F(x) is certainly invariant under SO(1,n). The gauge choice (2.1) is then

$$|X^o|^2 - X^a X^{*a} = 1 , \qquad (2.39)$$

which is precisely the SU(1,n)-variant constraint of non-compact CP^n models. The corresponding Kähler manifold is the coset space SU(1,n)/SU(n) x U(1). If follows from (2.4) that

$$Y = 1 - z^a z^{*a} = \frac{1}{1 + X^a X^{*a}} . \qquad (2.40)$$

Choosing the SO(2) gauge field to belong to the gravity multiplet, i.e. $g_a=0$, $g_o=g'$, we get the following scalar potential

$$V = -8g'^2[1 + 2(1 - z^a z^{*a})^{-1}]$$

$$+ g^2(1 - z^a z^{*a})^{-2}(C^a{}_{bc} z^b z^{*c})^2 \qquad (2.41)$$

deduced from the Kähler function [see Equation (2.13)]

$$G(z,z^*) = \ln\left(\frac{8g'^2}{1-z^a z^{*a}}\right) . \qquad (2.42)$$

Equivalently, using X variables in a matrix notation:

$$V = -24g'^2 - 16g'^2\mathrm{Tr}(XX^+) + g^2\mathrm{Tr}([X,X^+]^2) . \qquad (2.43)$$

The stationary points of this polynomial potential can be analyzed[32]. We first split the complex matrix X into

$$X = A + iB , \qquad (2.44)$$

with Λ and B hermitean. Then, from the minimum equation, we get:

$$\mathrm{Tr}(\langle A\rangle\langle B\rangle) = 0 , \qquad (2.45)$$

$$\mathrm{Tr}(\langle A\rangle^2) = \mathrm{Tr}(\langle B\rangle 2) , \qquad (2.46)$$

$$g^2\mathrm{Tr}([\langle X\rangle,\langle X^+\rangle]^2) = 16g'^2\mathrm{Tr}(\langle A\rangle^2) , \qquad (2.47)$$

$$\langle V\rangle = -24g'^2[1 + \tfrac{2}{3}\mathrm{Tr}(\langle A\rangle^2)] . \qquad (2.48)$$

All stationary points have negative cosmological constant. Equation (2.47) indicates that all supersymmetry breaking solutions break simultaneously the gauge symmetry. For the only supersymmetric vacuum, $\langle z\rangle = 0$, the potential can be written

$$V = -24g'^2(1 + \tfrac{2}{3} z^a z^{*a}) + O(z^4) . \qquad (2.49)$$

All scalar fields receive a negative "mass squared" term, $-2/3\ \Lambda$, in term of the cosmological constant Λ. This is precisely the conformal "mass term" required for effectively massless scalar fields in Anti-de Sitter space. This stationary point is, however, (classically and locally) stable due to the stability condition[33].

$$\frac{m^2}{\Lambda} < \frac{3}{4} , \qquad (2.50)$$

for the eigenvalues m^2 of the "mass matrix" obtained from the quadratic terms of the potential.

The gauge symmetry breaking patterns corresponding to stationary points (recall that N=2) is always broken into N=0) can be characterized in the following way[32]. For a gauge group G=SU(N), there is a stationary point of V with residual symmetry

$$SU(m_1) \times SU(m_2) \times \ldots \times SU(m_N) \times U(1)^{p-1},$$

where $m_1, \ldots, m_N$ are all possible sets of non-negative integers such that

$$\sum_{n=1}^{N} nm_n = N ,$$

and p is the number of non-zero m_n's. The case m_1=N corresponds to <Z>=0 and for m_N=1, all gauge symmetries and supersymmetries are broken. Stability of stationary points can, in general, be ensured, the condition (2.50) leading to a lower bound of order 1 on the ratio g'^2/g^2. In particular, for $N \geq 7$, $N \neq 10$, there is always an SU(3) x SU(2) x U(1)-invariant stationary point.

Notice that for the minimal coupling case, positivity of the kinetic energies is automatically satisfied due to the CP^n structure of the theory.

The minimally coupled theory does not possess any classical vacuum with zero cosmological constant. This is, however, not the case when more complicated couplings are considered. Demanding that, at a stationary point of the potential:

$$\langle \frac{dV}{dz^a} \rangle = 0 ,$$

the cosmological constant Λ = <V> is also zero leads to severe constraints on the possible coupling functions f(z), in particular when the stability requirements are also imposed. A general analysis of this problem can be found in Reference 31, which is the basis for the rest of this section.

Before coming to examples of couplings which exhibit N=2 super-Higgs effect with zero cosmological constant, let us mention a general property of the chiral potential, Equation (2.15). For all stationary points of the chiral potential, one finds that:

$$\left(\frac{d^2V}{dz^a dz^{*b}} - 2VG_{ab} \right) \left(H^6 - z^{*6} H^o \right) = o \tag{2.51}$$

($G_{ab} = d^2G/dz^a dz^{*b}$ is the metric of scalar kinetic terms). This simple result indicates that, for instance, all flat potentials which break supersymmetry are zero potentials, and that a stationary point with V = 0 is stable only if some scalars are massless. These two observations reinforce our interest in flat potentials, as a necessary ingredient for a "realistic" N=2 supergravity breaking. In the case of vanishing chiral potential, the trace of the (squared) mass matrix of the 2n spin ½ states in the n vector multiplets is

$$\mathrm{Tr}\, m^2_{1/2} = 4(n-1)m^2_{3/2} + \text{gauge terms.} \tag{2.52}$$

This indicates that these Majorana states receive a mean squared mass $2m^2_{3/2}$ as a consequence of the super-Higgs effect. (Recall that two spin ½ states are eaten up by the two massive gravitinos).

Couplings to N=2 supergravity of a single vector multiplet, with scalar field z, and giving rise to vanishing potentials are easily found. As we learned from N=1, we need a Kähler function G of the form

$$G = -\ln(Y/8g'^2)\ , \tag{2.53}$$

in the "gauge" $g_0 = g'$, $g_1 = 0$, with

$$Y = (\phi(z) + \phi^*(z^*))^3 \tag{2.54}$$

Thus, we have to find all function f(z) able to give such a form to the corresponding Y. The first remark is that

$$\begin{aligned}\frac{d^2Y}{dzdz^*} &= \frac{1}{4}\frac{d^2f(z)}{dz^2} + \text{c.c.}\\ &= 6\,\frac{d\phi}{dz}\frac{d\phi^*}{dz^*}(\phi+\phi^*)\ .\end{aligned} \tag{2.55}$$

This equation leads to only two solutions, which are apparent using:

$$\frac{d^4Y}{dz^2dz^{*2}} = 0 = 3\frac{d^2\phi}{dz^2}\frac{d^2\phi^*}{dz^{*2}}\left(\phi + \left(\frac{d\phi}{dz}\right)^2\Big/\frac{d^2\phi}{dz^2} + \text{c.c.}\right)\ . \tag{2.56}$$

We have either

$$\frac{d^2}{dz^2} = 0\ , \tag{2.57}$$

leading to

$$Y_1 = -ia(z-z^*)^3\ , \tag{2.58}$$

where a is a real constant, or

$$\frac{d}{dz}\left(\phi\frac{d\phi}{dz}\right) = \text{imaginary constant}\ , \tag{2.59}$$

leading to

$$Y_2 = \left(\sqrt{bz+c} + \sqrt{bz^*+c}\right)^3, \tag{2.60}$$

where b and c are real constants. These two solutions are obtained respectively from functions f(z) given by

$$f_1(z) = 4iaz^3 , \tag{2.61}$$

and

$$f_2(z) = 8(bz+c)^{3/2} , \tag{2.62}$$

up to a second order polynomial with imaginary coefficient which does not contribute to Y [see Equation (2.5)]. With a general gauging of SO(2), the two couplings with vanishing potentials are given by

$$F_1(X^o X^1) = i \frac{\left(\alpha_I X^I\right)^3}{g_I X^I} + ic_{IJ} X^I X^J , \tag{2.63}$$

and

$$F_2(X^o, X^1) = 4(g_I X^I)^{1/2} (\alpha_J X^J)^{3/2} + ic_{IJ} X^I X^J , \tag{2.64}$$

respectively, (α_I, g_I and C_{IJ} are real constants). In both cases, there exists a domain for which the kinetic energies are positive. Notice that, while in the N=1 case the constraint of zero potential was solved with an arbitrary function of z, the N=2 case is much more contrived: there are only two solutions, the only arbitrariness being the value of the real parameters a, b, c. The SU(1,1) invariance of the action found for N=1 is also present in the N=2 case, and the scalar manifold is also SU(1,1)/U(1).

Vanishing chiral potentials can also be found when an arbitrary number of vector multiplets are coupled to N=2 supergravity. We will consider here two classes of such couplings which make use of the results discussed in the case of N=1 supergravity.

For the first class, observe that if the function f(z) is the sum of functions $f_{(a)}$ of only one field z^a,

$$f = \sum_{a=1}^{n} f_{(a)}(z^a) , \tag{2.65}$$

then

$$Y = \sum_{a=1}^{n} Y_{(a)} , \tag{2.66}$$

with

$$Y_{(a)} = \frac{1}{2}\left[f_{(a)}+f^*_{(a)}-\frac{1}{2}\left(\frac{df_{(a)}}{dz^a}-\frac{df^*_{(a)}}{dz^{*a}}\right)(z^a-z^{*a})\right]$$

like in Equation (2.10). We can then use the result derived in Section 1 [see Equations (1.39) and (1.40)]: the chiral potential will vanish when each $f_{(a)}$ is one of the two solutions to the single field case, given in Equations (2.61) and (2.62), assuming $g_a=0$, $g_0\neq 0$. Alternatively, using X^I variables, the chiral potential will vanish when

$$F(X^I) = \sum_{a=1}^{n} F_{(a)}(X^o,X^a) , \tag{2.67}$$

each function $F_{(a)}$ being either Equation (2.63) or Equation (2.64).

We have also shown in Section 1 that zero potentials occur when the function exp(-G) is homogeneous of order 3 in the combinations $i(z^j-z^*_j)$. Such a function can easily be obtained in our N=2 case if we choose ($g_a=0$, $g_0\neq 0$):

$$f(z) = 4id_{abc}z^a z^b z^c , \tag{2.68}$$

where d_{abc} are real coefficients. We obtain

$$Y = -id_{abc}(z^a-z^{*a})(z^b-z^{*b})(z^c-z^{*c}) , \tag{2.69}$$

and the chiral potential is zero. For a general SO(2) gauging, this second class of couplings is generated by:

$$F(X) = \frac{id_{IJK}X^I X^J X^K}{g_L X^L} . \tag{2.70}$$

This form of F generalizes the single field solution given in Equation (2.63).

The coefficients d_{abc} are, in principle, arbitrary. However, f and Y have to be gauge invariant. There are different possibilities: if we underline indices corresponding to abelian vector multiplets, $d_{\underline{abc}}$ is arbitrary, $d_{\underline{ab}c}$ is zero, $d_{\underline{a}bc}$ is proportional to the Killing metric η_{bc} of the non-abelian group; d_{abc} can be non-zero only if the three indices correspond to gauge multiplets belonging to the adjoint representation of the same simple group. The adjoint representations of SU(N) groups only possess a cubic symmetric invariant. The coefficients d_{abc} exist then for each simple factor G_i=SU(N) of the gauge group. The invariances of f and

Y can, however, be larger than the gauge group, the vector multiplets being embedded in some representation possessing a cubic invariant.

This is in particular the case in the very restricted class of models in which the gauge choice (2.1), with a cubic function $F(X^I)$, gives rise to a non-linear σ-model for the scalars. The coset space G/H are obtained[34] from the requirement that G possesses an irreducible representations $\underset{\sim}{R}$ which decomposes into

$$\underset{\sim}{R} = \underset{\sim}{1} + \underset{\sim}{1} + \underset{\sim}{r}$$

with respect to the maximal compact subgroup H of G, $\underset{\sim}{r}$ being the representation of H classifying the elements of the coset G/H; $\underset{\sim}{r}$ is further required to possess a cubic H-invariant tensor. $\underset{\sim}{R}$ will classify the fields X^I. Interestingly enough, these models can be obtained by dimensional reduction of N=2 supergravity in 5 dimensions coupled to abelian vector multiplets[34]. The resulting four-dimensional theory has a scalar potential given by Equation (2.15), with Y as given in Equation (2.69). If a symmetric coefficient c^{abc} exist such that[31]

$$d_{i(ab}c^{ifg}d_{cd)f} = \delta^g_{(a}d_{bcd)} , \qquad (2.71)$$

then the potential can be written:

$$V = \frac{1}{3} c^{abc} g_a g_b h_c , \qquad (2.72)$$

with

$$h_c = d_{cab}(z^a - z^{*a})(z^b - z^{*b})/Y . \qquad (2.73)$$

Equation (2.72) is precisely the potential obtained in Reference 34. It can be made zero, if the condition

$$c^{abc} g_a g_b = 0 \qquad (2.74)$$

can be satisfied.

REFERENCES

1. For a review, see P. Fayet, Proc. 21st Int. Conf. on High-Energy Physics, Paris, 1982, P. Petiau and M. Porneuf, eds., J.Phys., 43:C3-673 (1982).
   R. Barbieri and S. Ferrara, Surveys in High-Energy Physics, 4:33 (1983).
   H. P. Nilles, Univ. of Geneva preprint UGVA-DPT 1983/12-412, to appear in Phys. Reports.
   H. E. Haber and G. L. Kane, Ann Arbor preprint (1984), to appear in Phys. Reports.

S. Ferrara, Supersymmetry and unification of particle interactions, to appear in Proc. "Ettore Majorana" Int. School of Subnuclear Physics; 20th Course: Gauge Interactions - Theory and Experiment, Erice, A. Zichichi, ed., Plenum Press, Inc., New York (1982).

2. E. Gildener and S. Weinberg, Phys.Rev., D13:3333 (1976).
E. Gildener, Phys.Rev., D14:1667 (1976).
L. Maiani, Proc. Summer School of Gif-sur-Yvette, 1980, IN2P3, Paris, 1980, p.3.
M. Veltman, Acta Phys.Pol., B12:437 (1981).
E. Witten, Nucl.Phys., B188:513 (1981).

3. S. Ferrara, Int. Europhysics Conf. on High-Energy Physics, Brighton, 1983, J. Guy and C. Costain, eds., Rutherford Appleton Laboratory, Chilton, Didcot, U.K., p.522.
P. Fayet, ibid., p.33.
D. V. Nanopoulos, ibid., p.38.
R. Barbieri, Univ. Pisa preprint (1983), in Proc. Int. Symp. on Lepton and Photon Interactions at High Energies, Cornell, to be published, 1983.
J. Ellis, CERN preprint TH 3718 (1983), in Proc. Int. Symp. on Lepton and Photon Interactions of High Energies, Cornell, to be published, 1983.

4. D. V. Volkov and V. A. Soroka, JETP Letters, 18:312 (1973).
S. Deser and B. Zumino, Phys.Rev.Lett., 38:1433 (1977).

5. E. Cremmer, B. Julia, J. Scherk, S. Ferrara, L. Girardello, and P. Van Nieuwenhuizen, Phys.Lett., 79B:231 (1978), and Nucl.Phys., B147:105 (1979).
E. Cremmer, S. Ferrara, L. Girardello and A. Van Proeyen, Phys.Lett., 116B:231 (1982), and Nucl.Phys., B212:413 (1983).

6. J. Bagger and E. Witten, Phys.Lett., 115B:202 (1982).
J. Bagger, Nucl.Phys., B211:302 (1982).

7. J. Polony, Budapest preprint KFKI-1977 (93) (unpublished).

8. R. Barbieri, S. Ferrara and C. A. Savoy, Phys.Lett., 119B:343 (1982).
R. Arnowitt, A. Chamseddine and P. Nath, Phys.Rev.Lett., 49:970 (1982).
H. P. Nilles, M. Srednicki and D. Wyler, Phys.Lett., 120B:346 (1983).
E. Cremmer, P. Fayet and L. Girardello, Phys.Lett., 122B:41 (1983),
L. J. Hall, J. Lykken and S. Weinberg, Phys.Rev., D27:2359 (1983).
S. K. Soni and H. A. Weldon, Phys.Lett., 126B:215 (1983).

9. E. Cremmer, S. Ferrara, C. Kounnas and D. V. Nanopoulos, Phys.Lett., 133B:61 (1983).

10. N. Chang, S. Ouvry and X. Wu, Phys.Rev.Lett., 51:327 (1983).

11. J. Ellis, A. B. Lahanas, D. V. Nanopoulos and K. Tamvakis, Phys.Lett., 134B:429 (1984).
J. Ellis, C. Kounnas and D. V. Nanopoulos, CERN preprints TH.3768 (1983), TH.3824 (1984) and TH.3848 (1984).

12. S. Ferrara and A. Van Proeyen, Phys.Lett., 138B:77 (1984).
13. M. Grisaru, M. Rocek and A. Karlhede, Phys.Lett., 120B:110 (1983).
14. E. Cremmer, S. Ferrara and J. Scherk, Phys.Lett., 74B:61 (1978).
15. N. Dragon, M. G. Schmidt and U. Ellwanger, Heidelberg preprint, HD-THEP-84-10 (1984).
16. S. Ferrara and P. Van Nieuwenhuizen, Phys.Rev.Lett., 37:1669 (1976).
17. D. Z. Freedman and A. Das, Nucl.Phys., B120:221 (1977).
E. S. Fradkin and M. A. Vasiliev, Lebedev Institute preprint 197 (1976).
18. B. de Wit, P. C. Lauwers, R. Philippe and A. Van Proeyen, Phys.Lett., 135B:295 (1984).
19. E. Cremmer and B. Julia, Phys.Lett., 80B:48 (1978), and Nucl. Phys., B159:141 (1979).
E. Cremmer, in: "Superspace and Supergravity," S. W. Hawking and M. Rocek, eds., Cambridge University Press, Cambridge, p.267 (1981).
20. P. H. Dondi and M. Sohnius, Nucl.Phys., B81:317 (1974).
R. J. Firth and O. J. Jenkins, Nucl.Phys., B85:525 (1975).
R. Grimm, M. Sohnius and J. Wess, Nucl.Phys., B133:275 (1978).
21. P. Fayet, Nucl.Phys., B113:135 (1976), and B149:137 (1979).
22. S. Ferrara and C. A. Savoy, Supergravity '81, S. Ferrara and J. G. Taylor, eds., Cambridge University Press, Cambridge, p.47 (1982).
23. J. P. Derendinger, S. Ferrara and A. Masiero, CERN preprint TH.3854 (1984), Phys.Lett., B to appear.
24. B. de Wit, P. G. Lauwers, R. Philippe, S. Q. Su and A. Van Proeyen, Phys.Lett., 134B:37 (1984).
25. P. Breitenlohner and M. Sohnius, Nucl.Phys., B187:409 (1981).
26. J. Bagger and E. Witten, Nucl.Phys., B222:1 (1983).
27. S. Ferrara and P. van Nieuwenhuizen, Phys.Lett., 127B:70 (1983).
28. S. Cecotti, L. Girardello and M. Porrati, Univ. of Pisa preprint (1984).
29. B. de Wit and A. Van Proeyen, preprint NIKHEF-H/84-4 (1984).
30. J. P. Derendinger, S. Ferrara, A. Masiero and A. Van Proeyen, CERN preprint TH.3813 (1984), Phys.Lett., B to appear.
31. E. Cremmer, J. P. Derendinger, B. de Wit, S. Ferrara, L. Girardello, C. Kounnas and A. Van Proeyen, Ecole Normale Superiéure preprint in preparation.
32. J. P. Derendinger, S. Ferrara, A. Masiero and A. Van Proeyen, Phys.Lett., 136B:354 (1984).
33. P. Breitenlohner and D. Z. Freedman, Phys.Lett., 115B:197 (1982), Ann.Phys., 144:249 (1982).
34. M. Günaydin, G. Sierra and P. K. Townsend, Ecole Normale Superiéure, preprint LPTENS 83/32 (1983), to appear in Nucl. Phys., B.
M. Günaydin, G. Sierra and P. K. Townsend, Cambridge preprint (1984).

DISCUSSION

CHAIRMAN : S. FERRARA

Scientific Secretaries: C. Burgess and S. Catto

DISCUSSION 1

- *BURGESS* :

Is there a non-renormalization theorem for locally supersymmetric theories ?

- *FERRARA* :

Yes, even if supergravity is non-renormalizable. In fact, one can still show that no F terms (or more generally terms which are not full superspace integrals) are generated in perturbation theory. Girardello and Grisaru also have investigated soft breaking and they have classified which terms do not destroy, for instance, the absence of quadratic divergences in supersymmetric theories.

- *BURGESS* :

Are the problems you mentioned in building models with $M_S \sim M_W$ particular to Fayet-Iliopoulos breaking, or do they apply more generally ?

- *FERRARA* :

Yes, I spoke only of the Fayet-Iliopoulos breaking. If you have O'Raifertaigh breaking, then the spontaneous breaking cannot occur at a tree-level because you get wrong mass relations between quarks, leptons and the scalar partners. If you have a fermion you get scalars which are split like

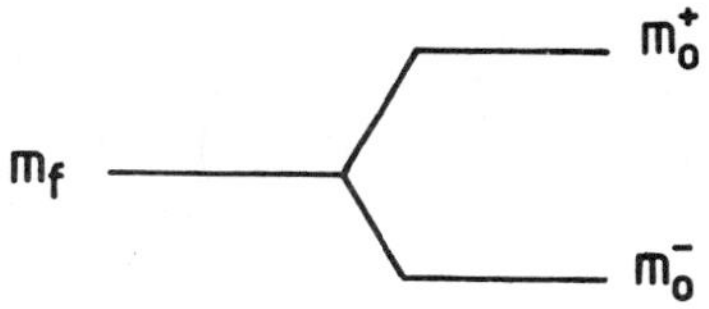

with respect to the fermions. So you always get some scalar partner of some fermion which is lighter than the fermion itself. So the breaking does not occur in the right way. You would like to have a breaking in which the scalar partners are both raised with respect to the fermion mass, i.e. the masses of the scalar partners are both higher than the fermion mass

which is only obtainable at the tree-level with Fayet-Iliopoulos supersymmetry breaking.

- *OHTA* :

The motivation for introducing supersymmetry is to solve the gauge hierarchy problem. But if you proceed to supergravity, even renormalizability of the theory is lost. What is your opinion about this ?

- *FERRARA* :

In these theories you certainly include supergravity in a very well defined way, namely, you neglect interactions which are suppressed by the inverse power of the Planck mass. What you do is the same as one does in general in ordinary theory. We know that gravity is coupled to ordinary Yang-Mills theories when interacting. So in any interaction when you include gravity it would become non-renormalizable. Even Yang-Mills theories when coupled to gravity become non-renormalizable. So that what is usuallys done is that you cut off these theories with some scale Λ which is assumed to be less than the Planck mass, and then you take the formal limit as $m_p \to \infty$, keeping the cut-off finite to give meaning to non-renomalizable terms. So what happens now is that you drop $1/m_p$ terms, but you keep the terms which are proportional to the gravitino mass.

itself, which cannot be neglected if the $m_{3/2} \sim O(m_W)$ which is exactly what happens when the supersymmetry breaking scale is in this energy range. So in this limit these theories become normal renormalizable theories with soft breaking terms which are proportional to the gravitino mass $m_{3/2}$. In this theory, guage hierarchy problem can be solved because soft breaking SUSY terms are sufficiently mild not to affect the ultraviolet improvement due to supersynmetry.

- *CATTO* :

In two dimensions you have shown that there exist supersymmetric equations depending on an arbitrary function f(S) of the scalar superfield S. In three dimensions however, a new invariant operator provides a generalization of analogous two-dimensional equations that are first order in the fermion field. In particular when one chooses $f(S) = \lambda S^3$, one gets the conformal invariant supersymmetric interacting field theories of Witten and Fronsdal. Do you know of any instanton solutions in such three-dimensional supersymmetric theories, and also could you tell me if the instantons are the origins of the supersymmetry breaking in four dimensions ?

- *FERRARA* :

I do not know of any solutions in three dimensions. But in four dimensions it has been shown very recently by Vainshtain, Voloshin and Zakharov that instantons can be incorporated in a supersymmetric theory without violating supersymmetry. They show that when you properly treat the zero modes in a supersymmetric theory and you consider the complete superconformal algebra in four dimensions, then you can restore supersynmetry.

- *HOU* :

Has there been any work combining supersymmetry and composite models ?

- *FERRARA* :

There have been attempts. The most promising in my opinion is to associate quarks and leptons to fermion partners of Goldstone bosons (quasi Goldstone fermions). This idea has been pursued by different groups at Max-Planck and CERN.

- *VAN DEN DOEL* :

Is it possible that naively unrenormalizable terms, such as

the four fermion couplings, might not spoil renormalizability in supersymmetric theories due to the cancellations or divergences ?

- *FERRARA* :

Such terms can improve the ultraviolet behavior of the theory. Supergravity is an example in which the four gravitino contact interactions improve the convergence of the theory to the extent that it becomes two-loop finite. In general, however, counterterms do arise at sufficiently high loops in supersymmetric non-renormalizable theories. Then the only remaining possibility is that perhaps their coefficients cancel and the theory is finite.

- *VAN DEN DOEL* :

Is it known whether or not local supersymmetry is spoiled by anomalies ?

- *FERRARA* :

As far as I know there is no regularization procedure which respects local supersymmetry. Although it is commonly believed that such a regularization scheme exists I do not think anyone has found it. Regularization schemes are known which respect global supersymmetry, however the question is still open in supergravity.

- *SIMIĆ* :

I would like to know if there are any two-loop calculations in N = 1 superconformal gravity, in particular I am interested in whether or not conformal invariance is broken by radiative corrections in this theory.

- *FERRARA* :

Superconformal gravity is widely believed to be a sick theory because of the appearance there of ghosts. N = 1 superconformal gravity is not ultraviolet finite but higher N superconformal gravities can be probably finite.

- *SIMIĆ* :

Recently Adler and Zee have revived the old idea of induced gravity originally due to Zakharov. I would like to know if it is possible to combine induced gravity with local supersymmetry.

- *FERRARA* :

It should be possible to combine these ideas. I think supersymmetry would help because it would improve the ultraviolet behavior. However, I do not think the renormalization of locally supersymmetric conformal gravity has been investigated much.

- *SALATI* :

If supersymmetry breaks the vacuum, energy becomes positive. How can this be reconciled with the known size of the cosmological constant ?

- *FERRARA* :

In global supersymmetry the potential for the scalars is positive, and supersymmetry is spontaneously broken if the potential is never zero. In local supersymmetry, however, this is no longer the case. In the presence of gravity the scalar potential is no longer positive definite and so it is possible to break supersymmetry and tune the cosmological constant to zero. Of course we do not have a natural reason to do so and must tune it to zero by hand.

- *BAGGER* :

In spontaneously broken local supersymmetry it is possible to cancel the cosmological constant. Is this cancellation preserved by radiative corrections ?

- *FERRARA* :

There are two types of radiative corrections to worry about. The non-renormalizable gravitational corrections will not preserve the cancellation. However in the softly broken theory arising in the limit of $m_p \to \infty$ with $m_{3/2}$ fixed the cosmological constant arises as a constant in the effective lagrangian and is of no consequence.

- *BAGGER* :

In global supersymmetry the connection between spontaneous supersymmetry breaking and vacuum energy density arises as a consequence of the supersymmetry algebra. In local supersymmetry there is no such connection. How does the argument change ?

- *FERRARA* :

In local supersymmetry the connection between the matter

vacuum energy and spontaneous SUSY breaking comes through the superhiggs effect. The order parameter of local SUSY is the coefficient of the spin 3/2 - spin 1/2 interference terms in the basic supergravity lagrangian. In absence of a cosmological constant this coefficient is related to the gravitino mass.

DISCUSSION 2 Scientific Secretaries: C. Burgess and P. Salati

- *OHTA* :

You said that in global supersymmetry all supersymmetric generators must break at the same time. How is this argument avoided in supergravity ?

- *FERRARA* :

The global argument does not apply because the local supersymmetry algebra is different. The algebra becomes field dependent in the sense that the parameters of the local transformations become field dependent. Further, there are explicit examples of N = 8 supergravity breaking down to N = 4, N = 2 or N = 1 supergravity.

- *LEURER* :

Why are the Higgs particles and leptons not placed in the same supermultiplet since they carry the same $SU_L(2) \times U(1)$ quantum numbers?

- *FERRARA* :

There are two reasons. If the Higgs were the superpartners of the neutrino it would carry lepton number and so lepton number would be broken when the Higgs develops a vacuum expectation value. It also turns out that two Higgs fields are required in supersymmetry and so there is a mismatch of degrees of freedom.

- *BURGESS* :

It is my understanding that model builders in supergravity usually do not use the most general coupling of matter to supergravity. For instance, terms proportional to four powers of the gauge boson superfield are dropped. Do such terms contribute to, say the scalar potential and would they have any effect on the phenomenology ?

- *FERRARA :*

These terms contribute non-renormalizable four-fermi and higher derivative interactions to the lagrangian. They do not survive when the Planck mass is sent to infinity with the gravitino mass fixed and so do not change the phenomenology of these models.

- *BURGESS :*

One of the original motivations for using N = 1 in globally supersymmetric theories was that it was impossible to have fermions in complex representation for $N \geq 2$. Does this argument still apply in supergravity theories ?

- *FERRARA :*

The arguments which forbid complex representations for fermions for $N \gtrsim 2$ rely on having the gauge group commute with the supersymmetry generators. This can be avoided, for example, in N = 8 supergravity by using the local SU(8) invariance of the action. Here the fermions can transform in complex representations.

In the Kaluza-Klein theories, in which the gauge invariance comes from coordinate invariance in the extra dimensions, there is again a difficulty in obtaining complex representations.

- *LAMARCHE :*

Are the coloured supersymmetric partners such as squarks, gluinos etc., confined ? Is there also an infrared divergence problem in supersymmetric QCD ?

- *FERRARA :*

In super-QCD, like QCD, all coloured particles are presumed to hadronize into coloured bound states. Consequently free gluinos should not be found, although there may be gluino balls. Various exotic hadrons become possible such as bound states of a quark and squark.

Nevertheless super-QCD could have rather different non-perturbative behaviour than QCD related to the possible patterns of chiral and supersymmetry breaking.

- *TAVANI :*

Can supergravity give rise to exotic macroscopic forces ?

- *FERRARA* :

In N = 1 supergravity you have only additional four-fermi interactions with a massive gravitino and so new long range forces do not arise. However, if you believe in extended supergravity theories the gravity multiplet can contain light spin zero and spin one bosons which could compete with the known long range forces.

- *D'AMBROSIO* :

Is supergravity a finite theory ?

- *FERRARA* :

At present there is no symmetry argument which shows supergravity is finite. There are calculations which show that various extended supergravity theories are finite at one or two loop order. Higher loops are under investigation. Supergravity has better ultra-violet behaviuor than Einstein gravity. However N = 1 supergravity coupled to matter is certainly non-renormalizable and should only be considered as a phenomenological low energy approximation to a more fundamental theory.

- *BAGGER* :

The models you discussed earlier today all use "minimal" kinetic energy terms. In the jargon of the trade they all correspond to flat Kähler manifolds. Will this flatness be preserved by radiative corrections ?

- *FERRARA* :

It would not be preserved by gravitational corrections. There is an argument, due to Weinberg, based on U(n) symmetry which implies that these corrections should be small at low energies.

- *BAGGER* :

What is your opinion about superstrings ?

- *FERRARA* :

They are interesting but do not correspond to normal lagrangian field theories as they are examples of non-local field theories. They have an infinite spectrum of states and suffer from many of the problems of Kaluza-Klein theories. The major difficulty is in obtaining is some way the observed low-energy spectrum of elementary particles.

- *VAN DEN DOEL* :

Does spontaneously broken supersymmetry get restored at high temperature like ordinary symmetries ?

- *FERRARA* :

This is a controversial point. Unlike most symmetries supersymmetry gets more and more broken as the temperature increases. This is due to the different behaviour of fermions and bosons at finite temperature. It is believed that supersymmetry is explicitly broken by finite temperature effects. However, the question is not settled yet.

- *SIMIĆ* :

How does proton decay differ in supersymmetric theories ?

- *FERRARA* :

New types of diagrams arise in supersymmetric proton decay due to the possibility of scalar superpartner exchange. The usual $p \to \pi^o e^+$ mode becomes swamped by new modes like $p \to \bar{\nu}X$. The lifetimes and branching ratios depend on the Yukawa couplings which are poorly known, so these calculations are quite uncertain. If the experimental proton lifetime were to be pushed above 10^{32} years supersymmetric grand unified theories would probably also be in trouble. However, in these theories the uncertainty in the Yukawa couplings and in the masses of the scalar superpartners makes it possible to push the proton lifetime above its experimental bound.

- *BALL* :

Why can we not go beyond N = 8 supersymmetry ?

- *FERRARA* :

It can be shown that any irreducible multiplet of N - extended supersymmetry must contain helicities greater than or equal to λ where

$$\lambda = N/4 \quad \text{for N even}$$

$$\lambda = (N+1)/4 \quad \text{for N odd.}$$

If we assume we cannot have mass particles of spin higher than 2 we must stop at N = 8. We stop at spin two because there is no known consistent lagrangian field theory of massless particles with helicity greater than 2.

- *BALL* :

Which supersymmetric theories are known to be ultraviolet finite ?

- *FERRARA* :

There are N = 2 and N = 4 globally supersymmetric theories which are finite. A class of soft-breaking terms can also be included. Unfortunately none of these contain gravity but are instead Yang-Mills theories. The particle content of these theories does not seem realistic.

- *CATTO* :

You earlier spoke of a class of supergravity GUT's in which $SU_L(2) \times U(1)$ is broken radiatively without the "sliding singlet" problem. How does this radiative breaking occur ?

- *FERRARA* :

This theory is renormalizable (once restricted to the $m_p \to \infty$ limit) and the radiative corrections to the Higgs mass-squared turn out negative. The terms which trigger this radiative breaking in the tree-level lagrangian are also those responsible for the soft-breaking of supersymmetry and are proportional to the gravitino mass.

TESTING SUPERSYMMETRY

G.L. Kane

Randall Laboratory of Physics
University of Michigan
Ann Arbor, MI 48109

ABSTRACT

Experiments which might detect supersymmetric partners of normal quarks, leptons, gauge and Higgs bosons, soon or eventually, are discussed. Current limits are examined.

1. INTRODUCTION

Among the ideas which are currently promising to explain, at least in part, the Standard Model and why it works, and to lead beyond the Standard Model, Supersymmetry is certainly the most popular.[1] And it is popular in spite of (because of?) a total absence of any experimental evidence which can be interpreted, even indirectly, as an indication that nature is Supersymmetric. Although finding any experimental evidence for Supersymmetry would have an extraordinary impact on particle physics, and although (as we will see below) it is not obvious how to look for most signatures of supersymmetry, very little effort has gone into this area. Fortunately, a number of ways exist to find evidence for Supersymmetry if it is present. These lectures will emphasize ways to find supersymmetric particles.

The expected states are listed in Table I. Each of the usual quarks, leptons gauge bosons, and Higgs bosons has its superpartner particle, displaced one half unit in spin but otherwise identical in quantum numbers to the familiar particle.

Table I

Standard Particles	Supersymmetric Partners
u_L, u_R, d_L, ...	$\tilde{u}_L$, $\tilde{u}_R$, $\tilde{d}_L$, ...
e_L^-, e_R^-, μ_L^-, ...	$\tilde{e}_L^-$, $\tilde{e}_R^-$, $\tilde{\mu}_L^-$, ...
ν_e, ν_μ, ...	$\tilde{\nu}_e$, $\tilde{\nu}_\mu$, ...
-·-·-·-·-·-·-·-·-·-·-·-·-·-	-·-·-·-·-·-·-·-·-·-·-·-·-·-
g	$\tilde{g}$
$W^\pm$	$\tilde{W}^\pm$
$\gamma, Z°$	$\tilde{\gamma}, \tilde{Z}°$
$\binom{H_1^+}{H_1^0}$, $\binom{H_2^0}{H_2^-}$	$\tilde{H}_1^+$, $\tilde{H}_1^0$, $\tilde{H}_2^0$, $\tilde{H}_2^-$

The set of standard particles (quarks, leptons, gauge bosons, Higgs) is given on the left. On the right are the superpartners, written in terms of the weak isospin eigenstates. Those superpartners above the dot-dashed line have spin zero, those below have spin 1/2. Since $\tilde{\gamma}, \tilde{Z}°$, $\tilde{H}_1^0$, $\tilde{H}_2^0$ all have spin 1/2 they can mix, so the mass eigenstates will be linear combinations of the weak eigenstates. The neutral mass eigenstates will be denoted as $\tilde{\chi}_i^o$ and called neutralinos (photinos, ziggsinos, or neutral higgsinos, depending on whether their couplings are photon-like, intermediate, or Higgs-like). Similarily, $\tilde{W}^\pm$ and $\tilde{H}_{1,2}^{\pm}$ will mix, and the mass mass eigenstates will be denoted $\tilde{\chi}_i^+$. The charged mass eigenstates will be called charginos (winos, wiggsinos, or charged higgsinos depending on whether their couplings are W-like, intermediate, or Higgs-like).

If an unbroken Supersymmetry held, the partners would have the same masses as the familiar particles, and they would not have escaped detection. It is customary to assume that the necessary breaking of Supersymmetry makes the superpartners heavier but does not affect their couplings or other properties, and that is presumably a good assumption for our purposes. One practical advantage of a Supersymmetric theory is that all the coupling strengths of the superpartners are determined in terms of those of the standard particles (essentially by replacing pairs of standard particles by their superpartners in any vertex). Thus, given a superpartner with some (unknown) mass but specific electric charge, weak isospin, and color one can see how it couples to quarks, gluons, and leptons and compute its production rates at any accelerator as a function of its mass. For the neutralino and chargino mass eigenstates, the unknown mass matrix affects the mixing, so the couplings of the mass eigenstates are not fully known.

In the simplest form of the theory the number of superpartners at a vertex is always even, so all superpartners will decay eventually into the lightest one. I will assume that is the photino, although it could be a higgsino or a scalar neutrino[4] -- then some modifications in detailed conclusions would occur, but not major changes.

What is needed for progress is experimental information, so I will concentrate on how to search for superpartners. As will be seen, that is sometimes subtle; most superpartners have detectable signatures, but often not ones that would be noticed without an explicit search. Background problems can be serious. Interestingly, some of the superpartners (gluinos, photinos and whiggsinos) could have been produced in experiments but not yet detected because the usual experimental cuts and procedures are not appropriate for the new states.

For some states [mainly scalar partners of leptons (sleptons) and partners of gluons (gluinos)] good lower limits exist on their masses. That is, if they were not heavier than a certain mass they would probably already have been detected. I will summarize these. Otherwise the emphasis will be on how a positive signal for supersymmetry might be detected at present machines and those of the next few years. Table II is a summary of the situation, meant to give a broad overview. Studies are needed because it is not just a matter of production rates (which are easily calculated) -- one must compare with backgrounds, make cuts, etc.

If supersymmetry were a spontaneously broken global symmetry, a Goldstone particle (a fermion called a Goldstino) would exist, coupled to every particle and its superpartner. Then various additional decays would be allowed, while production properties are unaffected. Goldstinos could also[5] be pair produced or produced associated with superpartners. I will not include the consideration of these in the present summary, because of space limitations; their role in gluino searches is described in Ref. 5, and they will be fully discussed in a forthcoming more complete review.[6]

It has been customary to assume photinos were rather light, with masses less than a few GeV. Recently Goldberg[7] has emphasized that cosmological arguments combined with the Majorana nature of the photino imply that photino masses may be rather large, up to of order 25 GeV. Various superpartner masses are correlated -- heavier sleptons implies heavier photinos. I will proceed here by giving the discussion for relatively light photinos, up to a few GeV (implying sleptons less than about 50 GeV in mass) and adding qualifying remarks for the situation with heavier photinos.

Various indirect methods exist to learn about restrictions on supersymmetric partners. Since these can never give a positive indication of supersymmetry, I will not cover them here. The relevant literature can be traced from the review of Fayet.[1] The status of models can be studied from the recent talk of Polchinski[2] or from the lectures of Ferrara in these proceedings. See also the review of Savoy[2].

Sometimes I will call various possibilities "excluded" by experiment. It should always be kept in mind that various experimental assumptions and cuts go into each reported result, and that it is very hard to find something for which one is not looking. Thus repeated examination of data for signals of new physics is very worthwhile.

Table II

SUSY Partners	Present Limits	Soon - - - - - -	- - - - - - - -	about 1989
$\tilde{\ell}^{\pm}$	$\tilde{m} \gtrsim 18$ GeV	~20 GeV	$[e^+e^- \to \tilde{\ell}^+\tilde{\ell}^-]$	~45 GeV
			$[W^{\pm} \to \tilde{\ell}^{\pm}\tilde{\nu}?]$	
$\tilde{\nu}$	$m(\tilde{\nu}_{\tau})+m(\tilde{\nu}_{\ell}) \gtrsim m_{\tau}$		$[e^+e^- \to \tilde{\nu}\tilde{\nu}]$	
$\tilde{q}$	$\tilde{m} \gtrsim 15$ GeV??		$[e^+e^- \to \tilde{q}\tilde{q}]$	~40 GeV
			$\bar{p}p, pp \to \tilde{q}\tilde{\gamma}X$	~75 GeV
$\tilde{g}$	$\tilde{m} > 2$-5 GeV	~25 GeV	$[\bar{p}p, pp \to \tilde{g}\tilde{g}X]$	~75 GeV
				[50-150 GeV if $\tilde{g} \to g\tilde{\gamma}$ large]
$\tilde{\chi}^0$	----	$m(\tilde{\gamma}_2)+m(\tilde{\gamma}_1) \lesssim 40$GeV		$m(\tilde{\gamma}_2)+m(\tilde{\gamma}_1) \lesssim 95$GeV
			$[e^+e^- \to \tilde{\gamma}_2\tilde{\gamma}_1]$	
$\tilde{\chi}^{\pm}$	$\tilde{m} \gtrsim 16$ GeV???	20 GeV	$[e^+e^- \to \tilde{\chi}_i^+\tilde{\chi}_i^-]$	~40 GeV
		$m(\tilde{w})+m(\tilde{\chi}^0) < 50$ GeV	$[W^{\pm} \to \tilde{w}_i^{\pm}\ \tilde{\chi}^0]$	$m(\tilde{w})+m(\tilde{\chi}^0) \lesssim 70$GeV
			$(\tilde{\chi}^0 = \tilde{\gamma}_1, \tilde{\gamma}_2, \tilde{h}^0)$	

Photino (and Other Supersymmetric Particle) Interactions

The detection of supersymmetric partners is always based on either (a) detecting the interaction of a light partner at the end of a decay chain (usually thought of as a photino) or, (b) missing $\vec{p}_T$ because such a particle escapes a detector, or both. The missing $\vec{p}_T$ can be distinguished from that of a neutrino because no charged lepton accompanies the photino.

If the particle involved is a higgsino or a Goldstino or a scalar neutrino there are some differences (see below) but the main points are the same, so we will emphasize the photino case. For photinos and Goldstinos the interaction was calculated by Fayet.

The incoming photino interacts by exciting or exchanging a scalar quark. The quark is in a hadron. The γ-hadron cross section is then, from the standard parton model formalism,

$$\sigma = \sum_q \int dx\, q(x)\, \hat{\sigma}(\hat{s})$$

where $\hat{\sigma}$ is the constituent ($\tilde{\gamma}q \to qx$) cross section, s is the $\tilde{\gamma}$-hadron c.m. energy squared, and $\hat{s}$=xs. The $\tilde{q}$ can decay to $q\tilde{g}$ if $\tilde{m}_g^2 < xs$, which may hold for relevant x,s; otherwise $q\tilde{\gamma}$ is available. If $\tilde{g}q$ is allowed it will dominate because the coupling is stronger. For beam dump analysis the mass of $\tilde{g}$ is, by the hypothesis being tested, sufficiently small that the $\tilde{g}q$ mode dominates.

The constituent cross section is of the form

$$\hat{\sigma} \sim \frac{\alpha\, \alpha_s}{\tilde{m}_q^4}\, \hat{s}$$

assuming $\hat{s} \ll \tilde{m}_q^2$ (if $\hat{s} \sim \tilde{m}_q^2$ considerable revision is needed). Putting α_s=0.15, the full interaction cross section is

$$\sigma_{Int}^{\tilde{\gamma}} \approx 20\times 10^{-38}\, \tilde{E}_\gamma \left(\frac{m_W}{\tilde{m}_q}\right)^4 \tilde{F}(\tilde{m}_g^2, s)\ \mathrm{cm}^2 .$$

$\tilde{F}$ is the analogue of the standard structure function F_2,

$$\tilde{F} = \sum_q \int_{\tilde{m}_g^2/s}^{1} dx\, q(x)\, (1-\tilde{m}^2/xs)^2\, (1+\tilde{m}^2/8xs).$$

For $\tilde{m}^2 \to 0$, $\tilde{F}=F_2 \simeq 0.15$. The phase space corrections for $\tilde{m}^2 \neq 0$ can be significant for $\tilde{m} \gtrsim 1$ GeV. The result for $\sigma_{Int}^{\tilde{\gamma}}$ has been written in the usual form of a neutrino cross section; $\tilde{E}_\gamma$ is the photino lab energy, i.e. $s=2\tilde{E}_\gamma m_{proton}$. Note s is the squared energy associated with the photino-hadron system.

If $m_{\tilde{W}} \simeq m_{\tilde{q}}$ and $\tilde{m}_g^2 << xs$, $\sigma_{Int}^{\tilde{\gamma}}$ is a few times larger than a neutrino charged current cross section. Thus photinos will interact in beam dump detectors, but will not interact in collider detectors. The interaction in a beam dump detector gives as a signature an event with no hard charged lepton and very little missing energy or p_T, so it is a neutral current candidate, but one with different characteristics from a neutrino event (a y distribution characteristic of scalar currents, and less missing energy, p_T). The signature in a collider is the p_T carried away (with no associated hard $\ell^\pm$).

With the above remarks in mind one can understand the main signatures discussed below. Now let us consider the entries in Table II, explaining them one row at a time.

Charged Scalar Leptons (sleptons) $\tilde{\ell}^\pm$

The charged sleptons $\tilde{e},\tilde{\mu},\tilde{\tau}$ are straightforward to produce at e^+e^- colliders, up to the beam energy in mass. All are produced by direct channel $\gamma, Z°$ as for any charged scalar, with $\beta^3/4$ units of R for each of $\tilde{\ell}_L^\pm$, $\tilde{\ell}_R^\pm$, and a $\sin^2\theta$ production distribution.

If photinos are light, $\tilde{\ell}^\pm$ decay rapidly, with essentially 100% branching ratio,

$$\tilde{\ell}^\pm \rightarrow \ell^\pm \tilde{\gamma}.$$

The $\tilde{\gamma}$ interacts too weakly to be detected in a normal collider detector, so it escapes. Thus on average 1/2 of the energy is missing. The ℓ^+ and ℓ^- are not colinear. Consequently the signature is quite good. There is some background from $\tau^+\tau^-$ events but the branching ratios reduce the background, and $\gamma\gamma$ events can be cut away.

All present detectors have results.[8] Combining them one can exclude $\tilde{e}$ and $\tilde{\mu}$ up to about 16 GeV and $\tilde{\tau}$ to 15.3 GeV, including[9] a retroactive analysis of SPEAR data at the lower end. One can do a little better on $\tilde{e}$, by singly producing[10] $\tilde{e}+\tilde{\gamma}$, giving $m_{\tilde{e}} > 19.5$ GeV.[11]

If photinos are not light one can still say something. Long-lived or stable charged hadrons are excluded[12] for m<14 GeV, so if

$\tilde{\gamma}$ is too heavy for $\tilde{\ell}^{\pm}$ to decay quickly via $\ell\tilde{\gamma}$, that is the limit. If $\tilde{m}_{\gamma}$ is less than about 10 GeV, the leptons from $\tilde{\ell}^{\pm}\to\ell^{\pm}\tilde{\gamma}$ would probably still have been seen, though experimenters need to examine that question with their cuts in mind. For $\tilde{m}_{\gamma}\geq 10$ GeV, but $\tilde{m}_{\ell^{\pm}}>\tilde{m}_{\gamma}$, the leptons from $\tilde{\ell}^{\pm}$ decay probably would not have been detected, so the limits on $\tilde{m}_{\ell^{\pm}}$ do not hold -- i.e., it appears that a 12 GeV $\tilde{\ell}^{\pm}$ and a 10 GeV photino would contradict no accelerator data. [For the $\tilde{\mu}$ there is a restriction[13] of order 15 GeV from g-2, but this is close enough so that it could probably be made acceptable.] The decay lifetime is still quite short, with a width at least in the MeV range unless $\tilde{\ell}^{\pm}$ and $\tilde{\gamma}$ are nearly degenerate.

Even if $\tilde{\gamma}$ were heavier than $\tilde{\ell}$, it is very unlikely that there could be a charged stable supersymmetric partner, since cosmological arguments suggest there would then be too much energy density in charged matter. There are several possibilities under consideration: (a) It could be that sneutrinos were the lightest stable partner[4]. Then $\tilde{\gamma}\to\tilde{\nu}\nu$ via one loop, and $\tilde{\ell}\to\ell\tilde{\nu}\nu$ with a lifetime on the weak scale. (b) If sneutrinos can get a vacuum expectation[14] value, then photinos will decay to normal particles via one loop, $\tilde{\gamma}\to\gamma\nu$. Then $\tilde{\ell}\to\ell\nu\gamma$. In such cases the final state signatures, including perhaps the decay of a long-lived charged object, are accessible at e^+e^- colliders and are generally somewhat different from standard signatures. Some relevant data is mentioned in the photino section.

Scalar Neutrinos (sneutrinos) $\tilde{\nu}$

The present limits on sneutrinos are not very strong. The best one can do so far[15] is to observe that there would be modifications to τ decay properties if the decay $\tau\to\ell\tilde{\nu}_{\ell}\tilde{\nu}_{\tau}$ could occur mediated by wino exchange. Depending on the wino mass, this gives limits which qualitatively imply $m(\tilde{\nu}_{\tau})+m(\tilde{\nu}_{\ell})\gtrsim m_{\tau}$ for $\ell=e,\mu$.

There are two ways to look for more massive $\tilde{\nu}$. (a) one can have $e^+e^-\to\tilde{\nu}\bar{\tilde{\nu}}$. This will occur if $m(\tilde{\nu})$ is not too large, with $\sigma<1$ pb. On the Z°, the branching ratio for $Z°\to\tilde{\nu}_x\bar{\tilde{\nu}}_x$ is 3% times phase space. The crucial question for detection is the signature, which requires a knowledge of the $\tilde{\nu}$ decay branching ratios. This has been studied by Barnett, Lackner, and Haber recently.[16]

One decay is via a wino, W, lepton loop, $\tilde{\nu}\to\nu\tilde{\gamma}$. Alternatively, one could have $\nu\to\nu\gamma\ell\,\ell$ with a number of contributions. If $\tilde{w},\tilde{\ell}$ are heavy, the conclusion of Ref. 16 is that the neutral mode $\tilde{\nu}\to\nu\tilde{\gamma}$ will dominate, but for certain ranges of parameters ($m(\tilde{w})$ or $m(\tilde{e})$)

not too large) the branching ratio for $\nu\tilde{\gamma}\ell'\bar{\ell}$ can be as large as 50%. If BR($\nu\tilde{\gamma}\ell'\bar{\ell}$) is large the signature is quite good, with only a very soft pair of leptons produced by one $\tilde{\nu}$ decay and only missing neutrals from the other. Under these conditions, unstable light $\tilde{\nu}$ could be found at present e^+e^- machines.

If such a signature were detected, it could be distinguished from other kinds of new physics by details of angular distributions, and by comparing e^+e^-, $\mu^+\mu^-$, μe modes -- here all are present, whereas for $\tilde{\gamma}_2\tilde{\gamma}_1$ final states (see below) one can have e^+e^- and $\mu^+\mu^-$ but not μe.

If the $\nu\tilde{\gamma}$ mode totally dominates, one can say very little here. Eventually, neutrino pair counting experiments could put a limit on the $\tilde{\nu}$ mass, and the absence of the branching ratio for $\nu\tilde{\gamma}\bar{\ell}\ell'$ constrains parameters.

(b) The second possibility is $W^{\pm} \to \tilde{\ell}^{\pm}\,\tilde{\nu}_\ell$ which can occur if it is kinematically allowed. The rate can be large, with

$$\frac{\Gamma(W\to\tilde{\ell}\tilde{\nu}_\ell)}{\Gamma(W\to\ell\nu)} = \frac{1}{2}\left\{[1-(m(\tilde{\ell})+m(\tilde{\nu}))^2/m_W^2]\,[1-(m(\tilde{\ell})-m(\tilde{\nu}))^2/m_W^2]\right\}^{3/2}$$

The signature is not too bad[16]. Assuming light photinos, presumably $\tilde{\ell}\to\ell\tilde{\gamma}$ and the photino escapes. If $\tilde{\nu}\to\nu\tilde{\gamma}$, only the ℓ can be detected, and it has an energy spectrum which depends on $m(\tilde{\ell})$, $m(\tilde{\nu})$. If $\tilde{\nu}\to\nu\tilde{\gamma}\bar{\ell}\ell'$ one has three soft charged leptons. The escaping particles give missing E_T,$\vec{p}_T$, and the single or three leptons are isolated, not in hadron jets and not accompanied by hadron jets.

The background from W decays and from QCD effects involving heavy quark semileptonic decays will be serious and needs to be studied.

Scalar quarks (squarks) $\tilde{q}$

Squarks can be looked for at both e^+e^- and hadron colliders. At e^+e^- machines the situation is somewhat similar to sleptons, but less good because (a) there is 1/9 or 4/9 the rate because of the smaller electric charge (in addition to the 1/4 for scalars), and (b) presumably $\tilde{q}\to q\tilde{\gamma}$ (with the same caveats for heavy $\tilde{\gamma}$) so one has non-colinear jets with missing energy (since the photinos escape). This is certainly detectable but much harder to see without careful study. Until this summer there have not existed any firm limits on $\tilde{q}$; now the JADE collaboration has given limits up to about 16 GeV. The details are presented in the talk of Min Chen at this school.

At hadron colliders one can have

$$\bar{p}p, pp \rightarrow \tilde{q}\tilde{\bar{q}}X$$

with reasonable cross sections, but combining rate and signature considerations perhaps the best way to search[17] is for $\bar{p}p, pp \rightarrow \tilde{q}\tilde{\gamma}X$. Then the signature is large missing p_T from the escaping $\tilde{\gamma}$, with a single quark jet at large p_T from $\tilde{q} \rightarrow q\tilde{\gamma}$. A full analysis in terms of useful variables has been done by Littenberg, Paige and collaborators, including background effects.

Photinos ($\tilde{\gamma}$)

There are no direct experimental constraints on photinos, nor is there any compelling argument about what mass they should have. Cosmological arguments analagous to the usual ones for neutrinos suggest[18] the mass should either be less than about 100 eV (actually, that the sum of all light particles masses, neutrinos + photinos + ..., should be less than about 100 eV) or above a GeV; the latter range has been studied quantitatively by Goldberg[7], who calculated a lower limit of about 2 GeV for scalar fermion masses $m(\tilde{f})$ below about 50 GeV and then increasing rapidly with $m(\tilde{f})$ to about 15 GeV at $m(\tilde{f}) \simeq m_W$. Theoretically, the photino starts out massless before the SUSY is broken, and can get mass directly or radiatively. In models radiative contributions to the mass are often related to gluino masses by $m(\tilde{\gamma}) \simeq k\alpha m(\tilde{g})/\alpha_s$ where k is a group factor of order unity. But in some models[19] the expectation is $m(\tilde{\gamma}) = m(\tilde{g})$ as all gauginos get a common mass. If photinos are heavy, many of the experimental tests of SUSY and ways to find SUSY particles get more difficult and subtle.

One direct test may be possible. Photinos can be produced in e^+e^- (or $q\bar{q}$) annihilation. It has been suggested[20] that the search analagous to neutrino counting is viable, and it may be attempted[21] in the near future. The rate is large enough to detect if $m(\tilde{e})$ is less than about 40 GeV. At future hadron colliders a gluon jet can be radiated from one of the quark lines to provide a recoil trigger.

If photinos decay to photon + anything, it is possible to search by using $e^+e^- \rightarrow \tilde{\gamma}\tilde{\gamma} \rightarrow \gamma\gamma X$. The mode $\tilde{\gamma} \rightarrow \gamma\tilde{G}$ is discussed in Ref. 20, 18, and $\tilde{\gamma} \rightarrow \gamma\tilde{\nu}$ in Ref. 14 and the slepton section above. A search with no candidate events based on 7.1 pb^{-1} at 34 GeV/c has been reported in Ref. 22, and further recent studies are reported in the talk of Min Chen.

Gluinos ($\tilde{g}$)

Gluinos are the partners of gluons and consequently they are a color octet of electrically neutral particles. When the SUSY is unbroken they are degenerate with the gluons, at zero mass. In models they acquire masses ranging from rather small ones below a GeV, to hundreds of GeV. Typical models give 15-100 GeV for $\tilde{m}_g$.

If photinos are lighter than gluinos, the gluinos will decay by $\tilde{q}$ exchange, $\tilde{g}\to q\bar{q}\tilde{\gamma}$, with a lifetime (for light $\tilde{\gamma}$)

$$\tau_{\tilde{g}} \approx 10^{-6} \left(\frac{m_\mu}{\tilde{m}_g}\right)^5 \left(\frac{\tilde{m}_q}{m_W}\right)^4 \text{ sec.}$$

This decay mode will dominate. An important two body mode, $\tilde{g}\to g\tilde{\gamma}$ is also present[23], with a branching ratio

$$\sum 3\alpha_s(\tilde{m}_R^2-\tilde{m}_L^2)^2/4\pi(\tilde{m}_R^4+\tilde{m}_L^4),$$

where $\tilde{m}_R,\tilde{m}_L$ are the masses of the partners of left-handed and right-handed scalar quark partners, and $\sum$ represents a sum over all families. This number probably ranges from 0.01 to 0.15, depending on the structure of the mass matrix. If it is not too small it can be important for detecting $\tilde{g}$ at colliders.

Gluinos, being colored, are produced by coupling to gluons in hadron collisions. The cross section for pair-producing gluinos is about an order of magnitude larger than that for quarks of the same mass.

If gluinos were very light they would be produced with mb cross sections, and then travel an observable distance, having been shielded by gluons to give a new, long-lived neutral (or charged) hadron. It is unlikely[5,24] such states would have escaped detection. The new hadrons are essentially like neutrons, perhaps with a somewhat smaller interaction cross section, and perhaps heavier, but hard to miss.

If $\tilde{m}_g \gtrsim 1$ GeV, one can do a perturbative QCD calculation[5] for the production cross section of $\tilde{g}$, and obtain σ as a function of $\tilde{m}_g$.

If gluinos are not detected, one assumes it is because the mass is large enough so that the production cross section is too small. [The mass of the hadron -- e.g. gluinoball -- contains some constituent mass, presumably of order a GeV, as well as the current algebra gluino mass.] Such arguments exclude $\tilde{m}_g \lesssim 1$ GeV.

For 1 GeV $\lesssim \tilde{m}_q \lesssim$ 6-8 GeV, beam dump experiments with $E_{beam} \lesssim$ 1 TeV are relevant. Gluino pairs are produced in the dump. The $\tilde{g}$ decay in a short distance. The photinos travel The photinos travel downstream until they interact in the beam dump detectors. In all cases, there are extra candidates for neutral current events in the detector because no charged lepton appears. [If such candidates appear, the y, E_{vis} distributions are different from those for neutrinos.] The absence of such events allows an experiment to set limits on the gluino mass; the limits depend on $\tilde{m}_q$ since the photino interaction probability varies as $\tilde{m}_q^{-4}$. The FNAL beam dump experiment[26] has the strongest limits, and the CHARM experiment[27] has similar limits, slightly less stringent since their detector was further from the dump.

If photinos are heavy, the gluinos do not decay, and there is no limit beyond the absence of stable new hadrons, under the assumptions discussed in the introduction.

At hadron colliders more massive gluino pairs can be produced. The signature is considerable missing $\vec{p}_T$ because of the escaped $\tilde{\gamma}$'s. An analysis has been done by Paige, Littenberg and collaborators[17] to make this quantitative, and to see the effect of detector efficiency. Basically they define variables such as $x_E = -\vec{p}_T \cdot \vec{p}_T\,'/|\vec{p}_T|^2$ where $\vec{p}_T$, $\vec{p}_T\,'$ are simply the transverse momenta of the particles in opposite hemispheres, whether clear jets are visible or not. If normal QCD jets were formed, one tends to find $\vec{p}_T \simeq -\vec{p}_T$, so $x_E \simeq 1$. For gluinos there is missing $\vec{p}_T$ so $|\vec{p}_T| \neq |\vec{p}_T\,'|$ and the clumps are at an angle, so x_E is less than unity, and almost flat down to $x_E \sim 0$. Then a cut at $x_E = 1/2$ eliminates almost all QCD background, but only about 2/3 of the signal. QCD background with $x_E < 1$ often arises from events with a heavier quark and a semileptonic decay, so a cut to eliminate events with a hard charged lepton further enhances the $\tilde{g}$ sample. Monte Carlo studies based on these procedures suggest that the SPS collider can produce and detect gluinos for $\tilde{m}_g \lesssim 25$ GeV, and FNAL can reach $\tilde{m}_g \lesssim 90$ GeV. Very crudely, future high energy hadron colliders with luminosities of $10^{30}/cm^3$sec can reach $x \simeq 2\tilde{m}_g/\sqrt{s} \lesssim 0.1$ and those with luminosities of $10^{33}/cm^2$sec can reach $x \lesssim 1/3$.

If the mode $\tilde{g}\to g\tilde{\gamma}$ is large enough, it may allow a search for individual events with no hard charged lepton and with large missing p_T, rather than requiring a statistical analysis as above. Then a signal could be detected at larger mass values, while no signal constrains $\sigma\times BR(\tilde{g}\to g\tilde{\gamma})$.

Although gluinos, being colored, but having no electric charge or weak isospin, are most easily produced by hadrons, they occur[28] with a reasonable signature in $Z°$ decay at one loop. The branching ratio depends on m_t, $m_{\tilde{q}}$, and $m_{\tilde{g}}$. For typical values one gets $BR\lesssim 10^{-5}$, which is detectable though not easily. Given the expected chronological order of new machines, it is not likely that gluinos will be discovered this way, but if they are found with $m_{\tilde{g}}<40$ GeV this is a nice check on the theory, and the branching ratio is sensitive to the scalar quark spectrum. In particular, every theoretical model has a definite prediction for $BR(Z°\to\tilde{g}\tilde{g})$.

Neutral, Spin 1/2, Color Singlet, SUSY Partners (neutralinos, $\tilde{\chi}^0$)

As shown above in Table I, there will be several neutral spin 1/2 SUSY partners, associated with the $Z°$, the photon, and the two neutral Higgs bosons H_1^0, H_2^0. The photino should really be considered in the context of the full set of neutral spin 1/2 SUSY partners. The crucial consideration is that they all have the same charge and spin, and they are coupled by the (SU(2) breaking) mass generating mechanism, so the mass eigenstates are mixtures of the weak eigenstates. They have been discussed in Ref. 29, and studied extensively in Ref. 30, 31.

The weak eigenstates are denoted $\tilde{Z}°,\tilde{\gamma},\tilde{H}_1^0,\tilde{H}_2^0$. The mass eigenstates will have couplings that are mixed gauge-like couplings and Higgs-like couplings and are denoted by $\tilde{\chi}_i^0$ in general; in special cases the names $\tilde{z}°$ for the heaviest, $\tilde{\gamma}_1$ for the lightest state with mainly photon-like couplings, $\tilde{h}_1^0$ for the lightest state with mainly Higgs-like couplings, and $\tilde{\gamma}_2^0$ for the state of intermediate mass are useful. The couplings of $\tilde{\gamma}_2$ will be part Higgs-like and part gauge-like (it is usually dominantly Higgs-like since $\tilde{z}°$ and $\tilde{\gamma}_1$ are more gauge-like, so it might be more appropriate to use $\tilde{h}_2^0$ for its name, but it is its gauge-like part which leads to its phenomenological importance so we will stick to $\tilde{\gamma}_2$ as a useful mnemonic name). The neutralinos will be denoted collectively by $\tilde{\chi}^0$

when it is useful to refer to them as a group or without specifying precisely which state is meant.

A neutralino with essentially gauge couplings is called a zino, one with essentially Higgs-like couplings is called a neutral higgsino, one with strongly mixed couplings a ziggsino, and one with photon-like couplings a photino. The masses of these neutralinos are (as all SUSY partner masses) very model dependent. Typically the $\tilde{z}^\circ$ mass is greater than m_Z, the $\tilde{\gamma}_1$ mass is 0.5-10 GeV, the $\tilde{h}^0$ and $\tilde{\gamma}_2$ masses in the 20-70 GeV range.

There are two especially good ways to look for neutralinos, both possibly applicable in the near future.

(a) In essentially all SUSY models the decay

$$W^\pm \to \tilde{w}_1^\pm \tilde{\chi}^\circ$$

will occur, with a potentially detectable branching ratio. I will consider it in detail in the next section, since a discussion of the charged spin 1/2 states is also required.

(b) Specific to the neutralinos is one very hopeful place to look for SUSY partners[30],

$$e^+e^- \to \tilde{\gamma}_2 \tilde{\gamma}_1.$$

Some models have $\tilde{m}_{\gamma_1} + \tilde{m}_{\gamma_2} < 35$ GeV, so this could be detected in the near future at PETRA/PEP. Almost all models have $\tilde{m}_{\gamma_1}+\tilde{m}_{\gamma_2}<100$ GeV, so this will be effectively (though not rigorously) a crucial test of whether supersymmetry is relevant to physics on the weak scale.

The cross section for $\tilde{\gamma}_2\tilde{\gamma}_1$ production is of the form

$$\sigma \simeq \pi(\lambda\lambda')^2 \alpha^2 s/\tilde{m}_e^4$$

for $s \lesssim \tilde{m}_e^2$. The factors λ,λ' arise at the vertices because of the mixing of the weak eigenstates into the mass eigenstates -- typically $0.1 \lesssim \lambda\lambda' \lesssim 0.7$. In models, $\tilde{m}_e$ may be of order $m_{W/2}$ or less, up to or above m_W. Thus typically $\sigma \gtrsim 0.1$ pb. Since e^+e^- colliders are currently collecting more than $\sim$pb^{-1}/day, they can be producing more than an event a week of $\tilde{\gamma}_2\tilde{\gamma}_1$.

The signature is quite good.[30] $\tilde{\gamma}_2$ can be seen to decay to any $f\bar{f}$ pair plus $\tilde{\gamma}_1$, where $f\bar{f}$ is e^+e^-, $\mu^+\mu^-$, $\tau^+\tau^-$, $q\bar{q}$. Since the primary and secondary $\tilde{\gamma}_1$ both escape, at least 2/3 of the energy of the event is missing, and there is a lepton pair or two jets with large $\vec{p}_T$ imbalance. [There is no standard model background for such events. Similar events could arise from $\tilde{\nu}\tilde{\nu}$ production or production of a fourth generation massive neutrino, either of which would be as exciting to detect; one can distinguish what kind of new physics is occurring by various properties of the branching ratios, energy distributions, etc.]
Thus,

$$e^+e^- \to \tilde{\gamma}_2\tilde{\gamma}_1$$

provides one of the most promising ways to produce, detect, and study SUSY partners, either soon or at the future colliders. At high luminosity hadron colliders the same analysis holds and even larger $\hat{s}$ can be reached for $q\bar{q} \to \tilde{\gamma}_2\tilde{\gamma}_1$, so one will eventually have essentially a definitive test of any supersymmetric approach which produces particles on the weak scale.

<u>Charged, Spin 1/2, Color Singlet SUSY Partners (charginos, $\tilde{\chi}^{\pm}$)</u>

The charged, spin 1/2 SUSY partners of W^+ and of H_1^+ are a pair of weak eigenstates that are mixed by mass terms in general, forming two mass eigenstates -- called "charginos". The mass eigenstates will have couplings that are a mixture of gauge-like couplings and Higgs-like couplings. I will denote the charginos as $\tilde{\chi}^{\pm}_{1,2}$. When a chargino has essentially gauge couplings it is called a wino, when it has essentially Higgs couplings it is called a charged higgsino, and when it has strongly mixed couplings it is called a wiggsino. The mass eigenstates have been discussed in References 29-35.

The chargino mass matrix will have the form

$$\begin{matrix} \tilde{w}^+ \tilde{H}_1^+ \end{matrix} \begin{pmatrix} M & g_2 v_2 \\ g_2 v_1 & \mu \end{pmatrix} \begin{matrix} \tilde{w}^- \\ \tilde{H}_2^- \end{matrix}$$

where $v_i = \langle 0|H_i^0|0\rangle$, $m_W = g_2\sqrt{(v_1^2+v_2^2)/2}$. The matrix is labeled in terms of two-component weak eigenstates. It may be that $v_1 \neq v_2$. M is a gaugino mass term and μ is from a Higgs sector coupling -- both are unknown and must originate in SUSY breaking physics. The eigenstates of the mass matrix are $\tilde{\chi}_1, \tilde{\chi}_2$. One of the mass eigenvalues can be smaller than m_W, and it will be smaller[32] if M=0 or if μ=0 [then the determinant of the mass matrix is $g_2^2 v_1 v_2$ in magnitude, and by comparison with m_W one can see one eigenvale is smaller], but it could easily happen that both eigenvalues are greater than M_W. In a number of present models, one eigenvalue is typically $M_W/3$ or $M_W/2$, as has been emphasized in Ref. 32, 33.

If the above mass matrix is mainly off-diagonal (M, μ small), then the eigenstates are Dirac particles

$$\tilde{\chi}^{\pm}_{1,2} = \begin{pmatrix} \tilde{W}^+ \\ \tilde{H}^{-*} \end{pmatrix}, \quad \begin{pmatrix} \tilde{H}^+ \\ \tilde{W}^{-*} \end{pmatrix}$$

with mixed couplings, i.e. wiggsinos. If a diagonal term is large compared to m_W, then the eigenstates are mainly wino or higgsino,

$$\tilde{\chi}^{\pm}_{1,2} = \begin{pmatrix} \tilde{W}^+ \\ \tilde{W}^{-*} \end{pmatrix}, \quad \begin{pmatrix} \tilde{H}^+ \\ \tilde{H}^{-*} \end{pmatrix} \quad ,$$

with gauge-like or Higgs-like couplings.

With these explanations, one can examine the production and decay of the charginos. One should think of the above Dirac eigenstates in the same way as one views an electron as

$$e^- = \begin{pmatrix} e^-_L \\ e^{+*}_R \end{pmatrix},$$

with e_L, e_R as degenerate in mass but different in weak interaction properties (SU(2) doublet for e_L, singlet for e_R). We speak of separate production of e_L, e_R by the weak interactions, while the electromagnetic interactions do not distinguish them. Thus above one can have separate production of upper or lower pieces of $\tilde{\chi}^{\pm}_i$.

The two most interesting ways to produce charginos soon are

(a) $e^+e^- \to \tilde{\chi}^+\tilde{\chi}^-$ (Ref. 29, 30, 31, 33)

(b) $\bar{p}p$, $pp \to W^{\pm}X$, $W^{\pm} \to \tilde{\chi}^{\pm}\tilde{\chi}^{\circ}$ (Ref. 30-33)

For (a), the cross section has the unit of R for production of a charged fermion via a photon, plus the Z° and $\tilde{e}$ contributions. The coupling to a Z° is large whether $\tilde{\chi}$ is wino or higgsino, and the $\tilde{e}$ contribution could be significant for lighter $\tilde{e}$.

The signatures[30,31,33] can be seen from the possible $\tilde{\chi}$ decays. One can have $\tilde{\chi}^{\pm} \rightarrow \tilde{\chi}^{0}\ell^{\pm}\nu$ via a $W^{\pm}$, $\tilde{\chi}^{\pm} \rightarrow \nu_{\ell}\ell^{\pm}\tilde{\chi}^{0}$ via a scalar lepton, $\tilde{\chi}^{\pm} \rightarrow \tilde{\chi}^{0}q'\bar{q}$ via a $W^{\pm}$, or $\tilde{\chi}^{\pm} \rightarrow \bar{q}q'\tilde{\chi}^{0}$ via a scalar quark. The relative rates of these contributions depend on the ratio of vacuum expectation values v_1/v_2, on $\tilde{m}_{\ell}$, $\tilde{m}_q$, $\tilde{m}_g$, $\tilde{m}_{\gamma}$ etc. All modes could be comparable in size.

Since $\tilde{\chi}^{\pm} \rightarrow \nu_{\ell}\ell^{\pm}\tilde{\gamma}$ looks like the decay of a heavy lepton if $\tilde{\gamma}$ is light, people have tended to say that $\tilde{\chi}$ is excluded for $\tilde{m}_W < 18$ GeV by PETRA and PEP data. However, because of the changes if $\tilde{\gamma}$ is massive, and various details of the SUSY couplings, published data does not provide firm limits on $\tilde{w}^{\pm}$ masses, and they still could be found[30] below 18 GeV. The essential point is that often enough energy is carried off undetected in these decays so that the events in question would not have passed typical experimental cuts. The charged lepton spectrum becomes much softer than for a typical heavy lepton decay. Recently MARK J has searched allowing for heavy photinos, and has given a limit of about 18 GeV for the case when the leptonic branching ratios of $\tilde{\chi}^{\pm}$ are not less than those of a sequential heavy lepton, and $\tilde{m}_{\gamma} \leq 4$ GeV (see the talk of Min Chen for details).

At higher $\sqrt{s}$ one can use the forward-backward asymmetry and the rate on the Z° to distinguish[29,30,31] among winos, wiggsinos, and higgsinos. For a heavy lepton $g_V^2+g_A^2=0.25$, for a wino for a $g_V^2+g_A^2=2.43$, higgsino $g_V^2+g_A^2=0.31$, and for a wiggsino $g_V^2+g_A^2=1.37$. Note the large couplings and the corresponding large effects expected for R. Intermediate mixing is possible, of course. The asymmetry[36] is different for a normal lepton, a pure wino or pure higgsino, a wiggsino with $\tilde{w}_L=\tilde{w}^-$, or a wiggsino with $\tilde{w}_L=\tilde{H}^-$. One can see that SLC, LEP will be very well suited to untangle many of the properties of SUSY partners, both charginos and neutralinos.

Next consider detecting SUSY partners via W decay,

$$W^{\pm} \rightarrow \tilde{\chi}^{\pm}\tilde{\chi}^{0}$$

Calculating the branching ratio is somewhat subtle. The interaction Lagrangian contains terms ($m_{\tilde{\chi}_2} < m_{\tilde{\chi}_1}$, $v_2 < v_1$)

$$\mathcal{L} \sim W^+_\mu \{ e\, \tilde{\gamma}_1\, \gamma^\mu R\, \tilde{\chi}_2^- - \frac{g_2}{\sqrt{2}} \frac{v_2}{v}\, \tilde{\gamma}_2 \gamma^\mu L\, \tilde{\chi}_2^-$$

$$+ \frac{g_2}{\sqrt{2}} \frac{v_1}{v}\, \tilde{h}^\circ\, \gamma^\mu L\, \tilde{\chi}_2^- + \frac{g_2}{\sqrt{2}}\, \bar{e}\, \gamma^\mu L\, \nu + \ldots \} .$$

Thus one expects three different decays if all are allowed by phase space,

$$W^\pm \to \tilde{\chi}_2 + \tilde{\gamma}_1, \tilde{\gamma}_2, \tilde{h}^\circ .$$

If all final masses were small, the sum of partial branching rates if

$$\Gamma(W^\pm \to \tilde{\chi}^\pm \chi^\circ)/\Gamma(W \to e\nu) \leqslant \frac{e^2 + \frac{g_2^2}{2}\left(\frac{v_1^2 + v_2^2}{v^2}\right)}{g_2^2/2} = 1 + 2\sin^2\theta_W \simeq 1.45 .$$

However, one expects a significant reduction by phase space. For W decay into final state masses m,μ the phase space factor is

$$[1-(m^2+\mu^2)/2M_W^2 - (m^2-\mu^2)^2/2M_W^4]\ [(1-(m+\mu)^2/M_W^2)\ (1-(m-\mu)^2/M_W^2)]^{1/2} .$$

The signatures[30,31,33] are quite different, and depend on which of the $\tilde{\chi}$ decay modes dominate. For $\tilde{\chi}^\pm$ $\tilde{\gamma}_1$ the $\tilde{\gamma}_1$ will escape, giving the large missing $\vec{p}_T$ siganture. $\tilde{\chi}^\pm$ may decay to $\ell^\pm \nu \tilde{\gamma}_1$, $\bar{q}q'\tilde{\gamma}'$, or $\bar{q}q'\tilde{g}$. The $\bar{q}q'\tilde{\gamma}$ events have been discussed in detail as a good signature (the "Zen" events) in Ref. 31. For the other modes both states decay, so one has $\tilde{\chi}^\pm$ decays on one side and $\tilde{\gamma}_2$ or $\tilde{h}^\circ$ decays on the other. There are various cases, all with some missing $\vec{p}_T$, and with some charged leptons or quark jets.

Because of the dilution of the signal by branching ratio effects (several decays with various signatures), and because of background effects [these have been studied in some detail by Littenberg and Paige[17] -- they find that with strong cuts, which reduce the signal, one can get ordinary standard model background that mimics the signal to the level where signal-to-noise is somewhat better than unity] it will not be easy to find a signal for $W \to \tilde{\chi}\tilde{\chi}$ unless we are very lucky about which modes occur and which branching ratios dominate. Nevertheless, it is such a good opportunity that experimenters should search mode-by-mode for a possible signal.

Acknowledgements

I am grateful to S. Raby, H. Haber, J. Hagelin, J. Leveille, and especially J.-M. Frere, for stimulating and instructive conversations and collaborations. I would like to thank Prof. Zichichi for providing great hospitality and a stimulating physics atmosphere at the school. This research was supported in part by the U.S.D.o.E.

References

1. For recent reviews of the theory, history, motivation, and models, see P. Fayet, Proceedings of the 21st International High Energy Physics Conference, Paris 1982, ed. P. Petiau and M. Porneuf, and Ref. 2. For earlier reviews of experimental aspects, see I. Hinchliffe ed L. Littenberg, Ref. 3, and G.L. Kane, Invited talk at the Fourth Workshop on Grand Unification, Phila., PA, April, 1983.
2. See J. Polchinski, Invited talk at the Fourth Workshop on Grand Unification, Phila., PA, April, 1983; C.A. Savoy, talks at Rencontre de Moriond, March 1983, SPhT/83/73 Saclay preprint.
3. Proceedings of the 1982 DPF study on Future Facilities, Snowmass, ed. R. Donaldson, R. Gustafson, and F. Paige.
4. J. Hagelin, G.L. Kane, and S. Raby, in preparation.
5. G.L. Kane and J.P. Leveille, Phys. Lett. 112B, 227 (1982).
6. H.E. Haber and G.L. Kane, in preparation.
7. H. Goldberg, Northeastern preprint NUB-2592 (1983).
8. A useful review is K.H. Lau, SLAC-PUB-3001, Oct. 1982. See also H. Behrend et al., Phys. Lett. 114B, 287 (1982) CELLO Collaboration; W. Bartel et al., Phys. Lett. 114B, 211 (1982) JADE Collaboration; D. Barber et al., Phys. Rev. Lett. 45, 1904 (1981) MARK J Collaboration; R. Brandelik et al., Phys. Lett. 117B, 365 (1982) TASSO Collaboration; C.A. Blocker et al., Phys. Rev. Lett. 49, 517 (1982) MARK II; D. Ritson, XXI International Conference on High Energy Physics, Paris, July 1982, MAC. See also the talks of Min Chen in these proceedings.
9. F.B. Heile et al., Nucl. Phys. B138 (1978) 189.
10. M.K. Gaillard, L. Hall, and I. Hinchliffe, Phys. Lett. 116B 279 (1982).
11. See Hinchliffe and Littenberg, Ref. 3, p. 248, for the Mark II result.
12. B. Adeva et al., MARK J Collaboration, Phys. Rev. Lett. 48 967 (1982).

13. P. Fayet, in "Unification of the Fundamental Particle Interactions (1980) ed, by. S. Ferrara, J. Ellis, and P. Van Nieuwenhuizen, p.587. See also, J.A. Grifols and A. Mendez, Barcelona preprint UABFT-84 (revised), and R. Barbieri and L. Maiani, Phys. Lett. 117B (1982) 203.
14. L. Hall and M. Suzuki, LBL preprint. See also J.-M. Frere and G.L. Kane, to be published. This question has also been discussed by Aulakh and Mohapatra, and by Polchinski and Wise.
15. G.L. Kane and W. Rolnick, Univ. of Michigan preprint UM-TH 83-14.
16. R.M. Barnett, K. Lackner, and H.E. Haber, SLAC-PUB-3066 (1983) and SLAC-PUB-3105 (1983).
17. L. Littenberg and F. Paige and collaborators, private communication and Ref. 3.
18. N. Cabibbo, G. Farrar, and L. Maiani, Phys. Lett. 105B (1981) 155.
19. E. Cremmer, P. Fayet, and L. Girardello, Phys. Lett. 122B (1983) 41.
20. P. Fayet, Phys. Lett. 117B (1982) 460; J. Ellis and J. Hagelin, Phys. Lett. 122B (1982) 303.
21. D. Burke, private communictaion.
22. H.J. Behrend et al., CELLO collaboration, Phys. Lett. 123B, 127 (1983).
23. H.E. Haber and G.L. Kane, Univ. of Michigan preprint UM-TH 83-18, July 1983.
24. G. Farrar and P. Fayet, Phys. Lett. 76B (1978) 575; 79B (1978) 442; P. Fayet, Phys. Lett. 78B (1978) 417; See also P.R. Harrison and C.H. Llewellyn Smith, Oxford preprint, 1982.
25. P. Fayet, Phys. Lett. 86B (1979) 272.
26. R.C. Ball et al., Univ. of Michigan preprint UM HE 82-21, Submitted to the XXI International Conference on High Energy Physics, Paris, July 1982.
27. F. Bergsma et al., Phys. Lett. 121B, 429 (1983).
28. G.L. Kane and W. Rolnick, Nucl. Phys. B217 (1983) 117.
29. J. Ellis and Graham G. Ross, Phys. Lett. 117B 397 (1982).
30. J.-M. Frere and G.L. Kane, Univ. of Michigan preprint UM TH 83-2, Feb. 1983.
31. J. Ellis, John S. Hagelin, D.V. Nanopoulos, and M. Srednicki, SLAC-PUB-3094, April 1983; J. Ellis, J.-M. Frere, John S. Hagelin, G.L. Kane, and S.T. Petcov, SLAC-PUB-3152, July 1983.
32. S. Weinberg, Phys. Rev. Lett 50 (1983) 387.
33. R. Arnowitt, A.H. Chamseddine, and Pran Nath, Phys. Rev. Lett. 50 232 (1983).
34. P. Fayet, Ecole Normale Superieure preprint LPTENS 83/16.
35. Luis Alvarez-Gaume, J. Polchinski, and Mark B. Wise, Harvard preprint HUTP-83/A063.

DISCUSSION

CHAIRMAN: G. KANE

Scientific Secretaries: J. Berdugo and M. Demarteau

- *BERNSTEIN* :

For the process $\tilde{g} \rightarrow q\bar{q}\tilde{\gamma}$ you showed us a plot for $d\sigma/dX_E$ versus X_E. The background in that plot comes in at about half the signal. Could you tell us what goes into that plot ?

- *KANE* :

The plot is due to Littenbirg, Paige and collaborators and was made by MC simulation and for the background standard QCD was used. If you make a cut at X_E=.6, for example, then you eliminate practically all the background and you have left about 1/3 of the signal, so you increased the signal to background ratio by orders of magnitude.

- *BERNSTEIN* :

How long would it take at Fermilab, for example, to see evidence for a gluino ?

- *KANE* :

If you are doing an experiment you just plot your data and see if there is a signal left after your cuts. If there is a signal left, you check by MC simulation if your background is reasonable and which process could reproduce the data. Up to a certain $\tilde{g}$ mass this will be done at the FNAL collider when it turns on; after a few months a signal could be seen if $\tilde{g}$ exists in the right mass region.

- *BERNSTEIN* :

What do you think is the most incisive and powerful test of SUSY at, say, the Fermilab or SSC ?

- *KANE* :

The best test for any collider would be to look for neutralinos where you see a pair of leptons in one hemisphere and nothing in the other. The Fermilab collider does not have enough luminosity to see such processes, so they should look at unbalanced p_T phenomena without charged leptons so that the unbalance cannot come from neutrinos; they are searching for $\tilde{g}$ or $\tilde{q}$.

- *BAGGER* :

This morning you didn't discuss gravitinos. Could one expect to see them, or is their coupling to matter too weak ?

- *KANE* :

Yes indeed, they couple very weakly and I don't know of any way to look for them experimentally. On the other hand, if goldstinos exist and become the spin 1/2 component of the spin 3/2 gravitinos then one could look at hadron colliders for the following process

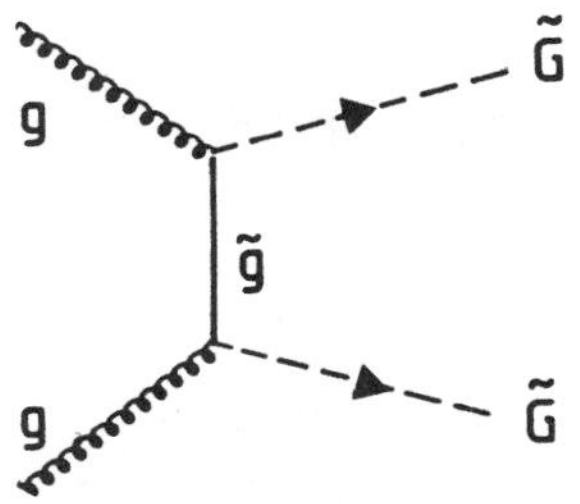

So then you have an event where the goldstinos escape the detector and you don't see anything in the central region. There is a whole set of cross-sections like this where things are missing at colliders.

If the scale of SUSY breaking is less than 1 TeV then the cross-sections for such processes become quite large, but the coupling of the goldstino goes inversely as the scale of the SUSY breaking, so for large scales of SUSY breaking these cross-sections become too small to be seen.

- *VAN DEN DOEL* :

Don't you destroy asymptotic freedom if you have scalar quarks?

- *KANE* :

No, in general not. In the minimal models you still have asymptotic freedom. In some models, however, extra gluinos are introduced to get rid of some symmetries and you can get non-asymptotically free theories. But in the models I am talking about you still have asymptotic freedom.

- *VAN DEN DOEL* :

Can you expect to find weird hadrons containing gluinos and/or squarks ?

- *KANE* :

If SUSY were not broken so badly you would be able to have, for example, baryons formed by two quarks and a squark because the color forces presumably commute with SUSY. If nature is indeed supersymmetric then you expect all the extra hadrons made from the standard hadrons by substituting squarks for quarks. But they will be very heavy.

- *VAN DEN DOEL* :

Are you going to look for those strange hadrons ?

- *KANE* :

As they are actually produced at colliders you won't see them in the form of hadrons but you'll see instead jet phenomena and missing p_T behaviour.

- *OGILVIE* :

Are there ways to be misled in a search for SUSY ? What is most likely to come along looking like a SUSY partner, but is not ? For example, we might find scalars carrying triplet color or fermions carrying octet color.

- *KANE* :

If new phenomena occur you have to find a pattern to find out what they are. You can't confirm SUSY by a single discovery.

You need a whole group of the minimal set of supersymmetric partners. It's the whole pattern that matters. Most likely other supersymmetric partners will show up in the same mass region, so to look for them is feasible. Alternative explanations can usually be tested but I have not had time to describe how, except for $\tilde{\gamma}_2$.

- *NEMESCHANSKY* :

You mentioned in your lecture that two identical Majorana fermions require p-wave annihilation. Can you explain this ?

- *KANE* :

A Majorana fermion is an eigenstate of C and you want to look at the scattering amplitude where it annihilates into a lighter pair of fermions. The annihilation must occur from a p-wave because in an s-wave the symmetry is wrong, since they are two identical fermions. It's just a matter of spin and statistics.

- *NEMESCHANSKY* :

What kind of effect does this have on cosmology ? Does this change the upper bound on the neutrino mass because p-wave annihilation is much slower than s-wave annihilation ?

- *KANE* :

For neutrinos you can have an s-wave annihilation as pointed out by Lee and Weinberg; the limit arises because otherwise you would have too much mass for the mass density of the universe. If you make it a p-wave annihilation you have a threshold factor. To compensate for this you must increase the mass of the neutrino because the cross-section is proportional to M^2. So, normally, you have to have neutrinos with masses heavier than about 2 GeV, and in the p-wave case they must be heavier.

- *RIBARICS* :

If we don't find the W-ino in the W decay, what would be our conclusion about SUSY ? I mean what kind of models would be excluded ?

- *KANE* :

Unfortunately you cannot draw any firm conclusion. The only restriction you get is that very specific models with a special kind of symmetry breaking can be excluded. No general statement however can be made. The masses and branching ratios are not sufficiently well determined.

- *SIMIĆ* :

Certain scalars in supersymmetric theories are very light so they might not have such short wavelengths. This however, would imply a long-range force. I would like to know whether those forces are long-ranged enough to have macroscopically measurable effects.

- *KANE* :

The present bounds on the masses already say that the particles that could give long-range forces are known to be too heavy.

- *ALTARELLI* :

I want to make a comment. In Rome we also considered the possibilities of observing W-inos (or Z-inos) in W decays. We found that the chances of observing them in this way are rather dim, even dimmer than you say. More promising seems to us the direct production of W-inos or Z-inos in association with a gluino leading to a signature consisting for example in one electron on one side and a very large transverse momentum hadron jet on the other side. Also, I decided that the branching ratio for a gluino decaying into a gluon and a photino was probably very small because of the cancellation between left and right scalars and also because of the partial cancellation of charges. Can you tell me how you evaluated this ratio Γ_2/Γ_3 and how much it can be ?

- *KANE* :

Both effects you mention do occur for $\tilde{g} \to g\tilde{\gamma}$. This expression for Γ_2/Γ_3 :

$$\Gamma_2/\Gamma_3 \simeq \sum_{\text{flavors}} \frac{\alpha_s}{4} \frac{\left(\tilde{m}_R^2 - \tilde{m}_L^2\right)^2}{m_R^4 + \tilde{m}_L^4}$$

is the full expression with some very minor approximations. If you put in values from some currently interesting models, you get a ratio on the order of 10^{-2} or more, which is in fact quite a big number because it allows you to identify a signal on an event-by-event basis rather than requiring a large statistical sample. Consequently it may be possible to more than compensate for the 10^{-2}.

- *OLEJNIK* :

What is the present experimental status of looking for light pseudo-Goldstone bosons proposed in some technicolor models and could their signals not be mixed up sometimes with that of supersymmetric particles ?

- *KANE* :

If the top quark were to be observed at the SPS collider with a mass of 25-35 GeV it would be possible to put much stronger constraints on the charged Higgs and technicolor models. If these particles exist, the t-quark would always decay semi-weakly into a charged Higgs which decays subsequently into a τ and ν_τ. The τ decays into other leptons which would be very soft so you would never be able to observe them as hard leptons. The moment you observe the standard decay $t \to b\ell\nu_\ell$ you have immediately excluded the existence of any charged Higgs particle up to the mass of M_t-M_b.

- *POHL* :

Do you see any way to derive an upper limit on the mass splitting between particles and sparticles beyond which SUSY gets unattractive theoretically ?

- *KANE* :

It could be that one reaches a stage where one says that only a SUSY field theory is a meaningful field theory because it can relate to gravity, improve the divergencies, and so forth.

But that could be a supersymmetry at the Planck mass and so doesn't give you any insight in the weak scale. That kind of theory need not be experimentally testable. However, if SUSY is to explain the origin of the weak scale, then it will produce other phenomena on the weak scale. If these phenomena are not observed, SUSY becomes uninteresting as for explaining the origin of the weak scale or the hierarchy problem.

- *VAN DEN DOEL* :

What is the supersymmetric partner of the magnetic monopole ?

- *KANE* :

I don't know.

- *BATTISTON* :

Supersymmetry is a very general theory and predicts many particles. Why are those particles always spin 0, 1/2 or 1 particles (apart from the gravitino) and not higher spin particles ?

- *KANE* :

When you insist on renormalisable field theories you don't allow spins greater than 2.

When you have N=2,4,8, supersymmetric theories you lose the direct connection with the Standard Model, because they do not commute anymore with $SU(2) \otimes U(1)$. So whatever the underlying theory, it is enough to consider N=1 SUSY, with one generator and thus one SUSY partner for each normal particle. To see exactly what spin particle is associated with a spin one particle, one must do a little technical analysis and write the superfields.

COMPOSITE W-BOSONS AND THEIR DYNAMICS

Harald Fritzsch

Sektion Physik der Universität München

Max-Planck-Institut für Physik und Astrophysik
- Werner Heisenberg Institut für Physik -
Munich (Fed. Rep. Germany)

Abstract Weak bosons are bound states. Approximately their properties are described within the standard SU(2) x U(1)-theory. Small deviations from predictions made using this theory for certain physical quantities as well as the discovery of new effects like unusual decays of W and Z might reveal soon the composite nature of the W- and Z- particle as well as of the leptons and quarks. The inverse size of the W-boson is expected to be in the range 0.1 ... 1 TeV.

+ Lecture given at the International School on Subnuclear Physics, Erice, Sicily (August 1983).
Supported by DFG- contract Fr 412/5-2

Recently evidence for the production of the weak intermediate bosons has been found in p-$\bar{p}$ collisions at high energies[1]. The observed properties of these bosons (masses, production- and decay characteristics) are in good agreement with the minimal theory of flavor dynamics based on the gauge group SU(2) x U(1). Nevertheless a number of questions remain to be answered, e.g.:

Is the SU(2) x U(1) gauge theory a microscopic theory of the electroweak interactions, or is it merely an effective theory, describing the low energy properties of a more fundamental theory?

Are the weak bosons basic gauge particles like the photon, or are they bound states of certain constituents?

Are leptons and quarks composite and are they made of the same constituents as the weak bosons?

Many authors have recently investigated the prospects for interpreting the observed leptons and quarks as bound states of yet smaller constituents[2]. Although at present there exists no indication whatsoever from experiment that leptons and quarks have a substructure on their own, one may strongly suspect that they do. It seems difficult to understand in any other way the rather complex structure of the observed spectrum of leptons and quarks. If they are composite objects, the question arises whether the weak bosons and possibly even the photon and the gluons are composite. Since photons and gluons are massless, we see no reason to interpret them as composite objects, although we do not exclude this possibility in a future unified theory including gravity. However the W- and Z- bosons are massive, and their mass might well be one of the manifestations of an internal structure.

The question, whether W-bosons are composite or not, is presumably related to the length scale at which the internal structure of leptons and quarks becomes apparent. If the latter arises at a length scale many orders of magnitude less than the present experimental limit of the order of 10^{-16} cm (e.g. a length scale close to the Planck length of the order of 10^{-33} cm), one may assume that the W-bosons are structureless objects and their dynamical behaviour is exactly described by the SU(2) x U(1) gauge theory. However if the radii of the leptons and quarks are related to the length scale given by the Fermi constant (300 $GeV^{-1} \sim 10^{-17}$cm), we find it very natural to assume that the W-bosons have a similar size. In that case no unification of the electromagnetic and weak interactions is achieved. The W-boson will be the ground state of a complicated spectrum of states, just like the ρ-meson is the ground state of the spectrum of infinitely many states in the $\bar{q}q(J^P = 1^-)$ channel. It is this possibility which we would like to address here.

During the progress of physics within the past hundred years it has happened twice that observed short range forces were recognized as indirect consequences of an underlying substructure of the objects considered. Thus the short-range molecular and van-der-Waals forces turned out to be indirect consequences of the substructure of atoms; they are remnants of the long range electromagnetic forces. Since 1970 something similar has happened to the short-range nuclear force, which has turned out to be a relict of the quark substructure of hadrons and the strong long range color forces between the quarks.

The only short range interaction left in physics which has not been traced back to a substructure and to a fundamental long range force between constituents is the weak interaction. Recently a number of authors has become interested in interpreting the weak force as some kind of "Van der Waals" remnant of an underlying lepton-quark substructure[3..8]. The lepton- quark constituents for which I will use the name "haplons"[x] are supposed to be bound together by very strong so-called hypercolor forces which are supposed to be confining forces, presumably described by a non- Abelean gauge theory (although other types of forces are not excluded). The short range character of the weak interaction arises since the leptons and quarks are hypercolor singlets, but have a finite size of the order of 1 TeV^{-1}.

[+] derived from Greek "haplos" (simple)

If the weak interaction turns out to be a remnant of the hypercolor force, a new interpretation of the relationship between the electromagnetic and weak interaction is required. The W- and Z-bosons cease to be fundamental gauge bosons, but acquire the less prestigious status of bound states of haplons. However the photon remains an elementary object (at least at the scale of the order of 10^{-17} cm, discussed here). As a whole, the SU(2) x U(1)-theory cannot be regarded anymore as a fundamental microscopic theory of the electroweak interactions, but at best can be interpreted as an effective theory, which is useful only at distances larger than the hypercolor confinement scale. It acquires a status comparable to the one of the σ-model in QCD, which correctly describes the chiral dynamics of π-mesons and nucleons at relatively low energies, but fails to be a reasonable description of the strong interaction at high energies.

However I would like to emphasize that at the present time no indication whatsoever comes from the experimental side that leptons, quarks and weak bosons may be bound states of yet smaller constituents. It may well be that the weak force will turn out in the future as a fundamental gauge force, as fundamental as the electromagnetic one and the color force. In fact, interpreting the weak forces as effective forces poses a number of problems which have not been solved in a satisfactory manner. First of all, the weak interactions violate parity, and they do that not in an uncontrolled way, but in a very simple one: only the lefthanded leptons and quarks take part in the charged current interactions. If we interpret the weak interactions as Van der Waals type interactions, the parity violation is a point of worry. How should one interpret the observed parity violation? Does it mean that the lefthanded fermions have a different internal structure than the righthanded ones? Or are we dealing with two or several different hypercolor confinement scales, for example one for the lefthanded fermions, and one for the righthanded fermions, such that the resulting effective theory is similar to the left - right symmetric gauge theory, based on the group $SU(2)_L \times SU(2)_R$?

Another point of concern is the fact that the weak interactions show a number of regularities, e.g. the universality of the weak couplings, which one would not a priori expect if the weak interaction is merely a hypercolor remnant. On the other hand it is well - known that the interaction of pions or ρ-mesons with hadrons shows a number of regularities which can be traced back to current algebra, combined with chiral symmetry or vector meson dominance. Despite the fact that both the ρ-mesons and the pions are quark - antiquark bound states for which one would not a priori expect that their interaction with other hadrons exhibits remarkable simple properties (e.g. the universality of the vector meson couplings), the latter arise as a consequence of the underlying current algebra, which is saturated rather well at low energies by the lowest lying pole (either the pion pole in the case of the divergence of the axial vector current, or the ρ- or A_1-pole in the case of the vector or axial vector current).

In the case of chiral SU(2) x SU(2) the pole dominance works very well - the predictions of current algebra and PCAC seem to be fulfilled within about 5 %. The universality of the weak interactions is observed to be valid within 1 % in the case of the couplings of the weak currents to electrons, myons, as well as u,d and s- quarks. Much weaker constraints exist for the heavy quarks. It remains to be seen whether the observed universality of the weak interactions will find an explanation along lines similar to the ones used in hadronic physics.

Here I shall concentrate on models in which the W-bosons consist of a haplon and an antihaplon. Since the observed weak interactions exhibit a symmetry $SU(2)_L$, we assume the existence of two haplons, denoted by the doublet $\binom{\alpha}{\beta}$, which carry hypercolor and electric charge. The electric charges are assumed to be: $Q(\alpha) = + 1/2$, $Q(\beta) = -1/2$ (see ref. (6)).

The spectral functions at energies much above the hypercolor confinement scale are supposed to be described by a continuum of haplon- antihaplon pairs. At low energies the weak amplitudes will be dominated by the lowest lying poles, which are identified with the W-particles. The latter form the triplet:

$$\begin{pmatrix} W^+ \\ W^3 \\ W^- \end{pmatrix} = \begin{pmatrix} \bar{\beta}\alpha \\ \frac{1}{\sqrt{2}}(\bar{\alpha}\alpha - \bar{\beta}\beta) \\ \bar{\alpha}\beta \end{pmatrix}$$

The experimental data on the neutral current interaction require a mixing between the photon and the W_3 boson (the neutral, isovector partner of W^+ and W^-), which in the standard SU(2) x U(1) scheme is caused by the spontaneous symmetry breaking. Within our approach this mixing is due to the $W_3 - \gamma$ transitions, generated dynamically like the ρ-γ transitions in QCD (for an early discussion, based on vector meson dominance, see ref. (9)). The magnitude of $\sin^2\theta_W$ is directly related to the strength of the γ-W_3 transition. The latter is determined by the electric charges of the W-constituents and by the W wave function near the origin. We suppose that in the absence of electromagnetism the weak interactions are mediated by the triplet (W^+, W^-, W^3), where $M(W^+) = M(W^-) = M(W^3) = 0\ (\Lambda_H)$.

After the introduction of the electromagnetic interaction the photon and the W^3- boson mix. We denote the strength of this mixing by a parameter λ, following ref. (6), which is related to g (W-fermion coupling constant) and the effective value of $\sin^2\theta_W$

$$\sin^2\theta_W = \frac{e}{g} \cdot \lambda$$

Furthermore one has:

$$M_W = g \cdot 123 \text{ GeV}$$
$$M_Z^2 = \frac{M_W^2}{1-\lambda^2}$$

The mixing parameter λ is determined by the decay constant F_W of the W-boson, which we define in analogy to the decay constants of the ρ_0-meson (F_ρ): $\langle o|j_\mu^{\ 3}|W^3\rangle = \varepsilon_\mu M_W F_W$. It is useful to express the decay constant in terms of the wave function at the origin:

$$|W^3\rangle = \frac{1}{\sqrt{2}}\frac{1}{\sqrt{n_h}}\sum_{j=1}^{n_h}(\bar{\alpha}_j\alpha_j - \bar{\beta}_j\beta_j)\ \phi(x)$$

($\phi(x)$: wave function in coordinate space, j: hypercolor index here we have assumed that the haplons carry hypercolor, but no ordinary color). The current matrix element can be written as

$$\langle o|j_\mu{}^3|W^3\rangle = \varepsilon_\mu\sqrt{n_h}\cdot\sqrt{2M_W}\cdot\phi(o) = \varepsilon_\mu\cdot M_W\ F_W$$

$$F_W = \sqrt{n_h}\cdot /2\sqrt{M_W}\ \phi(o)$$

$$\sin^2\theta_W = \frac{e^2}{g}\sqrt{n_h}\cdot\sqrt{2}/M_W{}^3\ \phi(o)$$

$$= e^2 / g\cdot F_W / M_W.$$

e.g. $\sin^2\theta_W$ is proportional to the coordinate space wave function of the W-boson at the origin. Taking for example G = 0.65 and M_W = 79 GeV, one obtains F_W = 123 GeV, a value which seems not unreasonable for a bound state of the size 10^{-16} cm.

In the SU(2) x U(1) gauge theory the SU(2) coupling constant g is related to e by the relation g = e/sin $_W$. In bound state models of the weak interactions discussed here this relation need not be true in general. However it has been emphasized recently[10)] that this relation is approximately fulfilled if the lowest lying W-pole dominates the weak spectral function at low energies. This leads to the relation

$$g = M_W / F_W = e/\sin\theta_W \approx 0.65$$

(we have used $\sin^2\theta_W = 0.22$).

It is interesting to note that many aspects of the bound state models can be derived from a local current algebra of the weak currents. We observe that the lefthanded leptons and quarks form doublets of the weak isospin. The weak isospin charges F_i^W (i= 1,2,3) obey the isospin charge algebra

$$[F_i^W, F_j^W] = i\,\varepsilon_{ijk}\,F_k^W$$

Let us assume that these charges can be constructed as integrals over local charge densities $F_{oi}^W(x)$, i.e.,

$$F_i^W(x^o) = \int F_{oi}^W(x)d^3x.$$

Furthermore we suppose that the charge densities obey at equal times the local current algebra

$$[F_{oi}^W(x), F_{oj}^W(y)]_{x^o=y^o} = i\varepsilon_{ijk}F_{ok}^W\delta^3(\vec{x}-\vec{y}).$$

The local algebra is trivially fulfilled in a model in which leptons and quarks are pointlike objects and the weak currents are simply bilinear in the lepton and quark fields. However, if leptons and quarks are extended objects, the situation changes. Currents, which are bilinear in the (composite) lepton and quark fields would not obey the local algebra, just like the currents, which are bilinear in nucleon fields, do not obey the local current algebra of QCD. The local algebra becomes a highly non-trivial constraint. It is fulfilled in the haplon models discussed above, in which the currents are bilinear in α and β.

We consider matrix elements of the weak currents between the various fermion fields. In order to do so, we shall assume that the higher families composed of μ, τ, ... etc. are dynamical excitations of the first family (ν_e, e^-, u, d), without specifying in detail the dynamical structure of these states.

Let us look at the form factors of the left-handed weak neutral current $F_{\mu 3}^L(x)$, i.e., the matrix elements of this current between different lefthanded lepton or quark states, e.g., $\langle e_L^- | F_{\mu 3}(0) | e_L^- \rangle$. Denoting these form factors by $F_e(t)$, $F_\mu(t)$, $F_\tau(t)$, etc., the weak isospin algebra requires a universal normalization at t = 0, i.e., $F_e = F_{\nu_e} = F_\mu = F_{\nu\mu} = \ldots = 1$. Assuming W dominance to be a reasonably good approximation, we may write for the dependence on the four-momentum transfer t,

$$F_f(t) = \frac{M_W^2}{f_W} \frac{f_W^{ff}}{M_W^2 - t},$$

where f denotes any one of the fermions e^-, ν_e, μ^-, ν_μ, etc. From $F_f(0) = 1$ we obtain the universality relation

$$f_W = f_W^{ff} \equiv g .$$

The neutral W and, because of the weak isospin algebra, also the charged W bosons couple universally to leptons and quarks, $f_W^{ff} \equiv g$. Thus the universality of the weak interactions follows from the W-dominance.

The saturation mechanism discussed here makes sense only if the constituents α, β of the W-bosons serve at the same time as constituents of the fermions. Thus both the weak bosons and the fermions should have comparable sizes.

A minimal scheme of haplons would involve only the W-constituents α and β. However it is easy to see that those fields are not enough for building up the quarks and leptons. (Here the situation is unlike QCD. The u and d- quarks as well as the corresponding antiquarks are sufficient to build up the (p, n) doublet and the (ρ^+, ρ^0, ρ^-)-triplet.) Thus further constituents are needed. For example, in some of the schemes discussed in ref. (6,7) new scalar fields of charges 1/2 and 1/6 are added: in this case the leptons and quarks consist of a fermion and a scalar. One may also add further spin 1/2 fields and interpret the leptons and quarks as composed of those objects.

If the weak bosons consist of a pair of constituents, a number of problems and questions arise:

a) Are the W- constituents scalars, vectors or spinors?

If they carry spin 1/2, the W-particles could be regarded (in a crude, nonrelativitic approximation) as s-wave bound states, like the ρ-mesons in QCD. If the W-constituents are scalar objects, the W-bosons would be p-wave states.

In the same nonrelativistic picture the parameter $\sin^2\theta_W$ is proportional to the wave function at the origin $\phi(0)$ (which, of course, vanishes in case of a p-wave). As we have seen, it is already difficult to obtain the observed value of $\sin^2\theta_W$ ($\sin^2\theta_W \approx 0.22$) in a model based on spin 1/2 haplons. In the case of scalar haplons $\sin^2\theta_W$ vanishes in the nonrelativistic limit; thus the observed value must be entirely due to relativistic corrections. We cannot exclude this possibility, although we find it difficult to accept. We conclude: The W- constituents are most likely spin 1/2 fermions. Scalar and spin 1 constituents are less likely, but not excluded.

b) <u>The W- fermion coupling constant</u>

The W- fermion coupling constant g is defined in the same way as in the SU(2) x U(1)-theory. Within the bound state picture it is related to F_W (W-decay constant) and M_W by:

$$g = F_W \;/\; M_W = e/\sin\theta_W \approx 0.65,$$

i.e. g is about twice as large as e. If the W-fermion vertex is a vertex among bound states like e.g. the $\rho\bar{N}N$-vertex in hadronic physics, there is no relationship between g and the electromagnetic coupling constant e. In strong interaction physics the ρ-nucleon coupling constant is of the order of five: $f_{\rho NN} \approx 5.6$. We conclude: If the W-vermion vertex is a bound state vertex, induced by the hypercolor interaction, the underlying dynamics must be quite different from the one in QCD. Of course, without knowledge about the dynamical details of the hypercolor interaction it is not possible to make precise statements about the strength of the W-fermion coupling. Being a bound state vertex, g should be of the order of 1. Whether it is slightly less than one (e.g. $g \approx 0.65$), or somewhat larger ($g \approx 5$), depends on dynamical details. Recently the meson-nucleon coupling constants have been calculated within the framework of QCD sum rules[11]. It remains to be seen whether similar methods can be applied to the W-fermion coupling constant.

c) Isoscalar Weak Bosons

If the weak bosons consist of a haplon- antihaplon pair, one expects the existence of an isoscalar partner W^o of the W^3 particle (composition $(\bar{\alpha}\alpha + \bar{\beta}\beta) / \sqrt{2}$). This particle would give an isoscalar contribution to the neutral current interaction which is not observed. If a W^o-particle exists, it must be heavier than about 250 GeV. It cannot be nearly degenerate in mass with the W-particle, i.e. a repetition of the ρ-ω-situation does not occur. In avoiding an isoscalar contribution to the neutral current (besides the one induced by electromagnetism), two possibilities are mentioned. The isoscalar singlet channel is distinct from the isotriplet channel by the possibility of hypergluon annihilation contributions. In QCD the gluonic annihilation channels in the isosinglet channel cause the η'-meson to be much heavier than the π^o-meson. Something similar could happen here in generating a large W^3-W^o splitting.

The second possibility works only if the W- constituents are scalar haplons[5]. Bose statistics requires the W^o or W^3 wave functions to be totally symmetric. Since the W-particles must be p-waves, the space wave function is antisymmetric. As a result the isospin wave function must also be antisymmetric, e.g. there is no low-lying W^o-state.

d) W-boson mass and the hypercolor energy scale

In the W-dominance approximation the magnitude of $\sin^2\theta_W$ is proportional to F_W / M_W:

$$\sin\theta_W = e \cdot \frac{F_W}{M_W} .$$

Furthermore the W-fermion coupling constant g is related to M_W and F_W:

$$g = M_W / F_W .$$

The smallness of g is therefore directly related to the relatively large value of F_W (about 123 GeV), the latter is even larger than the W-mass. This suggests that the W is lighter than what might be expected within a simple bound state models. It looks as if the W-mass is anomalously light, much like the pion mass in hadronic physics.

We emphasize that the smallness of the W-mass and the relatively small value of the W-fermion coupling constant are related to each other. If we can explain, why the W-mass is rather light, compared to the energy scale given by the W-decay constant F_W, the smallness of g will be understood simultaneously.

In a nonrelativistic approach one finds:

$$F_W/M_W = \left(\frac{2n_h}{M^3}\right)^{1/2} \cdot \phi(o)$$

($\phi(o)$: haplon wave function at the origin,
n_h: number of hypercolors). Thus there are two ways of obtaining a relatively large value of F_W/M_W:

1. $\phi(o)$ is large, i.e. the wave function is rather dense, compared to M^{-1}.
2. The number of hypercolors is large. If we suppose that the spatial structure of the W is identical to the spatial structure of the ρ-mesons, one finds

$$\frac{F_W \;/\; M_W}{F_\rho \;/\; M_\rho} = \sqrt{\frac{n_h}{3}} \;.$$

Thus for n_h of the order of 60 the desired order of magnitude for F_W/M_W is achieved.

In QCD the couplings of the hadronic vector mesons to the photon are not independent of the pointlike coupling of the photon to the quarks. They are related via duality: the vector meson peaks in the current spectral functions, when suitably averaged over, approximate the free quark spectral functions. Applying the same principle for W-bosons, one finds[12]:

$$\sin^2\theta_W \approx \frac{\alpha}{3\pi}\, n_h\, e_h^2 \,/\, M_W^2\, P(M^2_W)$$

(e_h: normalized haplon charge, $P(M^2_W)$: density of vector meson states, i.e. number of vector bosons per unit mass squared interval).

Taking for example $e^2_h = [\frac{1}{\sqrt{2}}(e_{\bar{\alpha}} - e_\alpha)]^2 = \frac{1}{2}(-\frac{1}{2}-\frac{1}{2})^2 = 1/2$, one finds for $n_h = 3$:

$$P(M^2_W) = 0.81\ \text{TeV}^{-2},$$

i.e. the large value of $\sin^2\theta_W$ or of F_W/M_W is made plausible,if there exists a large splitting of order 1 TeV between the W-boson and its first excited state. In other words: The natural energie scale of the hypercolor dynamics is of the order of 1 TeV; the W-mass is anomalously light (like the pion in QCD).

Also in the duality approach $\sin^2\theta_W$ is proportional to n_h, i.e. $P(M^2_W)$ could be significantly less than 0.8 TeV^{-2}, if n_h turns out to be very large. Obviously this question has immediate consequences for experimental physics. It makes a difference for experiments whether the natural scale of hypercolor dynamics is at 0.3 TeV or 2 TeV. At the present time nothing more precise can be said about the hypercolor energy scale.

The relatively large value of $\sin^2\theta_W$ is easily understood if the W-bosons are light in comparison to the typical hypercolor energy scale. In this case the photon and the W-boson are nearly degenerate; the mixing is naturally expected to be large, as in any quantum mechanical system with a small difference in energy levels.

e) Colored W-boson Constituents

Thus far we have treated the W-constituents as hypercolored objects, which in addition carry electric charge. The color (in the QCD sense) of the haplons inside the W-bosons was supposed to be zero. This is the case e.g. in the simplest versions of the models discussed in ref. (6,7). However it may well be that the W-constituents carry color besides hypercolor[6]. Of course, in this case the W-boson is a color singlet. Since the hypercolor force binds the haplons very densely, the W-bosons do not act as normal hadrons. The color of their constituents would only become apparent if the haplons are liberated. Nevertheless small changes are expected. For example, the ratio F_W / M_W is proportional to $\sqrt{n_h \cdot n_c} = \sqrt{n_h \cdot 3}$, i.e. $\sin^2\theta_W$ increases by a factor 3, and correspondingly for $n_h = 3$ the density function P is about 2.5 TeV^{-2}, i.e. between M_W and 1 TeV would lie about 2 excited states. Further changes apply to specific decays of the W-boson which will be discussed later.

Furthermore one expects the presence of color octet partners of the W-boson. These states would not be degenerate with the W-boson since the QCD force between the haplons inside the W shifts the W mass downward and the W-octet mass upward. In a crude nonrelativistic approximation one finds:

$$M(\text{color singlet } W) = M_o - \frac{4}{3} \cdot \alpha_s\, C$$
$$M(\text{color octet } W) = M_o + \frac{1}{6} \cdot \alpha_s\, C$$

where α_s is the QCD coupling constant at $E= \langle\frac{1}{R}\rangle$ ($\langle\frac{1}{R}\rangle$: expectation value of $\langle\frac{1}{R}\rangle$ for the W-state), and C is a mass parameter of the order of $\langle\frac{1}{R}\rangle$.

Taking $\alpha_s = 0.1$ and C = 0.3 TeV, one finds $1/6\ \alpha_s \cdot C =$ 5 GeV, i.e. ΔM= M(color octet W)- M(color singlet W) $\approx$ 50 GeV. Of course, this is a very crude estimate. But it shows that the typical mass range for the color octet partner of the W-boson is 120 ... 200 GeV.

Color octet partners of the weak bosons are interesting particles since they can easily be produced in $p\bar{p}$-collisions at high energies. They can decay either in a q-$\bar{q}$-pair, i.e. into two jets which presumably could easily be observed, or into a gluon and a color singlet W-boson. This process would lead to the emission of an highly energetic gluon jet and a W- or Z- boson. Both the gluon jet and the weak boson can have large transverse momenta, depending on the mass of the color octet W-boson.

f) The ρ-Parameter

The strength of the neutral current interaction can be described in terms of the so-called ρ-parameter, which in terms of the weak boson masses is given by:

$$\rho = \frac{M_W^2}{M_Z^2 \cos^2\theta_W}$$

In the SU(2) x U(1) gauge theory this parameter is equal to 1, apart from very small corrections[13] (typically of the order of 1 $^o/oo$ or less).

In bound state models for the weak bosons the electromagnetic interaction does not only give rise to the W^3-γ-mixing, but at the some time the charged and neutral W-bosons acquire different electromagnetic self energies. In a naive model (taking into account only the Coulomb attraction or repulsion of the haplons) one finds:

$$M_{W^\pm} = M_o + \frac{1}{4}\alpha \langle\frac{1}{R}\rangle = M_o + \Delta M$$

$$M_{W^3} = M_o - \frac{1}{4}\alpha \langle\frac{1}{R}\rangle = M_o - \Delta M$$

(The haplon charges are taken to be 1/2 and -1/2; $\langle\frac{1}{R}\rangle$ is the expectation value of $\frac{1}{R}$ for the W-state). One finds:

$$\rho = \frac{M_{W+}^2}{M_Z^2 \cos^2\theta_W} = \frac{M_{W+}^2}{M_{W3}^2} = \frac{(M_o - \Delta M)^2}{(M_o - \Delta M)^2} \cong 1 + \frac{4\Delta M}{M_W} .$$

(M_{W^3}: mass of W^3-boson before mixing with the photon). Taking $\langle 1/R\rangle$ to be 1 TeV, one finds ΔM = 0.9 GeV and ρ= 1.05. Such a value of ρ is not excluded by the experiments[14]. A general analysis carried out recently[15] shows that depending on the shape of the confining potential ρ varies between 1.01 and 1.05. We emphasize that ρ is always larger than one; this reflects the fact that $M(W^+) > M(W^3)$.

Thus a careful measurement of ρ or, alternatively, of the W-Z mass shift are important. For the W-Z mass shift one finds:

$$M(Z) - M(W) = [M(Z) - M(W)]_{SU(2) \times U(1)} - 2\Delta M.$$

This mass shift is slightly less than the one obtained within the standard SU(2) x U(1) theory. Using $\rho = 1.05$ as an upper limit allowed by experiment, i.e. $\Delta M < 1$ GeV, this deviation can be at most 2 GeV. If ρ turns out to be larger than 1.01, the composite nature of the W-boson must be taken very seriously.

g) Exotic Decays of W and Z

Many of the properties of the weak bosons are identical to the ones obtained within the standard SU(2) x U(1)-theory, due to the constraints, imposed by the algebra of the weak currents. However small deviations from the SU(2) x U(1) - results are supposed to be present, e.g. the deviation of ρ from one, discussed above. Especially one expects within a bound state model the occurrence of certain W- or Z- decays, which are strongly suppressed within the SU(2) x U(1) gauge theory. Examples of such decays are:

1. $Z \rightarrow \gamma\gamma\gamma$

This decay, which is analogous to e.g. $J/\psi \rightarrow \gamma\gamma\gamma$, is present, if the Z-particle consists of two electrically charged haplons, which can annihilate. One obtains[16)]

$$\Gamma(Z \rightarrow \gamma\gamma\gamma) \approx 100 \text{ KeV},$$

$$\text{i.e. } B(Z \rightarrow \gamma\gamma\gamma) \approx 0.3 \cdot 10^{-4}.$$

Such a decay can be observed e.g. at LEP. Within the standard SU(2) x U(1) gauge theory the decay $Z \rightarrow \gamma\gamma\gamma$ proceeds via fermion and gauge boson loops and is strongly suppressed ($B \sim 10^{-9}$)[17)].

Without knowing details about the dynamics of the constituents of the W-bosons, we cannot but speculate, how the spectrum of their bound states might look. One may suppose that this spectrum resembles at least in some of its properties the spectra observed in atomic, nuclear and hadronic physics. In particular there may exist spin zero states besides the spin one states (W-bosons) discussed above. If the W-constituents carry spin 1/2, these states would be the hyperfine partners of the W-bosons (analogous to the pions in hadronic physics, which are the hyperfine partners of the ρ-mesons). If the W-constituents have spin zero, those states are (in a non-relativistic picture) the s-wave analogs of the W-states, which must be p-wave states.

The dynamics of hypercolored W-constituents must differ in an essential way from the dynamics of QED or QCD: the effective couplings of the W-bosons to the fermions violate parity - only the lefthanded fermions couple to the weak bosons. The dynamical reason for this phenomenon is unclear; it is presumably related to the fact that the observed leptons and quarks are very light compared to their inverse sizes of order 1 TeV^{-1}. The lowest lying states of the haplon - composition $\bar{h}h$ ($h = \alpha, \beta$) are denoted as follows:

Spin 0:	X^+	X^0	X^-	U
$\bar{h}h$-content:	$(\bar{\beta}\alpha)$	$(\bar{\alpha}\alpha-\bar{\beta}\beta)/\sqrt{2})$	$\bar{\alpha}\beta$	$(\bar{\alpha}\alpha + \bar{\beta}\beta)/\sqrt{2}$
Spin 1:	W^+	Z	W^-	V

We discuss in particular the case of spin 1/2 haplons. The X- and W-states as well as the U- and V- states differ by their spin structure. Inside W,Z and V the haplon spins are aligned; inside X and U the spins are opposite to each other. In the absence of a bare mass term for α and β and of electromagnetism the X-particles are massless as a result of chiral symmetry. The U-particle acquires a mass due to the dynamical breaking of the axial U(1)-symmetry (this is analogous to the mass generation for the η'-meson in QCD). The V-particle is supposed to be heavier than the W/Z-particles, due to hypergluon annihilation effects as mentioned before. The charged X-particles must be heavier than about 15 GeV; otherwise they would have been observed in e^+e^--annihilation[18]. The charged X-particles can acquire a mass both via a violation of chiral symmetry by a bare mass term for α and β, and by electromagnetism (see ref. (19)).

The couplings of the fermions to the X-particles are probably highly suppressed (of order m_f / m_W), as a result of chiral symmetry and the corresponding PCAC relations analogous to the Goldberger-Treiman relation[19]. This would explain why the X-particle exchanges do not contribute to weak decay amplitudes (e.g. to the decay $\pi^+ \to e^+\nu_e$).

No such argument can be given for the neutral isoscalar boson U. This particle does not act as a Goldstone particle in the chiral limit $m_\alpha = m_\beta = 0$ (m_α, m_β: bare masses of the haplons). Its couplings to the fermions are expected to be finite in the chiral limit, hence presumably independent of the fermion masses. We shall assume that the U-fermion coupling constants are universal like the W-fermion coupling constants:

$$g_{U\bar{e}e} = g_{U\bar{\mu}\mu} = g_{U\bar{\nu}_e\nu_e} \cdots = g_{U\bar{u}u} = g_{U\bar{d}d} \cdots$$

Furthermore these couplings are supposed to be flavor diagonal, i.e. U-exchange will not contribute to decays like $K_L^o \to \mu^+\mu^-$.

The U-particle exchange can contribute to neutral current processes. In deep inelastic neutrino - hadron scattering it will cause a deviation in the y-distribution from the behaviour expected in the standard SU(2) x U(1)-theory. Contributions coming from U-exchange might be present on the level of 10 %. In case of the presence of a U-particle with a mass of 40 - 50 GeV this means that its coupling to the fermions must not be larger than about $\frac{1}{6}$ of the W-fermion coupling.

The U-particle can decay into two photons via the electromagnetic annihilation of its constituents. The decay amplitude depends on the haplon wave function at the origin. In the case of spin 1/2 haplons the latter is given in terms of F_U (see ref. (6,20)). One finds[20]

$$\Gamma(U \to \gamma\gamma) = \frac{\pi\alpha^2}{2 M_U} \cdot F_U^2$$

Taking for example $F_U = F_W \approx 100$ GeV and $M_U = 50$ GeV, one finds $\Gamma(U \rightarrow \gamma\gamma) \approx 17$ MeV.

The total width of the U-particle is given by:

$$\Gamma_U = \sum_f \Gamma_{U \rightarrow \bar{f}f} + \Gamma_{U \rightarrow \gamma\gamma}$$

(f: any lepton, quark).

In the range 20 GeV < M_U < 80 GeV the rate $\Sigma\Gamma_{U \rightarrow \bar{f}f}$ will depend on the mass of the t-quark (we assume the existence of three generations). For $M_U >> 2m_t$ one finds, taking into account the color quantum number:

$$\frac{\Gamma(U \rightarrow \bar{l}l)}{\sum_f \Gamma(U \rightarrow \bar{f}f)} = \frac{1}{24} \approx 4\ \% \qquad (= \frac{1}{21} \approx 5\ \%)$$

$$\frac{\Gamma(U \rightarrow \bar{q}q)}{\sum_f \Gamma(U \rightarrow \bar{f}f)} = \frac{1}{8} \approx 13\ \% \qquad (= \frac{1}{7} \approx 14\ \%)$$

In case $M_U < 2m_t$ one finds the ratios indicated in parantheses. It is useful to introduce the ratio $r = \sum_f \Gamma(U \rightarrow \bar{f}f) / \Gamma(U \rightarrow \text{all})$. Thus the leptonic branching ratio is given by

$$B(U \rightarrow \bar{l}l) = r/24 \approx 4r\ \% \ (r/21 \approx 5r\ \%)$$

($l = e, \mu, \tau;\ \nu_e, \nu_\mu, \nu_\tau$).

The ratio r is less than one, but may well be of the order of 1, in which case the leptonic branching ratio is of the order of 4 %.

If the X and U particles exist and have masses well below M_W, radiative transitions like $Z \rightarrow U(X^0) + \gamma$, $W^\pm \rightarrow X^\pm + \gamma$ will occur. In the case of spin 1/2 haplons these transitions will be magnetic transitions like $\rho \rightarrow \pi\gamma$. We estimate the radiative transitions by using a similar approach as used in the calculations of the radiative decays of vector mesons (see e.g. ref. (21)). One finds:

$$\Gamma(Z \to U\gamma) = \frac{16\alpha}{3} K^3 \mu^2$$

(K: photon momentum, : magnetic transition moment)

The magnetic moments of the haplons are described by $Q_i/2m_i$ (i:α, m_α, m_β denote the constituent haplon masses, for which we take m= $m_\alpha = m_\beta \sim M_U/2$. We obtain:

$$\Gamma(Z \to U\gamma) \approx \frac{4}{3} \alpha \left(\frac{1}{2m}\right)^2 K^3,$$

In the models discussed in ref. (6) the charges of the W/Z-constituents are $q_\alpha = + 1/2$ $q_\beta = -1/2$. In this case the transition $Z^o \to X^o\gamma$ is forbidden, likewise the transitions $W^\pm \to X^\pm\gamma$. In the following table the decay rates and leptonic branching ratios are given for various values of M_U[23].

Table 1.

M_U[GeV]	40	45	50	55	60
K [GeV]	39	37	34	32	29
$\Gamma(Z \to U\gamma)$[MeV]	360	243	154	106	66
$\Gamma(Z \to e^+e^-\gamma)$[MeV]	18	12.3	7.7	5.3	3.4
$\frac{B(Z \to e^+e^-\gamma)}{B(Z \to e^+e^-)}$[%]	20	13.6	8.4	5.9	3.7

Of course, these results are rather uncertain, since the decay rate for $Z \to U\gamma$ as well as the partial rates for $Z \to e^+e^-\gamma$, $Z \to \mu^+\mu$ depend strongly on the constituent haplon mass m; nothing is known about the dynamical rôle played by these masses. Furthermore the approach used by us based on nonrelativistic considerations is just as doubtful as its use in the calculation of decays like $\omega \to \pi\gamma$ or $\rho \to \pi\gamma$. Nevertheless it works approximately for those radiative decays and might work for the weak boson decays as well. Therefore the numerical results given in the table should not been taken too seriously. The rate for $Z \to U\gamma$ may well be larger or smaller as indicated in the table. Furthermore the rate for $U \to \gamma\gamma$ is uncertain as well within at least a factor three.

We add the following comments about the radiative decay of the Z-boson:

a) Both in the cases of spin 1/2 and spin 0 haplons the decay rate $\Gamma(Z \to U\gamma)$ is expected to be a sizable fraction of the total decay rate $Z \to$ all. Typically we expect $\Gamma(Z \to U\gamma) \sim 50 \ldots 500$ MeV.

b) The decay $Z \to X^o\gamma$ is suppressed in comparison to $Z \to U\gamma$. If the haplon charges are 1/2 and -1/2, the decay does not occur, likewise the decays $W^\pm \to X^\pm\gamma$.

c) Within the standard electroweak gauge theory one finds[22]:

$$\Gamma(Z \to e^+e^-) \approx 90 \text{ MeV}$$
$$\Gamma(W^- \to e^-\bar{\nu}_e) \approx 260 \text{ MeV}$$

Provided the rate for $Z \to U\gamma$ is of the order of hundred MeV as expected if $M_U \sim 40 \ldots 60$ GeV, the rate for $Z \to e^+e^-\gamma$ or $Z \to \mu^+\mu^-\gamma$ ($M(e^+e^-) = M(\mu^+\mu^-) = M_U$) is not small in comparison to $\Gamma(Z \to e^+e^-)$. The ratio $\Gamma(Z \to e^+e^-\gamma) / \Gamma(Z \to e^+e^-)$ may be of the order of 10 ...30%. The experimental data suggest that about 20 % of all decays $Z \to e^+e^-$ and $Z \to \mu^+\mu^-$ are accompanied by a photon. Furthermore the data are not inconsistent with the hypothesis that the lepton pairs originate from the decay of a spin 0 particle with a mass in the range 50 ... 60 GeV.

d) It is quite possible that the decay $U \to \gamma\gamma$ has a rate comparable to $U \to e^+e^-$. Therefore the decay $Z \to \gamma_1\gamma_2\gamma_3$ ($M(\gamma_2\gamma_3) = M_U$) should be present at the level of 1... 10%.

e) We expect $\Gamma(U \to l^+l^-) = \Gamma(U \to \bar{\nu}_l\nu_l)$ [$l = e,\mu,\tau$]. Thus the decay $Z \to \gamma$ + missing energy, ($[P_\mu(Z)-P_\mu(\gamma)]^2 = M_U^2$) should occur three times as often as the decay $Z \to \gamma\, e^+e^-$. This decay could be observed in the $\bar{p}p$-collider experiments. The decay $U \to \bar{q}q$($q = u,d,c,s,b,t$?) is expected to occur 18 (15, in case $U \to \bar{t}t$ does not occur) times more often than the decay $U \to e^+e^-$. Hence the decay $Z \to \gamma\,\bar{q}q$, the quarks manifesting themselves as two quark jets opposite to each other in the U rest frame, should be present.

f) The U-particle can be produced directly in hadronic or e^+e^--collisions. However the couplings $g(U\ e^+e^-)$ and $g\ (U\ \bar{q}\ q)$ are unknown, and the U-particle production cannot be safely estimated. Nevertheless one should look for the production of U and its subsequent decay into lepton, quark or photon pairs both in hadronic collisions and in e^+e^- annihilation.

The discovery of the radiative decay $Z \rightarrow U\gamma$ would be a major step towards establishing a substructure of the weak bosons and of the quarks and leptons[23]. Simultanously it would imply that the standard electroweak gauge theory is not a microscopic theory of the electroweak interactions but an effective field theory to describe in a good approximation the properties of the weak interactions at energies small compared to the inverse radius of the weak bosons, leptons and quarks.

REFERENCES

1. G. Arnison et al., Phys. Lett. 122 B (1983) 103, ibid. 126 B (1983) 398, ibid. 129 B (1983) 273. M. Banner et al., Phys. Lett. 122 B (1983) 476; P. Bagnaia et al., Phys. Lett. 129 B (1983), 130.
2. For a recent review see: H. Harari, Proceedings of the Banff Summerschool, Banff, Alberta (1981).
3. H. Harari and N. Seiberg, Phys. Lett. 98 B (1981) 269;
4. O. Greenberg and J. Šucher, Phys. Lett. 99 B (1981) 339.
5. L. Abbott and E. Farhi, Phys. Lett. 101 B (1981) 69; See also: R. Casalbuoni and R. Gatto, Phys. Lett. 103 B (1981)319.
6. H. Fritzsch and G. Mandelbaum, Phys. Lett. 102 B (1981) 319; and ibid. 109 B (1982) 224;
7. R. Barbieri, A. Masiero and R.N. Mohapatra, Phys. Lett. 105 B (1981) 369 B. Schrempp and F. Schrempp, preprint DESY 83-024 (1983)
8. H. Fritzsch, Proceedings of the Europhysics Conference on Electroweak Interactions, Erice February 1983
9. P. Hung and J. Sakurai, Nucl. Phys. B 143 (1978) 81;
10. H. Fritzsch, D. Schildknecht, and R. Kögerler, Phys. Lett. 114 B (1982) 157. See also: R. Kögerler and D. Schildknecht, CERN - preprint 3231 (

11. H. Rubinstein, private communication.

12. P. Chen and J.J. Sakurai, Phys. Lett. 110 B (1982) 481.

13. See e.g. M. Veltman, Nucl. Phys. B 123 (1977) 89.

14. See e.g.: M. Roos, Proceedings of the Europhysics Study Conference on Electroweak Interactions (Erice, 1983; H. Newman ed.).

15. U. Baur, Munich preprint MPI-PAE/PTh 72/83

16. F. Renard, Phys. Lett. 116 B, (1982) 269

17. M. Laursen, K. Mikaelian, and M. Samuel, Phys. Rev. D 23 (1981) 2795.

18. M. Althoff et al. DESY 82 - 069 (1982) preprint

19. S. Narison, LAPP preprint TH 68 (1982)
G. Girardi, S. Narison and M. Perrottet, LAPP preprint TH 82 (1983)

20. F.M. Renard, Phys. Lett. 126 B (1983), 59

21. J.D. Jackson in "Proceedings of the SLAC Summer Institute on particle physics August 2-13, 1976", p 147 and preprint LBL 5500 (1976)

22. See e.g. J. Ellis, M.K. Gaillard, G. Gierardi and P. Sorba, Ann. Rev. of Nucl. Part. Sci. 32 (1982), 443 for a review.

23. U. Baur and H. Fritzsch, to be published.

D I S C U S S I O N

CHAIRMAN : H. FRITZSCH

Scientific Secretaries : G. Hou and M. Leurer

- *WIGNER* :

Is it clear that the mass of quarks can be defined and has a meaning ?

- *FRITZSCH* :

In the field theory framework of QCD one can define quark masses. The trace of the energy-momentum-tensor has terms like $G_{\mu\nu}G^{\mu\nu}$ and $m_u\bar{u}u + m_d\bar{d}d + \ldots.$. Hadronic physics appears close to th (chiral) limit m_u, $m_d = 0$, and the nucleon mass is mostly due to the gluon term; the quark mass terms are regarded as perturbations. Then $m_\pi^2 \propto m_q$, and one determines ratios of quark masses, e.g.

$$m_q/m_s \approx m_\pi^2/2m_k^2 \sim \frac{1}{20}$$

As to the absolute value of m_q, there is the difficulty that on cannot go to the mass shell and measure m_q. Quarks are confined, and the mass of a quark has meaning only in propagators at short distances, as probed in eN scattering through a virtual photon. Also, it is not renormalization group independent, in that for each renormalization point μ, one has definite, μ-independent ratios, but the absolute values change logarithmically.

In this way quark masses are well defined (and so-called current algebra masses). One can also define constituent masses, used in certain bound state calculations.

- *WIGNER* :

Thank you. You seem to have redefined the concept of mass, but in a useful way.

- *FRITZSCH* :

This happens also in solid state physics. Electrons which are confined in a solid, have effective masses, which differ from the physical electron mass. Here also it is a scale dependent concept.

- *LEURER* :

If the W and Z are composite - how will you explain the fact that their masses are smaller than their compositeness scale by a factor of at least five ?

- *FRITZSCH* :

I do not know. A possible explanation could be a symmetry argument like supersymmetry. But, we do not see the supersymmetric partners of the W and Z, and moreover, I am not convinced that one has to evoke a symmetry in order to explain a factor of five or a little more. Another explanation may use the similarity between pions in QCD and W, Z in QHD. The pions are light on the QCD scale; the parameters f_π(~150MeV) and m_π(~140MeV) are almost equal. Likewise, W and Z are light on QHD scale and F_W(~123GeV) and M_W(~81GeV) have close values.

- *LEURER* :

In deriving the relation : $m_t = 4m_b$ you assumed that the QHD-excited u and d have equal masses. How do you justify this assumption?

- *FRITZSCH* :

These high energy states get essentially all of their masses through the hypergluon dynamics which is invariant under SU(2).

- *LEURER* :

Could you describe in more detail the constituents in your model and how you build the right- and left-handed fermions ?

- *FRITZSCH* :

The main problem to overcome here is the observed parity violation. In general there are several ways to proceed :

(a) The left-handed particles are composite, while the right-handed ones are fundamental.

(b) Both left- and right-handed particles are composite, but the radius of the right-handed fermions is smaller.

(c) Left-and right-handed fermions are composite with the same compositeness scale. Also, originally the left and right W's have the same mass (~ 1TeV) but, by some unknown effect of QHD, the W_L mass is pushed down to 80GeV, while the W(right)-masses remain heavy.

- *BURGESS* :

What are the main difficulties in constructing realistic composite models ?

- *FRITZSCH* :

The main difficulties are:

(a) to explain the observed parity violation as a dynamical effect,

(b) to fulfil the t'Hooft anomaly conditions,

(c) to suppress non-observed isoscalar neutral currents,

(d) to find a mechanism which will make the higher families excitations of the first one.

- *ALTARELLI* :

Actually there is no realistic mechanism that explains generations. For example, no such mechanism is consistent with the unobservation of $\mu \rightarrow e\gamma$ decay.

- *FRITZSCH :*

I agree that orbital excitations fail this experimental test, but the mechanism of excitation could be different, for example, pair excitation.

- *BURGESS :*

Is the prediction of vector dominance on the same footing as chiral dynamics, i.e., is it a theoretical consequence of symmetry breaking or is a certain amount of phenomenology required ?

- *FRITZSCH :*

The answer is no. In strong interactions current algebra and PCAC is good to 2 ~ 3%, whereas ρ-dominance is ~ 10%. The reason is the continuum above the pion pole depends on the quark mass and drops to zero as quark mass drops to zero. Only the π-pole remains. In ρ-dominance this is not so. For weak interactions we need a slightly different vector dominance. The continuum should be at least 3 ~ 4 times higher relative to the W pole compared to the ρ-dominance in QCD. We need to make it work to at least 2%. We probably need an analogous "σ-model". This is the reason why I say that W is lighter than one would have guessed naively.

- *BURGESS :*

In the haplon model the $SU_L(2)$ is a global symmetry. What happens to the Goldstone bosons when you spontaneously break this symmetry ?

- *FRITZSCH :*

The $SU_L(2)$ is not spontaneously broken, hence no Goldstone bosons are required. The W boson mass is generated in the same way as the ρ-mass in QCD.

- *HOU :*

In the analogy you made, the pion has vanishing mass (before EM) as a Goldstone particle. However, for haplondynamics you need to keep some chiral symmetry to protect composite fermions from acquiring a mass. QCD shows dynamical chiral symmetry breaking, how then are we to construct QHD as to respect chiral symmetry ?

- *FRITZSCH* :

Certainly QHD is not a copy of QCD, e.g. it has parity violation. QCD is a confining vector theory. Probably QHD is a confining "chiral" theory. What happens in a confining chiral theory is a well defined problem; just imagine QCD with only left-handed quarks. It may well be that in such a theory chiral symmetry is not realized in the Goldstone mode, but via a bunch of massless fermions (of course, with other massive ones in addition).

- *HOU* :

In this context let me mention that there is a recent approach viewing the first family of fermions as SUSY partners of Goldstone bosons corresponding to the breakdown of an internal symmetry group, say $SU(5) \rightarrow SU(3) \times U_{em}(1)$. Do you think this is a fruitful approach ? At least this is a way of evading breaking chiral symmetry or not.

- *FRITZSCH* :

I do not know the approach in detail. My impression is they have trouble in weak interactions. Making weak interactions effective in the way I described is more difficult in their case. Nevertheless this may be a useful way to proceed.

- *HOU* :

You separated quarks and leptons in your discussions. The charged lepton has more charge than quarks, how are you going to reconcile this in your scheme of EM origin of masses ? Do you have to invoke some similar argument of QCD induced mass splitting ?

- *FRITZSCH* :

I have thought of this problem, but have not solved it. In the model I showed, x carries color and y does not, so certainly one can imagine that it is QCD which makes the difference. However, the problem is that it cannot be additive, because then QED would only be a small correction and we would not be able to explain the mass-charge dependence as observed. We need a kind of multiplicative effect, but I have not found a way to accomplish this.

- *HOU* :

If we do see $K \rightarrow \mu e$, will we be able to distinguish between say, technicolor and compositeness ?

- *FRITZSCH* :

No, just our observation of flavor-nonconserving effects will not be good enough. Nevertheless the observation of such a decay, or a decay like $\mu \rightarrow e\gamma$, would be extremely interesting.

- *OGILVIE* :

What is the "Veltman's theorem" and do you think it could be proved ?

- *FRITZSCH* :

The so-called "Veltman's theorem" states that the effective low energy lagrangian of composite particles should be renormalizable. For composite vector mesons this implies that at low energies their interactions look like gauge-boson interactions. As I showed, this is approximately true when the vector boson pole dominates the low energy physics. As you see the theorem is only approximately true, and it is not really a mathematical theorem.

- *OGILVIE* :

Do you not find composite models unsatisfactory since they do not obey t'Hooft conditions ?

- *FRITZSCH* :

First, there are composite models (e.g. Abott-Fuhri model) or the model I discussed here, in which t'Hooft conditions are obeyed but do not say much. Second, though the t'Hooft conditions should be obeyed if the compositeness scale is much higher than the quark masses, the models I have in mind have quite a low compositeness scale. It is not absolutely clear that they have to obey t'Hooft conditions.

- *ZEPPENFELD* :

You said the second and third generations are "excitations" of the first family. I would expect these states to have mass of order $\sim \Lambda_{HC}$. How do you explain that these excitations remain massless before EM is turned on ?

- *FRITZSCH* :

By excitation I do not mean orbital excitations. For instance, you can have pair excitations. At the start these states are all massless. EM interactions will then produce a mass matrix, and after diagonalization you get the fermions of different families.

- *ZEPPENFELD* :

But then you need about 45 massless fermionic states before EM is turned on.

- *FRITZSCH* :

I expect fermions of different generations to come in replicas, so it is the number 3 that you should focus upon, not the number 45.

- *VAN DEN DOEL* :

It seems likely that chiral symmetry is always spontaneously broken in gauge theories. What do you think of the two possibilities of keeping it light :

(a) Spontaneous breaking of a global symmetry in a supersymmetric theory with quarks and leptons as SUSY partners of Goldstone bosons (as mentioned by Hou).

(b) Spontaneous breaking of extended supersymmetry.

- *FRITZSCH* :

I have already remarked on the first possibility. As far as the second one goes, I have not yet seen a realistic model in which the weak interactions arise as effective interactions.

- *VAN DEN DOEL* :

Can massless spin > ½ (photon, graviton ...) be composites ?

- *FRITZSCH* :

Probably not. There is a theorem (Case-Gasiorowicz-Weinberg-Witten) that forbids this.

- *VAN DEN DOEL* :

Why does the π^0 stay massless as you switch on EM ?

- *FRITZSCH* :

Essentially it follows from Dashen's theorem. The π^0 field commutes with the effective Hamiltonian which is responsible for the mass generation. It is a rigorous current algebra result.

- *MARUYAMA* :

How about experimental tests. Do you have any predictions for say neutrino oscillations, K decays ?

- *FRITZSCH* :

There are more model independent tests, e.g. searching for isoscalar components of the neutral current; it should be there. Neutrino oscillations are very model dependent; neutrinos do not get mass at order α, but they may get a mass in higher order, and it may well be consistent with several eV. As to mixing, I do not know how to calculate it. It might well be zero.

- *WIGNER* :

How is the intrinsic parity of the quarks related to the parity of the baryons ?

- *FRITZSCH* :

The parity of baryons is defined through the convention that the proton has positive parity, and all other baryons' parities are related to it through symmetry. Likewise, the parity of the quarks is defined to be positive. Of course, the parity of excited bound states (e.g. excited baryons) depends on the quark wave function.

- *SIMIĆ* :

What prevents the isoscalar particle from contributing substantially in the neutral current ?

- *FRITZSCH* :

Possible existing isoscalar W^o must be heavier than the W^+ - 140 GeV or more. This could arise by the annihilation of W constituents via hypergluons, thereby pushing up the mass, just as in QCD $m_{\eta'}$ is pushed up through the annihilation into two or more gluons.

I should also like to mention that in the case of spin 0 - W constituents there is no ground state isoscalar W-particle, due to Bose statistics.

- *ZICHICHI* :

Why do you expect $\rho > 1$ in composite models ?

- *FRITZSCH* :

QED breaking of SU(2) causes m_Z to go down and m_W to go up a bit, so $\rho > 1$ is expected.

- *SIMIĆ* :

You had small mass splittings compared to Λ_H. Would this not imply that mass splitting for W's are small also hence violate your single pole dominance assumption.

- *FRITZSCH* :

The splitting between generations is different from the excitations of W's. W excitations should be similar to pion excitations in QCD, e.g. $m_\pi \sim 140$ MeV $m_{\pi^*} \sim 1300$ MeV.

- *SIMIĆ* :

Why do you think W is composed of two constituents and not more ?

- *FRITZSCH* :

This is the simplest model you can construct. There are other models, e.g. in the rishon model the W^+ has six constituents. However, in all models in which the W consists of more than two constituents, the weak isospin cannot be interpreted as a flavor symmetry of an underlying hypercolor dynamics.

- *CHEN* :

In future accelerator experiments where we see hypercolor interactions, I am afraid of the case where we will see lots of charged particles (containing lots of leptons) concentrated in a pencil-like beam with the width of the electron mass (just like in QCD $\langle P_t \rangle \approx m_\pi$) and we will never be able to detect it.

- *FRITZSCH* :

The important scale is Λ_H and not m_e (it is practically zero). I expect these jets to have a width of about 10^{-16}cm. Certainly this will pose a problem for experimental studies.

INCLUSIVE DECAYS OF HEAVY FLAVOURS

G. Altarelli

CERN, Geneva, Switzerland
Dipartimento di Fisica, Università "La Sapienza", Rome
INFN - Sezione di Roma, Rome, Italy

1. INTRODUCTION

The study of weak decays of heavy flavoured particles is important as a source of information on many interesting aspects of the SU(3) ⊗ SU(2) ⊗ U(1) standard theory. Strong and electroweak effects all contribute to the determination of the decay properties. Much light can be shed on important new and old problems such as the understanding of parton dynamics, the riddle of non-leptonic weak decays of strange particles, and the determination of fundamental parameters in the theory such as the K-M mixing angles. The inclusive (semileptonic and total) decay rates are especially important in that they are the simplest and imply the least amount of model dependency. Many interesting problems are already met at this level, some of them not yet clarified, so that we shall concentrate on inclusive rates in this lecture.

2. PRELUDE: THE τ LEPTON

The decays of the τ lepton offer an example in a simpler context of a two-step strategy which will be developed in the following for heavy flavour decays.

Step one is the prediction of the leptonic rate (the analogue of the semileptonic rate for quark decay). Neglecting the final lepton masses this is simply obtained by rescaling the muon-decay rate:

$$\Gamma(\tau \to \nu_\tau \ell \nu_\ell) = \left(\frac{m_\tau}{m_\mu}\right)^5 \Gamma_\mu \ . \tag{1}$$

If one takes from experiment the value of the leptonic branching ratio[1]:

$$B_\ell(\tau)_{exp} = (0.176 \pm 0.016) \; , \tag{2}$$

the prediction of the leptonic rate can be translated into a value for the τ lifetime:

$$\tau(\tau) = \tau(\mu)\left(\frac{m_\mu}{m_\tau}\right)^5 B_\ell(\tau) = (2.81 \pm 0.25) \times 10^{-13}\ \mathrm{s} \tag{3}$$

to be compared with the measured value (presented at the Brighton Conference)

$$\tau(\tau)_{exp} = (3.20 \pm 0.41 \pm 0.35) \times 10^{-13}\ \mathrm{s} \qquad [\text{MARK II}] . \tag{4}$$

Step two is the theoretical evaluation of the leptonic branching ratio, or equivalently of the total rate, which is more difficult due to strong interactions and final quark mass effects. For massless final quarks and leptons one has:

$$B_\ell(\tau) = \left[2 + 3\left(1 + \frac{\alpha_s(m_\tau^2)}{\pi} + \ldots\right)\right]^{-1} \simeq 0.19 \; . \tag{5}$$

The factor of 3 in the quark channel is from colour (τ lepton decays thus measure the number of colours), $(1 + \alpha_s/\pi + \ldots)$ is the same Quantum Chromodynamics (QCD) correction that applies to $R_{e^+e^-}$ (both referring to the production of a quark pair from a current). Finally we observe that relatively large values for the u and d quark masses are disfavoured by the data, because they would lead to $B_\ell \gtrsim 0.2$. Light quarks are rather indicated by experiment which we interpret as due to the fact that in the final state mostly light mesons are found.

3. SEMILEPTONIC WIDTH OF A HEAVY QUARK

According to the strategy outlined above, one first tackles the problem of the semileptonic width of a heavy flavoured particle. When the momentum of the lepton pair is sufficiently large with respect to the hadron binding energy a parton description suggests that the rate for semileptonic decay is reduced to the corresponding quark decay: $Q \to q + \ell + \nu_\ell$. Deviations from the parton model associated with interference with spectators and "annihilation" are absent or suppressed in this mode, so that quark decay is appropriate in most cases. (One possible exception is the $F^{\pm}$ charmed meson: see in the following.) We thus concentrate on the decay of a heavy free quark. The semileptonic width is given by

$$\Gamma_{s\ell}(Q) = \sum_{q} |U_{Qq}|^2 \, I\left(\frac{m_q}{M_Q}, \frac{m_\ell}{M_Q}, 0\right)\left(\frac{M_Q}{m_\mu}\right)^5 \Gamma_\mu$$

$$\times \left\{1 - \frac{2}{3}\frac{\alpha_s}{\pi}\left(\pi^2 - \frac{25}{4}\right) f\left(\frac{m_q}{M_Q}\right) + \ldots\right\},$$

where U_{Qq} is the $Q \to q$ entry of the quark mixing matrix[2], $I(x,y,z)$ is the three-body phase-space factor[3], and the last bracket arises from the QCD leading correction for real and virtual gluon emission[4]. The function $f(x)$ decreases monotonically from $f(0) = 1$ to $f(1) = 0.41$. The QCD correction for $\Lambda = 250$ MeV is $\sim 25\%$ for $c \to s$ and $\sim 15\%$ for $b \to c$.

The main theoretical uncertainties of $\Gamma_{s\ell}(Q)$ arise from a) the value of M_Q [as $\Gamma_{s\ell} \sim M_Q^5$] and b) the incidence of bound-state effects. Information on both can be obtained from a study of the lepton spectrum through a bound-state model[5]. According to Fig. 1 for Q bound in a meson, $\Gamma_{s\ell}$ is given by a convolution of the decay width in flight of Q (taken as off shell, with virtual mass W and momentum $\vec{q}$) with the probability of finding Q with momentum q in the meson:

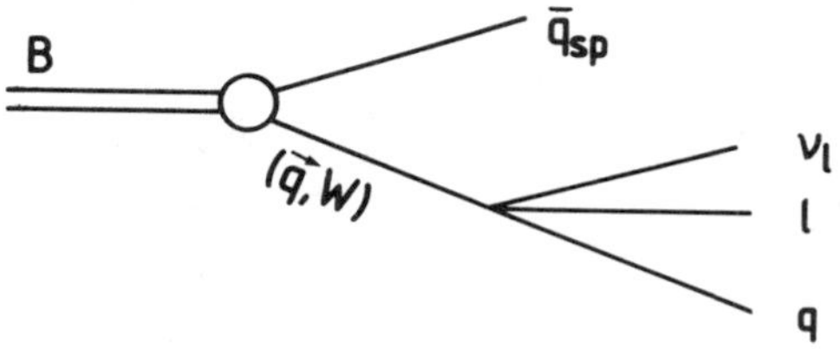

Fig. 1

$$\Gamma_{s\ell} = \int d^3q \, |\psi(q)|^2 \, \Gamma^Q_{s\ell}(\vec{q}, W) \, . \qquad (7)$$

Treating the spectator as an on-shell particle one immediately obtains

$$W^2 = M_B^2 + m_{sp}^2 - 2M_B \sqrt{\vec{q}^2 + m_{sp}^2} \, , \qquad (8)$$

where M_B and m_{sp} are the meson and spectator masses. Similarly, for the lepton spectrum:

$$\frac{d\Gamma_{s\ell}}{dE} = \int d^3q \; |\psi(q)|^2 \; \frac{d\Gamma^Q_{s\ell}}{dE} \; (\vec{q},W,E) \; . \tag{9}$$

The QCD correction of order α_s to $d\Gamma^Q_{s\ell}/dE$ is included, as computed by Cabibbo et al.[6] and Corbò[7].

The main virtue of this bound-state model is that it provides us with a parametrization of the Fermi motion of the heavy quark inside the meson in a way that fully respects the kinematics. We take a Gaussian form for the wave function:

$$\psi(q) = (\sqrt{\pi}\; p_F)^{-3/2} \; \exp\left(\frac{-\vec{q}^{\,2}}{2p_F^2}\right) \; . \tag{10}$$

p_F fixes at the same time the shape of the spectrum near the end point and the effective value of the Q mass: $M_Q \simeq \langle W \rangle$.

3.1 Charm Decay

For charm[8] the electron spectrum in D decay has been measured by the DELCO Collaboration at SLAC. The results[5] on p_F can be read from Figs. 2 and 3. The spectator and strange quark masses were taken in the ranges m_{sp} = (0-150) MeV, m_s = (300-500) MeV, which were chosen considering that the hadronic part of the final state of D semileptonic decays is mainly made up of a kaon plus pions. The value of α_s (0.38) is only important for the total width and not for the spectrum shape. Small values of p_F are indicated by the spectrum, in the range

$$p_F \lesssim 300 \text{ MeV} \; , \tag{11}$$

with preference for $p_F \simeq$ (0-200) MeV. The information obtained from the spectrum can be used to evaluate $\Gamma_{s\ell}$ for D mesons. With $\theta_C = 0$, m_{sp} = (100-150) MeV, m_s = (300-400) MeV, p_F = (0-300) MeV, α_s = 0.20-0.36 (corresponding to Λ = 100-400 MeV), one obtains:

$$\Gamma^D_{s\ell} = \cos^2\gamma \; (1.1\text{-}3.4) \times 10^{11} \; s^{-1} \; . \tag{12}$$

We follow the Maiani definition of mixing angles[9]:

$$\begin{aligned} u &\to \cos\beta \; d_c + \sin\beta \; b \\ c &\to \cos\gamma \; e^{i\delta} \; s_c + \sin\gamma \; (-\sin\beta \; d_c + \cos\beta \; b) \\ t &\to -\sin\gamma \; e^{i\delta} \; s_c + \cos\gamma \; (-\sin\beta \; d_c + \cos\beta \; b) \end{aligned} \tag{13}$$

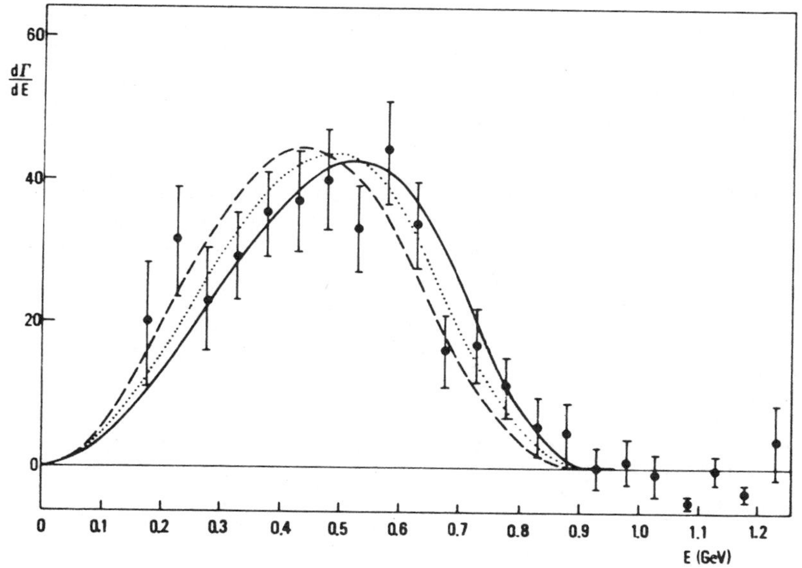

Fig. 2 Charged lepton spectrum in D decay for M_D = 1.866 GeV, m_s = 0.3 GeV, m_{sp} = 0.15 GeV; α_s = 0.38, p_{cm} = 0.26 GeV and p_F = 0 (solid), p_F = 0.15 GeV (dotted), p_F = 0.3 GeV (dashed). The normalization is fixed by the number of events.

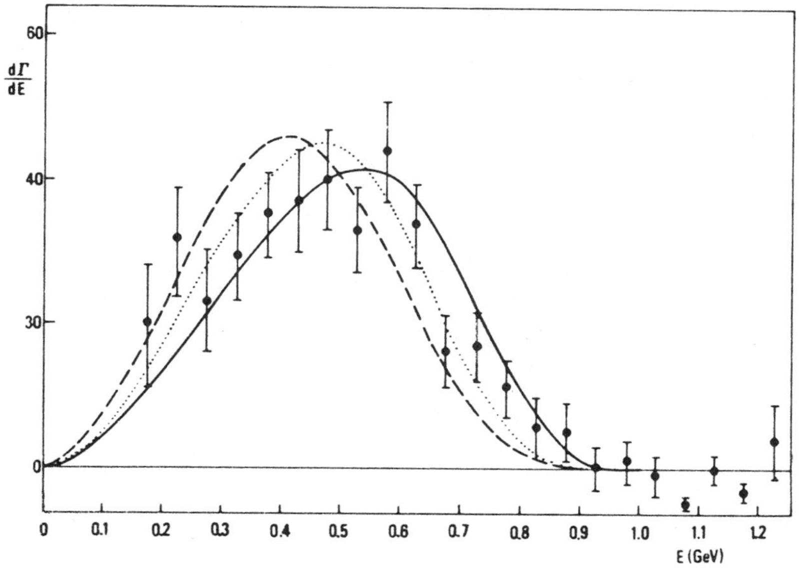

Fig. 3 Charged lepton spectrum in D decay for M_D = 1.866 GeV, m_s = 0.5 GeV, m_{sp} = 0, α_s = 0.38, p_{cm} = 0.26 GeV and p_F = 0 (solid), p_F = 0.15 GeV (dotted), p_F = 0.3 GeV (dashed). The normalization is fixed by the number of events.

leaving aside, for the time being, the recent SLAC results on the b lifetime (to be discussed later on), the existing limit on $\cos^2\gamma$ is derived by the CERN-Dortmund-Heidelberg-Saclay (CDHS)[10] Collaboration from the production of dimuons in ν and $\bar{\nu}$ deep inelastic scattering and is given by $\cos^2\gamma \gtrsim 0.6$.

If one knows the semileptonic branching ratio $B_{s\ell}$ of D mesons, then the estimate in Eq. (12) of the semileptonic width leads to a predicted range for the total lifetime. The average $B_{s\ell}$ of 56% D^0 and 44% D^+ is known with good accuracy from experiments at the ψ'' resonance[1]:

$$\bar{B}_{s\ell}(D^+,D^0) = (8.2 \pm 1.2)\% \; . \tag{14}$$

As the dominant transition $c \to s$ is isoscalar, $\Gamma_{s\ell}$ must be the same for D^+ and D^0 when isospin violations are neglected and $\theta_C \simeq 0$. Then from Eqs. (12) and (14) one obtains

$$\cos^2\gamma\ \bar{\tau}(D) = (2\text{-}8.5) \times 10^{-13}\ \mathrm{s} \qquad (p_F \lesssim 300\ \mathrm{MeV}) \; . \tag{15}$$

The present world averages of D^+, D^0, and F^+ lifetimes are given by[11]

$$\begin{aligned} \tau_{exp}(D^0) &= \left(4.4 \begin{array}{l} +\,0.6 \\ -\,0.5 \end{array}\right) \times 10^{-13}\ \mathrm{s} \\ \tau_{exp}(D^+) &= \left(8.8 \begin{array}{l} +\,1.8 \\ -\,1.0 \end{array}\right) \times 10^{-13}\ \mathrm{s} \\ \tau_{exp}(F^+) &= \left(2.1 \begin{array}{l} +\,1.3 \\ -\,0.6 \end{array}\right) \times 10^{-13}\ \mathrm{s} \; , \end{aligned} \tag{16}$$

which leads to

$$\bar{\tau}_{exp}(D) = \left(6.6 \begin{array}{l} +\,1.2 \\ -\,0.8 \end{array}\right) \times 10^{-13}\ \mathrm{s} \; . \tag{17}$$

Similarly for $B(D^+) \simeq 15\%$ (a value favoured by theory as we shall see) one obtains

$$\cos^2\gamma\ \tau(D^+) \simeq (4\text{-}14) \times 10^{-13}\ \mathrm{s} \; .$$

We also mention that from $B_{s\ell}(\Lambda_c) \simeq 5\%$, as known form experiment, if one takes, for purposes of illustration, $\Gamma_{s\ell}(\Lambda_c) \simeq \Gamma_{s\ell}(D)$ (as would be the case from free heavy quark decay) then one obtains

$$\cos^2\gamma\ \tau(\Lambda_c) \simeq (1.5\text{-}4.5) \times 10^{-13}\ \mathrm{s} \; ,$$

which is to be compared with the measured value[11]:

$$\tau_{exp}(\Lambda_c) \simeq \left(2.2 \begin{matrix} +\ 0.8 \\ -\ 0.5 \end{matrix}\right) \times 10^{-13}\ s\ . \qquad (20)$$

In conclusion, the theoretical estimates of $\Gamma_{s\ell}$ for charm lead to satisfactory agreement with the measured lifetime values, once the values of $B_{s\ell}$ are taken from experiment.

3.2 Beauty Decay

The electron spectrum of beauty decay is particularly interesting in that it is sensitive to the ratio of the $(b \to c) \simeq \simeq \sin\gamma \cdot \cos\beta \simeq \sin\gamma$ and the $(b \to u) \sim \sin\beta$ transitions:

$$\frac{d\Gamma^b_{s\ell}}{dE} = \sin^2\gamma\ \frac{d\Gamma^c}{dE} + \sin^2\beta\ \frac{d\Gamma^u}{dE}\ . \qquad (21)$$

The possibility of separating the two components depends on the theoretical control of the hard tail of the charged lepton spectrum. This is sensitive to QCD corrections and to bound-state effects. In particular, the case of a final massless quark is of special interest and is relevant for the $b \to u + e + \nu$ channel. The electron spectrum is given by

$$\frac{d\Gamma}{dx} = \frac{d\Gamma^0}{dx}\left[1 - \frac{2\alpha_s}{3\pi}\ G(x,0)\right], \qquad (22)$$

where $x = 2E/M_Q$ is the electron energy fraction in the rest frame of the heavy quark of mass M_Q, $d\Gamma^0/dx$ is the uncorrected spectrum, while $G(x,0)$ is a complicated function which describes the QCD correction[6,7]. Both $d\Gamma^0/dx$ and $G(x,0)$ are different for the $c \to s$ and the $b \to u$ transitions. In fact, as a consequence of the V - A structure of the current, the ν spectrum in the $c \to s$ transition is the same as the charged lepton spectrum in the $b \to u$ transition and conversely. In particular,

$$\frac{d\Gamma^0}{dx} = \begin{cases} \sim x^2\ (1 - x) & (c \to s) \\ \sim x^2\ (2 - x) & (b \to u)\ . \end{cases} \qquad (23)$$

In both cases[5] near $x \to 1$

$$G(x,0) \underset{x\to 1}{\to} \ln^2\ (1 - x)\ . \qquad (24)$$

While in the $c \to s$ case the singular behaviour of $G(x,0)$ at $x = 1$ is damped by the corresponding vanishing of $d\Gamma^0/dx$, it cannot be ignored in the $b \to u$ case, where the bare spectrum is non-vanishing at $x = 1$. In the latter case, first-order perturbation theory becomes inadequate near $x \to 1$, at values of $x = x_c$ such that

$$\frac{2\alpha_s}{3\pi} \ln^2 (1 - x_c) \simeq 1 \Rightarrow M_{QCD} = M_Q (1 - x_c)^{1/2} \simeq 0.55 \text{ GeV} \qquad (25)$$

(for $M_Q \simeq 5$ GeV). M_{QCD} is the minimum invariant mass of the recoiling hadronic system at which first-order perturbation theory can still be applied. The double logs are of infrared origin and can be resummed[5] at all orders leading to a Sudakov exponential:

$$\frac{d\Gamma}{dx} = \frac{d\Gamma^0}{dx} \exp\left[-\frac{2\alpha_s}{3\pi} \ln^2 (1 - x)\right]\left[1 - \frac{2\alpha_s}{3\pi} \tilde{G}(x,0)\right], \qquad (26)$$

where

$$\tilde{G}(x,0) = G(x,0) - \ln^2 (1 - x) . \qquad (27)$$

The resulting spectrum now vanishes for $x \to 1$ and one cannot distinguish a massless final quark from a quark with mass $m \sim M_{QCD}$. Since, however, the c quark mass is much larger than M_{QCD}, the possibility of distinguishing the $b \to c$ and the $b \to u$ transitions is not spoiled. In practice, these delicate effects are concealed by the Fermi motion smearing near the end point related to p_F.

Bound-state effects can be taken into account by the same model as already described[5]. The same values for p_F as found from the analysis of D decays can be tentatively assumed. This is justified in a crude non-relativistic picture of c and b flavoured mesons, where the wave function is determined by the reduced mass which is insensitive to the heavy quark mass provided it is sufficiently large.

The results obtained by the CUSB and CLEO Collaborations at Cornell can be summarized as follows. The $b \to c$ transition is found to be dominant with respect to the $b \to u$ transition. From the ratio of rates $\Gamma(b \to u)/\Gamma(b \to c) \lesssim 0.05$, we obtain:

$$\frac{\sin^2 \beta}{\sin^2 \gamma} \lesssim 0.04 . \qquad (28)$$

The total lifetime of a b flavoured meson can be written down in terms of the corresponding $B_{s\ell}$ as:

$$\tau(b) = \frac{B_{s\ell}}{0.131} \frac{10^{-14} \text{ s}}{\sin^2 \gamma \left[Z_c + (\sin^2 \beta/\sin^2 \gamma) Z_u\right]} . \qquad (29)$$

For $p_F \lesssim 300$ MeV and $\Lambda = 250$ MeV one finds

$$Z_c = 2.5\text{-}3.5, \quad Z_u = 5.6\text{-}7.2 . \tag{30}$$

From the experimental value of $B_{s\ell}$ measured at the Υ''', which refers to a weighted average of 60% B^+ and 40% B^0,

$$B_{s\ell}^{exp} = 0.131 + 0.012 , \tag{31}$$

and the bound in Eq. (28), one obtains

$$\bar{\tau}(B) = \frac{1}{\sin^2 \gamma} (2.4\text{-}4.4) \times 10^{-15} \text{ s} . \tag{32}$$

On the experimental side an upper limit found by the JADE Collaboration at DESY is known since last year:

$$\tau(B) < 1.4 \times 10^{-12} \text{ s} \qquad [\text{JADE}] . \tag{33a}$$

Quite recently two measured values have been reported from SLAC (presented at the Brighton and Cornell Conferences):

$$\tau(B) = (1.8 \pm 0.6 \pm 0.4) \times 10^{-12} \text{ s} \qquad [\text{MAC}] , \tag{33b}$$

which implies by Eq. (32):

$$0.03 \leq |\sin \gamma| \leq 0.07 ,$$

and

$$\tau(B) = \left(10.4 \begin{matrix} +\ 4.9 \\ -\ 3.9 \end{matrix}\right) \times 10^{-13} \text{ s} \qquad [\text{MARK II}] , \tag{33c}$$

corresponding to

$$0.4 \leq |\sin \gamma| \leq 0.08 .$$

Thus, if these results are confirmed, one is led to conclude that $\sin \gamma \sim O(\theta_C^2)$!

4. INCLUSIVE NON-LEPTONIC RATES

To proceed further toward a computation of $B_{s\ell}$ and of non-leptonic inclusive rates we shall divide the discussion into two conceptually different points: first the set-up of an effective Hamiltonian for non-leptonic weak amplitudes and then the attempts at evaluating its matrix elements. We shall see that the first part can be carried through on rather solid grounds, while the second step is more tentative.

Let us consider a non-leptonic weak process induced by charged currents. In lowest order in the weak coupling the transition matrix element is given by the time-ordered product of the two weak charged currents folded with the W propagator:

$$H_{fi} \simeq g_W^2 \int d^4x \, D_W(x^2, M_W^2) \, \langle F | T[J^\mu(x) \, J_\mu^+(0)] | I \rangle \, , \tag{34}$$

where M_W, g_W, and D_W are the W boson mass, coupling, and propagator, respectively. For flavour-changing amplitudes the leading contributions in the limit $M_W \to \infty$ arise from the four-fermion operators of dimension six in the short-distance operator expansion for the T-product[12]. In the particular example of charm-changing processes the relevant terms are of the form

$$H_{fi}^{\Delta c=1} = \frac{G_F}{\sqrt{2}} \Big\{ c_+(t,\alpha_s) \, \langle F | O_+(0) | I \rangle + $$

$$+ \, c_-(t,\alpha_s) \, \langle F | O_-(0) | I \rangle + \ldots \Big\} \, , \tag{35}$$

where $t = \ln M_W^2/\mu^2$ and μ is a reference mass scale, $\alpha_s = \alpha_s(\mu)$ being the renormalized QCD coupling at the scale μ, and

$$\begin{aligned} O_\pm &= \tfrac{1}{2} \left[(\bar{s}'c)_L \, (\bar{u}d')_L \pm (\bar{u}c)_L \, (\bar{s}'d')_L \right] \\ &= \frac{N \pm 1}{2N} (\bar{s}'c)_L \, (\bar{u}d')_L \pm \sum_A (\bar{s}'t^A c)_L \, (\bar{u}t^A d')_L \, . \end{aligned} \tag{36}$$

The shorthand notations for left-handed (right-handed) currents,

$$(\bar{q}_1 q_2)_{L,R} = \bar{q}_1 \gamma_\mu (1 \mp \gamma_5) \, q_2 \tag{37}$$

were used here; s' and d' are the Cabibbo-like quark mixtures coupled to the c and u quarks respectively; t^A are the SU(N) matrices in the quark (fundamental) representation, with the normalization $\mathrm{Tr}\,(t^A t^B) = \tfrac{1}{2}\,\delta_{AB}$. The second equality in Eq. (36) is obtained through Fierz rearrangement of $(\bar{u}c)\,(\bar{s}'d')$, according to the identity

$$(\bar{q}_1 q_2)_L \; (\bar{q}_3 q_4)_L = \frac{1}{N} (\bar{q}_1 q_4)_L \; (\bar{q}_3 q_2)_L + 2 \sum_A (\bar{q}_1 t^A q_4)_L \; (\bar{q}_3 t^A q_2)_L \; . \tag{38}$$

In Eq. (35) the dots stand for non-leading terms from operators of higher dimension and also from "penguin" operators[13] of dimension six, which are present in some channels for non-degenerate quark masses, and will be considered later on. In the free field limit $c_+ = c_- = 1$ and

$$H^{\Delta c=1} = \frac{G_F}{\sqrt{2}} (\bar{s}'c)_L \; (\bar{u}d')_L \qquad \text{(free fields)} \; . \tag{39}$$

In the leading logarithmic approximation 0_+ and 0_- are multiplicatively renormalizable, because anomalous dimensions are determined by the massless theory, and $0_\pm$ have definite and different transformation properties under the SU(N) flavour symmetry of the massless theory. c_+ and c_- deviate from the free field value by a known amount:

$$c_\pm = \left(\frac{\alpha_s(\mu^2)}{\alpha_s(M_W^2)} \right)^{\gamma_\pm} \tag{40}$$

$$\gamma_+ = \frac{-9(N-1)}{N(11N-2f)} \underset{\substack{N=3\\f=4}}{=} -\frac{6}{25}$$

$$(\gamma_+^2 = 1/\gamma_-) \tag{41}$$

$$\gamma_- = \frac{9(N+1)}{N(11N-2f)} \underset{\substack{N=3\\f=4}}{=} \frac{12}{25} \; ,$$

where N and f are the numbers of colours and of flavours, respectively. Recently a calculation of the effective Hamiltonian at the next to the leading log accuracy was completed[14] with the result:

$$c_\pm = \left(\frac{\alpha_s(\mu^2)}{\alpha_s(M_W^2)} \right)^{\gamma_\pm} \left\{ 1 + \frac{\alpha_s(\mu^2) - \alpha_s(M_W^2)}{\pi} \rho_\pm + \ldots \right\} \; , \tag{42}$$

where $\rho_\pm$ is independent of the renormalization scheme for the operators $0_\pm$, but only depends on the definition of α_s. Numerically, in the $\overline{\text{MS}}$ definition of α_s, for N = 3 and f = 4, 6, one obtains

$$\rho_\pm = \begin{cases} -0.469 \\ +1.36 \end{cases} \qquad f = 4$$

$$\rho_\pm = \begin{cases} -0.574 \\ +1.65 \end{cases} \qquad f = 6 \; . \tag{43}$$

It is important to observe that the next to the leading corrections are of "normal" size, i.e. $\rho_\pm \sim 0\ (1)$, and follow exactly the same pattern of enhancement and suppression of the leading term. Actual most of the correction could be reabsorbed in a redefinition of α_s (by scaling up $\Lambda_{\overline{MS}}$ by a factor of about 1.8). As a consequence, th above results considerably reinforce the leading log predictions an put the status of the effective Hamiltonian on a much more solid basis.

It is well known that for strange-particle decays the structur of the effective Hamiltonian works in favour of the $\Delta I = \frac{1}{2}$ rule[15]. In fact, for strangeness-changing amplitudes O_- is pure $\Delta I = \frac{1}{2}$ and c_- is enhanced by gluon effects, while O_+ also contains $\Delta I = \frac{3}{2}$ and c_+ is suppressed. However, $\mu \sim m_s$, where m_s is the strange quark mass, is too low to rely on massless perturbation theory in the descent from M_W to m_s. Below the charm mass m_c various effects con- nected with the no longer negligible c mass appear, such as penguin diagrams and separate anomalous dimensions for operators in differe $SU(3)_f$ representations[13,16]. On the other hand, the presence of still heavier quarks, such as b and t, is not of great relevance an can be effectively taken into account by adjusting the number f of excited flavours between 4 and 6. Penguin diagrams shown in Fig. 4 are zero for equal quark masses because of the Glashow-Iliopoulos-Maiani (GIM) mechanism. Their coefficient is rather small: $(c_+ + c_-)(\alpha_s/12\pi)\ \ln\ (m_c^2/m_s^2)$; (the argument of the log is the heav

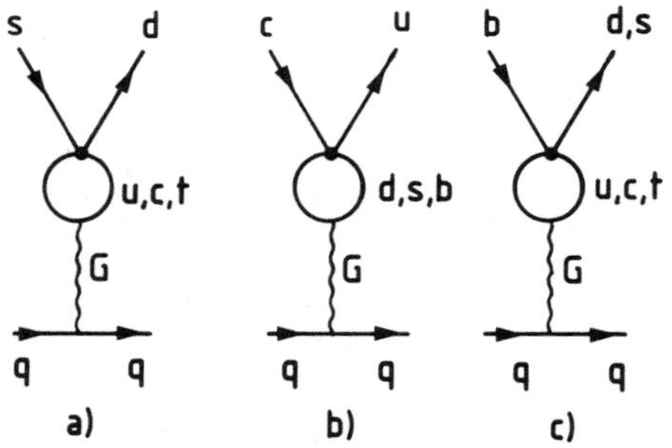

Fig. 4

quark mass in the loop divided by the energy scale for the process. The value of the K_L-K_S mass difference shows that the presence of the t quark is unimportant.). But the claim, originally made in Ref. 13, is that the matrix element is particularly large, owing to a low-energy effect connected with the smallness of the current mass of u and d quarks. In conclusion, the $\Delta I = \frac{1}{2}$ rule observed in K and hyperon decays corresponds to a factor $A_{1/2}/A_{3/2} \sim 20$ in amplitude. The QCD effect in the coefficients $c_\pm$ can account for the square root of this factor. The remaining factor is presumably due to low-energy effects in matrix elements, also including perhaps those of penguin diagrams. This view is to some extent supported by the fact that the $\Delta I = \frac{1}{2}$ rule is less pronounced in Ω^- decays, where low-energy effects are presumably less important.

For c decays the penguin diagrams, in the limit $m_b >> m_d \sim m_s \sim 0$, are given by

$$H = \frac{G_F}{\sqrt{2}} (c_+ + c_-) U_{cb} U_{ub} \frac{\alpha_s(m_c)}{12\pi} \ln \frac{m_b^2}{m_c^2} (\bar{c} t^A u)_L (\bar{q} t^A q)_{L+R} . \quad (44)$$

The existing bounds[17] on mixing angles give $|U_{cb}U_{ub}| \lesssim 0.04$ (even if we disregard Eqs. (33) for the b lifetime). Thus, as the matrix elements are not expected to be large, penguin diagrams are certainly irrelevant for charm decays. At most they can contribute in Cabibbo suppressed channels, but their contribution is expected to be small even at that level. For b decays, in the limit $m_t >> m_u \sim m_c \sim 0$, one similarly obtains

$$H = \frac{G_F}{\sqrt{2}} (c_+ + c_-) U_{tb} U_{ts} \frac{\alpha_s(m_b)}{12\pi} \ln \frac{m_t^2}{m_b^2} (\bar{b} t^A s)_L (\bar{q} t^A q)_{L+R} . \quad (45)$$

With little information on the mixing angles ($|U_{tb}U_{ts}| < 0.6$) one can only say that this contribution is presumably small, but could be not entirely negligible.

As a first conclusion, we have seen that the structure of the effective non-leptonic Hamiltonian is known in the form

$$\mathcal{H} = \frac{G_F}{\sqrt{2}} \left[c_+ O_+ + c_- O_- + \text{penguins} \right] \quad (46)$$

and that penguin diagrams are negligible in the $|\Delta c| = 1$ sector where the Hamiltonian reduces to

$$H = \frac{G_F}{\sqrt{2}} \left[\frac{c_+ + c_-}{2} (\bar{c}s')_L \, (\bar{d}'u)_L + \frac{c_+ - c_-}{2} (\bar{c}u)_L \, (\bar{d}'s')_L \right] . \tag{47}$$

Rather confusing is the status of matrix elements, in particular the relative importance of the quark decay with spectator(s) and the annihilation mechanism which we are now going to discuss.

Initially it was assumed, according to parton ideas, that the c quark decay was the main mechanism, with the other constituents in the hadron acting as spectators.

This obviously leads to approximately the same $B_{s\ell}$ or lifetimes for all charmed hadrons. Since the discovery of a longer lifetime for D^+ than for D^0 it was clear that quark decay cannot be the only effect. After the first data of the DELCO Collaboration followed by new evidence from MARK II, now there is also an increasing amount of data from emulsions and bubble chambers that led to the lifetimes quoted in Eqs. (16) and (20). In particular, the best value for the ratio of D^+ and D^0 lifetimes appears now to be:

$$\frac{\tau(D^+)}{\tau(D^0)} \simeq 2.0 \pm 0.4 \; . \tag{48}$$

This more recent value is much closer to 1 than previously reported and more understandable theoretically as we shall see.

Although quark decay cannot be, at least for charm, the only effect, however it certainly is there. It is thus important to summarize what would happen for the decay of a free heavy quark. We shall later discuss the possible role of the spectators.

In the leading log approximation the total non-leptonic width of a heavy quark Q arises from $Q \to q + q + \bar{q}$ and is given by[18]

$$\Gamma_{nL} = \frac{G_F^2 \, M_Q^5}{192\pi^3} \sum_{q, q_\alpha, \bar{q}_\beta} |U_{Qq}|^2 \, |U_{q_\alpha q_\beta}|^2$$

$$\times \, I\left(\frac{m_q}{M_Q}, \frac{m_{q_\alpha}}{M_Q}, \frac{m_{q_\beta}}{M_Q}\right) \frac{2c_+^2 + c_-^2}{3} \; , \tag{49}$$

where the last factor is larger than 1 as follows from Eqs. (40) and (41) for $c_\pm$. It arises as follows. By writing H in Eq. (47) in terms of one single ordering of quark fields, obtained by Fierz rearrangement of the second term according to Eq. (38), one gets

$$H \simeq \frac{G_F}{\sqrt{2}} \left\{ \left(\frac{2}{3} c_+ + \frac{1}{3} c_- \right) (\bar{c}s')_L \, (\bar{d}'u)_L \right.$$

$$\left. + (c_+ - c_-) \, (\bar{c}t^A s')_L \, (\bar{d}'t^A u)_L \right\} . \tag{50}$$

The incoherent sum of the rates into colour singlet and colour octet $d'u$ final states leads to

$$\left(\frac{2}{3} c_+ + \frac{1}{3} c_- \right)^2 + \frac{2}{9} (c_+ - c_-)^2 = \frac{2c_+^2 + c_-^2}{2} , \tag{51}$$

where the factor of $^2/_9$ arises from the ratio of Tr I · Tr I = 9 for the singlet-singlet contribution and $\Sigma_{AB} \left[Tr(t^A t^B) \right]^2 = 2$ for the octet-octet contribution. The overall normalization is chosen so as to reproduce the free field result for $c_+ = c_- = 1$.

Ignoring phase-space distortions and finite corrections of order α_s, this leads to:

$$B_{s\ell} \simeq \frac{1}{2 + 2c_+^2 + c_-^2} \simeq \begin{cases} 20 \text{ free fields} \\ (13 - 16)\% \text{ leading logs} \end{cases} . \tag{52}$$

Phase-space corrections are however of importance, because the three-body phase space is sensitive to masses. In particular, non-leptonic amplitudes could be somewhat suppressed compared to semi-leptonic modes, because of quark masses. It has been observed[19] that by setting $m_s \simeq 500$ MeV and $m_{u,d} \simeq 300$ MeV the effect is quite large and could bring $B_{s\ell}(D^+)$ back into the 20% range. On the other hand, the effect is quite small for $m_s \simeq 300$ MeV and $m_{u,d} \simeq 100$ MeV, which looks more reasonable in view of the final state being made up of π and K, and because no analogue effect is seen in τ semileptonic decay. Thus I think that this effect is probably marginal.

Corrections of order α_s to the non-leptonic rate have also been computed recently[14]. In the $\overline{MS}$ prescription for α_s they amount to the replacement

$$\frac{2c_+^2 + c_-^2}{3} \to \frac{2c_+^2 + c_-^2}{3} \left\{ 1 + \frac{2\alpha_s}{3\pi} \left(\frac{31}{4} - \pi^2 + \frac{19}{4} \frac{c_-^2 - c_+^2}{2c_+^2 + c_-^2} \right) \right.$$

$$\left. + 2 \frac{\alpha_s(m_c) - \alpha_s(M_W)}{\pi} \frac{2c_+^2 \rho_+ + c_-^2 \rho_-}{2c_+^2 + c_-^2} \right\} , \tag{53}$$

where $\rho_\pm$ are defined in Eq. (42). Numerically for $\Lambda \simeq 0.25$ GeV the curly bracket is 1.21. The size of the correction is normal and its sign is once more in the same direction as that of the leading log effect. The prediction of a depletion of $B_{s\ell}$ for quark decay below the free field value of 20% is thus confirmed by the corrections of order α_s, which lower the semi-leptonic width and increase the non-leptonic rate. We shall see that the annihilation mechanism cannot affect much the D^+ rate, and that spectator effects, which for D^+ could in principle be relevant, are found to be small. Thus the prediction of a small $B_{s\ell}$, obtained from the study of the decay properties of a heavy quark, is most probably relevant for D^+ decay.

Now naturally comes the question of possible spectator effects. An interesting point has been put forward[20]. At the Cabibbo allowed level a c quark decays according to $c \to su\bar{d}$. The D^+ meson is unique in that the spectator $\bar{d}$ finds an identical antiquark in the final state of c decay. Note parenthetically that also for Λ_c there would be a u spectator identical with a u quark from c decay. Assuming for a moment that the D^+ wave function was a δ-function at the origin, then the two $\bar{d}$'s would be emitted in the same space-time point. The final state would be the same as produced from the vacuum by an interaction where $\bar{c}$ is replaced by $\bar{d}$: for example $(cs)_L\ (\bar{d}u)_L \to (\bar{d}s)_L\ (\bar{d}u)_L$ because the D^+ is colour singlet and spin singlet, which means that colour and helicity are just right for the replacement of an incoming c with an outgoing $\bar{d}$. Thus the analogue of O_- would vanish and the non-leptonic rate of D^+ would be reduced by a factor

$$\frac{4c_+^2}{2c_+^2 + c_-^2} \sim \frac{1}{2} \, .$$

Now, of course this is an upper bound because the wave function is not a δ at the origin and the overlap between spectator and active antiquark must be far smaller. This matter can be quantitatively studied[21] by going back to the bound-state model developed in the analyses of semileptonic decays. We write the non-leptonic width of D^+ as a sum of a direct plus an interference term arising from the diagrams in Fig. 5 a and b, respectively:

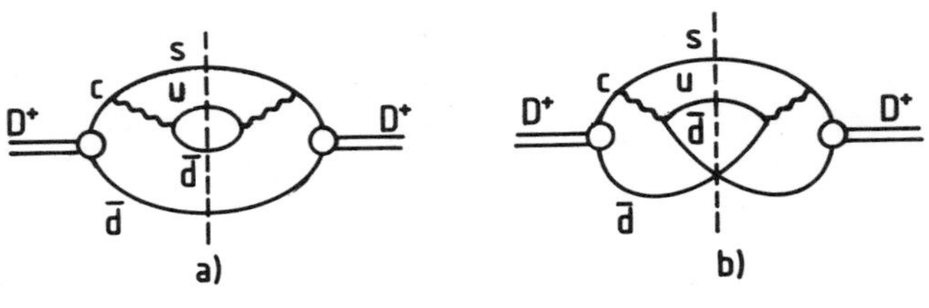

Fig. 5

$$\Gamma = \Gamma_Q + \Gamma_{int} \ . \tag{54}$$

Omitting mixing angles for simplicity we have

$$\Gamma_Q = (2c_+^2 + c_-^2)\ J_{2L} \int d^3q|\psi(q)|^2 \left(\frac{W}{E_W}\right) \frac{G_F^2\ W^5}{192\pi^3} \times I\left(\frac{m_s}{W}, \frac{m_{sp}}{W}, \frac{m_{sp}}{W}\right) , \tag{55}$$

where $(2c_+^2 + c_-^2)$ arises from QCD leading logs, J_{2L} is the two-loop QCD correction [Eq. (53)], W is defined in Eq. (8), W/E_W is the Lorentz factor for the decay in flight of a c quark of mass W, and I is the phase-space factor ($m_u \simeq m_d \equiv m_{sp}$). Correspondingly:

$$\Gamma_{int} = -(c_-^2 - 2c_+^2) \frac{G_F^2}{\pi} \int \frac{d^3q_1 d^3q_2}{(2\pi)^3} \psi(q_1)\psi^*(q_2) \times \left(\frac{W_1 W_2}{E_{W_1} E_{W_2}}\right)^{1/2} f(\vec{q}_1,\vec{q}_2) , \tag{56}$$

where $f(\vec{q}_1,\vec{q}_2)$ is a kernel arising from matrix elements and phase space which is given in Ref. 21. The limits on p_F obtained from the lepton spectrum lead to rather strong upper bounds for the ratio $R = -\Gamma_{int}/\Gamma_Q$ which are shown in Fig. 6. One realizes that the interference mechanism, although working in the right direction, can only play a marginal role in explaining a difference in lifetime between D^+ and D^0 of a factor of 2-3 as observed. The theoretical expectation for $B_{s\ell}(D^+)$ is plotted in Fig. 7 as a function of p_F and α_s. The resulting prediction turns out to be

$$B_{s\ell}(D^+) = \left(15 \begin{matrix} +\ 2 \\ -\ 3 \end{matrix}\right)\% \ . \tag{57}$$

We now consider the annihilation mechanism. First we recall that by this name we mean exchange of a W between constituents of a given hadron, in either the s- or t-channels. The name annihilation is, strictly speaking, only appropriate for an s-channel exchange.

The first point to recall is that annihilation without gluon emission cannot work, for two main reasons. The first is the helicity suppression factor. In the massless limit a pseudoscalar

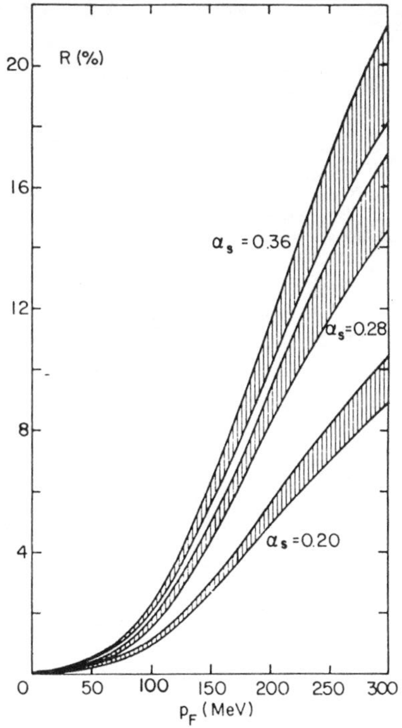

Fig. 6 The ratio $R = -\Gamma_I/\Gamma_Q$ of the interference and the direct term in D^+ non-leptonic decay, as a function of p_F for different values of $\alpha_s(M_D)$ and of final quark masses. The upper curve in each shaded band is for m_s = 300 MeV, m_{sp} = 150 MeV, while the lower curve corresponds to m_s = 400 MeV, m_{sp} = 100 MeV.

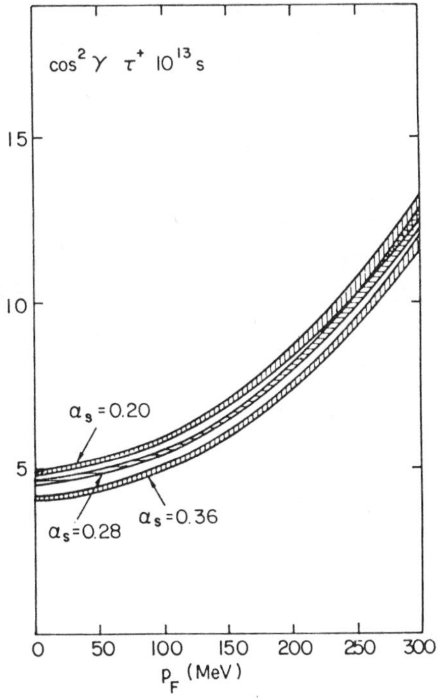

Fig. 7 The semileptonic branching ratio $B^+_{s\ell}$ of D^+ as a function of p_F for different values of $\alpha_s(M_D)$ and of final quark masses. The upper curve in each shaded band is for m_s = 300 MeV, m_{sp} = 150 MeV, while the lower curve corresponds to m_s = 400 MeV, m_{sp} = 100 MeV.

meson cannot decay into a spin-one $q\bar{q}$ pair as produced by a V - A interaction. Even for an s quark the suppression factor $m_s^2/m_c^2 \sim 1/10$ is too small. Also colour factors are completely unfavourable. The D^0 decay needs a $c\bar{u}$ pair in a colour singlet combination, which means an amplitude proportional to $X_- = 2/3c_+ - 1/3c_-$ [from Eqs. (38) and (47)], while F^+ and D^+ need $c\bar{s}$ or $c\bar{d}$ in a colour singlet state, with amplitude $X_+ = 2/3c_+ + 1/3c_-$. Since $X_+^2/X_-^2 \simeq 20$ this would make the lifetime of F^+ much shorter than that of D^0 and destroy the Cabibbo suppression pattern for D^+. Thus annihilation without gluon emission cannot work. However it exists and can provide a sizeable rate for $F^+ \to \tau^+\nu_\tau$ and contribute an important fraction of the D^+ Cabibbo-suppressed modes.

With single gluon emission the helicity suppression factor is removed and the colour factors look all right. In fact, in the D^0 case the amplitude for a $c\bar{u}$ pair in a colour octet is proportional to $(c_+ + c_-)$, while in the D^+ and F^+ cases the amplitude corresponding to colour octet $c\bar{s}$ and $c\bar{d}$ is proportional to $(c_+ - c_-)$. Other signatures for the relevance of the annihilation diagrams are also observed. For example, final states with no $s\bar{s}$ quarks in F^+ decay were seen, as well as dominance of isospin ½ amplitudes in D^0 decays and rather abundant Cabibbo-suppressed modes in D^+ decays[8].

If on qualitative grounds the annihilation mechanism with gluon emission is a convincing candidate for explaining the shorter lifetime of D^0, F^+, and Λ_c, compared to D^+, the quantitative estimate of the relative importance of annihilation versus quark decay for charm is a much more difficult problem[22]. A crude non-relativistic estimate leads[12] to:

$$\Gamma_{ann}^{D^0} \simeq \cos^2\gamma\,(c_+ + c_-)^2 \frac{G_F^2 M_D^2}{6\pi} |\psi^{(8)}(0)|^2 P^{(8)} , \tag{58}$$

where $P^{(8)}$ is the probability of finding a colour octet spin-one $c\bar{u}$ pair in the D^0 and $\psi^{(8)}$ is the corresponding wave function. A perturbative evaluation of $P^{(8)}$ leads to (see the first of Refs. 22)

$$P^{(8)} \simeq \frac{\alpha_s}{18\pi}\left(\frac{M_P}{m_u}\right)^2 \sim 0.2 \qquad \begin{pmatrix} \alpha_s \simeq 0.3 \\ m_u \simeq 300\text{ MeV} \end{pmatrix} . \tag{59}$$

The dependence on m_u^{-2} is a warning that perturbation theory might be misleading and $P^{(8)}$ could perhaps be larger. What about $|\psi^{(8)}(0)|^2$?

If one assumes that $\psi^{(8)}_{J=1} \simeq \psi^{(1)}_{J=0,1}$ then $\psi(0)$ can be related to f_D, the analogue of f_π for the D meson:

$$f_D = \frac{12}{M_D} |\psi(0)|^2 \ . \tag{60}$$

Theoretical estimates[12] of f_D from potential models[23], QCD sum rules[24], and bag models[25] all lead to values in the range $f_D = (150\text{-}300)$ MeV. From the hyperfine splitting of D^* and D one obtains[12]

$$\Delta M^2 = \frac{64\pi}{9} \alpha_s \frac{|\psi(0)|^2}{m} \Rightarrow f_D = 380 \text{ MeV} \quad \begin{pmatrix} \alpha_s = 0.3 \\ m = 300 \text{ MeV} \end{pmatrix} . \tag{61}$$

In conclusion, from f_D estimates it is difficult to imagine a violation of the bound $\Gamma_{ann}/\Gamma_Q \lesssim 1\text{-}2$ for D^0, so that one expects $\tau(D^+)/\tau(D^0) \lesssim 2\text{-}3$. The D^+ is almost unaffected, the c decay mechanism is dominant, and the prediction for $B_{s\ell}(D^+)$ in Eq. (57) is confirmed. For F^+ the decay $F^+ \to \tau^+\nu_\tau$ is a good f_F-meter. For $f_F \sim 150$ MeV one expects $B(F \to \tau\nu) \simeq 1\%$. In the non-leptonic channels annihilation with one-gluon exchange is proportional to $(c_- - c_+)^2\, P^{(8)}$ for F^+. If $P^{(8)}$ is large, then the probability of two-gluon emission could also be non-negligible. In this case the relevant factor is $3X_+^2\, P^{(1)}$ (from a colour singlet, spin 1 $c\bar{s}$ pair), which is colour-wise more favourable. In this case the lifetime of F^+ could be shortened near or below the D^0 level, and there could be a further sizeable contribution from a final state made up of leptons plus gluons to the semileptonic width of F^+ beyond quark decay.

In Λ_c decays the cd pair can have spin one even with no gluon emission and the annihilation rate for Λ_c depends on the baryon wave function, which is unrelated to the meson one. The observed value of the Λ_c lifetime indicates a sizeable annihilation term.

For beauty decay, violations of the parton model are expected to go down compared to charm decay. The spectator effects are certainly completely negligible (below 1%) because more phase space leads to far less wave-function overlap. Mass effects are also negligible unless two or more τ and/or c are present in the final state at the same time. Finally, the annihilation mode is decreasing with some power of m_Q. In the non-relativistic approximation, if one neglects the variation of $P^{(8)}$ (which is bound by one), then the ratio Γ_{ann}/Γ_Q decreases as $1/m_Q^3$. But this is very speculative. Neglecting the annihilation contribution one expects:

$$B_{s\ell}(B^0) \simeq B_{s\ell}(B^-) \simeq (13\text{-}15)\% \qquad \text{(no annihilation)}$$

while with annihilation the B^0 lifetime is shortened and

$$B_{s\ell}(B^-) \gtrsim B_{s\ell}(B^0) \qquad \text{(annihilation)} \ .$$

5. CONCLUSION

The picture is getting clearer and a reasonable understanding of heavy flavour decays is slowly emerging. On the one hand the experimental information is getting into focus and the initial indications of a giant ratio of D^+ and D^0 lifetimes are now replaced by a more reasonable (for theory) value of two or so. On the other hand, this crucial factor is now attributed, by general consensus, to the contribution of the annihilation term, while all other proposed mechanisms can presumably be discarded. Although the annihilation term cannot be precisely predicted, it is convergently estimated to be for charm of the same order as the quark-decay contribution. Furthermore, since the D^+ rates are almost unaffected by the annihilation term they are calculable and allow a test of the QCD-improved parton model. In particular, the experimental verification of the prediction for the semileptonic branching ratio of D^+, $B_{s\ell}(D^+) \simeq 15\%$, would be a striking confirmation of the QCD enhancement factor for non-leptonic rates arising from short-distance effects. For this interpretation to be unambiguous the accompanying signatures on the lifetimes of all charmed particles and on the quantum numbers of the exclusive modes and their relative intensities should eventually be detected. It is also of the utmost importance to measure beauty decays and verify that in this case the parton approximation is working far better than in the case of charm decay. Finally, we observe that once heavy flavour decays can be considered as understood, which we hope now to be not so far in the future, then one can probably reconsider the unsolved aspects of strange-particle decays and take advantage of the valuable lessons which have been learnt in order to tackle this more difficult problem.

REFERENCES

1. Particle Data Group, Phys. Lett. 111B:1 (1982).
2. M. Kobayashi and T. Maskawa, Progr. Theor. Phys. 49:652 (1973).
3. J.L. Cortes, X.Y. Pham and A. Tounsi, Phys. Rev. D25:188 (1982).
4. N. Cabibbo and L. Maiani, Phys. Lett. 79B:109 (1978); M. Suzuki, Nucl. Phys. B145:420 (1978).
5. G. Altarelli, N. Cabibbo, G. Corbò, L. Maiani and G. Martinelli, Nucl. Phys. B208:365 (1982).
6. N. Cabibbo, G. Corbò and L. Maiani, Nucl. Phys. B155:93 (1979).
7. G. Corbò, Phys. Lett. 116B:298 (1982), Nucl. Phys. B212:99 (1983).
8. For a review of charm decay see, for example, G. Trilling, Phys. Rep. 75:57 (1982).
9. L. Maiani, Phys. Lett. 62B:183 (1976).
10. H. Abramowicz et al., Z. Phys. C 15:19 (1982).
11. C. Jarlskog, Rapporteur's talk at Int. Europhysics Conf. on High Energy Physics, Brighton (UK), 1983.
12. K. Wilson, Phys. Rev. 179:1499 (1969).

13. M.A. Shifman, A.I. Vainshtein and V.I. Zakharov, Nucl. Phys. B120:316 (1977); Sov. Phys.-JETP 45:670 (1977).
14. G. Altarelli, G. Curci, G. Martinelli and S. Petrarca, Phys. Lett. 99B:141 (1981); Nucl. Phys. B187:461 (1981);
See also: G. Altarelli, Phys. Rep. 81:1 (1982).
15. B.W. Lee and M.K. Gaillard, Phys. Rev. Lett. 33:108 (1974);
G. Altarelli and L. Maiani, Phys. Lett. 52B:351 (1974).
16. For recent analyses see:
F.I. Gilman and M.B. Wise, Phys. Rev. D20:2392 (1979);
B. Guberina and R.D. Peccei, Nucl. Phys. B163:289 (1980);
F. Buccella, M. Lusignoli, L. Maiani and A. Pugliese, Nucl. Phys. B152:461 (1979).
17. L. Maiani, Proc. 21st Int. Conf. on High Energy Physics, Paris, 1982, J. Phys. 43:Suppl. 12, C3-631 (1982).
18. B.W. Lee, M.K. Gaillard and G. Rosner, Rev. Mod. Phys. 47:277 (1975).
G. Altarelli, N. Cabibbo and L. Maiani, Nucl. Phys. B88:285 (1975); Phys. Lett. 57B:277 (1975);
S.R. Kingsley, S. Treiman, F. Wilczek and A. Zee, Phys. Rev. D11:1914 (1975).
J. Ellis, M.K. Gaillard and D. Nanopoulos, Nucl. Phys. B100:313 (1975).
19. J.L. Cortes et al., Phys. Rev. D25:188 (1982).
U. Baur and H. Fritzsch, Phys. Lett. 109B:402 (1982).
See also: Q. Hokim and X.Y. Pham, Univ. Paris VI and VII preprint PAR-LPTHE 83/05 (1983).
20. B. Guberina, S. Nussinov, R.D. Peccei and R. Rückl, Phys. Lett. 98B:111 (1979);
See also: R.D. Peccei and R. Rückl, MPI-PAE/PTH 75/81/1981;
T. Kobayashi and N. Yamazaki, Progr. Theor. Phys. 65:775 (1981).
21. G. Altarelli and L. Maiani, Phys. Lett. 118B:414 (1982).
See also: H. Sawayanagi et al., Phys. Rev. D27:2107 (1983).
22. M. Bander, D. Silverman and A. Soni, Phys. Rev. Lett. 44:7 (1980); Errata 44:962 (1980);
H. Fritzsch and P. Minkowsky, Phys. Lett. 90B:455 (1980); Nucl. Phys. B171:413 (1980);
W. Bernreuther, O. Nachtmann and B. Stech, Z. Phys. C4:257 (1980).
23. H. Krasemann, Phys. Lett. 961B:397 (1980).
24. V.A. Novikov et al., Phys. Rev. Lett. 38:626 (1977);
E.V. Shuriak, Nucl. Phys. B198:83 (1982).
25. E. Golowich, Phys. Lett. 91B:271 (1980).

DISCUSSION

CHAIRMAN : G. ALTARELLI

Scientific Secretaries : G. D'Ambrosio and C. Ogilvie

- *KLEVANSKY* :

Could you clarify the presence of penguin diagrams and explain their significance ?

- *ALTARELLI* :

First of all, penguin diagrams superficially look one particle reducible; if they were, they should not be included in the operator product expansion. Actually, the $1/q^2$ gluon pole is cancelled by the vertex, which is proportional to $(\nabla_\mu F_{\mu\nu})\cdot[\bar{\psi}\gamma_\nu(1-\gamma^5)\psi]$. Thus penguin diagrams amount to an additional four-fermion interaction, which has the characteristic property of being V - A at one end and pure vector at the other end.

In the limit of equal masses for the quarks, penguin diagrams would vanish by GIM mechanism.

They are much less important for heavy flavors than for strange particle decay. For charm, they are completely irrelevant, while for b decays they presumably contribute a small effect, even if not completely negligible.

- *BERNSTEIN* :

The semi-leptonic branching ratio of D^+ seems to require M_s = 500 MeV and M_u = 300 MeV. Doesn't this conflict with successful prediction of $BR(\tau \rightarrow e\,\nu_\tau\,\bar{\nu}_e)$?

- *ALTARELLI* :

Let me summarize the situation with respect to D^+ decays. The Hamiltonian for non-leptonic decay is more or less known as a four-fermion interaction plus a penguin part. The penguins are not important in charm decay. The four-fermion operators are multiplied by coefficients which have been computed. The picture for charm is not terribly good in the parton approximation. This approximation would in fact imply equal lifetimes for all charmed particles, which is not the case.

Instead, there is an additional contribution from annihilation. The D^+ is special, because it is a $c\bar{d}$ state, and annihilation is Cabibbo suppressed. So it is interesting to precisely measure the D^+ semileptonic branching ratio, because in that case you measure the c-quark semileptonic branching ratio to a good approximation.

Mass effects have been considered in that they tend to enhance B_{SL} by suppressing the non-leptonic channels. In my opinion there is no need for this effect.

- *GASPARINI* :

Could you quantify the expected decay rate for F meson going into a final state with no strange quark via the annihilation mechanism ?

Can you give me some more explanation about the computation of the annihilation diagram amplitude when one or more gluons are radiated ? Particularly in which way the non-perturbative nature of the interaction is taken into account ?

- *ALTARELLI* :

Essentially, we have for F meson :

$$\frac{\Gamma(F \to \text{no } s\bar{s})}{\Gamma(F \to s\bar{s})} \sim \frac{\Gamma \text{ ann.}}{\Gamma \text{ spect.}}$$

which I have shown in my previous discussion on D^o lifetime to be of the order of unity. So we expect about 50% of F mesons to decay in non-strange mesons.

The presence of a gluon in the wave function of the initial state is a non-perturbative effect which we are not able to calculate precisely. But we can estimate the radiation of a gluon from a quark by perturbation theory. There is then a non-perturbative factor from the wave function at the origin of the quark-antiquark system which can be estimated in different ways : potential models, QCD sum rules and bag models; all converge towards values which do not differ more than a factor 2 or 3. So we can make a reasonable estimate of the annihilation amplitude.

- *LAMARCHE* :

What would change if the b lifetime was found to be shorter than reported at SLAC ?

- *ALTARELLI* :

We know $\sin^2\gamma < \frac{1}{2}$ roughly. Thus provided τ_B is greater than $(5 \div 9)10^{-15}$sec., there is no theoretical revolution.

- *MILOTTI* :

Why does the decay $b \rightarrow c+\ldots$ look different from $b \rightarrow u+\ldots$ near the end point ?

- *ALTARELLI* :

In $b \rightarrow c+\ldots$ there is less energy available for the electron than in the other case, because $m_c > m_u$.

- *BALLOCCHI* :

I would like to know more about the ambiguities in Γ_{sl} ?

- *ALTARELLI* :

The main uncertainty is the value for the masses. For semi-leptonic decays in general, the quark decay approximation is a good one; annihilation is not important for the semileptonic width (possible exception : the F, if two gluons can be emitted with sizeable probability).

The semileptonic width depends on the 5-th power of the quark masses. A small error in the masses means a large error in the width (a 30% error gives a factor of 4 in the width).

Our idea was to try to connect the effective mass to something measurable; to relate the effective mass to the electron spectrum. By a study of the electron spectrum we can deducethe characteristic momentum spread in the wave function. Once we know this, we derive an effective value for the quark mass.

- *BALLOCCHI* :

I think $p_F = 0$ gives a good fit for $\frac{d\Gamma}{dE}$ versus E in D decays. What do you think about that ?

- *ALTARELLI* :

It is good that p_F turns out to be small because it means that the free quark decay is not a terribly bad approximation for Γ_{sl}.

- *BAILEY* :

Would you comment on the possibility of $B^o - \bar{B}^o$ mixing and the consequence of B^o_{long} and B^o_{short} lifetimes ?

- *ALTARELLI* :

There could be interesting effects from mixing, but this could not explain large lifetime differences. The different lifetimes of the K^o_S and K^o_L are because the K^o_L must decay into three pions and this is phase space suppressed compared to the two-pion decay of the K^o_S. No such suppression should exist for the decays of heavy mesons. I would thus expect the lifetimes of the B^o_L and B^o_S to be almost degenerate.

SPIN DEPENDENCE AND TESTS OF QCD

Elliot Leader

Westfield College, London N.W.3

1) INTRODUCTION

It is presently believed that QCD is the correct theory of the strong interactions.

Since the theory, ultimately, will provide detailed information on the distribution of the spin and momentum of the constituents within a hadron, it is somewhat meaningless to concentrate soley on the momentum distribution and to regard the spin distribution as something separate and disconnected. Thus deep inelastic lepton scattering measurements with polarised beams and polarised targets are vital; just as important, in the long run, as the unpolarised experiments are.

There are two very distinctive features of QCD. One, the non-Abelian nature of its structure, is much emphasised, and indeed is crucial for its non-perturbative behaviour. But the other, the vector nature of the coupling between the fundamental fermions and gluons, is just as important and specific, yet is often overlooked as a crucial feature of the theory. The fact that the coupling is of the form

$$\psi \gamma_\mu \psi A^\mu$$

(just as it is in QED) is a non-trivial element and has powerful consequences, especially in controlling the spin-dependence of scattering amplitudes.

It is therefore very important indeed to test this structure of the theory and the best way to do that is via large p_T hadronic reactions involving polarised beams and/or targets and/or final-state hadrons, such as the hyperons, whose polarisation can also be measured. The perturbative QCD predictions for these processes are rather clean, but depend upon a knowledge of both the spin and momentum distributions of the constituents of the hadrons, which can be deduced from deep inelastic lepton scattering experiments, of the type mentioned above.

This lecture is intended to be pedagogical, and certainly not comprehensive. We shall thus sample the situation by examining two typical questions, one in the realm of non-perturbative QCD, the other relevant to perturbative QCD. We shall firstly concentrate upon one typical, but key non-perturbative question:

In a proton of helicity Λ what is the number density of "up" quarks with momentum fraction x and with helicity λ either parallel (P) or anti-parallel (A) to Λ?

The calculation of these number densities $\mu_P(x)$, $\mu_A(x)$ is a classic problem in non-perturbative QCD. It is exceedingly difficult but not beyond hope with sophisticated lattice field theory approaches. This sort of calculation provides the ultimate test as to whether QCD truly describes the dynamics of hadrons - it gives quantitatively the quark structure of the hadrons.

We shall discuss which quantities are of importance and how they can be measured experimentally.

Secondly, we shall ask a typical key question in perturbative QCD: What is the polarisation of a Λ-hyperon produced at large p_T in the collision of unpolarised protons? We shall see that we need as input for the calculation of $p + p \to \Lambda + X$ just these non-perturbative quantities mentioned above. Thereafter the predictions are fairly clean and restrictive and should provide a fertile area for testing QCD. We shall end with a rapid survey of the present phenomenological situation.

It should be noted that we will work at the level of the simple parton model and shall not here discuss scale breaking and other questions of Q^2 dependence.

2) The Non-Perturbative Element

In the "classical" parton model a hadron is just a container carrying the quarks and gluons into the reaction; it is like a source providing a beam of partons which interact. Being classical probabilistic in structure, the model for a proton in a definite helicity state is described in terms of just two density functions for each type of constituent. Thus for say the "up" quark the two number densities $\mu_P(x)$ and $\mu_A(x)$ defined in Section 1 are the whole story. The usual number density $\mu(x)$ is given by

$$\mu(x) = \mu_P(x) + \mu_A(x) \tag{1}$$

This "classical" picture is too simplistic and we shall return to this question later.

Let us turn now to the problem of measuring these number densities. Eventually we would hope to confront the experimental results with calculated values, but the non-perturbative computations are not yet refined enough. For the present our measurements are simply telling us about the structure of a hadron and we shall utilise the information for the perturbative calculations in Section 3.

The principal and most direct source of information is deep inelastic lepton scattering using polarised lepton beams and polarised hadron targets. (That both must be polarised is essential, as will become obvious later. Experimentally the lepton beams present no problem - they are already highly polarised, ν and $\bar{\nu}$ 100% and μ's somewhat less. The technology of polarised targets has been improving steadily and a recent breakthrough[1] suggests that a dramatic improvement is imminent). For purposes of discussion we shall talk about electromagnetic processes say $\mu p \to \mu X$. The process is, as usual, visualised to proceed via one-photon exchange.

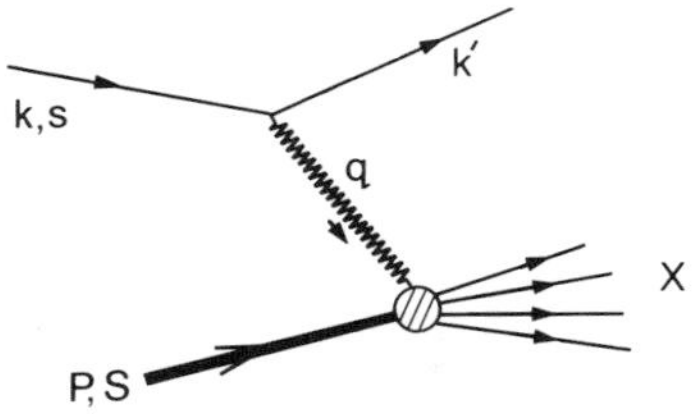

Figure 1.

Besides the obvious momentum labels, s and S are covariant spin vectors describing the spin states of the initial lepton and hadron. All final state spins are unobserved and therefore summed over.

The differential cross-section has the form:[3]

$$\frac{d^2\sigma}{d\Omega dE'}(s,S) = \frac{\alpha^2}{2Mq^4}\left(\frac{E'}{E}\right) L_{\alpha\beta}(s)W^{\alpha\beta}(S) \qquad (2)$$

where E,E' are the initial and final lepton Lab. energies, M is the hadron mass, and $L_{\alpha\beta}, W^{\alpha\beta}$ are respectively the leptonic and hadronic tensors which now depend on the spin vectors s and S.

The leptonic tensor is exactly known

$$L_{\alpha\beta}(s) = L^{SYM}_{\alpha\beta} + iL^{ANTI}_{\alpha\beta}(s) \qquad (3)$$

where $L_{\alpha\beta}^{SYM}$, the symmetric piece that appears for unpolarised leptons, is given by

$$L_{\alpha\beta}^{SYM} = 2[k_\alpha k'_\beta + k'_\alpha k_\beta - g_{\alpha\beta}(k.k' - m_\ell^2)] \qquad (4)$$

and the anti-symmetric piece is, for a lepton of helicity h,

$$L_{\alpha\beta}^{ANTI}(h) = -4h\varepsilon_{\alpha\beta\gamma\delta}k^\gamma k'^\delta \qquad (5)$$

The hadronic tensor, whose structure it is the aim of these experiments to discover, has the general form:

$$W_{\alpha\beta}(S) = W_{\alpha\beta}^{SYM} + iW_{\alpha\beta}^{ANTI}(S) \qquad (6)$$

with the usual symmetric piece that occurs for unpolarised hadrons given by

$$\frac{1}{2M} W_{\alpha\beta}^{SYM} = \left(\frac{q_\alpha q_\beta}{q^2} - g_{\alpha\beta}\right) W_1 + \frac{1}{M^2}\left(P_\alpha - \frac{P\cdot q}{q^2} q_\alpha\right) x$$

$$\left(P_\beta - \frac{P\cdot q}{q^2} q_\beta\right) W_2 \qquad (7)$$

where P is the hadron momentum and $W_{1,2}$ the usual spin-independent structure functions.

The anti-symmetric piece can be written

$$\frac{1}{2M} W_{\alpha\beta}^{ANTI}(S) = \varepsilon_{\alpha\beta\gamma\delta}\, q^\gamma \left\{ S^\delta \left[MG_1 + \frac{P\cdot q}{M} G_2 \right] - P^\delta \frac{S\cdot q}{M} G_2 \right\} \qquad (8)$$

where $G_{1,2}$ are often referred to as the spin-dependent structure functions, though they themselves do not depend upon S. [4)]

All the structure functions $W_{1,2}$, $G_{1,2}$ are scalar functions that can in principle depend upon two scalar variables

$$Q^2 \equiv -q^2 \quad \text{and} \quad \nu \equiv \frac{P\cdot q}{M} \qquad (9)$$

It is the content of Bjorken scaling that in the limit $Q^2 \to \infty$, $\nu \to \infty$ in such a way that

$$x \equiv \frac{Q^2}{2M\nu} \quad \text{remains finite,} \tag{10}$$

the structure functions scale in a specially simple fashion:

$$MW_1\ (\nu, Q^2) \rightarrow F_1(x) \qquad M^2\nu G_1\ (\nu, Q\ ) \rightarrow g_1(x)$$

$$\nu W_2\ (\nu, Q^2) \rightarrow F_2(x) \qquad M\nu^2 G_2\ (\nu, Q\ ) \rightarrow g_2(x) \tag{11}$$

The experimental determination of $G_{1,2}$ is carried out as follows. Using a longitudinally polarised lepton beam (indicated by $\rightarrow$) on a polarised proton target one compares the differential cross-section for the target protons polarised longitudinally i.e. along or opposite to the beam direction ($\rightarrow$ or $\leftarrow$) or polarised perpendicularly, in the reaction plane ($\rightarrow$ or $\leftarrow$).

The important cross-section differences are:

$$\frac{d\sigma^{\overset{\rightarrow}{\leftarrow}}}{d\Omega dE'} - \frac{d\sigma^{\overset{\rightarrow}{\rightarrow}}}{d\Omega dE'} = \frac{4\alpha^2 E'}{Q^2 E}\left\{(E + E'\cos\theta)\, MG_1 - Q^2 G_2\right\} \tag{12}$$

and

$$\frac{d\sigma^{\rightarrow\downarrow}}{d\Omega dE'} - \frac{d\sigma^{\rightarrow\uparrow}}{d\Omega dE'} = \frac{4\alpha^2 (E')^2}{Q^2 E}\sin\theta \left\{MG_1 + 2EG_2\right\} \tag{13}$$

from which, clearly, one can determine G_1 and G_2.

Experiment therefore can give us the four scaling functions $F_{1,2}(x)$, $g_{1,2}(x)$.

In the classical parton model these are given in terms of the two basic number densities $q_P^j(x)$ and $q_A^j(x)$ for each species of quark j. One has [3]

$$F_1(x) = \tfrac{1}{2} \sum_j Q_j^2 \left\{q_P^j(x) + q_A^j(x)\right\} \tag{14}$$

$$g_1(x) = \tfrac{1}{2} \sum_j Q_j^2 \left\{q_P^j(x) - q_A^j(x)\right\} \tag{15}$$

Clearly, in this model, two relations must hold between the scaling functions. These are the usual Callan-Gross relation

$$F_2(x) = 2xF_1(x) \tag{16}$$

and also

$$g_2(x) = -g_1(x) \tag{17}$$

The latter relation can be tested directly. It follows from (17) that for the differential cross-sections

$$\frac{(1 - y)\sin\theta}{y} \cdot \frac{d\sigma^{\overset{\rightarrow}{\Leftarrow}} - d\sigma^{\overset{\rightarrow}{\Rightarrow}}}{d\sigma^{\rightarrow\Uparrow} - d\sigma^{\rightarrow\Downarrow}} = 1 \tag{18}$$

where, as usual, $y \equiv \nu/E$ and we have neglected terms of order M^2/Q^2 on the R.H.S. This prediction is remarkably strong, since in principle the L.H.S. can depend on E,y and x.

I believe that these predictions are unrealistic because the classical parton model, dealing only with probabilities is too restrictive. From a modern point of view, given a QCD Lagrangian, we have a well defined set of Feynman diagrams, and partonic type results will emerge from a subset of these. It is no longer necessary to invoke infinite momentum frames, which is any case are a nuisance when several hadrons are involved. A more general approach was suggested by Ralston and Soper [5] in their analysis of Drell-Yan reactions, though they still clung to infinite momentum frames. We shall outline very briefly an analogous approach which seems interesting for spin dependence. It can be regarded as a generalisation of the classical parton model.

In lowest order QCD and at large q^2 the process of interest is given by the following diagram:

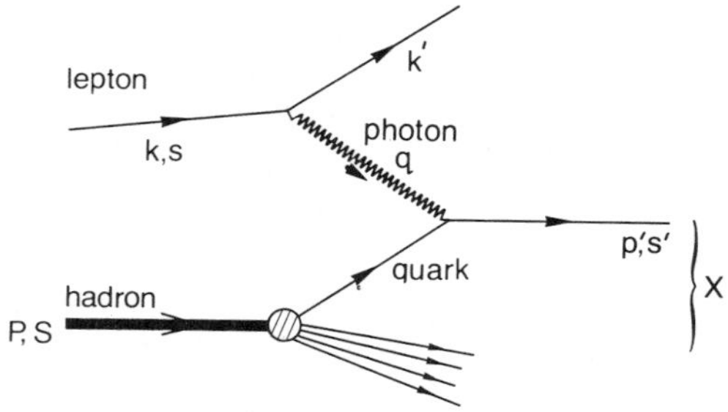

Figure 2.

where p and p' label quark lines.

After taking the modulus squared one is to sum over all momenta and spins in X, including p' and s'. The result then depends upon the Green's function (strictly its absorptive part) $\hat{\Phi}_{\alpha\beta}$

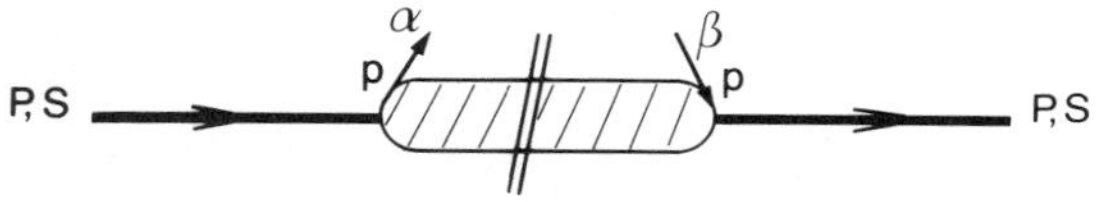

Figure 3.

which is the generalisation of the quark number densities. (α,β are Dirac indices). In detail

$$\hat{\Phi}_{\alpha\beta} \equiv \int d^4z e^{iz\cdot p} \langle P,S|\bar{\Psi}_\beta(0)\bar{\Psi}_\alpha(z)|P,S\rangle \quad (19)$$

where Ψ is the quark field.

One then finds the following:-

i) $\hat{\Phi}$ depends on NINE independent scalar functions!

ii) $W_{\mu\nu}$, when calculated using the most general form for $\hat{\Phi}$ is not gauge invariant.

Clearly this is too far from the parton model. So we introduce the condition that the partons are essentially free during the collision by taking only the mass-shell contribution i.e. we restrict $p^2 \approx m^2$. The nine functions immediately reduce to THREE and $\hat{\Phi}$ collapses to the form of a standard covariant density matrix for a Dirac particle [2]:-

$$\hat{\Phi} = N\,(\not{p} + m)(I + \gamma_5\ \not{s}\ ) \tag{20}$$

where N is a normalisation function (it is the number density of quarks produced with momentum $\not{p}$) and s is "effective spin vector" of the emitted quark:

$$s_\mu = B\,(S_\mu - \frac{p.S}{p.P} P_\mu) + C\,\frac{p.S}{m\ M}(p_\mu - \frac{m^2}{p.P} P_\mu) \tag{21}$$

where B,C are (unknown) scalar functions controlled by the basic dynamics.

Using (20) yields a perfectly gauge-invariant $W_{\mu\nu}$ and the four scaling functions are now expressed in terms of THREE independent partonic functions. The Callan-Gross relation (16) still holds, but now (17) is not valid, i.e.

$$g_1(x) \neq -g_2(x) \tag{22}$$

so (18) is not expected to hold.

It is most interesting that the sum $g_1(x) + g_2(x)$ differs from zero only by terms proportional to m^2/M^2 and $p_T^2\,/\,M^2$. It thus provides a possible method of studying these quantities in a situation where they are not damped by Q^2 factors.

In the generalised parton description given by (20) a hadron acts as a source of partons, on-shell, and described by a non-diagonal density matrix. This is more general than the classical parton model where the parton density matrix is diagonal. Equation (20) forms a useful basis from which to attack spin dependence in other reactions.

3) Perturbative QCD - Large p_T Reactions

We turn now to spin-dependent observables which can be calculated perturbatively and which, as will be seen, provide quite a severe test for QCD. Clearly we must restrict ourselves to large p_T reactions where the perturbative treatment is justified. Our discussion will be based on the classical parton model only, so will be subject to some modification, but the principal points will not be altered significantly.

As input we require the number density functions $q_P(x)$ and $q_A(x)$ for at least "up" and "down" quarks. But what is measured at present is only the cross-section asymmetry (12) for protons. Assuming the validity of (17) we can derive from this the function $g_1(x)$ for protons. The figure below shows essentially the x-dependence of

$$\frac{2xg_1(x)}{F_2(x)} = \frac{\sum_j Q_j^2 \, [q_P^j(x) - q_A^j(x)]}{\sum_j Q_j^2 \, [q_P^j(x) + q_A^j(x)]} \tag{23}$$

which is something like the mean polarization of the quarks at fixed x, as measured by the SLAC-YALE group [6].

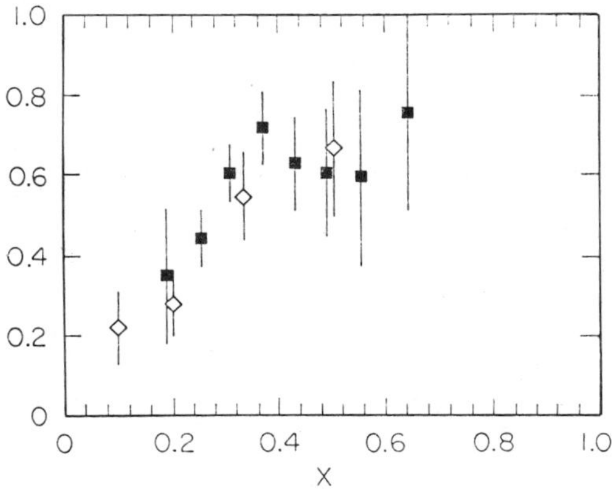

Figure 4.

It is clear that there is a significant polarisation as $x \to 1$ and correspondingly almost none as $x \to 0$. We shall not discuss phenomenological models of these distributions (they are not very profound) and we simply use this data as a rough guide to the shape of the number density functions.

We stress, however, that much more data is needed, especially on neutrons, and also for the perpendicular spin orientation, before we can hope to extract reliable information to use as input for the perturbative calculations.

A hadronic interaction at large p_T, say

$$A + B \to C + X$$

is typically viewed as follows:

and thus consists of non-perturbative distribution and fragmentation input functions convoluted with the high p_T, QCD controlled parton-parton interaction calculated perturbatively.

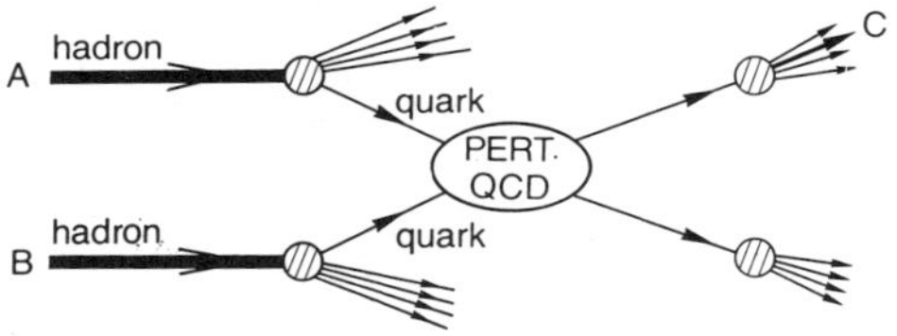

Figure 5.

The spin structure of the perturbative QCD reaction is very restrictive. Suppose for a moment that quarks are massless. In that case the Dirac spinors for helicity $\pm\frac{1}{2}$ (i.e. R or L) are eigenspinors of γ_5:-

$$\gamma_5 U_R = U_R \qquad \gamma_5 U_L = -U_L$$
$$\overline{U}_R \gamma_5 = -\overline{U}_R \qquad \overline{U}_L \gamma_5 = \overline{U}_L \tag{24}$$

Suppose now that a quark line with helicity λ enters an <u>arbitrarily complicated</u> Feynman diagram and emerges with helicity λ' as shown:

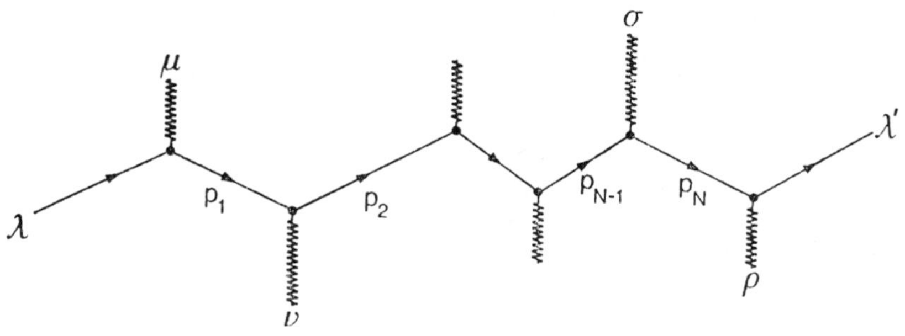

Figure 6.

All internal quark lines will have propagators of the form $\frac{\not p}{p^2}$ so the γ-matrix structure will look like

$$M_{\lambda'\lambda} \equiv \overline{U}_{\lambda'}\gamma_\rho \not p_N \gamma_\sigma \;\cdots\; \gamma_\nu \not p_1 \gamma_\mu U_\lambda \tag{25}$$

Now replace U_λ by $\pm\gamma_5 U_\lambda$ according to (24), anti-commute the γ_5 step by step towards the left until it hits $\overline{U}_{\lambda'}$ and replace $\overline{U}_{\lambda'}\gamma_5$ by $\pm\overline{U}_{\lambda'}$ using (24). It will be found that there are always an odd number of steps and that one ends up with

$$M_{\lambda'\lambda} = M_{\lambda'\lambda} \quad \text{if} \quad \lambda' = \lambda$$

but (26)

$$M_{\lambda'\lambda} = -M_{\lambda'\lambda} \quad \text{if} \quad \lambda' = -\lambda$$

So the amplitude vanishes if $\lambda' \neq \lambda$. In other words the helicity of an incoming quark is unchanged in any finite order diagram. The above assumed massless quarks. It will be invalid to the extent that m/E for the quark is not negligible in the particular reference frame that one is untilising. For the large p_T reactions under discussion m/E should always be <<1 and the conclusion that the helicity does not flip should be essentially valid.

This sort of argument does not work for gluons. So, even though they are massless, they can emerge from a diagram with altered helicity. But this is not important since our primary question is how to flip the helicity of a hadron and the sea of gluons in the hadron is quite insensitive to the spin state of the parent hadron.

By and large then hadrons entering a large p_T reaction will tend to emerge with their helicities unaltered. This means that the polarisation will be essentially zero; but at the same time certain other spin parameters or correlations will be large.

Indeed the statement about polarisation is even stronger, since to get a non-zero polarisation requires <u>both</u> the flipping of a helicity <u>and</u> the existence of an imaginary part to the amplitude[7]. To get an imaginary part one must go to higher order than Born approximation and thus one expects a small result also on account of the smallness of the coupling constant.

It seems therefore a 3-star prediction that

$$P \approx 0 \text{ at large } p_T \tag{27}$$

in both exclusive and inclusive reactions.

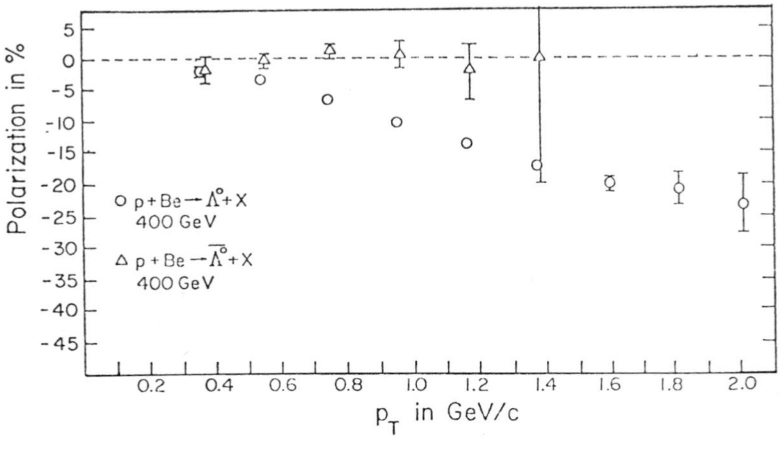

Figure 7.

The present data, for the hyperon polarisation in $p + p \to$ hyperon $+ X$ is summarised in the following graphs:-

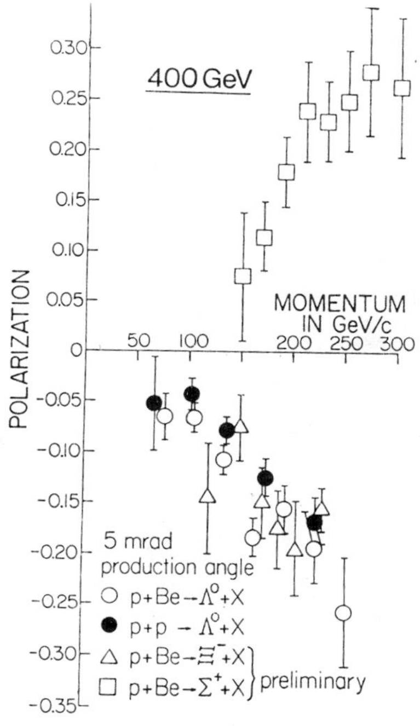

Figure 8.

It is seen that the polarisation is far from zero and that it displays interesting dependence upon the hyperon type. The crucial question is whether p_T is large enough to justify comparison with the perturbative conclusion - presumably not. The maximum p_T in these experiments is around 2 GeV/c, whereas one might demand $\gtrsim$ 4 to 5 GeV/c for a valid comparison.

So we must suspend judgement and await with great interest the result of future experiments at larger p_T. Their importance cannot be overemphasised. The basic vector nature of QCD is under scrutiny!

Time does not permit me to discuss other reactions, or spin correlations. Access to the literature can be obtained from my paper in the CERN Workshop on SPS Fixed Target Physics in the years 1984-1989 [8].

4) Conclusions

The ultimate test of QCD will be to compare the measured momentum and spin distributions of the partons inside a hadron with the results of non-perturbative calculations. The latter are not yet ready, so in the meantime it should be a major experimental aim to gather as much data as possible on both the momentum and the spin dependence of the partons. This is best attacked in Deep Inelastic and Drell-Yan reactions using polarised beams and targets.

With these distributions known from experiment one can use them as input in order to calculate perturbatively the QCD predictions for various reactions at large p_T. The measurements of polarisations and other spin-dependent correlations will then provide a critical test for the vector nature of the quark-gluon coupling in QCD.

Present day experiments seem, tantalizingly, to disagree with the perturbative QCD predictions, but this may only be a consequence of too low p_T in the reactions thus far investigated. Measurements at larger p_T are vitally important.

References

1) See Proc. High Energy Spin Physics Conference, Brookhaven Natl. Lab. 1982.

2) J.D. Bjorken and S.D. Drell: "Relativistic Quantum Mechanics", Chapter 1. (McGraw-Hill, 1964).

3) E. Leader and E. Predazzi: "An Introduction to Gauge Theories and the 'New Physics'", Chapters 12 and 13. (Cambridge Univ. Press, 1982).

4) Equations (2), (3) and (6) show that both beam and target must be polarised to get information on $G_{1,2}$.

5) J.P. Ralston and D.E. Soper, N.P. B152, 109(1979). Note that these authors found only 8 of the 9 invariant functions.

6) See talk by R.F. Oppenheim in ref (1), p.255.

7) For example in $\pi p \to \pi p$ the polarisation P of the proton is given by $\text{Im} f^*_{++} f_{+-}$. Even if $f_{+-} \neq 0$, P will only be non-zero if the phase of f_{+-} differs from that of f_{++}.

8) E. Leader: Proc. Workshop on SPS Fixed Target Physics in the Years 1984-1989, Vol II, pp 23 and 384, CERN, 1983.

DISCUSSION

CHAIRMAN : E. LEADER

Scientific Secretary : S.P. Klevansky

- *BERNSTEIN* :

Why is the Λ^0 produced polarized ? Could you give me a physical explanation ?

- *LEADER* :

That is easy, I cannot! Within QCD there is no decent explanation to date, and I would argue that the polarization is zero, provided that the perturbative argument is valid. You can say that $p_\perp$ is too small in those experiments; you should not trust perturbative QCD. So I cannot do a calculation for you. There are other models on the market, but these are extremely artificial and unconvincing.

- *BERNSTEIN* :

Do you understand the polarization of the pp elastic cross-section measured at Argonne and Brookhaven ?

- *LEADER* :

No. The reason is that we can identify at least three classes of diagrams, so called end-point diagrams of Swed, Landshoff and Brodsky-Lepage. All three classes have different spin properties, and the relative normalization of these is unknown. Although there are many quantitative statements in the literature, I would regard them with scepticism, since they are always based on results from one type of diagram.

- *BERNSTEIN* :

What should we measure experimentally, as the cleanest way to understand these proeesses ?

- *LEADER* :

Measure the polarizations in inclusive reactions. The cleanest prediction of perturbative QCD is that the polarization should vanish as $p_\perp$ increases in inclusive reactions.

- *LAMARCHE* :

What is the meaning of the symbols $d\sigma^{\rightarrow\uparrow}$ and $d\sigma^{\rightarrow\downarrow}$; I do not see how nature can make any distinction between the upward and downward arrows.

- *LEADER* :

Yes, nature can: let me explain what the arrows mean. A horizontal thin arrow means the lepton has its helicity along its motion $\rightarrow$ or against its motion $\leftarrow$. The arrows $\rightarrow$ or $\leftarrow$ indicate proton spin along or opposite to the beam direction. Since the reaction plane is defined by the placing of the experimental counters, it is possible to define "up" or "down" with respect to this plane. By convention "up" corresponds to the direction $P_i \times P_f$, where $P_{i,f}$ are the initial and final beam momenta respectively. The arrows $\uparrow$, $\downarrow$ indicate proton spin along these directions.

- *MOUNT* :

My question concerns the use of relationships between structure functions to study $p_\perp$ effects. You said that the breakdown of $g_1 = -g_2$ was due to $p_\perp$. In what way is this better than using existing measurements of breakdown of the Callan-Gross relation ?

- *LEADER* :

At the level of the classical parton model, there is no breakdown of the Callan-Gross relation. The breakdown comes from scaling violations. The breakdown of $g_1 = -g_2$ may be proportional to $p_\perp^2/M^2$ and would therefore be much larger than the $p_\perp^2/Q^2$ deviations from the Callan-Gross relation.

- *ARNEODO* :

In order to get a non-zero polarization in a reaction, you

mentioned that one needs an imaginary part. Why is this so ?

- *LEADER* :

Consider a hadronic reaction A+B $\rightarrow$ C+D, i.e. an elastic or pseudoelastic reaction. Suppose A and B are unpolarized, and you ask for the polarization of C, i.e. the mean spin vector of particle C. Parity conservation forces that vector to lie perpendicular to the reaction plane. Using the usual convention, perpendicular refers to the y-axis, and the mean value of spin in the y-direction involves taking the expectation value of the Pauli matrix σ_y, which contains i. That is the simplest way to say it. (You might object and ask me to relabel the axes. But then you find other things become artificially complex, i.e. their imaginary parts would not be related to physical intermediate states.) One knows that the formula for a polarization always involves an interference of two amplitudes. Consider for example $\pi p \rightarrow \pi p$. There are two scattering amplitudes, f++ in which the proton helicity does not flip, and f+- in which it does. Typically $P \alpha \, \mathrm{Im}(f^*_{++} f_{+-})$. This means that even if the theory can flip the spin, i.e. change the helicity of the p, it does not guarantee a polarization. It must not only flip spin, it must flip it with a different phase from the non-flip amplitude. So you see, there are two requirements to get polarization. This is a nice way of telling you that if in this reaction you found polarization zero, you would not know whether the theory cannot flip spin, or whether it can but does so with the same phase as the non-flip.

Polarization is the simplest spin quantity that you can measure, but there are many others which are independent of it. And to really sort out what is happening to the amplitudes in the reaction, you have to measure several different spin parameters, for example, in pp $\rightarrow$ pp, at least 9 parameters at each energy and scattering angle.

- *FRITZSCH* :

I do not understand why there is a difference between Λ and $\bar{\Lambda}$.

- *LEADER* :

Nor do I. I know of no convincing mechanism to predict the polarization. But all models that exist, and there are about three, converge to the QCD statement, that as p increases, the polarization goes to zero.

- *ZICHICHI* :

I would like to make a comment. The production of Λ follows the leading effect. For $\bar{\Lambda}$, the number of quarks propagating would be exactly zero. It would be very interesting to measure the polarization of the Λs produced in association with the $\bar{\Lambda}$s, because most of the cross-section in hadronic interactions producing Λs, going essentially in the leading way, occurs with the antihyperons. I'm sure if you do measure the polarization of the $\bar{\Lambda}$s and the coupled Λs, you find exactly the same result. The leading effect has been neglected in the present analyses and must be accounted for. All the data should be reanalysed.

- *LEADER* :

The question of leading quarks is an interesting one, but as you can see from this table the situation is not so simple. The table compares different reactions summarized as pp → hyperon + X. It lists how many new quarks, i.e. those not in the incoming p, are needed to build the hyperon that is detected.

Table 1

n = 0	n = 1	n = 2	n = 3
p(uud)	Λ(uds)	Ξ^o(uss)	$\bar{\Lambda}(\bar{u}\bar{d}\bar{s})$
	Σ^o(uds)	Ξ^-(dss)	$\bar{\Xi}^o(\bar{u}\bar{s}\bar{s})$
	Σ^+(uus)	Σ^-(dds)	Ω^-(sss)
	n(udd)		

For example, if you start with a p(uud), then in each column you see the number of new quarks needed. You see that to end up with Λ(uds) or n(udd), you need n = 1 new quark, and the same for Σ^+(uus). To produce Σ^- or higher cascades, you need n = 2 or n = 3 new quarks, etc.

Starting from a general state of ignorance, this might be a nice way to begin to analyse the problem. Compare the polarizations and categorize according to the number of new quarks you need to produce the final particle. But it is not so clear cut as can be seen from this rough diagram :

You see here the strange fact that the ps and $\bar{\Lambda}$s are both unpolarized. This is very odd, because the ps have n = 0, whereas the $\bar{\Lambda}$s require three new quarks to construct them. It's a very

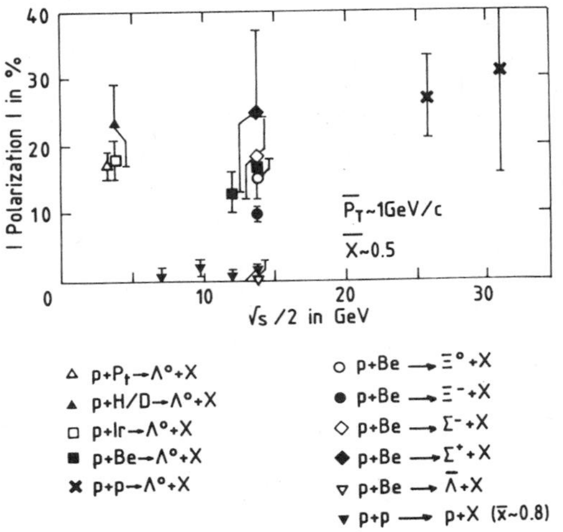

surprising and unobvious result. It's not clear that there is any pattern. We do not at this stage understand the situation, but it is certainly very interesting. However these experiments may be at too low $p_\perp$ to be relevant to the QCD prediction. It is vital to push out to larger $p_\perp$ (say $p_\perp >$ 5 GeV/c) to really test QCD.

PHYSICS RESULTS OF THE UA1 COLLABORATION

AT THE CERN PROTON-ANTIPROTON COLLIDER

Carlo Rubbia

Cern
Geneva
Switzerland

1. THE CERN SUPER PROTON SYNCHROTRON (SPS) AS A PROTON-ANTIPROTON COLLIDER

The conversion of the SPS into a $\bar{p}p$ collider[1] and the associated physics programs of the UA1 collaboration were motivated by three very specific physics goals, namely:

(i) the observation of jets and a detailed comparison with predictions of QCD.
(ii) The discovery of the charged Intermediate Vector Boson (IVB) $W^{\pm}$ in the electron and muon decay modes and the measurement of its fundamental charge asymmetry in the decay.
(iii) The discovery of the neutral IVB, Z° both in the electron and muon decay channels.

These goals are now essentially fulfilled. In addition and perhaps more surprising has been the extreme cleanliness of the events. A modest P_t threshold cut can easily separate the interesting phenomena removing the background due to spectator quarks.

The first collisons between protons and antiprotons at $\sqrt{s}$=540 GeV in the SPS accelerator operating as a storage ring[2] were observed in the early summer of 1981, not even three years after approval of the project. Two years later, a very large amount of information has become available which will be the subject of these lectures. Before discussing these results, it may be worth recalling that the collider has been the first and so far the only storage ring in which bunched protons and antiprotons collide head-on. Although the $\bar{p}p$ collider uses bunched beams like e^+-e^- colliders, the phase-space damping due to synchrotron radiation is now absent. Furthermore, since antiprotons are scarce one has to operate the collider in conditions of relatively large tune shift which is not the case for the continuous proton beams of the ISR. Therefore the machine itself has been breaking new grounds.

One of the most remarkable results of the $\bar{p}p$ collider has been probably the fact that it has operated at such high luminosity, which in turn means a large beam-beam tune shift. Most serious concern had been voiced in the early days of the construction about the stability of the beams due to beam-beam interactions. The beam-beam force can be approximated as a periodic δ-function of extremely non-linear potential kicks and it is expected to excite a continuum of resonances, in principle with the density of rational numbers. Reduced to bare essentials, we can consider the case of a weak antiproton beam colliding head-on with a strong bunched proton beam. The increment of the action invariant $W=\gamma x^2+2\alpha xx'+\beta x'^2$ of an antiproton due to the angular kick $\Delta x'$ is $\Delta W=\beta(\Delta x')^2+2(\alpha x+\beta x')\Delta x'$ and can be expressed in terms of the "tune shift" ΔQ as $\Delta x'=4\pi\Delta Qx/\beta$. If now we assume that the successive kicks are randomized, the second term of ΔW averages to zero and we get:

$$< \frac{\Delta W}{W} > = \frac{1}{2} (4\pi\Delta Q)^2$$

For the design luminosity we need $\Delta Q \sim 0.003$ leading to $(\Delta W/W)=7.1 \times 10^{-4}$. This is a very large number indeed giving an e-fold increase of W in only $1/7.1 \times 10^{-4}=1.41 \times 10^3$ kicks! Therefore the only way in which the antiproton motion remains stable is because these strong kicks are not random but periodic and the beam has a long "memory" which allows one to add them coherently rather than at random. Off-resonances, effects of these kicks then cancel on average, giving a zero overall amplitude growth. The beam-beam effects are very difficult, albeit impossible to evaluate theoretically, since this, a priori purely deterministic problem can exhibit stochastic behaviors and irreversible diffusion-like characteristics.

An old measurement at the electron-positron collider SPEAR had further increased the general concern about the viability of the $\bar{p}p$ collider scheme. Reducing the energy of the electron collider (Figure 1) resulted in a smaller value of the maximum allowed tune shift, interpreted as due to the reduced synchrotron radiation damping. Equating the needed beam lifetime for the $\bar{p}p$ collider (where damping is absent) with the extrapolated damping time of an e^+e^- collider gives a maximum allowed tune shift $\Delta Q=10^{-5}-10^{-6}$ which is catastrophically low. This bleak prediction did not find itself confirmed by the experience at the collider, where $\Delta Q=0.003$ per crossing and six crossing is routinely achieved with a beam luminosity lifetime approaching one day. What is then the cause of such a striking contradiction between the experiments with protons and electrons? The difference is related to the presence of synchrotron

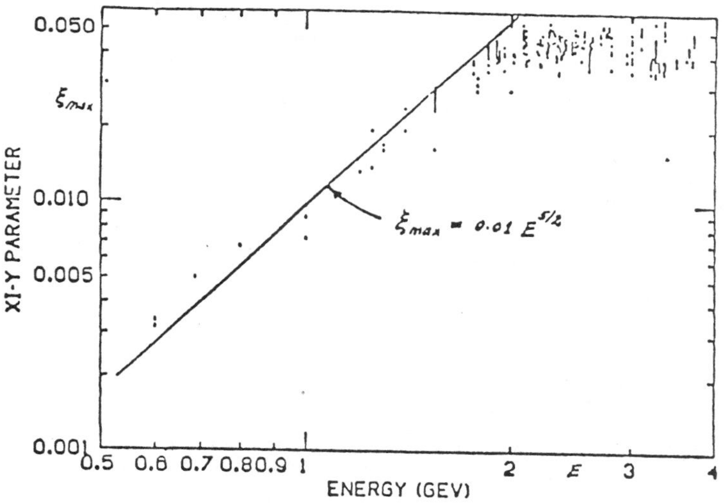

Fig. 1. Maximum allowed tune shift ΔQ at SPEAR as a function of the energy.

radiation in the latter case. The emission of synchrotron photons is a main source of quick randomization between crossings and it leads to a rapid deterioration of the beam emittance. Fortunately the same phenomenon provides us also with an effective damping mechanism. The proton-antiproton collider works since both the randomizing and the damping mechanisms are absent. This unusually favorable combination of effects ensures that proton-antiproton colliders become viable devices. They are capable of substantial improvements in the future. Accumulating more antiprotons would permit us to obtain a substantially larger luminosity. A project is on its way at CERN which is expected to be able to deliver enough antiprotons to accumulate in one single day approximately the integrated luminosity on which the results of these lectures have been based ($\sim$100 nb^{-1}). On a longer time scale, a $\bar{p}p$ collider built in the LEP tunnel with superconducting magnets of high field (10 T) is also conceivable. The luminosity will be further increased because of adiabatic beam damping with energy ($L \sim \gamma/R$). If the proton and antiproton bunches are transferred from the SPS collider to a 10 TeV + 10 TeV collider in the LEP tunnel, a further increase of luminosity of about one order of magnitude is gained. Luminosities of order of 10^{32} cm^{-2} sec^{-1}, which are likely to be at the limit of a general purpose detector are therefore quite conceivable. A further, important advantage is provided by the emergence of a significant amount of synchrotron damping, which at 10 TeV has an e-folding time of the order of several hours. This could be very helpful in improving even further the beam lifetimes and to increase significantly the attainable luminosities.

2. JETS

2.1. Introduction

Jets appear as the dominant, new phenomenon at the S$\bar{p}$pS collider, thus confirming the earlier cosmic ray observations and predictions of QCD. In this lecture we shall make use mostly of UA1 results. Very similar results and analogous conclusions are in general given by the parallel experiment UA2.

As realized very early in the experimentation around the collider a threshold in the transverse energy $E_T = \Sigma_i E_T^i$ summed over calorimeter cells can be used to trigger on an essentially 100% pure jet sample. The energy flow around the jet axis shows a striking sharp peak on a relatively low background due to other particles. The identification of the jet parameters is therefore very clean.

The energy spectrum covered by jets at the collider greatly exceed the one explored so far with the e^+e^- collider. Invariant masses in excess of 200 GeV have been observed. Also the nature of these jets is different, since the projectiles now are made both of quarks and gluons. In spite of these differences, however, fragmentation distribution of charged particles appear remarkably similar to the one measured for e^+e^- jets. A significant fraction of jet events contain more than two jets. For instance for events with $E_T^{(1)} > 20$ GeV, $E_T^{(2)} > 20$ GeV, about 30% have $E_T^{(3)} > 4$ GeV and $\sim$ 10% have $E_T^{(3)} > 7$ GeV. The presence of the third jet strongly suggests the gluon events with bremsstrahlung mechanism (roughly α_s times smaller in cross-section) and very similar to the familiar observation at the e^+e^- colliders. Indeed, the acoplanarity distribution for these events is in excellent agreement with QCD predictions which take precisely this effect into account. Appearance of jets at the collider is interpreted as hard scatterings amongst constituents of the proton and the antiproton. Kinematics of this "elementary" process can be derived from the energies and angles of the jets. There are several processes which can concurrently occur, due to the presence of quarks and gluons:

$$gg \to gg \qquad gq \to gq$$
$$g\bar{q} \to g\bar{q} \qquad qq \to qq$$
$$q\bar{q} \to q\bar{q}$$

Fortunately, in the center of mass of the parton collision, all processes have almost identical angular distributions. Only cross-sections differ significantly.

2.2. The UA1 Detector and the Trigger Conditions

The UA1 detector has been described in detail elsewhere[3], so only the aspects specifically concerned with this study will be presented. The central part of the detector consists of a large cylindrical tracking chamber centered on the collision point, surrounded by a shell of electromagnetic (e.m.) calorimeters and then by the hadronic calorimeter, which also serves as the return yoke for the 0.7 T dipole magnetic field. There are also tracking chambers and calorimeters in the more forward regions but these were not used in this study. The central detector (CD) and central calorimetry has almost complete geometrical coverage down to 5° to the beam axis. In the variables commonly used for such descriptions, this translates to -3.0 to 3.0 in pseudorapidity ($\eta=-\ell n[\tan\theta/2]$, where θ is the polar angle from the beam axis), and nearly 2π coverage in azimuth about the beam axis (ϕ).

The central tracking chamber consists of a 5.8 m long and 2.3 m diameter cylindrical drift chamber. This chamber provides three-dimensional coordinate information, enabling efficient track reconstruction. This, combined with the 0.7 T magnetic field, results in accurate momentum measurement for nearly all charged tracks.

The central calorimetry consists of lead/scintillator sandwich e.m. shower calorimeters surrounded by iron/scintillator sandwich hadronic calorimeters. These calorimeters are highly segmented in order to obtain position information of the energy deposition. Details of the geometry are given in Table 1.

Making use of the knowledge gained from the previous (1981) collider run[4], a localized transverse energy hardware trigger was implemented to select jet-like events for the 1982 run. This trigger required that the transverse energy (E_T) measured within a calorimeter "block" be greater than 15 GeV. A "block" was defined in the central region as two hadron calorimeter units plus the e.m. calorimeter elements in front of them. A "block" in the end-cap region was defined as the hadronic and e.m. elements comprising one quadrant of an end-cap. With this trigger, a data sample of $\int L\, dt=14\ nb^{-1}$ was obtained in the 1982 collider run, which constitutes the sample used for the jet studies reported in this paper. In the 1983 run, approximately 118 nb^{-1} of data were collected. Only the inclusive jet cross-section will include results from the 1983 data sample.

2.3. Definition of Jets

Jets are defined as clusters in pseudorapidity/azimuth (η/ϕ) space by the following procedure[4]. An energy vector is associated to each calorimeter cell. For hadronic cells, the vector points from the interaction vertex to the center of the cell. For electromag-

Table 1. UA1 Central Calorimeters

Calori-meters	Type	Thickness at normal incidence	Rapidity range	No of cells	Cell size $\Delta\eta \times \Delta\phi$ (approx.)	Comments
Gondolas	Electromag.	$26.4\ X_0/1.1\lambda$	$\lvert\eta\rvert < 1.5$	48	$0.11 \times 180°$	Azimuth from light attenuation
C's	Hadronic	$5.0\ \lambda$	$\lvert\eta\rvert < 1.5$	232	$0.30 \times 15°$	
Bouchons	Electromag.	$27.0\ X_0/1.2\lambda$	$1.5 \leq \lvert\eta\rvert < 3$	64	$1.5 \times 11°$	Rapidity from position detector
I's	Hadronic	$7.1\ \lambda$	$1.5 \leq \lvert\eta\rvert < 3$	128	$0.40 \times 15°$	

(x_0 are radiation lengths, λ absorption lengths).

netic cells, the vector points to the energy centroid determined by pulse height measurements (Gondolas) or by position detectors (Bouchons).

In the subsequent clustering, cells are treated differently depending on their E_T being above or below 2.5 GeV:

- Among the cells with $E_T \geq 2.5$ GeV, the highest E_T cell initiates the first jet. Subsequent cells are considered in order of decreasing E_T. Each cell in turn is added vectorially to the jet closest in (η,ϕ) space, i.e. with the smallest $d \equiv \sqrt{(\Delta\eta^2 + \delta\phi^2)}$ (with ϕ in radians), if $d \leq 1.0$. If there is no jet with $d \leq 1.0$, the cell initiates a new jet.
- Cells with $E_T<2.5$ GeV are finally added vectorially to the jet nearest in (η,ϕ) if their transverse momentum relative to the jet axis is less than 1 GeV and if they are not further than 45° in direction from the jet axis.

The cut at $d = 1.0$ has been derived from the jet energy profile (see below).

All triggers selected previously were fully reconstructed in the central detector, and calorimeter depositions were corrected using detailed light attenuation maps and an average response correction factor of 1.13 to allow for the hadronic energy in the electromagnetic calorimeters[4]. After reconstruction of jets according to the procedure above over the full pseudorapidity range $|\eta|<3$, an E_T threshold was applied requiring at least one jet with $E_T>30$ GeV in the Gondola region. This jet is called the trigger jet. A threshold was applied to the fraction of jet energy deposited in the electromagnetic calorimeters ($\geq$10%), to discard signals from beam halo in the hadronic calorimeters. This leaves <1% of events with a potential halo problem, while reducing the good data sample by <2%. To avoid edge effects due to the calorimeter geometry, the trigger jet axis has to be contained in $\eta = \pm 1.2$ and in $\phi = \pm 60°$ from the horizontal plane. These cuts ensure good energy containment and minimal particle losses due to the narrow dead zones between the calorimeters in the vertical plane.

To understand possible trigger biases, we used a sample of events taken with a global (i.e. $|\eta|<1.5$) $\Sigma|E_T| > 20$ GeV trigger and studied the efficiency of jet finding. We found that the localized energy trigger and the selection procedure is ∿80% efficient for jets with $E_T = 30$ GeV, and more than 95% efficient for jets with $E_T \geq 35$ GeV.

2.4. Jet Axis and Jet Energy

The above jet finding procedure results in jets with energy and axis defined from calorimeter depositions. It is therefore necessary to discuss the precision of these parameters.

The definition of a jet axis is rather independent of the algorithm used for finding the jet, as it is mostly dependent on large energy depositions in few calorimeter cells. To assess the precision of the jet axis we have simulated in our detector high E_T jets using a Monte Carlo program with fragmentation according to a cylindrical phase space (CPS) model defined below. The result of this study is that the jet axis given by the jet finding procedure agrees with the vectorial sum of the momenta of fragments from the generated jet to within ±6° in ϕ and ±0.04 in η (rms). This shows that the granularity of our calorimeters does not introduce any appreciable error in the jet axis definition. In using the same jet finding algorithm on the charged tracks given by the central detector, determined with superior angular resolution, one obtains a charged jet axis which coincides with the calorimetric jet axis to within ±0.1 in η and ±10° in ϕ (rms). The difference reflects mostly the fluctuations between the charged and neutral parts of jets, and constitutes a lower limit to the precision of the jet axis definition.

The definition of the jet energy, on the other hand, is directly related to the cutoff parameter d in (η,ϕ) space. We use the energy profile and Monte Carlo studies to obtain better understanding of the jet energy.

Given the axis of a jet, the average values per jet of deposited transverse energy and of charged particle transverse momentum can be studied as function of $\Delta\eta$ and $\Delta\phi$ referred to the jet axis. We restrict ourselves here to the pseudorapidity projection, where the granularity is best. We define an average jet profile by superimposing many jets, leaving out from the average any low-acceptance regions in η or ϕ. The hemisphere opposite to the jet axis in ϕ ($|\Delta\phi| > \pi/2$) is not included. All jets found in our event sample are included, if their transverse energy is at least 20 GeV and if their axis lies within the same (η,ϕ) limits as used for the trigger jet.

The transverse energy flow as a function of $\Delta\eta$ is shown in Figure 2 (a-c) for three ranges of jet E_T. A clear enhancement is observed on top of a flat energy plateau. The full width of the enhancement at the base is given by $\Delta\eta = \pm 1.0$, independent of the jet energy (this justifies the cutoff value $d \leq 1.0$ in the jet finding algorithm). We distinguish two distinct regions in the average jet, the hard core of the jet ($|\Delta\eta| < 0.2$), and the wings ($0.2 \leq |\Delta\eta| < 1.0$). The relative amount of transverse energy contained in the core increases with the jet E_T. The E_T content of the wings seems rather independent of the jet E_T. Outside the enhancement, for $|\Delta\eta| > 1.0$, a

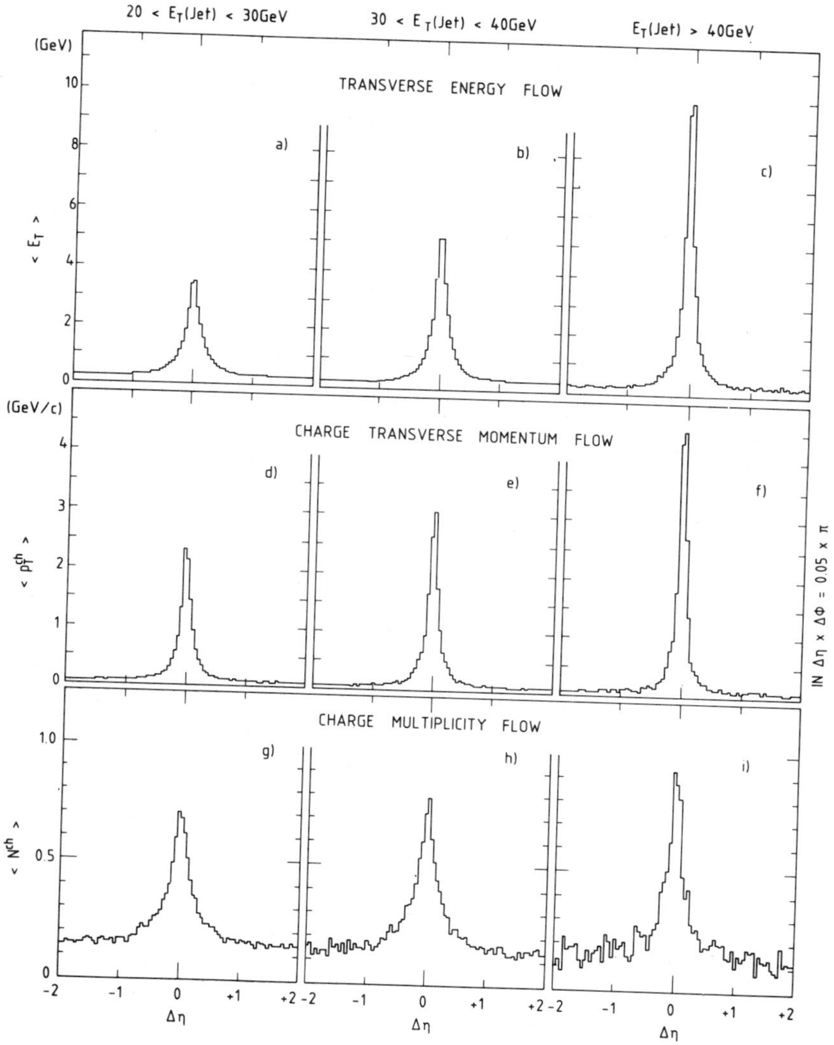

Fig. 2. (a-c): Transverse energy flow as function of $\Delta\eta$, i.e. pseudorapidity distance from the jet axis, for 3 slices of jet E_T. Cells inside $\Delta\phi = \pm 90°$ are used. (d-e): Charged transverse momentum flow as function of $\Delta\eta$. (g-i): Charged multiplicity flow as function of $\Delta\eta$.

constant E_T plateau is observed, whose height is independent of the jet E_T. Its value is substantially higher than the one observed for minimum bias events[5]. When the events with more than two jets are excluded the plateau height lies between the values for minimum bias events and for all events. Hence the origin of the plateau is in two distinct phenomena. The first is due to the debris of the beam particles, namely spectator jets. The second one is the production of multijet events, clearly observed in the present data sample. In this case more than one jet can be emitted in the same azimuthal hemisphere, thus increasing the height of the plateau. It should be noted, though, that an absolute measurement of this plateau is difficult: the particles outside the hard core of the jet are mostly of low momentum, and are not measured with good precision due to the non-linear response of the calorimeters at low energy[6].

We have therefore used the information coming from the central detector for a more quantitative assessment of the jet and plateau composition. The jet profile is now studied with respect to the charged jet axis. Again, particle losses in the small regions of low acceptance in the central detector are corrected for by our averaging procedure. Figures 2 (d-f) show the mean charged transverse momentum flow in the same slices of jet E_T as before. The same conclusions are reached as for the transverse energy flow. This demonstrates in particular that the granularity of our calorimeters does not influence our measurement of the jet width. The distributions also show that calorimetric measurements for energetic jets are not significantly distorted by the magnetic field.

2.5. Multiplicities

The multiplicity flow is obtained by the same averaging procedure as the energy flow, counting tracks reconstructed in the central detector and excluding regions of low acceptance. In Figures 2 (g-i) the average multiplicity flow of charged tracks around the charged jet axis is shown, again as function of the jet E_T. The multiplicity in the jet increases only slowly with E_T, implying that the leading particles of the jet carry an increasing transverse momentum.

Outside the jet a flat plateau is observed. The mean multiplicity for the plateau depends on the topology of the events. A definite increase in the average multiplicity is observed over the minimum bias level[7]. For pure 2-jet events the rise is 40%, and is almost a factor of 2 when the multijet events are included. A similar increase is observed in the mean p_t per charged particle over the minimum bias value, 14% for 2-jet events and 25% when including the multijet events.

The increase observed in the transverse energy flow plateau over minimum bias events can therefore be attributed to an increase of both the average multiplicity and the mean p_t per particle, due to the multijet production. The precise determination of charged particle multiplicities inside the jet is difficult. It is not possible to separate low-momentum fragments of the jet which contribute sizeably to the jet multiplicity, from the multiplicity background arising from spectator or additional jets, on an event-by-event basis. To estimate losses of low-momentum fragments from the jet by the jet finding algorithm, we rely upon a fragmentation model and a full Monte Carlo simulation of the detector.

2.6. Monte Carlo Studies

Two Monte Carlo programs with different hadronization models have been compared with the data. The first one is a naive parton-parton hard scattering model without QCD radiative effects, in which the systems of hard scattered partons and spectator jet fragment independently according to cylindrical phase space (CPS). The mean multiplicity of fragments is obtained as a function of s (square of c.m.s. parton-parton energy) from a phenomenological fit to multiplicity measurements in hadronic events from e^+e^- collisions[8]:

$$\langle n(s)\rangle = 2.0 + 0.027 \exp\{2\sqrt{[\ln(s/\Lambda^2)]}\} \qquad (1)$$

with $\Lambda = 0.3$ GeV.

The second program is ISAJET[9], which incorporates QCD radiated gluons in the final state. The fragmentation model used for the spectators is ISAJET tuned to reproduce minimum bias events. For both models, generated particles are tracked through the magnetic field and the showers in the calorimeters are simulated[10]. In Figure 3 the experimental transverse energy flow for jets with $E_T >$ 35 GeV is compared to the Monte Carlo results. ISAJET gives a better description of the jet shape than CPS: Both programs fail to reproduce the plateau region as they do not include a complete multijet production, in particular through initial state bremsstrahlung. The charged multiplicity flow is shown in Figure 4, again for jets with $E_T > 35$ GeV. The multiplicity given by ISAJET is low and coincides with the minimum bias level. CPS gives a good overall description of the multiplicity flow.

To measure charged jet multiplicities, we sum up average track multiplicities in the window $\Delta\eta = \pm 1$, apply a flat background subtraction obtained from the region $1 \leq |\Delta\eta| < 2$, and correct for the loss of jet-associated particles outside $\Delta\eta = \pm 1$ according to CPS. As in our data, low acceptance regions are excluded and corrected for. A further correction is applied for unreconstructed low-momentum particles. The final corrected results on charged track multiplicities in jets are shown in Table 2.

It should be noted that ISAJET when tuned to fit the plateau multiplicity results in larger corrections to the multiplicity inside the jet window, as more fragments are generated far from the jet axis. The 10% systematic error in Table 2 is an attempt to account for this model dependence of the correction.

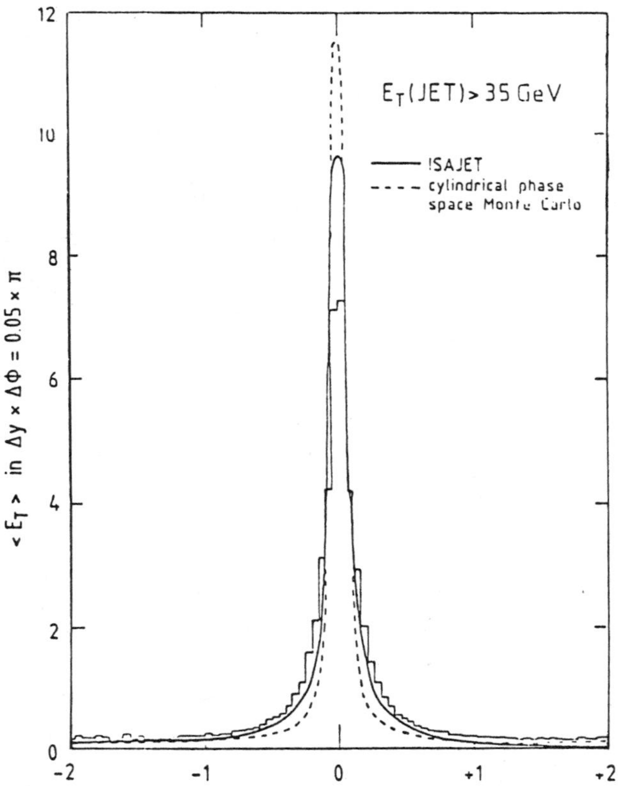

Fig. 3. Transverse energy flow for jets with $E_T > 35$ GeV as function of $\Delta\eta$. Dashed curve: Cylindrical Phase Space model. Full curve: ISAJET model.

Using the multiplicity extrapolation (1) from e^+e^- for our energy range would also result in higher multiplicities, of the order of 9.0 charged particles for a jet of 25 GeV. A more detailed comparison with e^+e^- jets can be found in the discussion of jet fragmentation.

2.7. Jet Cross-sections

To measure the inclusive jet yield, all events are selected as above, imposing additional constraints on the vertex position (±40 cm from the detector's center, for a measured rms spread of ±12 cm) and on the jet axes (±1 in pseudorapidity, ±60° in ϕ from the horizontal plane). All jets with $E_T > 35$ GeV are considered. Events containing jets with transverse energy above 55 GeV were, in addition, inspected visually on a MEGATEK 3-dimensional display, and a small number of badly reconstructed or background events were rejected, typically events in which cosmic showers or beam halo overlap with collisions.

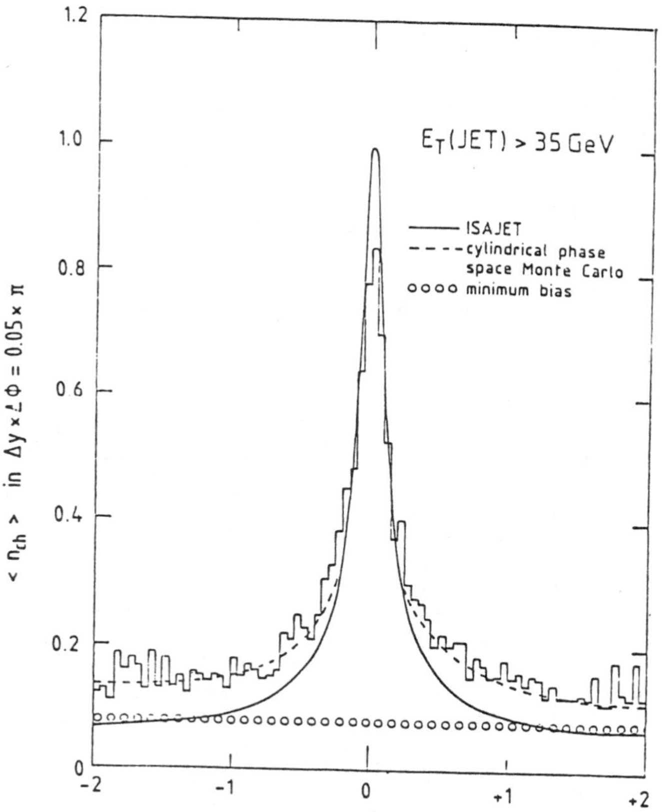

Fig. 4. Charged Multiplicity flow for jets with E_T > 35 GeV as function of $\Delta\eta$. Dashed curve: Cylindrical Phase Space model. Full curve: ISAJET model.

Table 2. Jet population used for the present study after the cuts mentioned in the text for different E_T(jet) ranges

N(jets)	E_T(jet) range (GeV)
853	30-35
323	35-40
195	40-50
64	> 50

The inclusive E_T distribution thus obtained is further corrected for geometrical acceptance, experimental procedure and detector resolution using a Monte Carlo method. Jets are simulated using the models discussed above. The comparison between the reconstructed E_T spectrum and the Monte Carlo input results in a correction factor for each bin of E_T. The global correction is not strongly dependent on E_T and does not exceed a factor 1.25. The inclusive cross-section $d\sigma/dE_T d\eta$ obtained after correction is shown in Figure 5, together with the data from 1981 published in Reference 4. In the small region of overlap the measurements agree. All errors shown are purely statistical. The following systematic errors are not included:

- The uncertainty of the absolute luminosity (±15%);
- Uncertainties in geometrical acceptance due to the choice of the fiducial region, of the jet finding algorithm and of the fragmentation model used in the Monte Carlo programs (±5%);
- An uncertainty in the jet energy which for a given event varies when changing the angular aperture in the jet finding algorithm. This translates into a ±20% uncertainty in cross-section;
- The dependence of the jet acceptance and jet finding efficiency on the production and fragmentation models used in the Monte Carlo programs (±20%);
- The uncertainty arising from the definition of the energy scale: the absolute energy calibration is known to ±5%[4];
- The additional uncertainty in energy scale due to the poorly known particle composition of jets: electromagnetic and hadronic calorimeters exhibit a different response to electromagnetic particles and hadrons, and this response difference is strongly dependent on the particle momenta for $p < 5$ GeV[6]. The uncertainty in the average response correction used is estimated at ±5%. We combine these energy effects into an overall error of ±7.5% of the energy scale, which translates into an uncertainty of a factor 1.5 in the cross-section.

Our overall systematic uncertainty in the cross-section can therefore be given as a factor of 1.65, essentially independent of the jet transverse energy. The measured cross-section is compared in the figure to a band of QCD predictions with a width corresponding to uncertainties in the theory[11]. Note that QCD predicts transverse momentum distributions for partons, while our measurements refer to the sum of the jet fragment momenta. Despite large experimental and theoretical uncertainties, the agreement observed is excellent. The large increase in jet cross-section from ISR to collider energies, by three orders of magnitude, is thus confirmed. In order to obtain the cross-sections for producing several jets in an event, we now consider, in addition to the trigger jet, all jets with $E_T > 15$ GeV reconstructed with a jet axis inside $|\eta| < 2.5$, again attempting to avoid edge effects and jets that might be faked by spectator background.

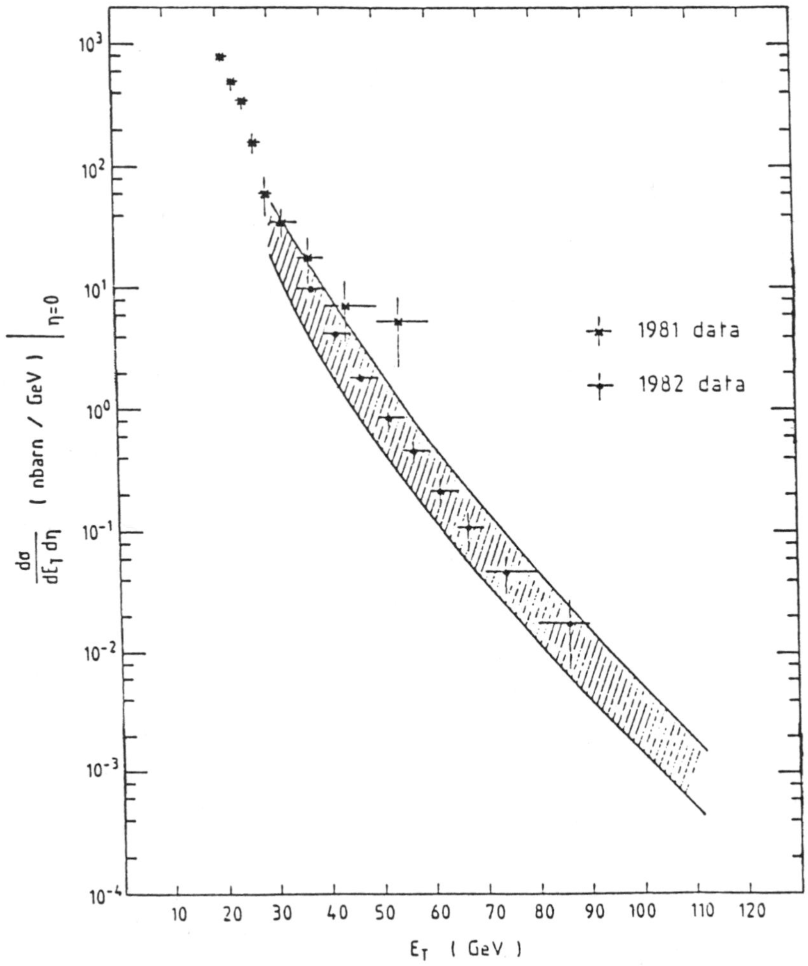

Fig. 5. Inclusive Jet cross section $d\sigma/dE_T d\eta$ (η=0) as function of E_T. 1981 data (x) and 1982 data (+). The hatched band corresponds to possible QCD predictions[15].

In Figure 6 we show the fraction of events with 1, 2 and 3 jets (trigger jets included) as function of the E_T of the trigger jet. The 2-jet topology dominates over the full range in E_T at a level of $\sim$ 80-85%. The fraction of 1-jet events becomes negligible at high trigger jet E_T, whereas the fraction of 3-jet events rises in the region of low E_T and levels off at $\sim$ 15%. We should stress that our jet finding algorithm, with the window $\Delta\eta = \pm 1$, and the additional requirement $E_T > 15$ GeV, can be expected to have a direct influence on the number of jets found, and that the topological cross-sections as presented here have to be understood in relation to a given jet finding procedure. We also have not corrected these cross-sections in any way for geometrical acceptance.

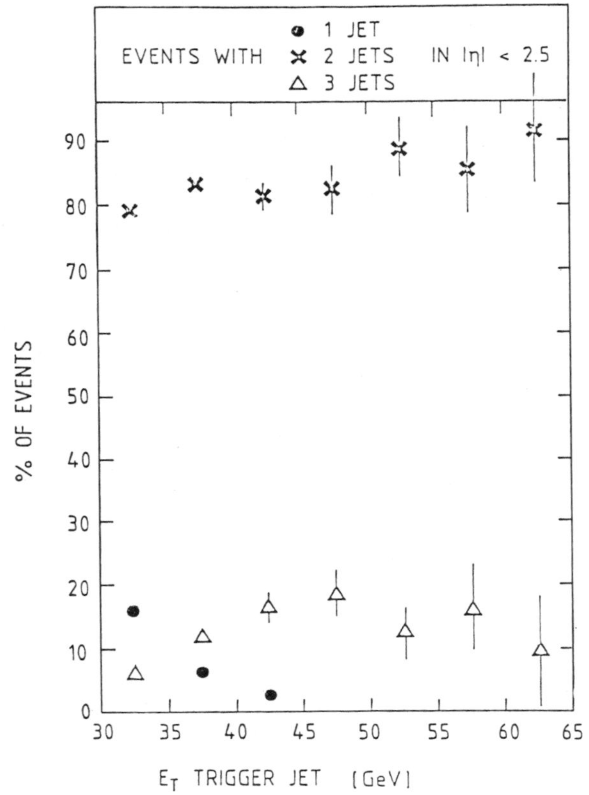

Fig. 6. Fraction of events with 1, 2 and 3 jets of $E_T > 15$ GeV, found by the jet algorithm in a pseudorapidity window $|\eta| < 2.5$, as function of the trigger jet E_T.

The occasional presence of a third jet strongly suggests a gluon bremsstrahlung mechanism similar to what has been observed in hadronic e^+e^- events. QCD predicts multijet events due to quark-gluon and gluon-gluon couplings with rates that are proportional to the products of coupling constants appearing in the bremsstrahlung processes of the original parton. 3-jet events would then, for instance, be produced with a cross-section roughly α_s times the cross-section for 2-jet production. The rate of multijet events can be estimated by measuring the differential cross-section in terms of some suitable parameter describing the non-coplanarity. One such parameter is p^{out}, the momentum perpendicular to the plane defined by the trigger jet and the beam momentum. For large enough p^{out} the 3-jet production rate can be calculated perturbatively from QCD[14, 15] and be compared with the data.

For the study of non-coplanarity we used events with a trigger jet as defined in 2.3 above. In order to minimize any effects coming from problems in jet finding we first calculate p^{out} directly from all E_T-vectors not belonging to the trigger jet. To avoid contamination from the spectator jets we require these E_T-vectors to have $|\eta| < 2.5$. The p^{out} is reconstructed by adding the E_T-vector components perpendicular to the plane defined by the energy axis of the trigger jet and the beam direction, separately on both sides of the plane. If the energies of the jets are balanced, the two $|p^{out}|$ values should be the same. The difference shows a width which is consistent with the experimental resolution. We therefore take as p^{out} the average value of $|p^{out}|_{left}$ and $|p^{out}|_{right}$. The resulting p^{out} distribution is shown in Figure 7a. This distribution has a contribution from the p_t of the background, i.e. from particles not belonging to the jets, which is difficult to estimate accurately. From the transverse energy flow in minimum bias events one can roughly estimate the contribution to be about 5 GeV. To understand further the size of the background we have repeated the analysis using the jets as found by the jet algorithm. The results are shown in Figure 7b. One sees that the p^{out} distributions obtained have the

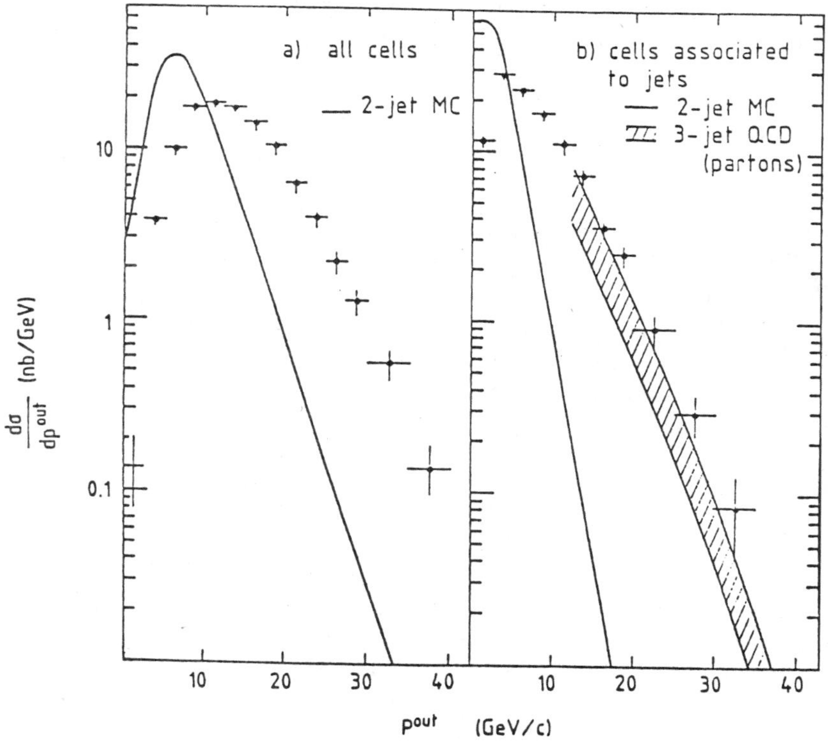

Fig. 7. p^{out} distribution from calorimeter cells (a) and from jets (b), compared to a 2-jet model and to a perturbative QCD 3-jet calculation[16].

same slope but are shifted by about 5 GeV down in p^{out} when compared with the distribution obtained from all E_T-vectors. This shift is consistent with the estimate from minimum bias events. In the same figure is also shown the p^{out} distributions obtained from events with two back-to-back jets generated with our CPS Monte Carlo. In this model the partons are strictly coplanar, the non-zero p^{out} comes from the hadronization model and the resolution of the apparatus. It is clear from the figure that this 2-jet model cannot reproduce the observed large p^{out} tail.

Instead the distribution obtained from a complete QCD calculation for 3-jet production[12] as shown in Figure 7b agrees much better with the data. The band reflects again the theoretical uncertainties, which are of the same origin as for the inclusive jet cross-section. Note that the QCD calculation is for partons, i.e. there is no broadening due to fragmentation and no background from spectator partons either. We conclude from our study that we observe multijet events with a rate that is roughly consistent with the expectation from QCD.

2.8. Studies on Jet Fragmentation in Charged Particles

So far the jets are found by using exclusively the information coming from the electromagnetic and hadronic calorimeters, via the clustering method explained above and in References 4, 14, and 18. The axis and the total energy of the jet are known from the vectorial sum of all its cell energy vectors. Charged particles are then associated with a particular jet, provided they are confined inside a cone centered around the jet axis. The jet axis and directions of charged particles show obvious correlation up to an angle of 35°, as shown in Figure 8. The half aperture of the cone is then fixed to that value. The dashed curve is an estimate of the background of the beam fragments (spectators) not associated with the jet, by a simple Monte Carlo where the particles are generated uncorrelated according to a cylindrical phase-space model.

Owing to the fixed opening angle of the cone, only the central region of the calorimeters is used. This region extends from -1.5 to +1.5 in pseudorapidity η. To minimize edge effects, we restrict the jet-axis direction to be contained in $-1.25 \leq \eta \leq 1.25$. In order to consider only charged particles pointing to the central calorimeter cells, and consequently depositing their energy in those cells, the pseudorapidity range of charged particles is within the limits $-1.5 \leq \eta \leq +1.5$. If the jet axis is close to its pseudorapidity limit of 1.25, charged particles can be emitted in the region $|\eta| > 1.5$. The loss of particles due to this situation is small, of the order of 7%.

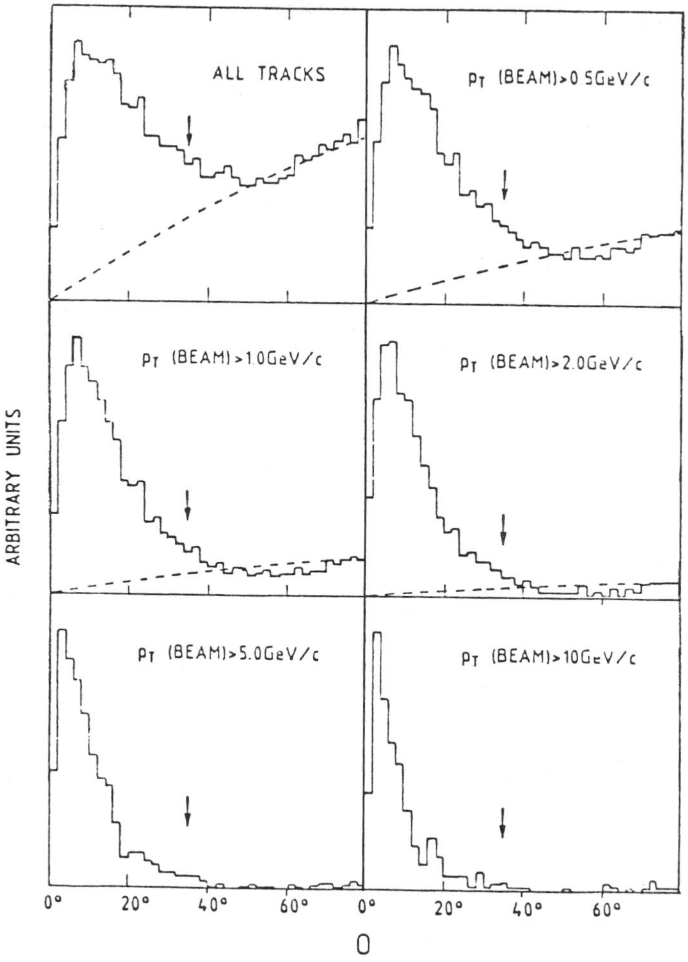

Fig. 8. Distribution of the spatial angle between the jet axis as given by the calorimeters and the directions of charged particles. Arrows indicate the 35° cut. The dashed line is the result of a Monte Carlo assuming no correlation between jet axes and charged particles.

Because of the horizontal orientation of the magnetic field and the presence of gaps between the two halves of the calorimeters in the vertical plane, the azimuthal angle about the beam line of the jet axis is restricted to lie within four sectors of a 15° half opening angle centered on ±45° axes. This restriction in solid angle does not affect any result of the present fragmentation study, as everything is normalized to the final jet population obtained after the above-mentioned cuts.

A loss of 13% is estimated for charged particles emitted outside the sectors. Including the rapidity acceptance, the overall geometrical acceptance is, then, of the order of 80%. A systematic uncertainty of 10% has been added to the statistical errors. No correction has been introduced for track-finding efficiency within jets. However, visual scanning of a reduced population of events did not show any evidence of unfitted straight tracks. The percentage of tracks giving an ionization larger than 1.7 times that of minimum ionizing particles has been measured in the central drift chamber volumes. The small percentage observed, 4.5%, is compatible with the tail of the dE/dx distribution. We deduce that no undetected multi-tracks are contained in the data sample. A small fraction of the tracks within the 35° cone (< 5%) has been discarded because of non-association with the main vertex of the interaction. After the acceptance cuts listed above, the jet population used for the present analysis is given in Table 2.

2.9. Jet Fragmentation Function

The inclusive variable used for the jet fragmentation study has been chosen to be:

$$z = p_L^{\pm}(\text{jet axis})/E(\text{jet}), \qquad 0 \leq z \leq 1,$$

where $p_L^{\pm}$(jet axis) is the momentum of the charged particle projected on the jet axis. The energy of the jet, E(jet), is defined as the modulus of the vectorial sum of all energy cell vectors belonging to the jet.

The distribution of the variable z - the fragmentation function - is defined as:

$$D(z) = (1/N_{jets})(dN_{ch}/dz)$$

The integral of D(z) is then the mean charged multiplicity contained in the jet:

$$N_{ch}(\text{jet}) = \int_0^1 D(z)dz$$

Several corrections have been applied to D(z). These are discussed below.

The form of the background below D(z) is obtained in the control region $52.5° \leq \theta \leq 70°$, where θ is the spatial angle between the jet axis and the charged particles. Background normalization is given by the Monte Carlo prediction of the number of particles uncorrelated with the jet for $\theta < 35°$ (dashed curve in Figure 8).

For relatively high z's, all the charged particles are contained in the 35° cone around the jet axis. At low values of z this is no longer true. Many soft particles are emitted at large angles. In particular, below 0.02 the emission of particles looks so isotropic that the correlation of the particles with the jet is no longer obvious. Loss of jet particles emitted outside the cone varies from 35% for z between 0.02 and 0.03, to 5% for z around 0.07.

The jets we are dealing with in this paper are very energetic; they have a total energy above 30 GeV. Associated charged particles with a high z value (therefore a high momentum) have then a large momentum uncertainty. The smearing in z which creates an important deformation of D(z) for values above 0.7 has been removed from the data, assuming an exponential law for D(z) at large z.

Monte Carlo studies with full calorimeter reconstruction indicate that the uncertainty in the measurement of the jet energy is more or less constant between 30 and 50 GeV, and is of the order of 15%. Five percent of the resolution comes strictly from cell association of the jet cluster algorithm, the rest of it being due to the finite resolution and granularity of the calorimeters. No correction has been applied to the data in order to take into account the jet energy smearing.

Figure 9 shows the plot for D(z) with z > 0.02. This distribution falls rapidly with z at low z values. At higher z values its form is approximately exponential.

We can compare the shape and the normalization of D(z) for the present experiment with $(1/\sigma_{tot}) \times (d\sigma/dx_L)$ obtained by the TASSO Collaboration for jet energies of 17 GeV, where $x_L = p_L^{\pm}/p_{beam}$, and $p_L^{\pm}$ is the momentum of the charged particles projected on the jet axis whose direction is determined from minimizing the sphericity of the e^+e^- events[8]. The energies of the jets are of course different for both cases. However, the comparison is meaningful because scaling violations in e^+e^- annihilations are known to be small[13]. No striking differences can be observed between these two sets of data, as can be seen in Figure 9. This means that quark-dominated and gluon-dominated fragmentation functions are, on the whole, not different from each other, at least for values of z > 0.02.

Within our own data we can look for possible variations of D(z) as a function of the transverse energy of the jet. After background subtractions and corrections, D(z) is plotted in Figure 10 for three E_T bands: 30-35 GeV, 40-45 GeV, and > 50 GeV; D(z) is approximately independent of the jet energy. A possible tendency for D(z) to shrink at low z with increasing E_T(jet) cannot be excluded. This is not observed in the high-z region, probably on account of the very large uncertainties in the data introduced by the track momentum smearing, which are difficult to remove entirely owing to the lack of statistics[14].

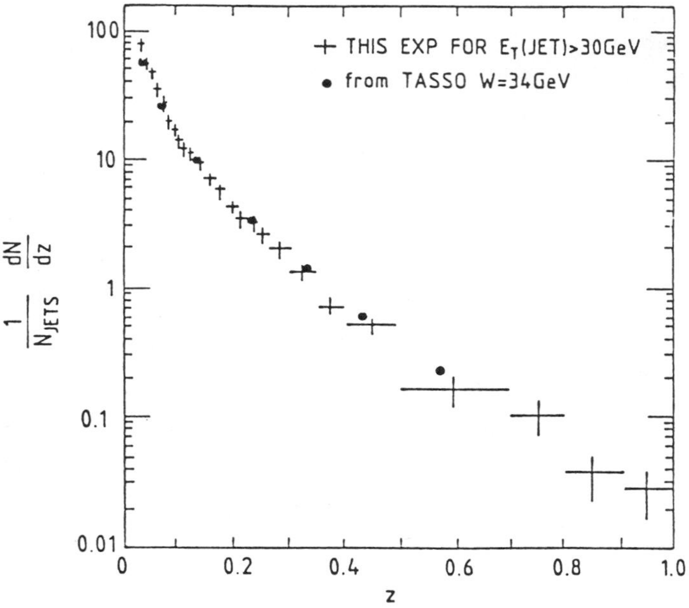

Fig. 9. Charged-particle fragmentation function for E_T(jet) > 30 GeV, compared with similar results from the TASSO detector at PETRA at W = 34 GeV.

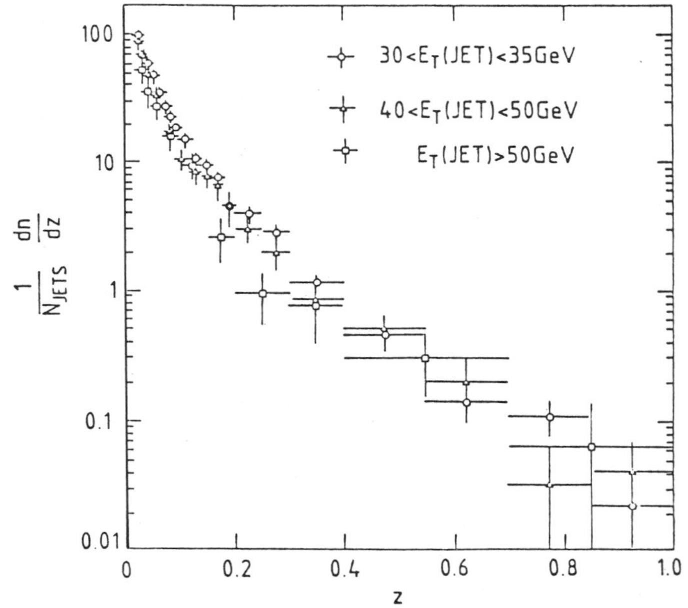

Fig. 10. Charged-particle fragmentation functions for E_T(jet) = 30-35 GeV, 40-50 GeV, and > 50 GeV.

2.10. Transverse Momentum with Respect to the Jet Axis

The jet axis given by the calorimeters is not precise enough for studying the transverse momentum p_t of the charged particles with respect to the jet axis. For this reason it is replaced by a charged jet axis whose direction is given by the vectorial sum of all charged-particle momenta. The charged particles used to define this axis are inside the cone of 35° half aperture around the calorimeter jet axis. Of course, the charged jet axis is correct only if we assume that the charged and neutral axes are aligned evenly. If this assumption is not valid on an event-by-event basis, it is probably true statistically.

As we have seen before, the association of particles with the jet is questionable at lower z values. For this reason a cut $z > 0.1$ is applied to select particles unambiguously associated with the jet. Owing to the "seagull effect" discussed below (Figure 11), this cut will result in a higher mean p_t within the jet, compared with a mean p_t value obtained for all particles belonging to the jet regardless of their z value. For all jets with $E_T > 30$ GeV, the variation in the average transverse momentum of charged particles measured with respect to the jet axis is plotted in Figure 11 as a function of z. A "seagull effect" is observed, showing the increase of $\langle p_t \rangle$ from a value of 0.5 GeV/c at a z value around 0.1 to a value approaching 1 GeV/c for z values above 0.5.

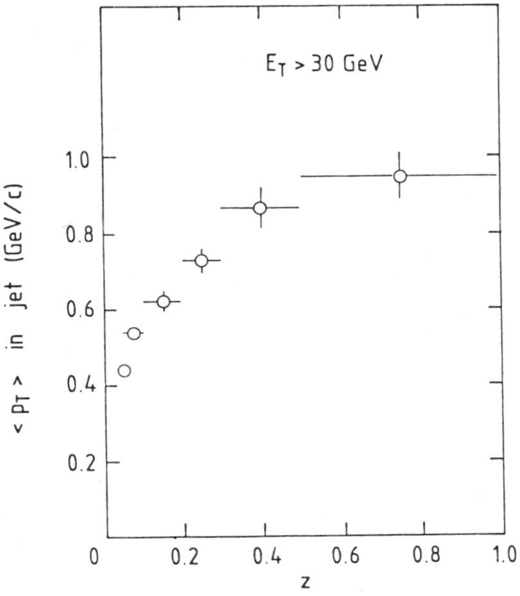

Fig. 11. Variation of $\langle p_T \rangle$ with respect to the jet axis for charged particles as a function of z. Errors are due to statistics only.

The invariant p_t spectrum, $(1/p_t)(dN/dp_t)$, is shown in Figure 12 together with the results of a fit:

$$(1/p_t)(dN/dp_t) = A/(p_t + p_{t_0})$$

for all jets with $E_T > 30$ GeV. The above function was shown to reproduce well the p_t spectrum of charged particles in minimum bias events[15]. The p_t spectrum is well fitted by the values $p_{t_0} = 4$ GeV/c, n = 14.8. The observed mean p value internal to the jet is $\langle p_t \rangle = 600$ MeV/c, after having applied the cut z > 0.1 on all particles. A large p_t tail is observed up to $p_t = 4$ GeV/c. This tail could well be an indication of gluon bremsstrahlung. On the other hand, it could also be due to an experimental misalignment of the jet axis, or to events whose leading particles are neutrals.

Evolution of the mean p_t within the jet has been studied for the following regions of E_T(jet): 30-35, 40-50, and > 50 GeV. Figure 13 shows the p_t spectrum obtained for each of these transverse energy bands. The mean p_t increases from 600 MeV/c at $E_T = 30$ GeV to 700 MeV/c for $E_T > 50$ GeV.

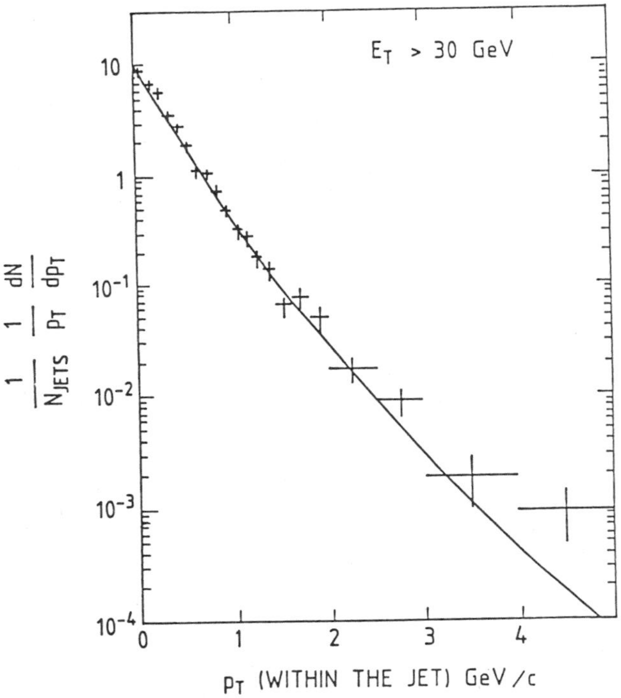

Fig. 12. $(1/p_T)(dN/dp_T)$ spectrum (p_T with respect to the jet axis) for charged particles with z > 0.1. The solid line is the result of a fit $1/(P_T + P_{T_0})^N$ with $P_{T_0} = 4.0$ GeV/c, N = 14.8.

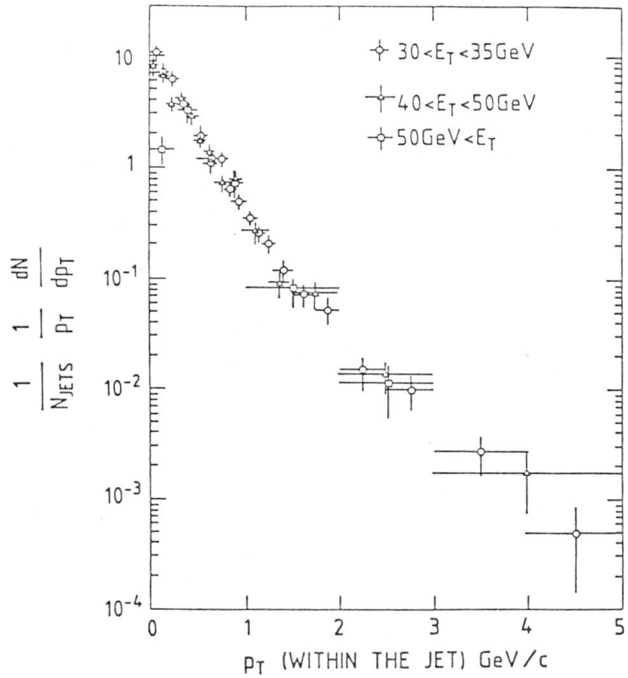

Fig. 13. $(1/p_T)(dN/dp_T)$ spectra (p_T with respect to the jet axis) for charged particles with $z > 0.1$ and for three ranges of E_T(jet) = 30-35 GeV, 40-45 GeV, and > 50 GeV.

2.11. Structure Functions

So far nucleon structure functions were the exclusive domain of lepton-hadron scattering experiments. The observation of well defined two-jet events in proton-antiproton collisions at high energy opens up the possibility of proton structure function measurements at values of four-momentum transfer squared (Q^2) in excess of 2000 GeV^2, far higher than previously accessible using lepton beams. In the parton model two-jet events result when an incoming parton from the antiproton and incoming parton from the proton interact with each other to produce two outgoing high transverse momentum partons which are observed as jets.

If $d\sigma/d\cos\theta$ is the differential cross-section for a particular parton-parton subprocess as a function of the c.m.s. scattering angle θ, the corresponding contribution to the two-jet cross-section may be written:

$$d^3\sigma/dx_1dx_2d\cos\theta = [F(x_1)/x_1][F(x_2)/x_2]d\sigma/d\cos\theta \qquad (1)$$

where $[F(x_1)/x_1][F(x_2)/x_2]$ is a structure function representing the number of density of the appropriate partons in the antiproton [proton] as a function of the scaled longitudinal momentum $x_1[x_2]$ of the partons.

The differential cross-sections for the possible subprocesses have been calculated to leading order in QCD[16]. The elastic scattering subprocesses [gluon-gluon, gluon-quark(antiquark) and quark(antiquark)-quark(antiquark)] have a similar angular dependence and become large as $\cos\theta \rightarrow 1$ [like$(1-\cos\theta)^{2}$] as a consequence of vector gluon exchange. In the approximation that the elastic subprocesses dominate and have a common angular dependence, the total two-jet cross-section may be written[17] in the form of Equation (1). In particular, if $d\sigma/d\cos\theta$ is taken to be the differential cross-section for gluon-gluon elastic scattering:

$$d\sigma/d\cos\theta = (9/8)[\pi\alpha^2/2x_1x_2s](3+\cos^2\theta)^3(1-\cos^2\theta)^{2} \qquad (2)$$

where α_s is the QCD coupling constant and s is the total c.m.s. energy squared, then the structure function F(x) becomes:

$$F(x) = G(x) + (4/9)[Q(x) + \bar{Q}(x)] \qquad (3)$$

where G(x), Q(x), and $\bar{Q}(x)$ are respectively the gluon, quark, and antiquark structure functions of the proton. In Equation (3) the factor 4/9 reflects the relative strengths of the quark-gluon and gluon-gluon couplings predicted by QCD.

The experimental angular distribution of jet pairs produced in pp collisions is analyzed as a test of vector gluon exchange and results are presented on the structure function F(x) defined by Equations (1) to (3).

A set of Monte Carlo events[10] generated using ISAJET [9] has been analyzed in parallel with the real data. Isajet generates two jet events and simulates the fragmentation of each jet into hadrons including the effects of QCD bremsstrahlung. The Monte Carlo program simulates in detail the subsequent behavior of the hadrons in the UA1 apparatus. The Monte Carlo events are used to calculate various corrections which are discussed below, and to estimate the jet energy resolution and the uncertainty in the determination of the jet direction.

After full calorimeter reconstruction jets are defined using the UA1 jet algorithm[18]. The energy and momentum of each jet is computed by taking respectively the scalar and vector sum over the associated calorimeter cells. A correction is applied to the measured energy ($\sim$ + 10%) and momentum ($\sim$ + 6%) of each jet, as a function of the pseudorapidity and azimuth for the jet, on the basis of the Monte Carlo analysis, to account for the effect of uninstrumented material and containment losses.

After jet finding, events are selected with $\geq$ 2 jets, within the acceptance of the central calorimetry $|\eta| < 3$. While the majority of these events have a topology consistent with two balanced high E_T jets, some 10-15% of the events have additional jets with $E_T > 15$ GeV[18]. A preceding analysis has shown that multijet events are largely accounted for in terms of initial-and final-state bremsstrahlung processes[19]. For this analysis, in order to compare with theoretical expectations for the two-jet cross-section, additional jets, apart from the two highest E_T jets, are ignored. The r.m.s. transverse momentum of the two-jet system (taking account of resolution) is then $\sim$ 10 GeV. A further correction is then applied, on the basis of the Monte Carlo analysis, to the energy ($\sim$ + 12%) and momentum ($\sim$ + 7%) of the two highest E_T jets in order to account for final state radiation falling outside the jet, as defined by the jet algorithm. After all corrections, averaged over the full acceptance, the jet energy resolution $\delta E/E \sim \pm 26\%$ and the uncertainty in the jet direction (in pseudorapidity) $\delta\eta \sim \pm 0.05$.

For each event the x_1 and x_2 of the interacting partons are computed as follows:

$$x_1 = [x_F + \surd(x_F^2 + 4\tau)]/2 \qquad (4)$$

$$x_2 = [-x_F + \surd(x_F^2 + 4\tau)]/2$$

where

$$x_F = (p_{3L} + p_{4L})/(\surd s/2) \qquad (5)$$

$$\tau = (p_3 + p_4)^2/s$$

In Equation (6), p_3 and p_4 are the 4-momenta of the final two jets and p_{3L} and p_{4L} are longitudinal momentum components measured along the beam direction in the laboratory frame.

The c.m.s. scattering angle is computed in the rest frame of the final two jets $[(p_3 + p_4)]$ relative to the axis defined by the interacting partons $[(p_1 - p_2)]$ assumed massless and collinear with the beams[20]:

$$\cos\theta = (p_3-p_4).(p_1-p_2)/(|p_3-p_4||p_1-p_2|). \qquad (6)$$

The finite angular acceptance of the apparatus and the trigger E_T threshold requirement discriminates against events with small scattering angles (i.e. large $\cos\theta$), and restricts the range of $\cos\theta$ over which the trigger is fully efficient. Events which are close to the limits of the acceptance in $\cos\theta$ are rejected by applying a fiducial cut in $\cos\theta$. This fiducial cut ($\cos\theta_{max}$) is defined for each event as the maximum value of $\cos\theta$ for which both the final two jets would fall in the acceptance region $|\eta| < 2.5$ (with at least one jet having $|\eta| < 1.0$) and for which the mean of the transverse energies of the final two jets would exceed (20)35 GeV [in the (un)filtered data set]. Events with azimuthal angle ϕ (defined by the final two jets) within ±45° of the vertical, where the two halves of the calorimeter are joined, are also rejected. The final event sample consists of 2432 events satisfying the above cuts, for which the acceptance is essentially uniform.

Figures 14 (a-d) show the raw histograms in $|\cos\theta|$ for various x_1, x_2 intervals. In each case a cut has been applied on $\cos\theta_{max}$ as indicated. The curve, which has been normalized to the number of events in each histogram, shows the expected angular dependence (Equation (2)) assuming vector gluon exchange (QCD). The data are consistent with the curve independent of x_1, x_2 in accordance with Equation (2).

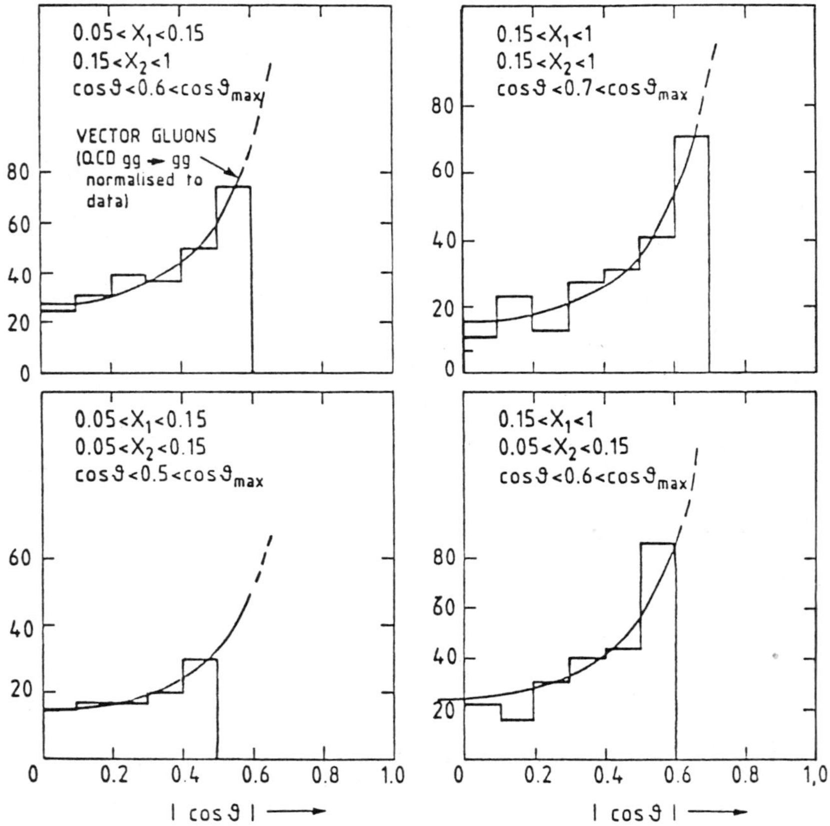

Fig. 14. (a-d): Histograms in $\cos\theta$ for various x_1, x_2 intervals. A cut on $\cos\theta_{max}$ has been applied as indicated. The curve, which has been normalized to the number of events in each histogram, is the expected angular dependence (Equation (2)) assuming vector gluon exchange (QCD).

Figure 15 shows the angular distribution obtained using all the events in the sample, on the assumption that the angular dependence is independent of x_1 and x_2. The distribution is computed starting from the raw distributions in cosθ, classified in intervals of $\cos\theta_{max}$($\Delta \cos\theta_{max}$ = 0.1), normalizing one distribution to another in the region of overlap. The final distribution is divided by the number of events with cosθ<0.1 to obtain $(d\sigma/d\cos\theta)(d\sigma/d\cos\theta)_{\cos\theta=0}$. Monte Carlo studies demonstrate that, for the range of cosθ shown, the angular distribution is not appreciably modified by the effects of resolution smearing. In Figure 15 the data are compared with theoretical forms for gluon-gluon, gluon-quark(antiquark)-quark(antiquark) elastic scattering in QCD (i.e. assuming vector gluon exchange). Theoretical forms for quark(antiquark)-quark(antiquark) and gluon-quark(antiquark) scattering in an Abelian scalar gluon theory[21] are also shown. The data are clearly consistent with the predictions of the vector gluon theory (QCD) and the Abelian scalar gluon theory is excluded. The data are not yet sufficiently accurate, however, to discriminate between the various subprocesses. A fit to the data of the form $(1-\cos\theta)^{-n}$ for cosθ > 0.4 yields n = 2.08±0.10. This result is consistent with n = 2 and, in analogy with the case of Rutherford scattering at low energy, may be regarded as a test of the inverse square law for the interaction between the partons.

In order to extract the structure function F(x) (Equation (3)) the quantity $S(x_1x_2)$ is defined in terms of the measured differential cross-section $d^2\sigma/dx_1dx_2$ as follows:

$$S(x_1,x_2)=x_1x_2(d^2\sigma/dx_1dx_2)/\int_0^{\cos\theta max} K(d\sigma/d\cos\theta)d\cos\theta. \quad (7)$$

If Equation (1) is valid then:

$$S(x_1,x_2)=F(x_1)F(x_2)$$

is determined from the data by weighting the events individually as a function of x_1, x_2 and $\cos\theta_{max}$.

The parton-parton differential cross-section dσ/dcosθ is taken from Equation (2) with $\alpha_s=12\pi/[23\ \ell n(Q^2/\Lambda^2)]$, i.e. assuming five effective quark flavors and Λ = 0.2 GeV. For this analysis Q^2 is defined by $Q^2 = -t$, where $t = (p_1-p_3)^2$ (or $t = (p_1-p_4)^2$, whichever is numerically smaller) in analogy with deep inelastic scattering. A factor K has been introduced in Equation (7) to allow for the effect of higher order QCD corrections, which may change the effective Q^2 scale and hence the normalization of the parton-parton cross-section. These corrections have been computed theoretically[22,23], and are expected to be appreciable even at the energy of the SPS Collider. The results given below have been computed assuming K = 2.

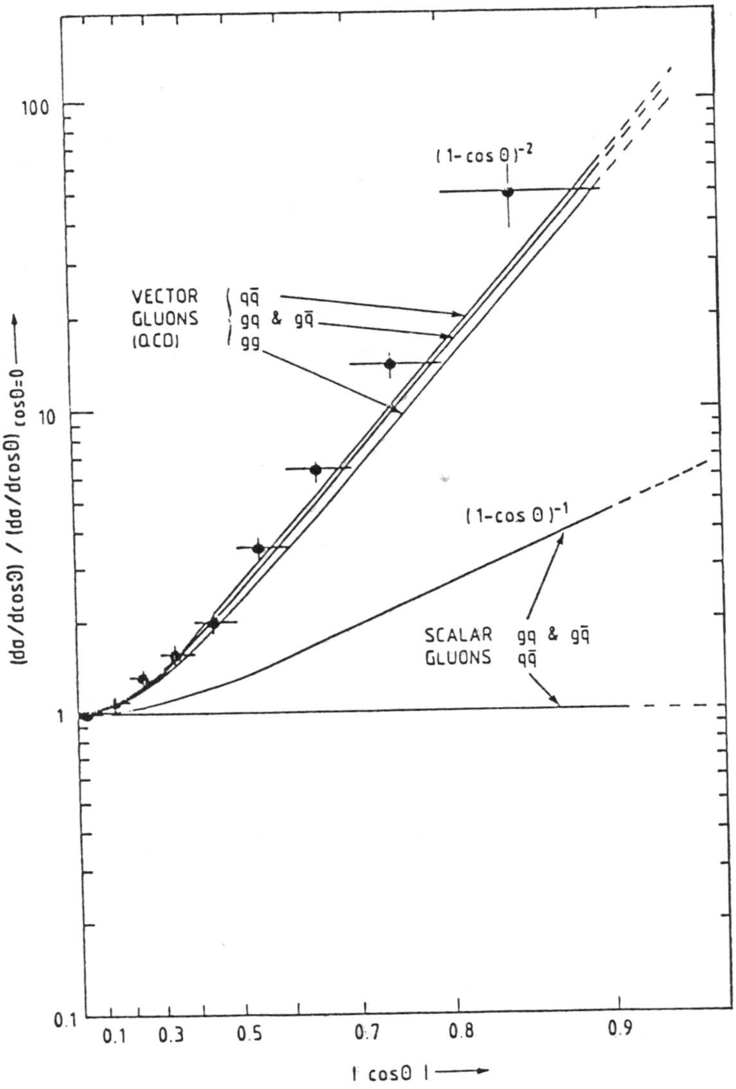

Fig. 15. The angular distribution obtained using all the events in the sample on the assumption that the angular dependence is independent of x_1 and x_2. The data are compared with theoretical forms for parton-parton elastic scattering in QCD and in an Abelian scalar gluon theory[9].

The experimental results for $S(x_1,x_2)$ are tabulated in Table 3. The data are symmetric in x_1 and x_2 and have been folded appropriately. The raw event numbers in each bin are given. The data for $\sqrt{\tau} < 0.2$ are based entirely on the unfiltered data set. The data

Table 3. The experimental result for $S(x_1,x_2)$ (see text). The raw event numbers in each x_1,x_2-bin are also given. The quoted errors are statistical only: the systematic error affecting mainly the overall normalization $\sim \pm 65\%$. The corresponding results for $F(x)$ are also tabulated, together with the mean value of $-\hat{t}$ in each bin.

			$S(x_1,x_2)$				
x_1 or x_2	0.05-0.10	0.10-0.20	0.20-0.30	0.30-0.40	0.40-0.50	0.50-0.60	0.60-0.80
0.05-0.10	-	-	3.053±0.276 (340)	1.224±0.207 (225)	0.449±0.122 (145)	0.298±0.134 (66)	0.064±0.038 (38)
0.10-0.20	-	2.541±0.247 (263)	0.842±0.118 (392)	0.373±0.061 (202)	0.218±0.047 (115)	0.055±0.027 (30)	0.024±0.015 (16)
0.20-0.30	-	-	0.391±0.080 (75)	0.117±0.044 (28)	0.038±0.029 (13)	0.032±0.038 (11)	0.005±0.011 (1)
0.30-0.40	-	-	-	0.042±0.053 (3)	0.022±0.040 (3)	0.009±0.028 (1)	0.000±0.000 (0)
			$F(x)$				
$F(x)$	5.497±0.578	1.607±0.082	0.542±0.049	0.222±0.030	0.112±0.027	0.038±0.017	0.012±0.006
$\langle -\hat{t} \rangle$ GeV²	1500	2200	3000	4100	4100	4600	5100

have been corrected for the effects of resolution smearing. This correction varies from -24% at small x to -70% at the largest x values. The quoted errors are statistical only: the systematic error affecting mainly the overall normalization is $\sim \pm 65\%$, due largely to the uncertainty in the jet energy scale. If factorization holds then $S(x_1,x'_2)/S(x_1,x_2)$ will be independent of x_1 for any choice of x_2 and x'_2. This ratio is plotted in Figure 16a for $x_2 = 0.1-0.2$, $x'_2 = 0.05-0.1$ and in Figure 16b for $x_2 = 0.1-0.2$, $x'_2 = 0.2-0.3$. Since the data are symmetric in x_1 and x_2 the folded data (Table 3) have been used to compute the ratio. In each case the ratio is consistent with a constant (broken line) independent of x_1. The data are therefore consistent with factorization in x_1 and x_2.

Figure 17 shows the structure function F(x) (Equation (3)) obtained from $S(x_1,s_2)$ assuming factorization in x_1 and x_2. The results for F(x) are also tabulated in Table 3, together with the mean value of -t in each x-bin. The measured integral of F(x) over the

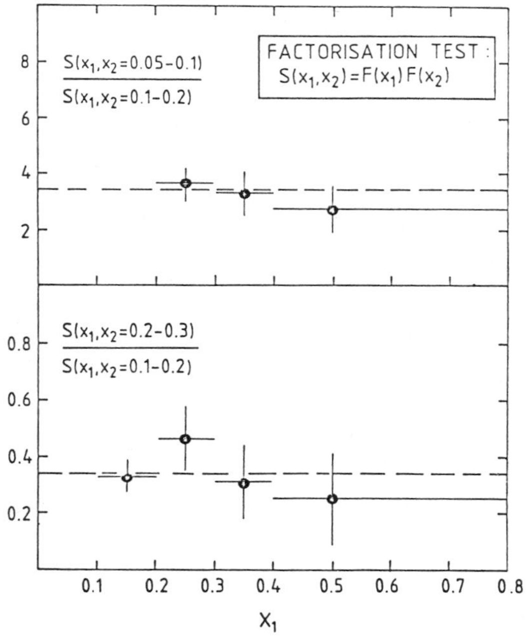

Fig. 16. The ratio $S(x_1,x_2')/S(x_1,x_2)$ for a) $x_2 = 0.1-0.2$, $x_2' = 0.05-0.1$ and b) $x_2 = 0.1-0.2$, $x_2' = 0.2-0.3$. In each case the data are consistent with a constant (broken line) independent of x_1.

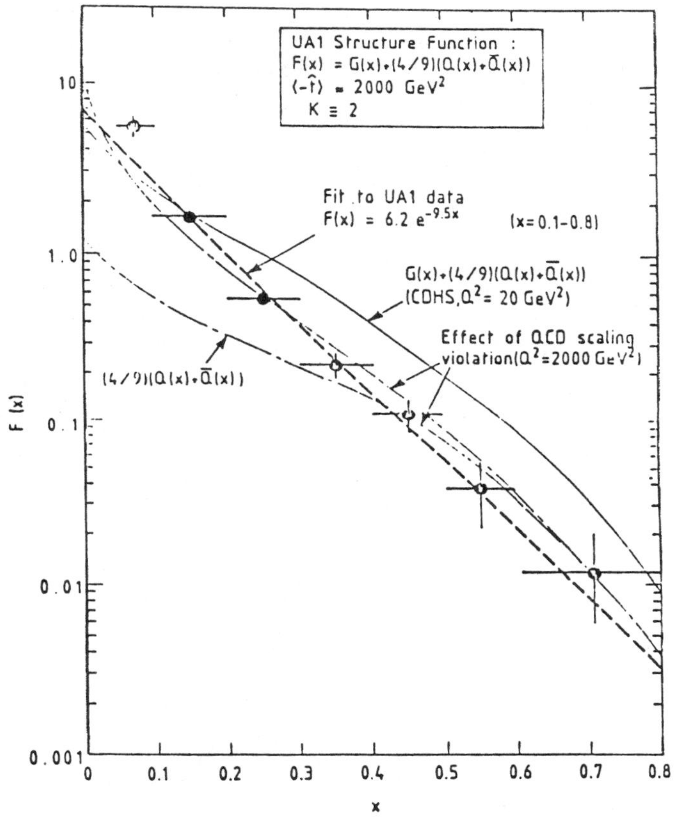

Fig. 17. The structure function F(x) (Equation (3)). The errors are statistical only: the systematic error affecting mainly the overall normalization is ∿ ±30%. The broken line represents the parametrization $F(x) = 6.2^{-9.5x}$ (see text). The solid curve represents a QCD parametrization of the structure function $G(x) + (4/9)\ [Q(x) + \bar{Q}(x)]$ at $Q^2 = 20\ GeV^2$ based on the CDHS[13] measurements at low $Q^2 (Q^2 = 2\text{-}200\ GeV^2)$. The broken curves show the expected modification of the structure function at the values of Q^2 appropriate in this experiment. The expected contribution of quarks and antiquarks is shown separately.

range x = 0.05-0.80 is 0.53±0.03. The errors quoted are statistical only. Since the cross-section takes the form of a product of structure functions the errors in the structure function are roughly one half the errors in $S(x_1,x_2)$. In particular, the systematic uncertainty in the overall normalization of the structure function is ∿ ±30%. Over the range x = 0.1-0.8, the data show an exponential x dependence and may be parametrized by the form $F(x) \sim 6.2\ e^{-9.5x}$ (broken line). Over the same range the data cannot be described by a single term of the form $(1-x)^n$ for any choice of n but may be parametrized as the sum of two such terms e.g. $F(x) \sim 8.0(1-x)^{13} + 1.1(1-x)^4$. Changing the assumed value of Λ by a factor of two (up or down) changes the structure function by $\sim \pm 15\%$. Changing the assumed value of the K-factor (Equation 7)) from K = 2 to K = 1 increases the structure function by a factor of $\sqrt{2}$, i.e. by an amount comparable to the systematic error.

In Figure 17 the measured F(x) is compared with expectations based on QCD fits to deep inelastic scattering data[24]. The solid curve represents the structure function $G(x) + (4/9[Q(x)+\bar{Q}(x)]$ at $Q^2 = 20$ GeV^2 based on a QCD parametrization of CDHS measurements of G(x) and $[Q(x)+\bar{Q}(x)]$ at low $Q^2(Q^2 = 2\text{-}200$ GeV^2) using a neutrino beam on an iron target. The broken curve shows the expected modification of the structure function, at the value of Q^2 appropriate in this experiment, due to QCD scaling violations (assuming $\Lambda = 0.2$ GeV). The expected contribution due to the quarks and the antiquarks is shown separately. The two-jet data measure directly, for the first time, the very large flux of gluons in the proton at small $x(x \lesssim 0.3)$ and also demonstrate the existence of QCD scaling violations at large Q^2 relative to the low Q^2 data. The present data suggest somewhat softer parton distributions than would be expected based on the CDHS parametrization.

Finally, our result for the integral of F(x) given above may be reinterpreted to yield information of σ and K. Assuming the validity of the parton model momentum sum-rule we have $\int_0^1 F(x)dx \lesssim 1$, from which we obtain $\alpha_s\sqrt{K} \gtrsim 0.12$ (±30% systematic error, due to the uncertainty in the energy scale). It is emphasized that this result is independent of the comparison with deep inelastic scattering data.

3. OBSERVATION OF CHARGED INTERMEDIATE VECTOR BOSONS

3.1. Introduction

The observation of the charged intermediate vector boson (IVB) was reported by the UA1 Collaboration in January 1983[25] on the basis of the observation of five high energy isolated electrons. These events all exibited a very large missing energy, interpreted as neutrino emission. The transverse vector momenta of the electron and neutrino share a strong correlation in angle and energy, indicating the same physical origin and are consistent with the assumption of the process:

$$p\bar{p} \rightarrow W^{\pm} + X \quad (1)$$
$$\qquad |\!\rightarrow e^{\pm}\nu_e$$

Since no other background could be formulated they were interpreted as evidence for IVB. Mass values were also given, $m_W = (80\pm5)$ GeV/c^2 (UA1) and $m_W = (80^{+10}_{-6})$ GeV/c^2 (UA2).

Since then the CERN sample has been considerably increased:

UA1[26]	53	$W^{\pm} \to e^{\pm}\nu$ events
	14	$W^{\pm} \to \mu^{\pm}\nu$ events
UA2[27]	36	$W \to e\nu$ events
Total	103	events

Such a comparatively large sample of events can now be used to proceed further in understanding the phenomenon. In particular the assignment of reaction (1) can now be "proven" rather than "postulated".

(i) The transverse decay kinematics is incompatible with emission of two or more neutrinos and it indicates the two body decay ($\nu_e e$). A greatly improved value for the mass is also obtained.

(ii) A subset (80% of the sample) of events from UA1 can be completely reconstructed, the twofold ambiguity for the unmeasured longitudinal momentum of neutrino being removed by the overall calorimetric information from the event. Using the well-known relation $x_W = x_p - x_{\bar{p}}$ and $x_p x_{\bar{p}} = m_W^2/s$ we can determine the distribution of the partons participating in the process. One can see that the distributions are in excellent agreement with the assumptions of quarks and antiquarks for the proton and antiproton respectively and exclude "sea" effects and gluons. Therefore the production mechanism is proven to be valence quark-antiquark annihilation.

(iii) A subsample of the fully reconstructed events have the sign of the lepton determined by magnetic curvature. Weak interactions should act as a longitudinal polarizer of the W particles since quarks(antiquarks) are provided by the proton(antiproton) beam. Likewise decay angular distributions from a polarizer are expected to have a large asymmetry, which acts as a polarization analyser. A strong backward-forward asymmetry is therefore expected, in which electrons(positrons) prefer to be emitted in the direction of the proton(antiproton). In order to study this effect independently of W-production mechanism, one has looked at the angular distribution of the emission angle θ^* of the electron(positron) with respect to the proton(antiproton) direction in the W center of mass. According to expectation of V-A theory the distribution should be of the type $(1+\cos\theta^*)^2$, in excellent agreement with the experimental data. More generally it has been shown by Jacob that for a particle of arbitrary spin J produced with average polarization $\langle\mu\rangle$ and decaying with an asymmetry parameter $\langle\lambda\rangle$ one expects:

$$\langle\cos\theta^*\rangle = \frac{\langle\lambda\rangle\langle\mu\rangle}{J(J+1)}$$

For V-A theory one expects $\langle\lambda\rangle = \langle\mu\rangle = -1$ and $J = 1$, leading to the <u>maximal</u> value $\langle\cos\theta^*\rangle = 0.5$. For $J = 0$ one obviously expects $\langle\cos\theta^*\rangle = 0$ and for any other spin value $J > 2$, $\langle\cos\theta^*\rangle \leq 1/6$. Experimentally, we find $\langle\cos\theta^*\rangle = 0.5 \pm 0.1$ close to its maximal value, which supports both the J = 1 assignment and maximal helicity states at production and decay. Note that the choice of sign $\langle\mu\rangle = \langle\lambda\rangle = \pm 1$ cannot be separated, i.e. right- and left-handed currents at production and decay cannot be resolved without a polarization measurement.

(iv) Finally the production cross-section can be determined: $(\sigma.B)_W = 0.53 \pm 0.08\ (\pm 0.09)$ nb, where the last error takes into account systematic errors. This value is in excellent agreement with the expectations of the V-A theory $(\sigma.B)_W = 0.39$ nb.

All these results can be used now to perform the interesting exercise of deducing classic weak interactions from $W^{\pm}$ particle observation. Indeed we know that $W^{\pm}$ must couple to valence quarks at production and to $(e\nu)$ pairs at decay, which implies the existence of the beta decay processes $n \rightarrow p + e^- + \nu_e$ and $(p) \rightarrow (n) + e^+ + \nu_e$. The mass value m_W and the cross-section measurement can be then used to calculate G_F, the Fermi coupling constant. The interaction must be Vector since J = 1 and parity is maximally violated since $\langle\mu\rangle = \langle\lambda\rangle = \pm 1$. The only missing element is the separation between V + A and V - A alternatives. For this purpose a polarization measurement is needed. It may be accomplished in the near future by studying for instance the decay $W \rightarrow \tau + \nu_\tau$ and using the τ decay as polarization analyzer or producing IVB with longitudinally polarized protons. Isn't this a nice way of teaching weak interactions to young students?

3.2. Event Selection for $W \rightarrow e\nu$ Decays

Results are based on an integrated luminosity of 0.136 pb^{-1}, which is corrected for dead-time and other similar losses and which includes the exposure for which results have been initially reported[25]. The trigger selection used throughout the investigation required the presence of an electromagnetic cluster at angles larger than 5°, with transverse energy in excess of 10 GeV. After on-line filtering and complete off-line reconstruction, about 1.5×10^5 events had at least one electromagnetic (e.m.) cluster with $E_T > 15$ GeV. By requiring the presence of an associated, isolated[25] track with $p_t > 7$ GeV/c in the central detector, we reduce the sample by a factor of about 100. Next, a maximum energy deposition (leakage) of 600 MeV is allowed in the hadron calorimeter cells after the e.m. counters, leading to a sample of 346 events. We then classify events according to whether there is prominent jet activity. We find that in 291 events there is a clearly visible jet within an azimuthal angle cone $|\Delta\phi| < 30°$ opposite to the "electron" track. These events are strongly contaminated by jet-jet events in which one jet fakes the electron signature and must be rejected. We are left with 55 events without any jet or with a jet not back-to-back with the "electron" within 30°.

The bulk of these events is characterized by the presence of neutrino emission, signalled by a significant missing energy (see Figure 18). According to the experimental energy resolutions, at most the three lowest missing energy events are compatible with no neutrino emission. They are excluded by the cut $E_T^{miss} > 15$ GeV. We are then left with 52 events. These events have a very clean electron signature (Figures 19 a-c) and a perfect matching between the point of electron incidence and the centroid in the shower detectors, further supporting the absence of composite overlaps of a charged track and neutral π^0's expected from jets.

In order to ensure the best accuracy in the electron energy determination, only events in which the electron track hits the electromagnetic detectors more than ±15° away from their top and bottom edges have been retained. The sample is then reduced to 43 events.

We have estimated, in detail, the possible sources of background coming from ordinary hadronic interactions with the help of a sample of isolated hadrons at large transverse momenta and we conclude that they are negligible (< 0.5 events). For more details on background we refer the reader to Reference 25. We may, however, detect some background events from other decays of the W, namely:

$W \to \tau\nu_\tau$ (< 0.5 events)
$\quad|\to \pi^\pm (\pi^0) \nu_\tau$

or $W \to \tau\nu_\tau$ (= 2 events)
$\quad|\to e\nu_e \nu_\tau$

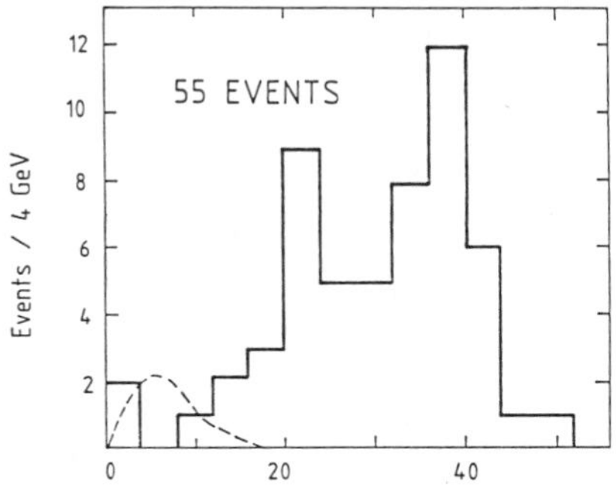

Fig. 18. The distribution of the missing transverse energy for those events in which there is a single electron with $E_T > 15$ GeV, and no co-planar jet activity. The curve represents the resolution function for no missing energy normalized to the three lowest missing-energy events.

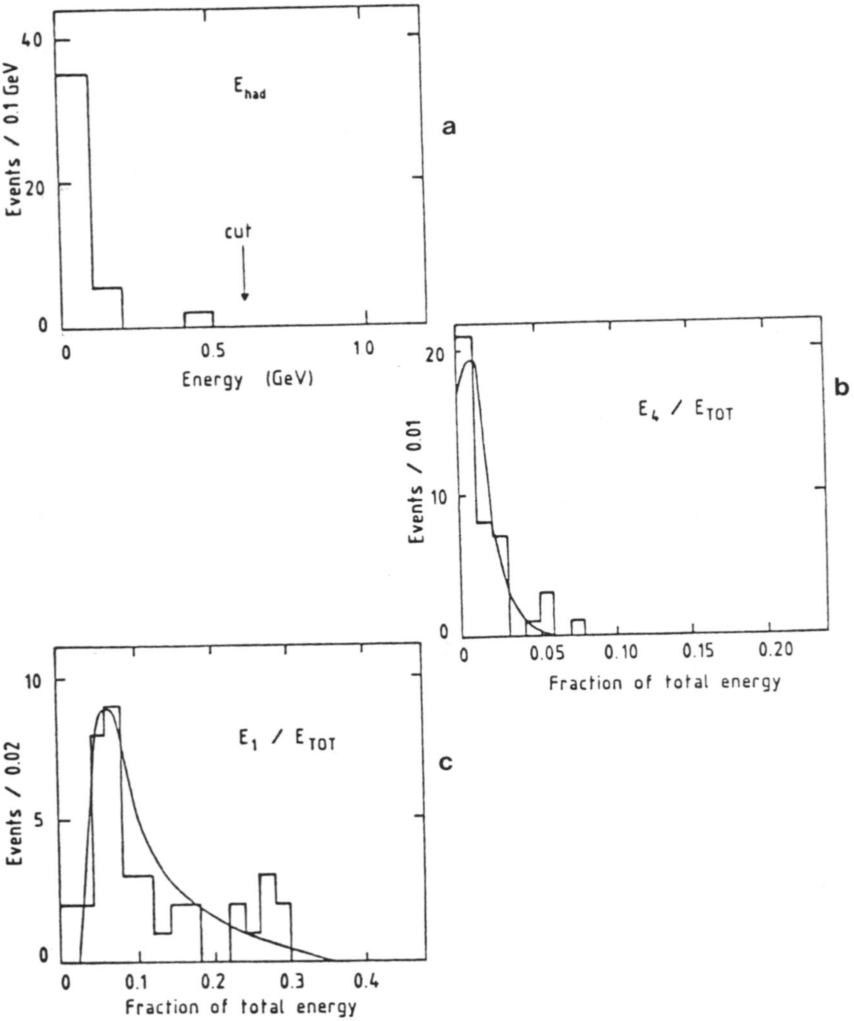

Fig. 19. Distributions showing the quality of the electron signature: a) The energy deposition in the hadron calorimeter cells behind the 27 radiation lengths (r.l.) of the e.m. shower detector. b) The fraction of the electron energy deposited in the fourth sampling (6 r.l. deep, after 18 r.l. convertor) of the e.m. shower detector. The curve is the expected distribution from test-beam data.

These events are expected to contribute only at the low p_t part of the electron spectrum and they can be eliminated in a more restrictive sample.

3.3. Origin of the Electron-neutrino Events

We proceed to a detailed investigation of the events in order to elucidate their physical origin. The large missing energy observed in all events is interpreted as being due to the emission of one or of several non-interacting neutrinos. A very strong correlation in angle and energy is observed (in the plane normal to the colliding beams, where it can be determined accurately), with the corresponding electron quantities, in a characteristic back-to-back configuration expected from the decay of a massive, slow particle (Figures 20a and b). This suggests a common physical origin for the electron and for

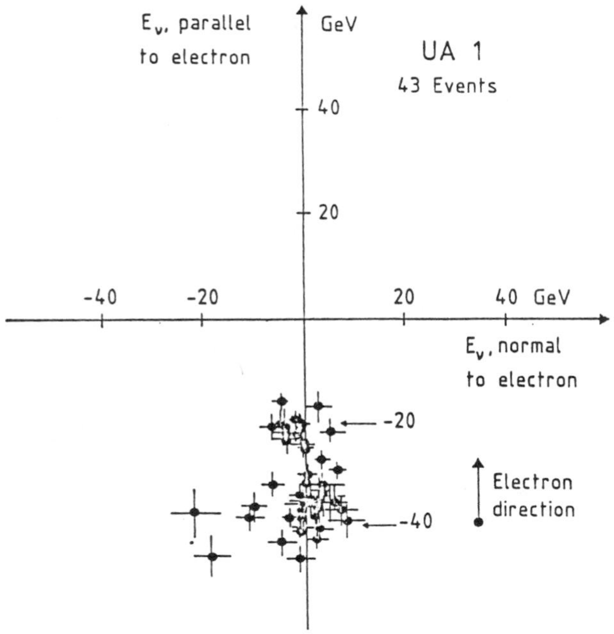

Fig. 20. a) Two-dimensional plot of the transverse components of the missing energy (neutrino momentum). Events have been rotated to bring the electron direction pointing along the vertical axis. The striking back-to-back configuration of the electron-neutrino system is apparent.

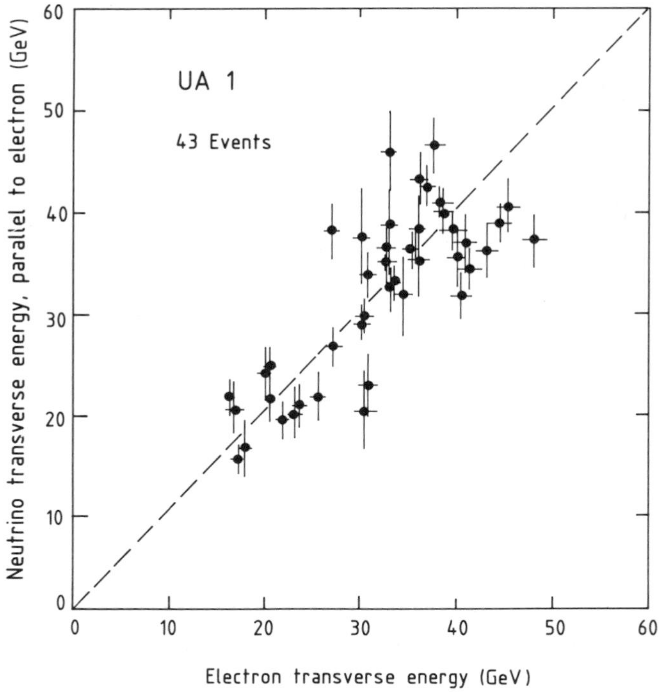

Fig. 20. b) Correlation between the electron and neutrino transverse energies. The neutrino component along the electron direction is plotted against the electron transverse energy.

one or several neutrinos. In order to understand better the transverse motion of the electron-neutrino(s) system one can study the experimental distribution of the resultant transverse momentum $p_t^{(W)}$ obtained by adding neutrino(s) and electron momenta (Figure 21). The average value is $p_t^{(W)}$ = 6.3 GeV/c. Five events which have a visible jet have also the highest values of $p_t^{(W)}$. Transverse momentum balance can be almost exactly restored if the vector momentum of the jet is added. The experimental distribution is in good agreement with the many theoretical expectations from QCD for the production of a massive state via the Drell-Yan quark-antiquark annihilation[28]. The small fraction (10%) of events with a jet are then explained as hard gluon bremsstrahlung in the initial state[29].

Several different hypotheses on the physical origin of the events can be tested by looking at kinematical quantities constructed from the transverse variables of the electron and the neutrino(s). We retain here two possibilities, namely (i) the two-body decay of a massive particle into the electron and one neutrino, $W \rightarrow e\nu_e$; and (ii) the three-body decay into two, or possibly more, neutrinos and

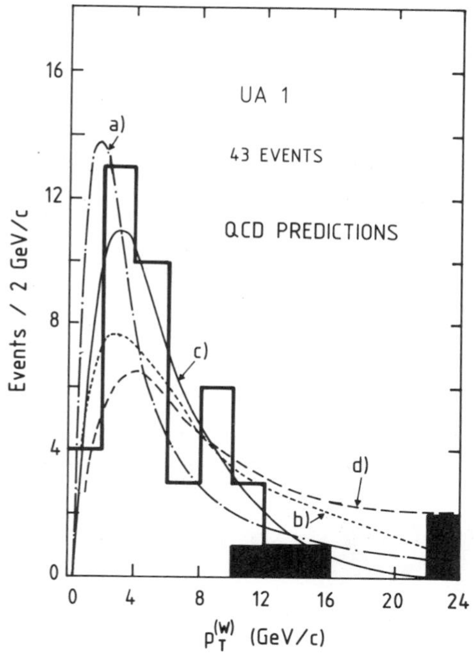

Fig. 21. The transverse momentum distribution of the W derived from our events using the electron and missing transverse-energy vectors. The highest $p_t^{(W)}$ events have a visible jet (shown in black in the figure). The data are compared with the theoretical predictions of Halzen et al., for W production (a) without $[O(\alpha_s)]$ and (b) with QCD smearing; and predictions by (c) Aurenche et al., and (d) Nakamura et al.[4].

the electron. One can see from Figures 22a and b that hypothesis (i) is strongly favored. At this stage, the experiment cannot distinguish between one or several closely spaced massive states.

3.4 Determination of the Invariant Mass of the $(e\nu_e)$ System

A (common) value of the mass m_W can be extracted from the data in a number of ways, namely:

i) It can be obtained from the inclusive transverse-momentum distribution of the electrons (Figure 22a). The drawback of this technique is that the transverse momentum of the W particle must be known. Taking the QCD predictions[28], in reasonable agreement with experiment, we obtain $m_W = (80.5 \pm 0.5)$ GeV/c^2.

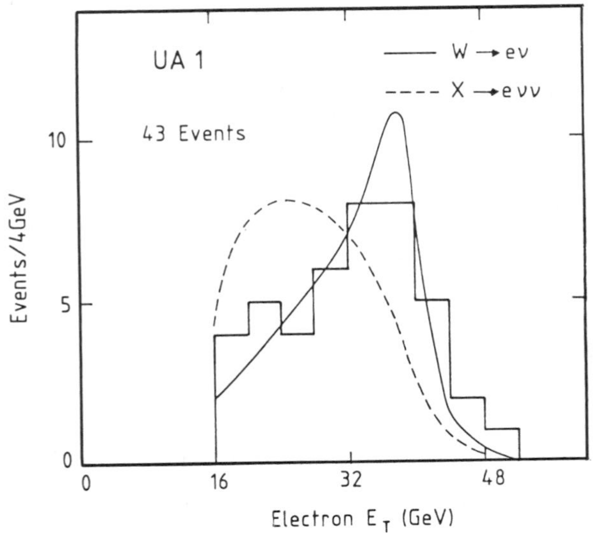

Fig. 22. a) The electron transverse-energy distribution. The two curves show the results of a fit of the enhanced transverse mass distribution to the hypothesis $W \to e\nu$ and $X \to e\nu\nu$. The first hypothesis is clearly preferred.

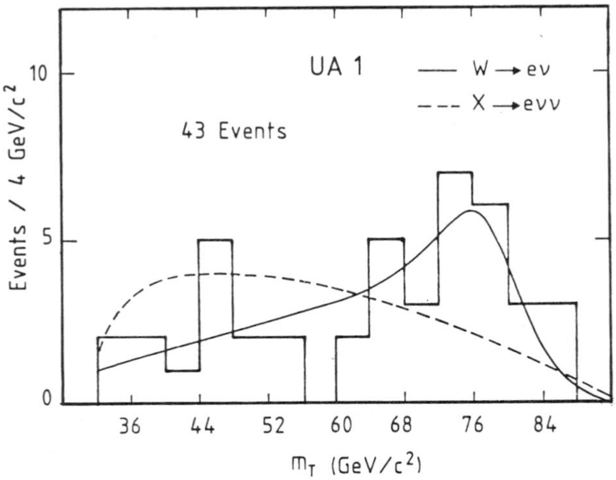

Fig. 22. b) The distribution of the transverse mass derived from the measured electron and neutrino vectors. The two curves show the results of a fit to the hypothesis $W \to e\nu$ and $X \to e\nu\nu$.

ii) We can define a transverse mass variable, $m_T^2 = 2p_t^{(e)}p_t^{(\nu)}(1 - \cos\phi)$, with the property $m_T \le m_W$, where the equality holds only for events with no longitudinal momentum components. Fitting Figure 22b to a common value of the mass can be done almost independently of the transverse motion of the W particles, $m_W = (80.3^{+0.4}_{-1.3})$ GeV/c². It should be noted that the lower part of the distribution in $m_t^{(W)}$ may be slightly affected by $W \to \tau\nu_\tau$ decays and other backgrounds.

iii) We can define an enhanced transverse mass distribution, selecting only events in which the decay kinematics is largely dominated by the transverse variable with the simple cuts $p_t^{(e)}$, $p_t^{(\nu)} > 30$ GeV/c. The resultant distribution (Figure 22c) shows then a relatively narrow peak, at approximately 76 GeV/c². Model-dependent corrections contribute now only to the difference between this average mass value and the fitted m_W value, $m_W = (80.0 \pm 1.5)$ GeV/c². An interesting upper limit to the width of the W can also be derived from the distribution, namely $\Gamma_T \le 7$ GeV/c² (90% confidence level).

The three mass determinations give very similar results. We prefer to retain the result of method (iii), since we believe it is the least affected by systematic effects, even if it gives the largest statistical error. Two important contributions must be added to the statistical errors:

i) Counter-to-counter energy calibration differences. They can be estimated indirectly from calibrations of several units in a beam of electrons; or, and more reliably, by comparing

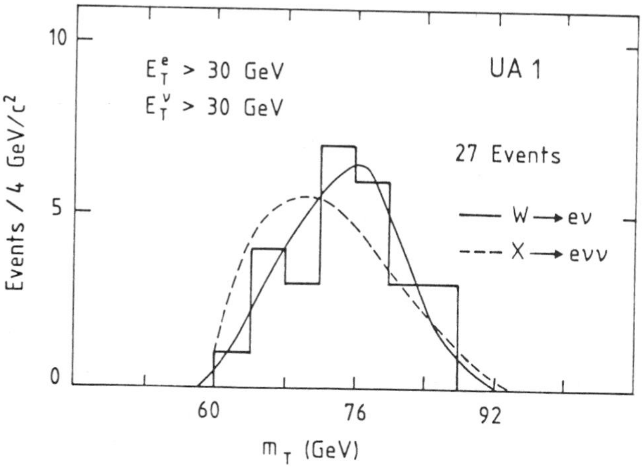

Fig. 22. c) The enhanced electron-neutrino transverse-mass distribution (see text). The two curves show the results of a fit to the hypotheses $W \to e\nu$ and $X \to e\nu\nu$.

the average energy deposited by minimum bias events recorded periodically during the experiment. From these measurements we find that the r.m.s. spread does not exceed 4%. In the determination of the W mass this effect is greatly attenuated, to the point of being small compared to statistical errors, since many different counter elements contribute to the event sample.

ii) Calibration of the absolute energy scale. This has been performed using a strong ^{60}Co source in order to transfer test-beam measurements to the counters in the experiment. Several small effects introduce uncertainties in such a procedure, some of which are still under investigation. At the present stage we quote an overall error of ±3% on the energy scale of the experiment. Of course this uncertainty influences both the $W^{\pm}$ and Z° mass determinations by the same multiplicative correction factor.

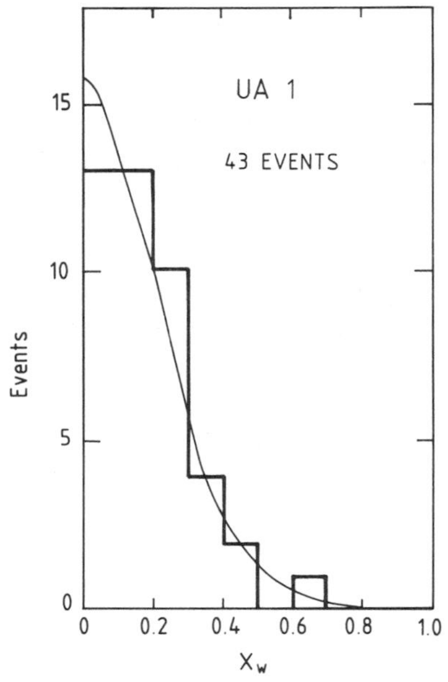

Fig. 23. a) The fractional beam energy x_W carried by the W. The curve is the prediction obtained by assuming the W has been produced by $q\bar{q}$ fusion. Note that in general there are two kinematic solutions for x_W (see text), which are resolved in 70% of the events by consideration of the energy flow in the rest of the event. Where this ambiguity has been resolved the preferred kinematic solution has been the one with the lowest x_W. In the 30% of the events where the ambiguity is not resolved the lowest x_W solution has therefore been chosen.

3.5 Longitudinal Motion of the W Particle

Once the decay reaction $W \to e\nu_e$ has been established, the longitudinal momentum of the electron-neutrino system can be determined with a two-fold ambiguity for the unmeasured longitudinal component of the neutrino momentum. The overall information of the event can be used to establish momentum and energy conservation bounds in order to resolve this ambiguity in 70% of the cases. Most of the remaining events have solutions which are quite close, and the physical conclusions are nearly the same for both solutions. The fractional beam energy x_W carried by the W particle is shown in Figure 23a and it appears to be in excellent agreement with the hypothesis of W production in $q\bar{q}$ annihilation[30]. Using the well-known relations $x_W = x_p - x_{\bar{p}}$ and $x_p \cdot x_{\bar{p}} = m_W^2/s$, we can determine the relevant parton distribution in the proton and antiproton. One can see that the distributions are in excellent agreement with the expected x distributions for quarks and antiquarks respectively in the proton and antiproton (Figure 23b and c). Contributions of the u and d quarks can also be neatly separated, by looking at the charges of produced W events, since $(\bar{u}d) \to W^+$ and $(\bar{u}d) \to W^-$ (Figures 23d and e).

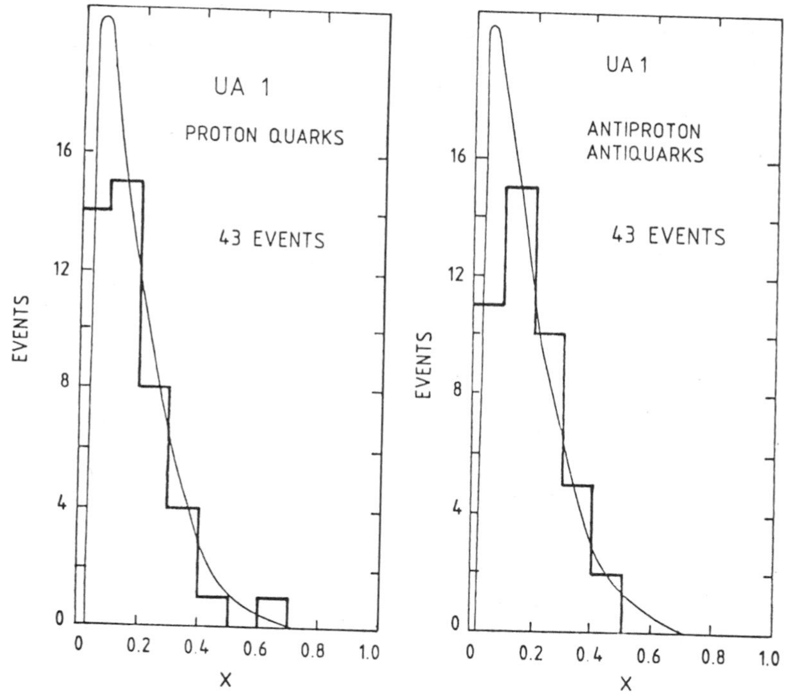

Fig. 23. b) The x-distribution of the proton quarks producing the W by $q\bar{q}$ fusion. The curve is the prediction assuming $q\bar{q}$ fusion. c) The same as (b) for the antiproton quarks.

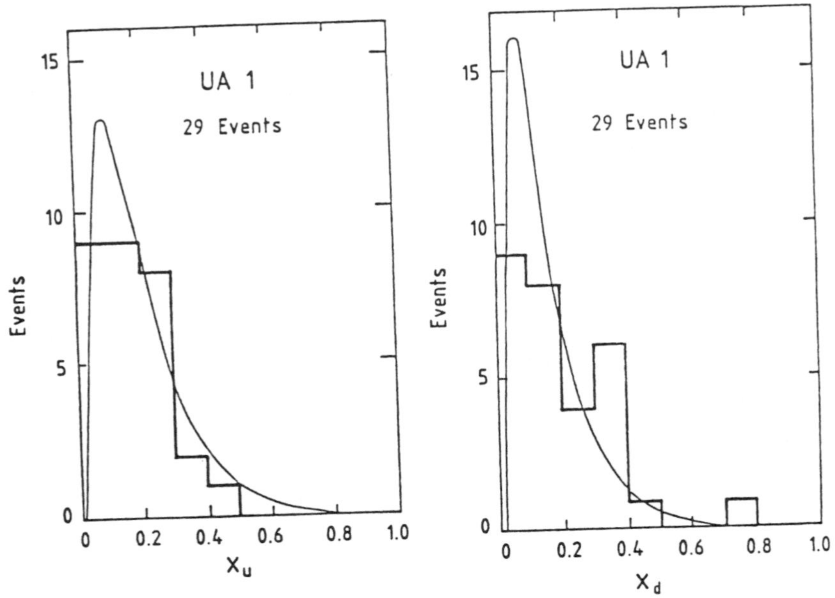

Fig. 23. d) The same as Fig. 23b) but for u($\bar{u}$) quarks in the proton (antiproton). e) The same as Fig. 23b) but for d($\bar{d}$) quarks in the proton (antiproton).

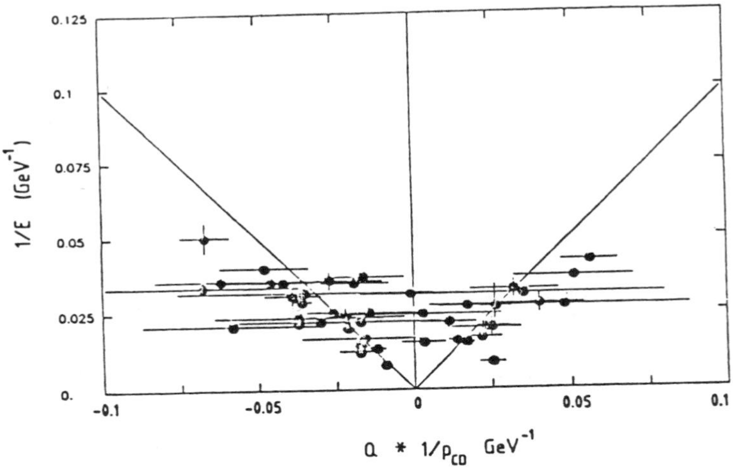

Fig. 24. a) 1/E plotted against Q/P_{CD} where E is the electron energy determined by the calorimeter, p_{CD} the momentum determined from the curvature of the central detector track, and Q the charge of the track.

3.6. Effects Related to the Sign of the Electron Charge

The momentum of the electron is measured by its curvature in the magnetic field of the central detector. Out of the 52 events, 24 (14) have a negative (positive) charge assignment; 14 events have a track topology which makes charge determination uncertain. Energy determinations by calorimetry and momentum measurements are compared in Figure 24a, and they are, in general, in quite reasonable agreement with what is expected from isolated high-energy electrons. A closer examination can be performed, looking at the difference between curvature observed and expected from the calorimeter energy determination, normalized to the expected errors (Figure 24b). One can observe a significant deviation from symmetry (corresponding to $p < E$), which can be well understood once the presence of radiative losses of the electron track (internal and external bremsstrahlung), is taken into account[31].

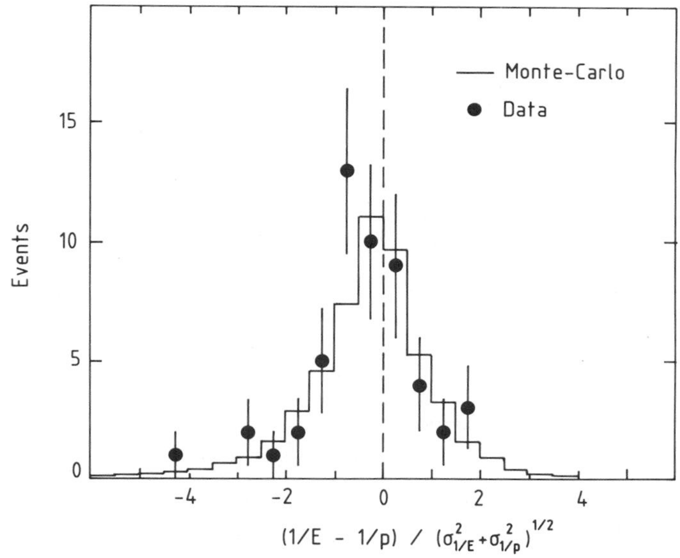

Fig. 24. b) $(1/E - 1/p)$ normalized by the error on the determination of this quantity. The curve is a Monte Carlo calculation, in which radiative losses due to internal and external bremsstrahlung have been folded with the experimental resolution[6].

Weak interactions should act as a longitudinal polarizer of the W particles since quarks(antiquarks) are provided by the proton(antiproton) beam. Likewise decay angular distributions from a polarizer are expected to have a large asymmetry, which acts as a polarization analyzer. A strong backward-forward asymmetry is therefore expected, in which electrons(positrons) prefer to be emitted in the direction of the proton(antiproton). In order to study this effect independently of W-production mechanisms, we have looked at the angular distribution of the emission angle θ^* of the electron(positron) with respect to the proton(antiproton) direction in the W center of mass. Only events with no reconstruction ambiguity can be used. It has been verified that this does not bias the distribution in the variable $\cos\theta^*$. According to the expectations of V-A theory the distribution should be of the type $(1 + \cos\theta^*)^2$, in excellent agreement with the experimental data (Figure 25).

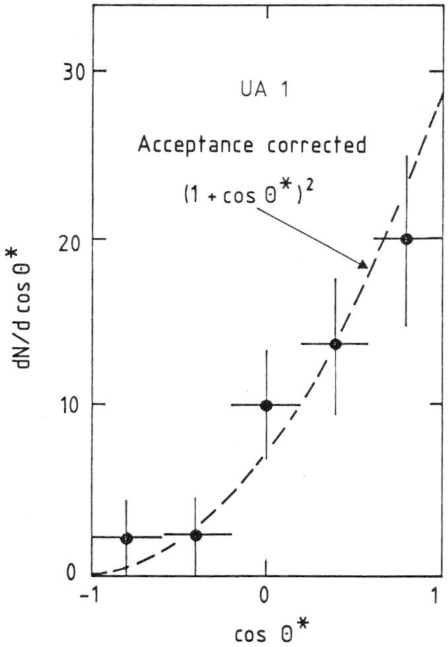

Fig. 25. The angular distribution of the electron emission angle θ^* in the rest frame of the W after correction for experimental acceptance. Only those events in which the electron charge is determined and the kinematic ambiguity (see text) has been resolved have been used. The latter requirement has been corrected for in the acceptance calculation.

3.7. Determination of the Parity Violation Parameters and of the Spin of the W-particle

It has been shown by Jacob[32] that for a particle of arbitrary spin J one expects:

$$\langle \cos\theta^* \rangle = \frac{\langle\lambda\rangle\langle\mu\rangle}{J(J+1)} ,$$

where $\langle\mu\rangle$ and $\langle\lambda\rangle$ are, respectively, the global helicity of the production system (ud) and of the decay system $(e\nu)$.

The detailed derivation follows closely the paper of reference. Let θ be the angle between the direction of the electron and the spin of the W particle in the rest system of the W-particle. The decay amplitude of W into a ν is proportional to:

$$D^{J^*}_{\mu\lambda}(\phi,\theta,-\phi)$$

The decay angular distribution is given by:

$$I(\theta,\phi) = \sum_{\mu\mu'} \ell_{\mu\mu'} \frac{2J+1}{4\pi} \sum_{\lambda_1\lambda_2} |c_{\lambda_1\lambda_2}|^2 D^{J^*}_{\mu\lambda}(\phi\theta-\phi) D^{J}_{\mu'\lambda}(\phi\theta,-\phi)$$

with

$$\lambda = \lambda_1 - \lambda_2 \text{ and } \rho_{\mu\mu'} = \delta_{\mu\mu'}$$

p_μ the matrix density for the W in absence of polarization of the p and p beams; p_μ is the probability that the W is the state of helicity μ.

Combining the two D functions making use of the Clebsch-Gordon series:

$$I(\theta,\phi) = \frac{2J+1}{4\pi} \sum_{\mu} (-1)^{\mu} p_{\mu} \sum_{\lambda_1\lambda_2} (-1)^{\lambda_1\lambda_2} |c_{\lambda_1\lambda_2}|^2$$
$$\sum_{\ell=0}^{2J} c(JJ\ell|\mu-\mu)c(JJ\ell|\lambda-\lambda)P_{\ell}(\cos\theta)$$

We now take the average value of $\cos\theta$. This gives a particularly simple expression:

$$\langle\cos\theta\rangle = \frac{1}{J(J+1)} (\sum_{\mu} \mu p_{\mu})(\sum_{\lambda_1\lambda_2} \lambda |c_{\lambda_1\lambda_2}|^2)$$

The two terms within parenthesis are the average helicities, therefore:

$$\langle\cos\theta\rangle = \frac{\langle\mu\rangle\langle\lambda\rangle}{J(J+1)}$$

For V-A one then has $\langle\lambda\rangle = \langle\mu\rangle = -1$, $J = 1$, leading to the maximal value $\langle\cos\theta\rangle = 0.5$. For $J = 0$ one obviously expects $\langle\cos\theta^*\rangle = 0$ and for any other spin value $J \geq 2$, $\langle\cos\theta\rangle \leq 1/6$. Experimentally, we find $\langle\cos\theta^*\rangle = 0.5 \pm 0.1$, which supports both the $J = 1$ assignment and maximal helicity states at production and decay. Note that the choice of sign $\langle\mu\rangle = \langle\lambda\rangle = \pm 1$ cannot be separated, i.e. right- and left-handed currents both at production and decay cannot be resolved without a polarization measurement.

3.8. Total Cross-section and Limits to Higher Mass W's

The integrated luminosity of the experiment was 136 nb^{-1} and it is known to about ±15% uncertainty. In order to get a clean $W \to e\nu_e$ sample we select 47 events with $p_t^{(e)} > 20$ GeV/c. The $W \to \tau\nu_\tau$ contamination in the sample is estimated to be 2 ± 2 events. The event acceptance is estimated to be 0.65, due primarily to: (i) the $p_t^{(e)} > 20$ GeV/c cut (0.80); (ii) the jet veto requirement within $\Delta\phi = \pm 30°$ (0.96±0.02); (iii) the electron-track isolation requirement (0.90±0.07); and (iv) the acceptance of events due to geometry (0.94±0.03). The cross-section is then:

$$(\sigma \cdot B)_W = 0.53 \pm 0.08 \ (\pm 0.09) \ \text{nb},$$

where the last error takes into account systematic errors. This value is in excellent agreement with the expectations for the Standard Model[30] $(\sigma \cdot B)_W = 0.39$ nb.

No event with $p_t^{(e)}$ or $p_t^{(\nu)}$ in excess of the expected distribution for $W \to e\nu$ events has been observed. This result can be used in order to set a limit to the possible existence of very massive W-like objects (W') decaying into electron-neutrino pairs. We find $(\sigma \cdot B) \leq 30$ pb at 90% confidence level, corresponding to $m > 170$ GeV/c², if standard couplings and quark distributions are used to evaluate the cross-sections.

3.9 Observation of the Decay Mode $W \to \mu + \nu$

Muon-electron universality predicts an equal number of events in which the electron is replaced by its heavy counterpart, the muon:

$$p\bar{p} \to W^{\pm} X \ ; \ W^{\pm} \to \mu^{\pm} \overset{(-)}{\nu}_{\mu}$$

Although almost identical with decays with electrons (1) in theory, the muonic decay has a completely different experimental signature. Whereas an electron produces an electromagnetic shower (detected in the electromagnetic calorimeters), a high momentum muon traverses the whole detector with almost minimum energy loss. Muons are identified by their ability to penetrate many absorption lengths of material.

Thus potential backgrounds for muons are radically different from those for electrons. The observation of the same rate for processes $e\nu$ and $\mu\nu$ is therefore not only the most direct confirmation of muon-electron universality in charged-current interactions, but it also provides an important experimental verification of the previous results.

We now briefly describe the muon detection. A fast muon, emerging from the pp interaction region, will pass in turn through the central detector, the electromagnetic calorimeter and the hadron calorimeter, which consists of the instrumented magnet return yoke. After 60 cm of additional iron shielding (except in the forward region), it will then enter the muon chambers, having traversed about $8/\sin\theta$ nuclear interaction lengths, where θ is its emission angle with respect to the beam axis. The number of hadrons penetrating this much material is negligible; however there are two sources of hadron-induced background:

i) stray radiation leaking through gaps and holes;
ii) genuine muons from hadron decays, such as $\pi \rightarrow \mu\nu$, $K \rightarrow \mu\nu$, etc.

It is therefore essential to follow the behavior of all muon candidates throughout the whole apparatus. Tracks are recorded in the central detector. The momenta of muons are determined by their deflection in the central dipole magnet, which generates a field of 0.7 T over a volume of $0.7 \times 3.5 \times 3.5\ m^3$. The momentum accuracy for high-momentum tracks is limited by the localization error inherent in the system ($\leq$ 100 μm) and by the diffusion of electrons drifting in the gas, which is proportional to $\sqrt{\ell}$ and amount to about 350 μm after the maximum drift length of ℓ = 19.2 cm. This results in a momentum accuracy of about ±20% for a 1 m long track at p = 40 GeV/c, in the best direction with respect to the field. In general, the precision depends greatly on the length and orientation of the track. For the muon sample under discussion, the typical error is around ±30%.

In the present investigation the calorimeters have a fourfold purpose:

i) they provide enough material to attenuate hadrons, and constitute a threshold for muon detection of $p_t > 2$ GeV/c;
ii) they identify hadronic interactions and/or accompanying neutral particles by an excess in the energy deposition;
iii) they ensure a continuous tracking of the muon over six segments in depth;
iv) they provide an almost hermetically closed energy flow measurement around the collision point, which makes possible the determination of the transverse components of the neutrino momentum by transverse energy conservation.

Fifty muon chambers[33], nearly 4 m x 6 m in size, surround the whole detector, covering an area of almost 500 m^2. A graphical display of a $W \to \mu\nu$ event is shown in Figure 26, with an expanded view of the muon chambers shown as an insert. Each chamber consists of four layers of drift tubes, two for each projection. The tubes in adjacent parellel layers are staggered. This resolves the left-right drift time ambiguity and reduces the inefficiency from the intervening dead spaces. The extruded aluminium drift tubes have a cross-section of 45 mm x 150 mm, giving a maximum drift length of 70 mm. An average spatial resolution of 300 μm has been achieved through the sensitive volume of the tubes[34]. In order to obtain good angular resolution on the muon tracks, two chambers of four planes each, separated by 60 cm, are placed on five sides of the detector. This long lever-arm was chosen in order to reach an angular resolution of a few milliradians, comparable to the average multiple scattering angle of high-energy muons (3 mrad at 40 GeV/c). Because of space limitations, the remaining side, beneath the detector, was closed with special chambers consisting of four parallel layers of drift tubes.

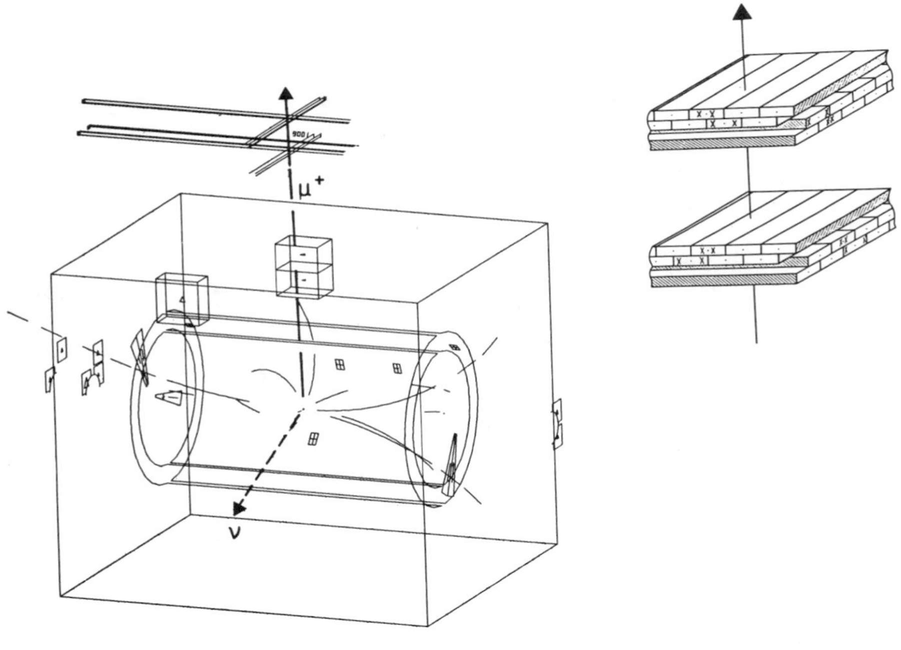

Fig. 26. A graphical display of a $W^+ \to \mu^+\nu$ event. The vertical arrow show the trajectory of the 25 GeV/c μ^+ up to the muon chamber while the other arrow shows the transverse direction of the neutrino. The curved lines from the vertex are the charged tracks seen by the central detector, and the petals and boxes illustrate the electromagnetic and hadronic energy depositions. An expanded view of a muon module is shown as an insert.

The track position and angle measurements in the muon chambers permit a second, essentially independent, measurement of momentum. The statistical and systematic errors in this second momentum determination were carefully checked with high-momentum cosmic-ray muons; Figure 27 compares the momentum measurements in the central detector and muon chambers. Because of the long lever-arm to the muon chambers, a significant increase in precision is achieved by combining the two measurements.

The presence of neutrino emission is signalled by an apparent transverse energy imbalance when the calorimeter measurement of missing transverse energy is combined with the muon momentum measurement. This determines the neutrino transverse momentum error perpendicular to the muon p whereas the error parallel to the muon p is dominated by the track momentum accuracy.

The muon sample is contaminated by several background sources such as leakage through the absorber, beam halo, meson decays, and cosmic rays. Some of the background can be eliminated by requiring a matching central detector track with sufficiently high momentum to penetrate to the muon chambers. All events were therefore passed through a fast filter program which selected muon candidates with $p_t > 3$ GeV/c or $p > 6$ GeV/c. This filter program reconstructed

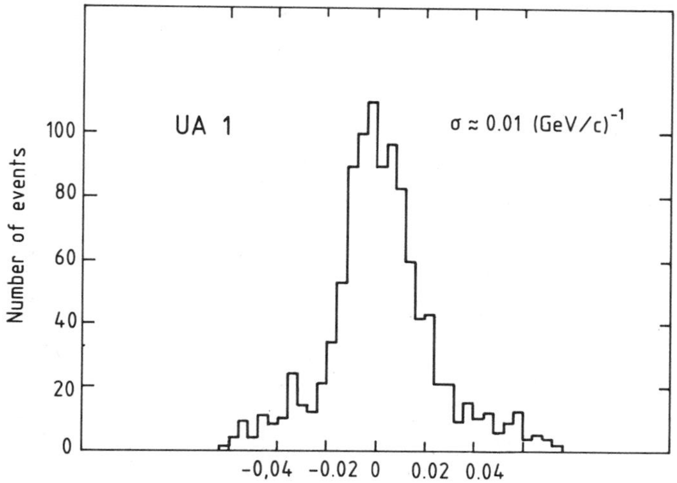

Fig. 27. Distribution of $1/p_\mu - 1/p_{CD}$ for vertical cosmic-ray muons with $p_{CD} > 10$ GeV/c, where p_μ and p_{CD} are the momenta measured in the muon chambers and central detector respectively.

tracks in the muon chambers. For each track pointing roughly towards the interaction region, the central detector information was decoded along a path from the muon chamber track to the interaction region. Track finding and fitting were performed in this path. Events were kept if a central detector track satisfied the above momentum cut and matched the muon chamber track within generous limits. The filter program selected about 72000 events. Since only limited regions of the central detector were considered, the program took about 10% of the average reconstruction time of a full event.

The 17326 events from the fast filter which contained a muon candidate with $p_t > 5$ GeV/c were passed through the standard UA1 processing chain. Of these, 713 events had a muon candidate with $p_t > 15$ GeV/c or $p > 30$ GeV/c. These events were passed through an automatic selection program which eliminated most of the remaining background by applying strict track quality and matching cuts. Independently of this, all events were examined on an interactive scanning facility. This confirmed that no W-candidate events were rejected by the selection program.

The selection program imposed additional requirements on event topology in order to reject events with muons in jets or back-to-back with jets.

Events were also rejected if the jet algorithm found a calorimeter jet with $E_T > 10$ GeV or a central detector jet with $p_t > 7.5$ GeV/c back-to-back with the muon to within ±30° in the plane perpendicular to the beam. Thirty-six events survived these cuts, and were carefully rescanned. After eliminating additional cosmics and probable $K \to \mu\nu$ decays, 18 events remained. The final W-sample of 14 events was obtained after the additional requirement that the neutrino transverse energy exceed 15 GeV. The effects of the different cuts are shown in Table 4.

The most dangerous background to the $W \to \mu\nu$ sample comes from the decay of medium-energy kaons into muons within the volume of the central detector such that the transverse momentum kick from the decay balances the deflection of the particle in the magnetic field. This simulates at the same time a high-momentum muon track and, in order to preserve momentum balance in the transverse plane, a recoiling "neutrino". Most of these events are rejected by the selection program. We have performed a Monte Carlo calculation to estimate the residual background. Charged kaons with $3 < p_t < 15$ GeV/c and decaying in the central detector were generated according to a parametrization of the transverse momentum distribution of charged particles[35], assuming a ratio of kaons to all charged particles of 0.25[36]. A full simulation of the UA1 detector was performed, and each track was subjected to the same reconstruction and selection procedures as the experimental data, including the scanning of these events. Normalizing to the integrated luminosity of 108 nb^{-1}, we

Table 4. Selection of $W \to \mu\nu$ candidates. The number of events is indicated at each stage

Events with a $p_t > 5$ GeV/c muon selected by the fast filter program	17,326
Fully reconstructed events with muon $p_t > 15$ GeV/c or $p > 30$ GeV/c	713
Events with a good qualtiy CD track which matches the muon chambers well	285
Events remaining after rejection of cosmic rays	247
Events remaining after a tight cut on the χ^2 of the CD track fit to remove decays	144
Events where the muon is isolated both in the CD and in the calorimeters	53
Events with no jet activity opposite the muon in the transverse plane	36
Events remaining as W candidates after scanning (see text)	18
Events with a neutrino transverse momentum > 15 GeV/c	14

found 4 events in which the K decay was recognized and simulated a muon with $p_t > 15$ GeV/c. Imposing the additional requirement of $p_t > 15$ GeV/c for the accompanying neutrino leaves less than one event as an upper limit to the background to $W \to \mu\nu$ from this source.

In addition, we expect about 5 events in our data sample with muons from decays of pions or kaons with $p_t > 15$ GeV/c. These will be similarly suppressed by the reconstruction and selection procedures; in particular such events will be characterized by jets which transversely balance the high-p_t hadrons and are therefore rejected by our topological cuts. The momentum measurements in the central detector and in the iron agree very well (Figure 28), as a good check of our procedure.

Eighteen events survive our selection criteria and contain a muon with $p_t > 15$ GeV/c. The muons are isolated, and there is no visible structure to compensate their transverse momenta, in contrast with what might be expected for background events from heavy-flavor decays. Including the muon in the transverse energy balance, all events exhibit a large missing transverse energy of more than 10 GeV, attributed to an emitted neutrino. For the final $W \to \mu\nu$ sample, we consider only those 14 events with a neutrino transverse momentum $p_t > 15$ GeV/c. As in the electron case, the transverse momentum of the neutrino is strongly correlated, both in magnitude and in direction, with the transverse momentum of the muon. Figure 29a shows this correlation in the direction parallel to the muon p_t. Similarly the component of the neutrino p_t perpendicular to the muon p_t is small. The characteristic back-to-back configuration and the high

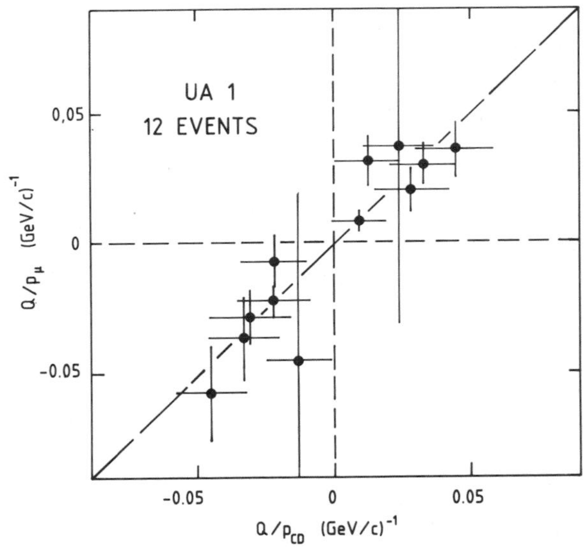

Fig. 28. Two-dimensional plot of Q/p_μ versus Q/p_{CD} for the $W \rightarrow \mu\nu$ events, where p_μ and p_{CD} are the momenta measured in the muon chambers and central detector respectively, and Q is the charge of the muon. The events with tracks in the bottom chambers are not shown.

momenta of both leptons, well above the threshold, are very suggestive of a two-body decay of a massive, slow particle. The large errors in the momentum determination of the muons smear the expected Jacobian peak of a two-body decay. However, the transverse momentum distribution agrees well with that expected from a $W^{\pm}$ decay, once it is smeared with the experimental errors (Figure 29b).

The transverse momentum $p_t^{(W)}$ of the decaying particle is well measured, because the muon momentum does not enter into its determination. In fact, $p_t^{(W)}$ is simply energy measured in the calorimeters, after subtraction of the muon deposition. The measured distribution is given in Figure 30a and agrees well with our previous measurement from the $W \rightarrow e\nu$ sample, shown in Figure 30b. Each of the two events with the highest $p_t^{(W)}$ has a jet which locally balances the transverse momentum of the W.

In order to determine the mass of the muon-neutrino system, we have used in a maximum likelihood fit the eight measured quantities for each event (momentum determination of the muon in the CD and in the muon chambers, angles of the muon, four-vector of the energy for the rest of the event) and their relevant resolution functions. We have taken account of the cust imposed on the measured muon and neutrino transverse momenta[37]. We obtain a fitted W mass of

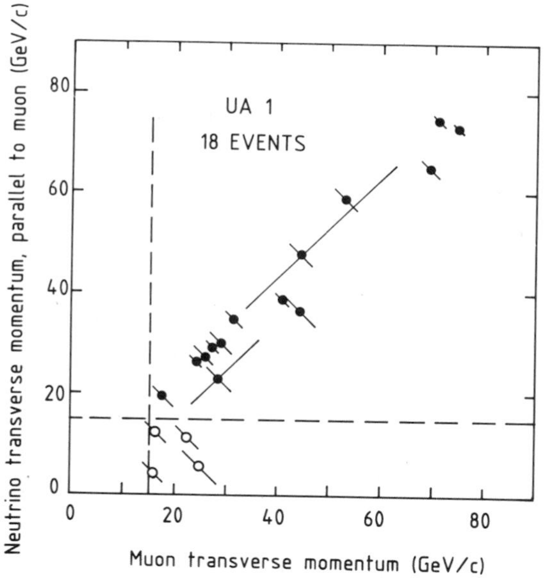

Fig. 29. a) Transverse energy of the neutrino parallel to the muon versus transverse momentum of the muon. Since the two quantities are correlated, error bars are shown for the difference and the sum. The difference in the transverse energy of the W parallel to the muon which is measured in the calorimetry and is therefore not correlated with the transverse momentum of the muon. For the errors in the sum only two error bars are shown for typical events. The filled circles correspond to the final sample of 14 W events, and the open circles to the 4 events with neutrino $p_t < 15$ GeV/c.

$m_W = 81^{+6}_{-7}$ GeV/c^2, in excellent agreement with the measured value from $W \rightarrow e\nu$. This result is insensitive to the assumed decay angular distribution of the W. If the mass is fixed at the electron value of 80.9 GeV/c^2, a fit of the decay asymmetry gives $\langle \cos \theta^* \rangle = 0.3 \pm 0.2$, fully consistent with our result from $W \rightarrow e\nu$ and with the expected V-A coupling. The asymmetry measurement is not very significant since the ambiguity due to the two possible solutions for the longitudinal momentum of the W could be resolved in only a few cases. This is due to the large momentum errors and the limited acceptance in pseudorapidity ($|\eta| < 1.3$) for the muons.

The overall acceptance for the final sample of 14 $W \rightarrow \mu\nu$ events is limited by two main factors, namely the geometrical acceptance of the muon trigger system for muons with $p_t > 15$ GeV/c (49%) and the influence of the track quality cuts applied to the muon. The latter

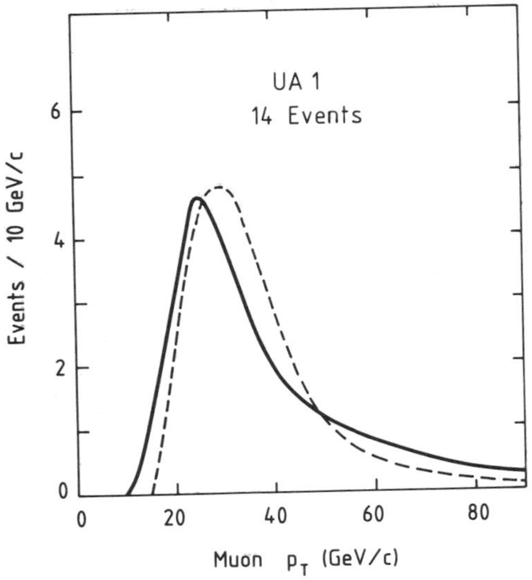

Fig. 29. b) The solid curve is an ideogram of the transverse momentum distribution of the muons in the final sample of 14 $W \rightarrow \mu\nu$ events. The dashed curve is a Monte Carlo prediction, based on the W production spectra measured in $W \rightarrow e\nu$ decays and a W mass of 80.9 GeV/c^2 smeared with errors.

has been estimated by applying identical cuts to an equivalent sample of 46 $W \rightarrow e\nu$ events from the 1983 data sample. 21 events remain, giving an acceptance of (46 ± 7)%. A further correction of (87 ± 7)% is included to account for the jet veto and track isolation requirement. These three factors give an overall acceptance of (20 ± 3.5)%.

The integrated luminosity for the present data sample is 108 nb^{-1}, with an estimated uncertainty of ±15%. The cross-section is then:

$$(\sigma \cdot B)_\mu = 0.67 \pm 0.17\ (\pm 0.15)\ \text{nb}$$

where the last error includes the systematics from both acceptance and luminosity. This value is in good agreement both with the standard model predictions[30] and with our result for $W \rightarrow e\nu$, namely $(\sigma \cdot B)_e = 0.53 \pm 0.08\ (\pm 0.09)$ nb.

A direct comparison between the electron and muon results has been made by selecting those $W \rightarrow e\nu$ events which are within the acceptance of the muon trigger system. Twelve events remain from the

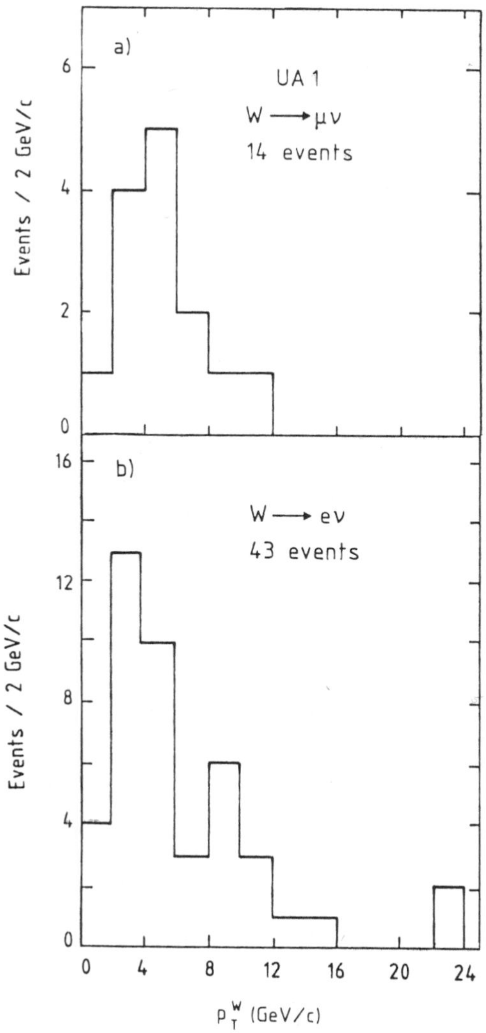

Fig. 30. a) The transverse momentum distribution of the W derived from the energy imbalance measured in the calorimetry.
b) The corresponding distribution from the W → eν data is shown for comparison.

21 which pass the muon track quality cuts. After correction for the difference in integrated luminosity (118 nb^{-1} in the electron case) this gives the following cross-section ratio, in which systematic errors approximately cancel:

$$R = (\sigma \cdot B)_\mu / (\sigma \cdot B)_e = 1.24^{+0.6}_{-0.4}$$

4. OBSERVATION OF THE NEUTRAL BOSON Z°

4.1. Event Selection

We now extend our search to the neutral partner Z°, responsible for neutral currents. As in our previous work, production of intermediate vector bosons is achieved with proton-antiproton collisions at $\sqrt{s}$ = 540 GeV in the UA1 detector, except that now we search for electron and muon pairs rather than for electron-neutrino coincidence. The process is then:

$$\bar{p} + p \rightarrow Z^\circ + X$$
$$\quad\quad |\rightarrow e^+ + e^- \text{ or } \mu^+ + \mu^-$$

The reaction is then approximately a factor of 10 less frequent than the corresponding $W^\pm$ leptonic decay channels. A few events of this type are therefore expected in our muon or electron samples. Evidence for the existence of the Z° in the range of masses accessible to the UA1 experiment can also be drawn from weak-electromagentic interference experiments at the highest PETRA energies, where deviations from point-like expectations have been reported.

Events for the present paper were selected by the so-called "express line," consisting of a set of four 168E computers[38] operated independently in real time during the data-taking. A subsample of events with $E_T \geq 12$ GeV in the electromagnetic calorimeters and dimuons are selected and written on a dedicated magnetic tape. These events have been fully processed off-line and further subdivided into four main classes:

i) single, isolated electromagnetic clusters with $E_T > 15$ GeV and missing energy events with $E_{miss} > 15$ GeV, in order to extract $W^\pm \rightarrow e^\pm\nu$ events;
ii) two or more isolated electromagnetic clusters with $E_T >$ 25 GeV for $Z^\circ \rightarrow e^+e^-$ candidates;
iii) muon pair selection to find $Z^\circ \rightarrow \mu^+\mu^-$ events; and
iv) events with a tract reconstructed in the central detector, of transverse momentum within one standard deviation, $p_t \geq$ 25 GeV/c, in order to evaluate some of the background contributions.

We will discuss these different categories in more detail.

4.2. Events of Type $Z^\circ \rightarrow e^+e^-$

An electron signature is defined as a localized energy deposition in two contiguous cells of the electromagnetic detectors with $E_T > 25$ GeV, and a small (or no) energy deposition ($\leq$ 800 MeV) in the hadron calorimeters immediately behind them. The isolation requirement is defined as the absence of charged tracks with momenta adding up to more than 3 GeV/c of transverse momentum and pointing towards the electron cluster cells. The effects of the successive cuts on the invariant electron-electron mass are shown in Figure 31. Four e^+e^- events survive cuts, consistent with a common value of (e^+e^-) invariant mass. They have been carefully studied using the interactive event display facility MEGATEK. One of these events is shown in Figures 32a and 32b. The main parameters of the four events are listed in Table 5 and 7. As one can see from the energy deposition plots (Figure 33), their dominant feature is two very prominent electromagnetic energy depositions. All events appear to balance the visible total transverse energy components; namely, there is no evidence for the emission of energetic neutrinos. Except for

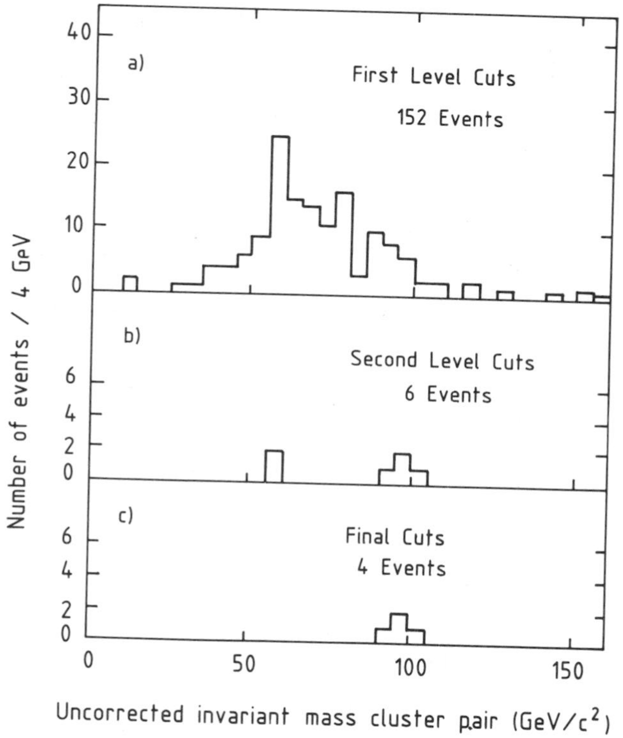

Fig. 31. Invariant mass distribution (uncorrected) of two electromagnetic clusters: a) with $E_T > 25$ GeV; b) as above and a track with $p_t > 7$ GeV/c and projection length > cm pointing to the cluster. In addition, a small energy deposition in the hadron calorimeters immediately behind (< 0.8 GeV) ensures the electron signature. Isolation is required with $\Sigma p_t < 3$ GeV/c for all other tracks pointing to the cluster. c) The second cluster also has an isolated track.

Table 5. Properties of the Individual Electrons of the Pair Events

	Drift chamber measurement						Shower counter measurement					
								Electromagnetic samples (Gev)				Had. energy
Run event	p (Gev)	Δp a) (Gev)	Q	dE/dx b)	y c)	ϕ (deg)	E (Gev)	S_1	S_2	S_3	S_4	H
A 7433 1001	33	+9 -6	+	1.8±0.3	1.01	144	44	14	27	3	0.0	0.0
	63	+23 -13	-	1.7±0.2	-1.19	-31	48	6	37	4	0.2	0.0
B 7434 746	27	+19 -8	+	1.6±0.3	-0.36	131	42	2	18	20	1.3	0.1
	93	+66 -28	-	1.8±0.2	-1.45	-60	102	42	56	4	0.2	0.0
C 6059 1010	32	+11 -6	+	1.3±0.2	0.64	67	61	1	37	22	0.6	0.0
	9	+1 -1	-	1.4±0.1	0.24	-121	48	1	23	23	1.3	0.0
D 7739 1279	d)	d)	d)	d)	-0.19	169	51	1	13	34	2.4	0.0
	50	+50 -17	-	1.5±0.2	-0.79	-9	55	8	38	9	0.0	0.1

a) $\pm 1\sigma$ including systematic errors.

b) Ionization loss normalized to minimum ionizing pion.

c) The rapidity y is defined as positive in the direction of outgoing $\bar{p}$.

d) Unmeasured owing to large dip angle.

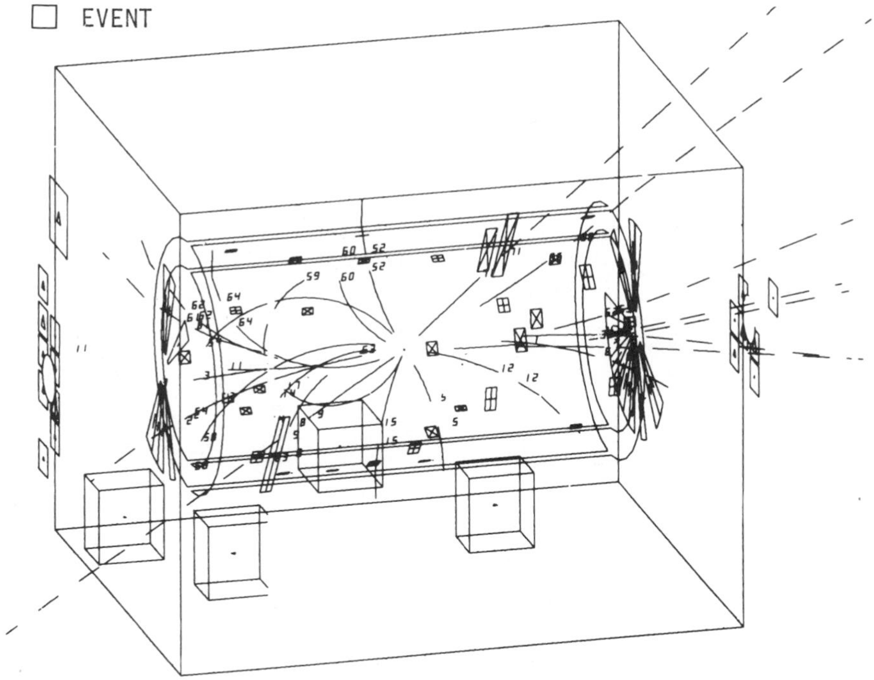

Fig. 32. a) Event display. All reconstructed vertex associated tracks and all calorimeter hits are displayed.

one track of event D which travels at less than 15° parallel to the magnetic field, all tracks are shown in Figure 34, where the momenta measured in the central detector are compared with the energy deposition in the electromagnetic calorimeters. All tracks but one have consistent energy and momentum measurements. The negative track of event C shows a value of (9 ± 1) GeV/c, much smaller than the corresponding calorimeter deposition of (48 ± 2) GeV. One can interpret this event as the likely emission of a hard "photon" accompanying the electron. Subsequent calibrations of the electromagnetic (e.m.) counters indicated that the centroid of the energy deposition in the calorimeters was significantly displaced with respect to the incident electron track, indicating an angle of $\Delta\phi$ = (14 ± 4) degrees between the charged and neutral components. This excluded the possibility of an external bremsstrahlung in the vacuum pipe and detector window. The estimated probability of an internal bremsstrahlung exceeding the angle and energy observed is, according to Berends[31], about 0.005, or 2% for the sample of four events. A rather similar event has been reported by the UA2 Collaboration. In this case the photon and electron hit separate cells, thus directly indicating a finite e-γ separation. More recently, an event of the type $\mu^+\mu^-\gamma$ has been observed in the UA1 study of $Z^\circ \to \mu^+\mu^-$ decays. Also in this

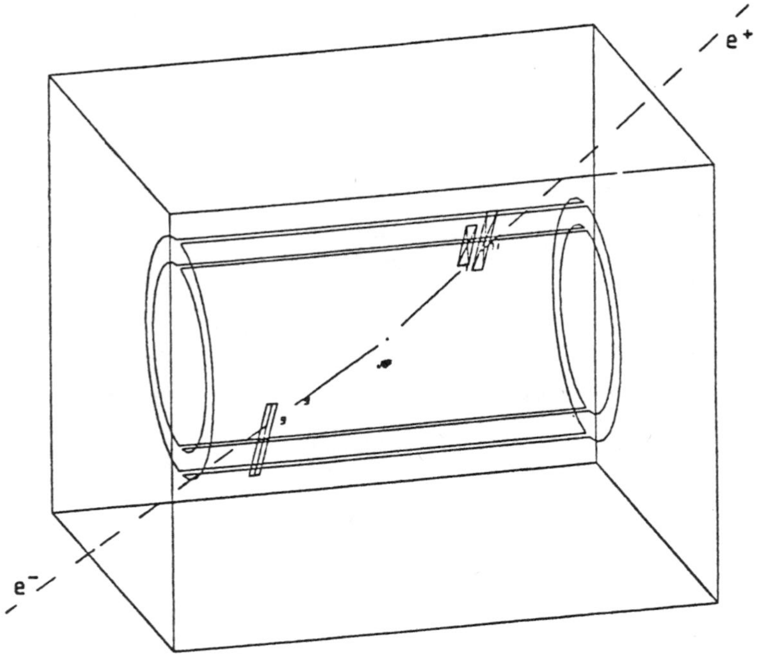

Fig. 32. b) The same, but thresholds are raised to $p_t > 2$ GeV/c for charged tracks and $E_T > 2$ GeV for calorimeter hits. We remark that only the electron pair survives these mild cuts.

event, a small but finite angle (∿8°) is observed between the muon and the hard photon (E∿30 GeV). Event parameters are summarized in Table 8. It is certainly premature to draw any conclusion about the origin of such events. It shows however how entirely new phenomena may be occurring in the collider energy range.

The average invariant mass of the pairs, combining the four consistent values, is (95.2±2.5) GeV/c² (Table 7).

4.3 Events of Type $Z^\circ \rightarrow \mu^+\mu^-$

Events from the dimuon trigger flag have been submitted to the additional requirement that there is at least one muon track reconstructed off-line in the muon chambers, and with one track in the central detector of reasonable projected length ($\geq$ 40 cm) and $p_t \geq 7$ GeV/c. Only 42 events survive these selection criteria. Careful scanning of these events has led to only one clean dimuon event, with two "isolated" tracks (Figure 35). Most of the events are due to cosmics. Parameters are given in Table 6 and 7. Energy losses in

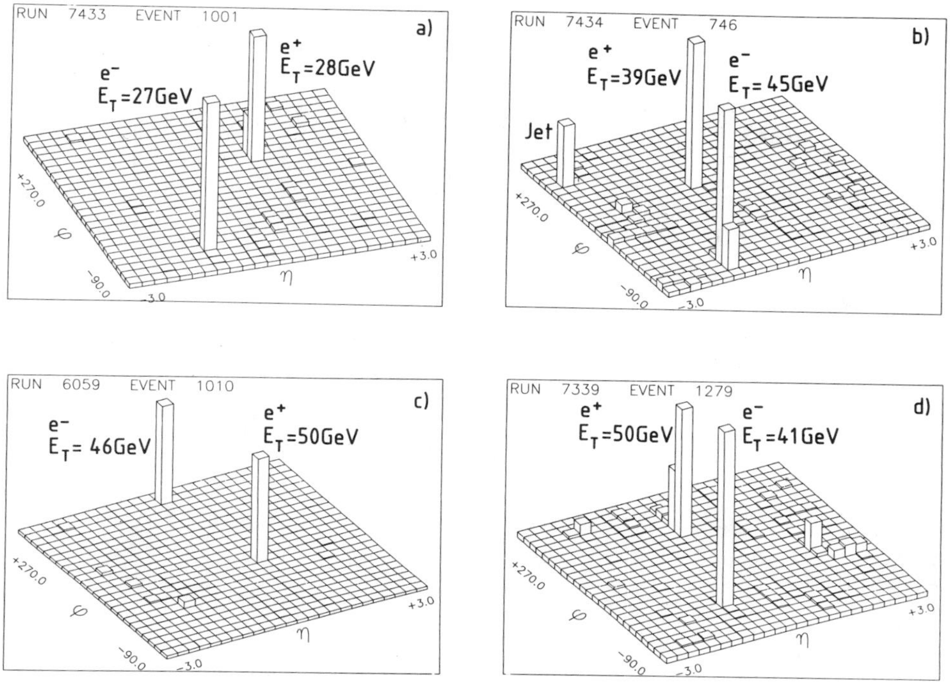

Fig. 33. Electromagnetic energy depositions at angles > 5° with respect to the beam direction for the four electron pairs.

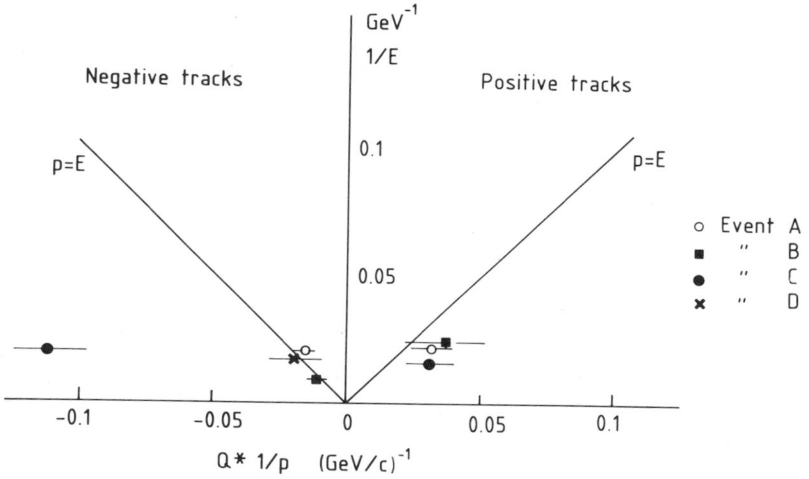

Fig. 34. Magnetic deflection in 1/p units compared to the inverse of the energy deposited in the electromagnetic calorimeters. Ideally, all electrons should lie on the 1/E = 1/p line.

Table 6. Properties of the Muons of the Dimuon Event 6600-222

	Track parameters				Normalized ionization I/I_o			μ/CD matching[d)]	
Q	p (Gev/c)	ℓ (cm)	y	φ (°)	e.m. calorimeter	Hadron calorimeter	abs	Position (cm)	Angle (mrad)
	58.8^{+8}_{-6} a)								
+	60.3±10.8 b)								
	$59.2^{+6.4}_{-5.2}$ c)	170	1.19	-27.6	0.8±0.5	1.2±0.5	10.2	ΔX_1 = -1.3±1.9 ΔX_2 = 11.6±10.7	$\Delta\phi$ = -2±6 $\Delta\Lambda$ = 11±14
	63.6^{+30}_{-15} a)								
-	43.1±6.2 b)								
	$46.1^{+6.1}_{-5.7}$ c)	80	-0.28	119.1	1.2±0.9	1.6±0.8	11.1	ΔX_1 = -0.1±8.0 ΔX_2 = -8.0±8.5	$\Delta\phi$ = -6±3 $\Delta\Lambda$ = -9±14

Momentum determination: a) Central detector and μ chamber (statistical errors only); b) Transverse momentum balance;
c) Weighted average of (a) and (b);

μ-CD matching: d) Difference between the extrapolated CD track measured in the μ chambers (see Fig.6).

Remarks: The acceptance of the single muon trigger starts at a transverse momentum of about 2.5 GeV/c and reaches its full efficiency of 97% at 5.5 GeV/c. The geometrical acceptance of the dimuon trigger used in this analysis, reduced the acceptance for Z° events to about 30%.

Table 7. Mass and Energy Properties of Lepton Pair Events

	Lepton pair properties			General event properties			
Run event	Mass [a] (Gev/c²)	p_t (Gev/c)	x_F [b]	E_{tot} (Gev)	$\Sigma\|E_T\|$ (Gev)	Missing E_T (Gev)	Charged tracks
A 7433 1001	91±5	2.9±0.9	0.02±0.01	274	82	2.1±3.6	27
B 7434 746	97±5	7.9±1.2	0.39±0.01	494	149	9.3±5.0	67
C 6059 1010	98±5	8.0±1.5	0.17±0.01	412	143	3.3±4.8	38
D 7339 1279	95±5	8.4±1.4	0.17±0.01	493	157	0.8±5.0	54
E 6600 222 (μμ)	95±8	24±5	0.14±0.02	278 [c]	128 [c]	3.4±5.9 [c]	28

[a] These errors have been scaled up arbitrarily to 5 GeV to represent the present level of uncertainty in the overall calibration of the e.m. calorimeter which will be recalibrated completely at the end of the present run. This scale factor is not included in the error bars plotted in Fig.8.

[b] x is defined as the longitudinal momentum of the dilepton divided by beam energy.

[c] Includes the muon energies.

Table 8. Energy, Angle, and Mass Properties of the eeγ Events

		UA1	UA2
$E\gamma$	(GeV)	38.8±1.5	24.4±1.4
Ee_1	(GeV)	61.0±1.2	68.5±1.6
Ee_2	(GeV)	9±1	11.4±0.9
$\Delta\phi(e_2\gamma)$	(^{0})	14.4±4.0	31.8
$m(e_1e_2)$	(GeV/c^2)	42.7±2.4	49.8
$m(e_1e_2\gamma)$	(GeV/c^2)	98.7±5.0	89.7±2.8
$m(e_1\gamma)$	(GeV/c^2)	88.8±2.5	74.1
$m(e_2\gamma)$	(GeV/c^2)	4.6±1.0	9.1

Notes: For the UA1 event $e_1 = e^+$ and $e_2 = e^-$.
The UA2 numbers are calculated from ref.[4].

the calorimeters traversed by the two muon tracks are well within expectations of ionization losses of high-energy muons (Figure 36a). The position in the coordinate and the angles at the exit of the iron absorber (Figure 36b) are in agreement with the extrapolated track from the central detector, once multiple scattering and other instrumental effects have been calibrated with $p > 50$ GeV cosmic-ray muons traversing the same area of the apparatus. There are two ways of measuring momenta, either in the central detector or using the muon detector. Both measurements give consistent results. Furthermore, if no neutrino is emitted (as suggested by the electron events which exhibit no missing energy), the recoil of the hadronic debris, which is significant for this event, must be equal to the transverse momentum of the $(\mu^+\mu^-)$ pair by momentum conservation. The directions of the two muons then suffice to calculate the momenta of the two tracks. Uncertainties of muon parameters are then dominated by the errors of calorimetry. As shown in Table 6 this determination is in agreement with magnetic deflection measurements. The invariant mass of the $(\mu^+\mu^-)$ pair is found to be $m_{\mu\mu} = 95.5\pm7.3$ GeV/c^2, in excellent agreement with that of the four electron pairs (see Table 7).

4.4. Backgrounds

The most striking feature of the events is their common value of the invariant mass (Figure 37); values agree within a few percent and with expectations from experimental resolution. Detection efficiency is determined by the energy thresholds in the track selection, 15 GeV/c for $e^\pm$ and 7 GeV/c for $\mu^\pm$. Most "trivial" sources of background are not expected to exhibit such a clustering at high masses. Also, most backgrounds would have an equal probability for $(e\mu)$ pairs, which are not observed. Nevertheless, we have considered

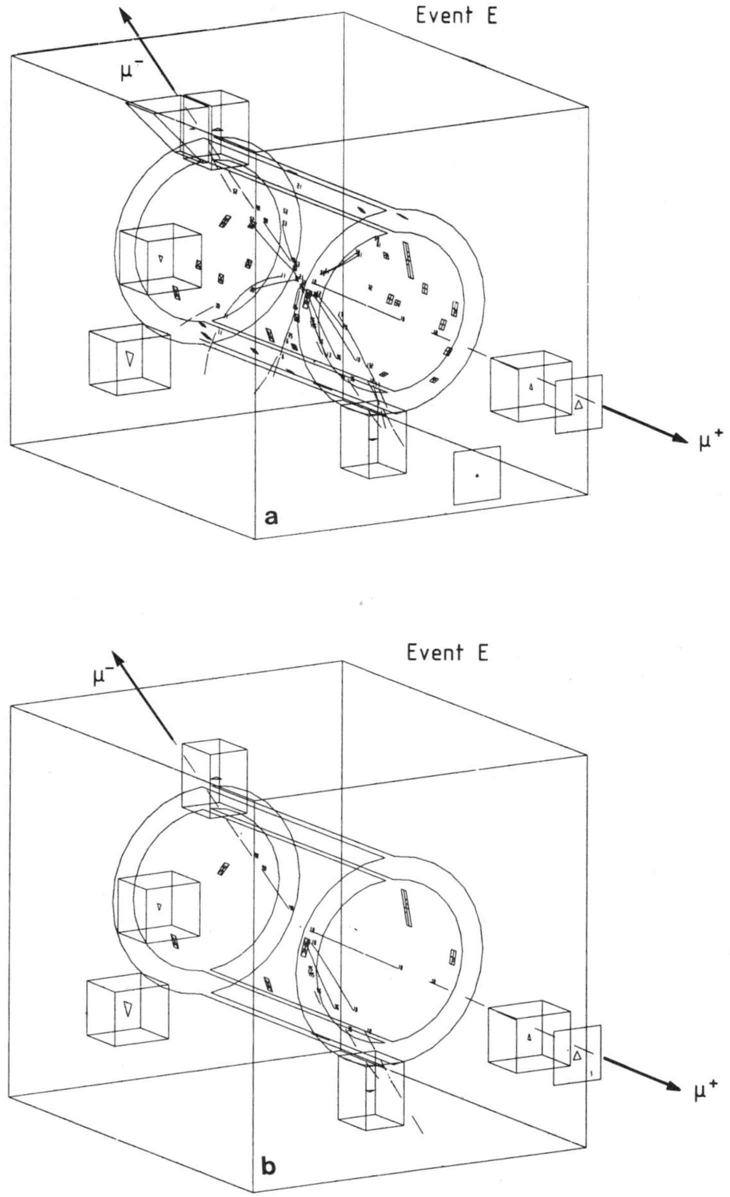

Fig. 35. Display for the high-invariant-mass muon pair event: a) without cuts and b) with $p_t > 1$ GeV thresholds for tracks and $E_T > 0.5$ GeV for calorimeter hits.

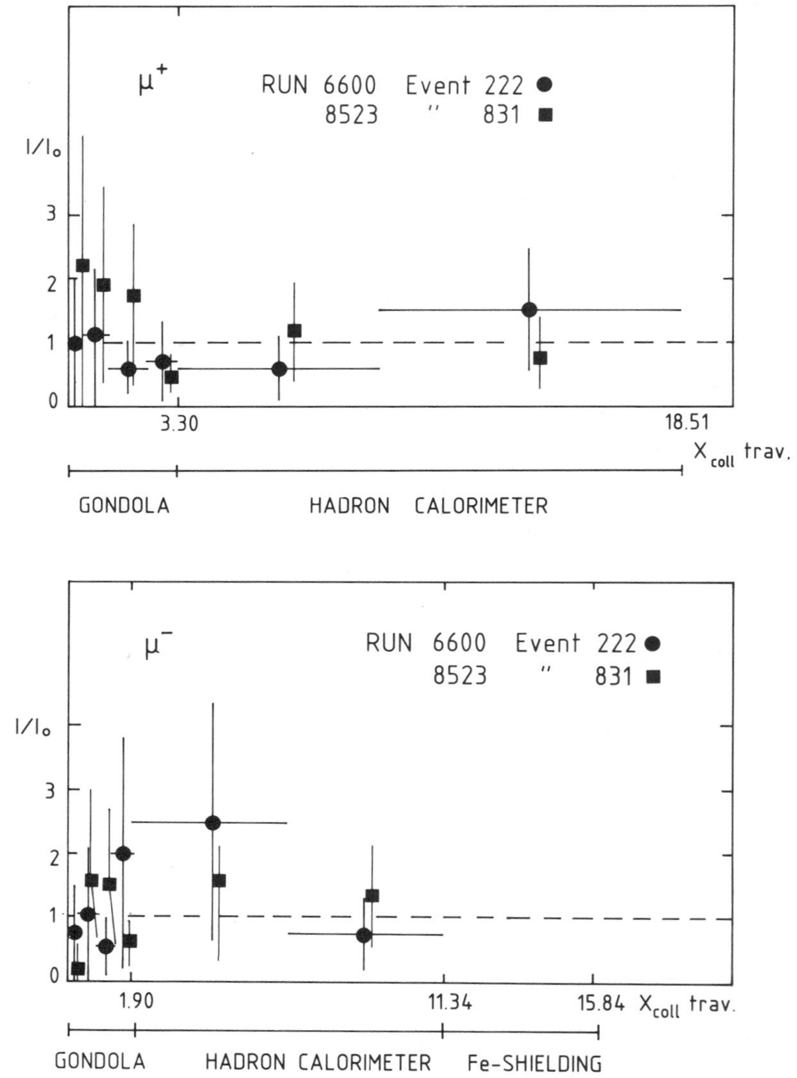

Fig. 36. a) Normalized energy losses in calorimeter cells transversed by the two muon tracks.

several possible spurious sources of events:

i) Ordinary large transverse momentum jets which fragment into two apparently isolated, high-momentum tracks, both simulating either muons or electrons. To evaluate this effect, events with (hadronic)tracks of momenta compatible with $p_t > 25$ GeV/c were also selected in the express line. After requiring that the track is isolated, one finds one surviv-

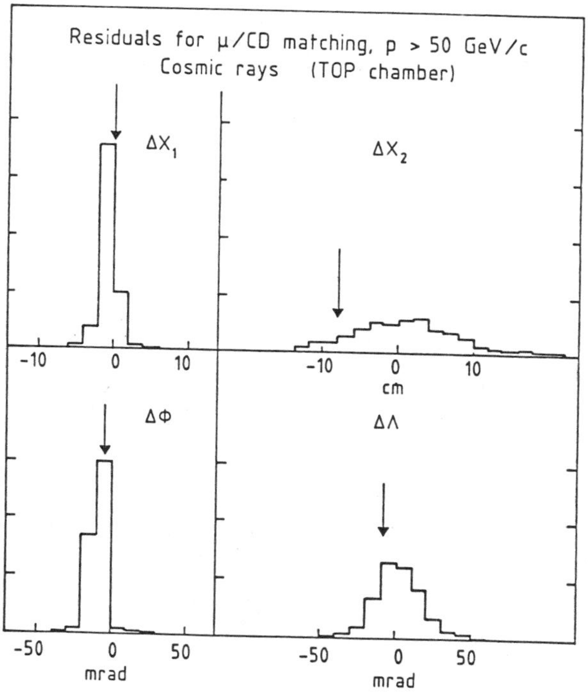

Fig. 36. b) Arrows show residuals in angle and position for muon track. Distribution come from cosmic-ray calibration with $p > 50$ GeV/c.

ing event with transverse energy ∿ 25 GeV in a sample corresponding to 30 nb^{-1}. Including the probability that this track simulates either a muon (∿ 2 x 10^{-3}) or an electron (∿ 6 x 10^{-3}), we conclude that this effect is negligible[39]. Note that two tracks (rather than one) are needed

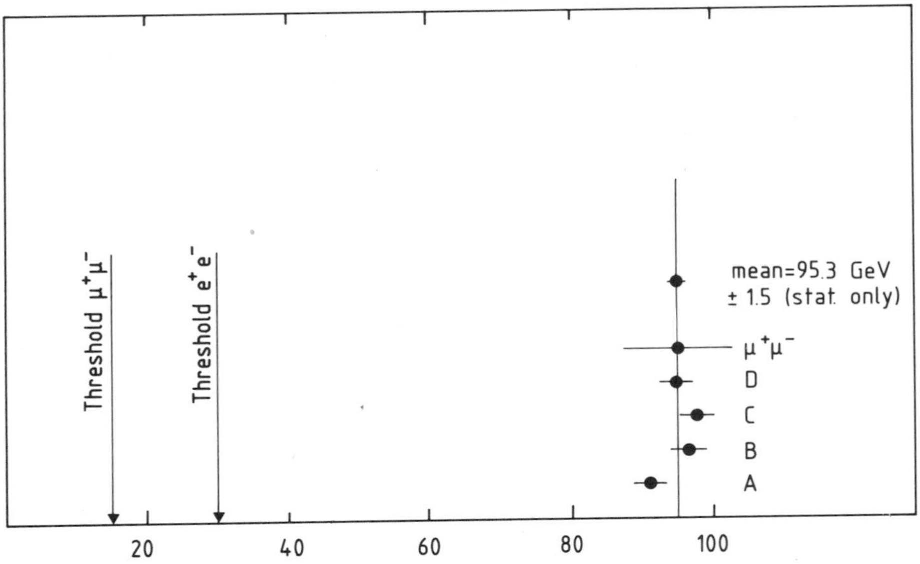

Fig. 37. Invariant masses of lepton pairs

to simulate our events (probabilities must be squared!) and that the invariant mass of the events is much higher than the background. The background is expected to fall approximately like m^{-5} according to the observed jet-jet mass distributions[14].

ii) Heavy-flavored jets with subsequent decay into leading muons or electrons. In the 1982 event sample (11 nb^{-1}), two events have been observed with a single isolated muon of $p_t > 15$ GeV and one electron event with $p_t > 25$ GeV/c. Some jet activity in the opposite hemisphere is required. One event exhibits also a significant missing energy. Once this is taken into account they all have a total (jet+jet+lepton+neutrino) transverse mass of around 80 GeV/c^2, which indicates that they are most likely due to heavy-flavor decay of W particles. This background will be kinematically suppressed at the mass of our five events. Nevertheless, if the fragmentation of the other jet is also required to give a leading lepton and no other visible debris, this background contributes at most to 10^{-4} events. Monte Carlo calculations using ISAJET lead to essentially the same conclusion[30].

iii) Drell-Yan continuum. The estimated number and the invariant mass distribution make it negligible[40].

iv) W^+W^- pair production is expected to be entirely negligible at our energy[41].

v) Onium decay from a new quark, of mass compatible with the observation (∿ 95 GeV/c^2). Cross-sections for this process have been estimated by different authors[42], and they appear much too small to account for the desired effect.

In conclusion, none of the effects listed above can produce either the number or the features of the observed events.

4.5. Mass Determination

All the observations are in agreement with the hypothesis that events are due to the production and decay of the neutral intermediate vector boson Z° according to reaction (1). The transverse momentum distribution is shown in Figure 38, compared with the observed

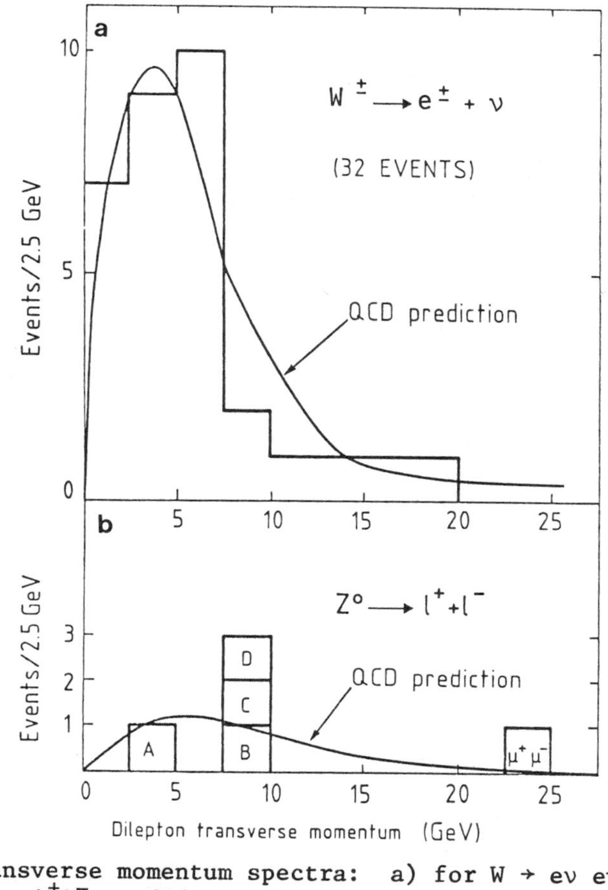

Fig. 38. Transverse momentum spectra: a) for W → eν events, and b) $Z° \rightarrow \ell^+\ell^-$ candidates. The lines represent QCD predictions.

distributions for the $W^{\pm} \rightarrow e\nu$ events and with QCD calculations. The muon events and one of the electron events (event B) have visible jet structure. Other events are instead apparently structureless.

From our observation, we deduce a mass value for the Z° particle.

$$m_{Z^\circ} = (95.2 \pm 2.5) \text{ GeV}/c^2$$

The half width based on the four electron events is 3.1 GeV/c^2 (< 5.1 GeV/c^2 at 90% c.l.), consistent with expectation from the experimental resolution and the natural Z° width, $\Gamma_{Z_\circ} = 3.0$ GeV. At this point it is important to stress that the final calibration of the electromagnetic calorimeters is still in progress and that small scale shifts are still possible, most likely affecting both the $W^{\pm}$ and Z° mass values. No e.m. radiative corrections have been applied to the masses.

5. COMPARING THEORY WITH EXPERIMENT

The experiments discussed in the previous paragraph have shown that the W-particle has most of the properties required in order to be the carrier of weak interactions. The presence of a narrow di-lepton peak has been observed around 95 GeV/c^2. Rates and features of the events are consistent with the hypothesis that one has indeed observed the neutral partner of the $W^{\pm}$. Statistics at present is not sufficient to test experimentally the form of the interaction, neither has parity violation been detected. The precise values of the masses of Z° and $W^{\pm}$ now available constitute however a crucial test of the idea of unification between weak and electromagnetic forces and in particular of the predictions of SU(2) x U(1) theory of Glashow, Weinberg and Salam. Careful account of systematic errors is needed in order to evaluate an average between the UA1 and UA2 mass determination.

The charged vector boson mass given in the present work is

$$m_{W^{\pm}} = (80.9 \pm 1.5) \text{ GeV}/c^2 \quad \text{(statistical errors only)}$$

to which a 3% energy scale uncertainty must be added. In the present report a value for the Z° mass, $m_{Z^\circ} = (95.1 \pm 2.5)$ GeV/c^2 has been given. Neglecting systematic errors, a mass value is found with somewhat smaller errors:

$$m_{Z^\circ} = (95.6 \pm 1.4) \text{ GeV}/c^2 \quad \text{(statistical errors only)}$$

to which the same scale uncertainty as for the $W^{\pm}$ applies. The quoted errors includes: i) the neutral width of the $Z_\circ$ peak, which is found to be $\Gamma < 8.5$ GeV/c^2 (90% confidence level), ii) the

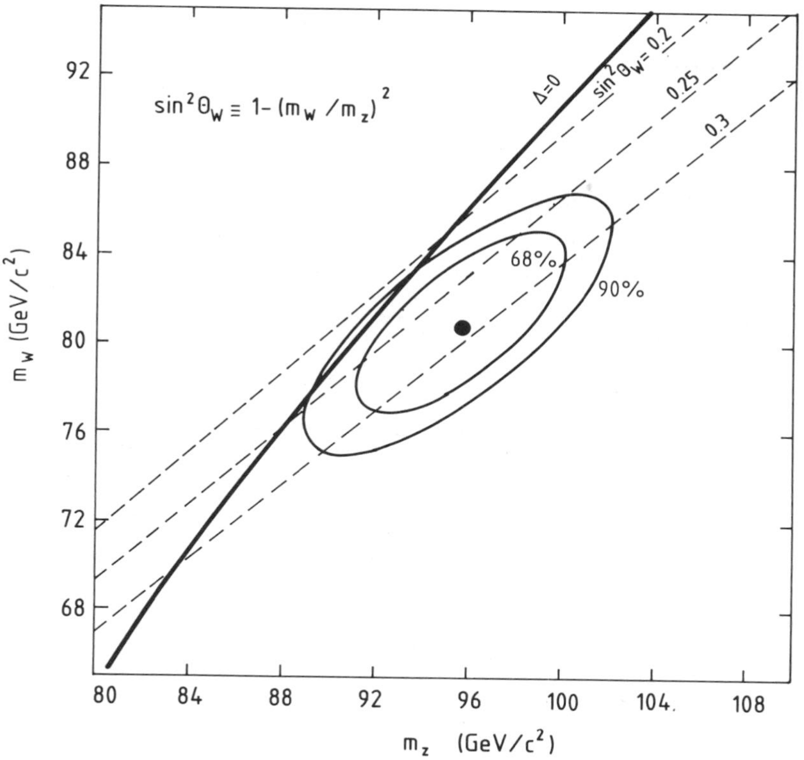

Fig. 39. a) m_Z plotted against m_W determined by the UA1 experiment. The elliptical error curves reflect the uncertainty in the energy scale at the 68% and 90% confidence levels. The heavy curve shows the Standard Model prediction for $\rho = 1$ as a function of the Intermediate Vector Boson (IVB) masses.

experimental resolution of counters, and iii) the r.m.s. spread between calibration constants of individual elements. In Figure 39a we have plotted m_Z against m_W. The elliptical shape of the errors reflects the uncertainty in the energy scale. One can see that there is excellent agreement with the expectations of the SU(2) x U(1) Standard Model. One can also determine the classic parameters:

$$\sin^2 \theta_W = \frac{38.5 \text{ GeV/c}^2}{m_W} = 0.226 \pm 0.008 \ (\pm 0.014) \ ,$$

$$\rho = \frac{m_W^2}{m_Z^2 \cos^2 \theta_W} = 0.925 \pm 0.05 \ ,$$

where the number in parenthesis is due to systematic errors. The result is shown in Figure 39b again in good agreement with expectations and published results[43].

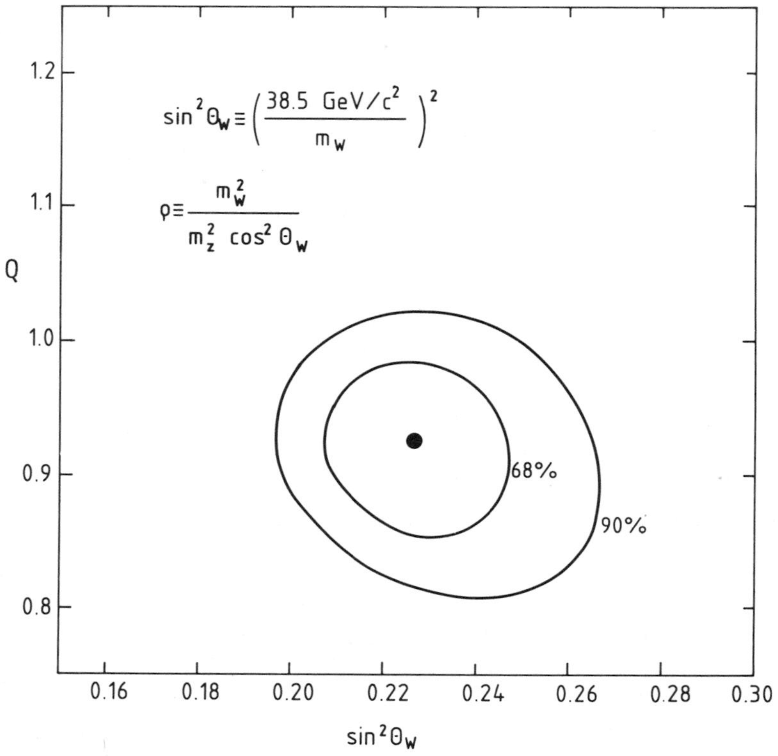

Fig. 39. b) ρ plotted against $\sin^2 \theta_W$ as determined from the measurements of the IVB masses. The 68% and 90% confidence level limits are shown.

Acknowledgements

These lectures are based on the work of the UA1 Collaboration team, to which I want to express all my appreciation for the remarkable work which has led to so many results. At present the following individuals are members of the collaboration:

G. Arnison, A. Astbury, B. Aubert,
C. Bacci, A. Bezaguet, R. K. Bock, T. J. V. Bowcock,
M. Calvetti, P. Catz, P. Cennini, S. Centro, F. Ceradini,
S. Cittolin, D. Cline, C. Cochet, J. Colas, M. Corden,
D. Dallman, D. Dau, M. DeBeer, M. Della Negra, M. Demoulin,
D. Denegri, A. Diciaccio, D. DiBitonto, L. Dobrzynski,
J. D. Dowell,
K. Eggert, E. Eisenhandler, N. Ellis, P. Erhard,
H. Faissner, M. Fincke, G. Fontaine, R. Frey, R. Frühwirth,

J. Garvey, S. Geer, C. Ghesquiere, P. Ghez, K. L. Giboni,
W. R. Gibson, Y. Giraud-Heraud, A. Givernaud, A. Gonidec,
G. Grayer,
T. Hansl-Kozanecka, W. J. Haynes, L. O. Hertzberger, C. Hodges,
D. Hoffman, H. Hoffman, D. J. Holthuizen, R. J. Homer, A. Honma,
W. Jank, G. Jorat,
P. I. P. Kalmus, V. Karimaki, R. Keeler, I. Kenyon, A. Kernan,
R. Kinnunen, W. Kozanecki, D. Kryn,
F. Lacava, J. P. Laugier, J. P. Lees, H. Lehmann, R. Leuchs,
A. Leveque, D. Linglin, E. Locci,
J. J. Malosse, T. Markiewicz, G. Maurin, T. McMahon,
J. P. Mendiburu, M. N. Minard, M. Mohammadi, M. Moricca,
K. Morgan, H. Muirhead, F. Muller,
A. K. Nandi, L. Naumann, A. Norton,
A. Orkin-Lecourtois,
L. Paoluzi, F. Pauss, G. Piano Mortari, E. Pietarinen, M. Pimiä,
J. P. Porte,
E. Radermacher, J. Ransdell, H. Reithler, J. P. Revol, J. Rich,
M. Rijssenbeek, C. Roberts, J. Rohlf, P. Rossi, C. Rubbia,
B. Sadoulet, G. Sajot, G. Salvi, G. Salvini, J. Sass,
J. Saudraix, A. Savoy-Navarro, D. Schinzel, W. Scott,
T. P. Shah, D. Smith, M. Spiro, J. Strauss, J. Streets,
K. Sumorok, F. Szoncso,
C. Tao, G. Thompson, J. Timmer, E. Tscheslog, J. Tuominiemi,
B. Van Eijk, J. P. Vialle, J. Vrana, V. Vuillemin,
H. D. Wahl, P. Watkins, J. Wilson, R. Wilson, C. E. Wulz,
Y. G. Xie,
M. Yvert,
E. Zurfluh.

REFERENCES

1. C. Rubbia, P. McIntyre, and D. Cline, Study Group, Design study of a Proton-Antiproton Colliding Beam Facility, CERN/PS/AA 78-3, reprinted in Proc. Workshop on Producing High-Luminosity, High Energy Proton-Antiproton Collisions (Berkely, 1978), report LBL-7574, UC34, p.189 (1978); Proc. Inter. Neutrino Conf. (Aachen, 1976) (Vieweg, Braunschweig, 1977) p.683.
2. The Staff of the CERN "Proton-Antiproton Project," Phys.Lett., 107B:306 (1981).
3. UA1 Proposal: A 4π solid-angle detector for the SPS used as a proton-antiproton collider at a centre-of-mass energy of 540 GeV, CERN/SPSC 78-06 (1978); M. Barranco Luque, Nucl.Instrum. Methods, 176:175 (1980); M. Calvetti, Nucl.Instrum.Methods, 176:255 (1980); K. Eggert, Nucl.Instrum.Methods, 176:217, 233 (1980); A. Astbury, Phys.Scr., 23:397 (1981); UA1 Collaboration, the UA1 detector (presented by J. Timmer), in: "Proc. 18th Rencontre de Moriond, Antiproton-Proton Physics and the W discovery," J. Tran Thanh Van, ed., p.593 (1983).

4. UA1 Collaboration, G. Arnison, Phys.Lett., 123B:115 (1983).
5. UA1 Collaboration, Transverse Energy Distributions in the Central Calorimeters, preprint CERN EP/82-122 (1982).
6. M. J. Corden, et al., Physica Scripta, 25:468 (1982); C. Cochet, et al., UA1 Tech. Note TN 82-40.
7. G. Arnison, et al., Phys.Lett., 107B:320 (1981); 123B:108 (1983).
8. G. Wolf, DESY Report EP/82-122 (1982).
9. F. E. Paige and S. D. Protopopescu, ISAJET, BNL 31987.
10. M. Della Negra, Physica Scripta, 25:468 (1982); R. K. Bock, et al., Nucl.Inst.Meth., 186:533 (1981).
11. Z. Kunszt and E. Pietarinen, CERN preprint TH 3584 (1983).
12. K. Kunszt and E. Pietarinen, Nucl.Phys., B164:45 (1980); T. Gottschalk and D. Sivers, Phys.Rev., D21:102 (1980); F. Berends, Phys.Lett., 103B:124 (1981).
13. R. Brandelik, Phys.Lett., 114B:65 (1982) [16].
14. UA1 Collaboration, Jet fragmentation at the SPS pp collider - UA1 experiment (presented by V. Vuillemin), in: "Proc. 18th Rencontre de Moriond, Antiproton-Proton Physics and the W discovery 1983," J. Tran Thanh Van, ed., p.309.
15. G. Arnison, et al., Phys.Lett., 118B:173 (1982).
16. B. Combridge, et al., Phys.Lett., 70B:234 (1977).
17. B. Combridge and C. Maxwell, preprint RL-83-095 (1983).
18. UA1 Collaboration, G. Arnison, et al., Phys.Lett., 132B:214 (1983).
19. F. Berends, et al., Phys.Lett., 103B:124 (1981).
20. J. C. Collins and D. E. Soper, Phys.Rev., D16:2219 (1977).
21. D. Drijard, et al., Phys.Lett., 121B:433 (1983).
22. R. K. Ellis, et al., Nucl.Phys., B173:397 (1980).
23. N. G. Antoniou, et al., Phys.Lett., 128B:257 (1983).
24. H. Abramowicz, et al., Z.Phys., C12:289 (1982).
25. G. Arnison, et al., Phys.Lett., 122B:103 (1983).
26. G. Arnison, et al., Phys.Lett., 129B:273 (1983).
27. UA2 Collaboration in Proceeding of the International Europhysics Conference, Brighton, July 1983, p.472.
28. A. Nakamura, G. Pancheri, and Y. Srivastava, Frascati preprint LFN-83/43 (R) (June 1983).
29. P. Aurenche and J. Lindfors, Nucl.Phys., B185:274 (1981).
30. F. E. Paige and S. D. Protopopescu, ISAJET program, BNL 29777 (1981). All cross-sections are calculated in the leading log approximation assuming SU(2) x U(1).
31. F. Berends, et al., Nucl.Phys., B202:63 (1982), and private communications.
32. M. Jacob, to be published. We thank Professor M. Jacob for very helpful comments on the subject.
33. K. Eggert, et al., Nucl.Instrum.Methods, 176:217, 233 (1980).
34. For more detailed information, see for example: UA1 Collaboration, Search for Isolated Large Transverse Energy muons at $\sqrt{s}$=540 GeV, in Proc. 18th Recontre de Moriond on Antiproton-Proton Physics, La Plagne 1983 (Editions Frontières, Gif-sur-Yvette, 1983), p.431.

35. G. Arnison, et al., Phys.Lett., 118B:167 (1982).
36. Calculation based on: M. Banner, et al., Phys.Lett., 122B:322 (1983).
37. In the maximum likelihood fit, the measured quantities of each event are compared with computed distribution functions, smeared with experimental errors. A Breit-Wigner form is assumed for the W mass (with a width (FWHM) of 3 GeV/c^2), and Gaussian distributions are used for the transverse and longitudinal momenta of the W (with r.m.s. widths of 7.5 GeV/c and 67.5 GeV/c, respectively). In the W centre of mass, the angle θ^* of the emitted positive (negative) lepton with respect to the outgoing antiproton (proton) direction is generated according to a distribution in $\cos\theta^*$ of $(1+\cos\theta^*)^2$ as expected for V(±A) coupling.
38. J. T. Carrol, S. Cittolin, M. Demoulin, A. Fucci, B. Martin, A. Norton, J. P. Porte, P. Ross, and K. M. Storr, Data Acquisition using the 168E, Paper presented at the Three-Day In-Depth Review on the Impact of Specialized Processors in Elementary Particle Physics, Padua 1983, ed. Istituto Nazionale di Fisica Nucleare, Padova, p.47 (1983).
39. Electron-pion discrimination has been measured in a test beam in the full energy range and angles of interest. The muon tracks have the following probabilities: i) no interaction: 2×10^{-5} (4×10^{-5}); ii) interaction but undetected by the calorimeter and geometrical cuts: 10^{-4} (4×10^{-4}); iii) decay: 10^{-3} (0.7×10^{-3}).
40. S. D. Drell and T. M. Yan, Phys.Rev.Lett., 25:316 (1970); F. Halzen and D. H. Scott, Phys.Rev., D18:3378 (1978). See also ref.6; S. Pakvasa, M. Dechantsreiter, F. Halzen and D. M. Scott, Phys.Rev., D20:2862 (1979).
41. R. Kinnunen, Proc. Proton-Antiproton Collider Physics Workshop, Madison, 1981 (univ. Wisconsin, Madison, 1982); R. W. Brown and K. O. Mikaelian, Phys.Rev., D19:922 (1979).
42. T. G. Gaisser, F. Halzen, and E. A. Paschos, Phys.Rev., D15:2572 (1977); R. Baier and R. Rückl, Phys.Lett., 102B:364 (1981); F. Halzen, Proc. 21st Int. Conf. on High Energy Physics, Paris, 1982 (J.Phys.(France), No. 12 t 43: (1982)), p.C3-381; F. D. Jackson, S. Olsen, and S. H. H. Tye, Proc. AIP Dept. of Particles and Fields Summer Study on Elementary Particle Physics and Future Facilities, Snowmass, Colorado, 1982 (AIP, New York, 1983), p.175.
43. J. E. Kim, Rev.Mod.Phys., 53:211 (1981).

DISCUSSION

CHAIRMAN : C. RUBBIA

Scientific Secretaries : M. Demarteau and C. Mana

- *POHL* :

You have shown very impressive jet data. Why didn't you show any data on $W \to q\bar{q}$ and $Z \to q\bar{q}$?

- *RUBBIA* :

I will explain you one thing. You must be a theoretician.

- *POHL* :

No, I'm not.

- *RUBBIA* :

Well in genearl $W \to$ jets and $Z \to$ jets are hard to detect because the background from normal (QCD produced) jets is 10 to 15 times more frequent.

- *POHL* :

But you could find top by $W \to t\bar{b}$ that way.

- *RUBBIA* :

It is very different for heavy flavors, because they can be tagged by leptons. But due to the relatively low mass difference between the W and the t-quark they are not really the pencil-like jets we were talking about, namely, the ones you get from light flavors. But of course $W \to t\bar{b}$ will be the next thing we will concentrate on.

- *BENCHOUK* :

Can you compare the fraction of 3-jet events in $p\bar{p}$ to 3-jet events in e^+e^- ? I would expect from the strong gluon-gluon coupling in $p\bar{p}$ collisions a higher fraction of jet events.

- *RUBBIA* :

$p\bar{p}$ colliders and e^+e^- machines are not comparable, because they are very far separated in energy, even if they show similar results on the fragmentation function distribution. Besides, the jets we get push QCD to a region where higher order QCD corrections become important.

- *ZICHICHI* :

It seems to me that the only way to compare hadronic machines with e^+e^- machines is to compare the P_T distribution of the jets with respect to the beam axis. At the ISR we found no discrepancy with the e^+e^- machines besides the fact that the P_T distribution with respect to the jet axis is a little bit softer. When we use, however, the normalised variable P_T / $<P_T>$ the curves become degenerate.

At the collider you can compare the jets with high P_T with respect to the beam axis with the jets with low P_T. Do you find the same distribution of the produced particles as the e^+e^- machines ?

- *RUBBIA* :

For the moment we have taken the more naive point of view and tried to consolidate more obvious things. It is certainly our next item in the list of our understanding of nature.

- *EREDITATO* :

Do you apply topological criteria in order to distinguish between quark and gluon jets ? Namely, does a study of the distribution of particles show any significant difference ?

- *RUBBIA* :

I have shown you the transverse momentum with respect to the beam axis and the longitudinal energy carried by single charged particles and we cannot tell clearly any difference between quark and gluon jets.

Another thing you can look at is the multiplicity. Jet events have an associated multiplicity that is higher than for non-jet events which is not concentrated in a jet. From the event itself you can tell if there was a jet around or not. This may be due to color rearrangement. The associated multiplicity in W events, however, is down to a minimum bias level.

- *MARUYAMA* :

Can you explain why you have about four times more electron decays than muon decays in the W-events ?

- *RUBBIA* :

The detector efficiency of our detector for $W \to \mu\bar{\nu}$ is about one half of the efficiency for $W \to e\bar{\nu}$. Another factor two comes from the event selection. Up to now W's and Z^o's have been searched for and analysed in a so-called "express line". Triggering on di-muon, di-electron and single electron events is much easier than triggering on single muon events because we trigger calorimetrically. For single muons you have to reconstruct the whole track to make sure it is high energetic or not and this, of course, takes some time.

- *POSCHMANN* :

In one of your plots you have shown the number of e^+ and e^- from the W-decay. There seems to be an imbalance. Do you have an explanation for this ?

- *RUBBIA* :

This has been changing up and down. In this plot only those events which have a good measured momentum were selected. If one looks only for the charge of the W's or if one loosens the cuts the asymmetry disappears, so I don't think that we have to be worried about it, but we will keep an eye on that.

- *BERNSTEIN* :

How do you estimate the contribution to the fragmentation function from quark jets as compared to gluon jets ?

- *RUBBIA* :

Sorry, I didn't say that they were gluon jets, I said that our prejudice is that they are dominated by gluons because, firstly, the gluon-gluon cross-section is bigger than the quark-quark cross-section by a factor $(4/9)^2$ and, secondly, because we have chosen particularly low q^2 jets ($x \approx 0.1$) and from the neutrino experiments we know that gluon jets dominate. It's true that we are going from a gluon dominated region at low energies to a quark dominated region at high energies. My own opinion is that gluon jets are supposed to be the dominant contributors but I have not given any mechanism whereby you can separate a gluon jet from a quark jet.

- *BERNSTEIN* :

Please discuss the systematic errors in more detail in the mass determination of the W and the Z. Specifically aging effects in the calorimeter and the differences in the systematic errors for the W and Z mass.

- *RUBBIA* :

Aging effects due to radiation damage in the calorimeter and also angular effects due to particles hitting near cracks in the spectrometer are quite important and we are redesigning the apparatus to avoid this. At the moment the systematic errors dominate the statistical errors. We have estimated the systematic errors to be about 3% and we will try to improve it.

On the other hand, the collider is the only machine (unfortunately CBA is gone) which can provide us with a simultaneous measurement of the W and the Z^0 mass and so, for the determination of the mass difference, most of the systematic errors will disappear. A very accurate measurement of one of the two is not enough. You need both in order to compare the theory with something else.

- *VAN DEN DOEL* :

Is there a limit on the size of the W ?

- *RUBBIA* :

Not that I know. Everything is consistent with the standard model up to 20%. So within this limit the W is pointlike.

- *FRITZSCH* :

From this kind of experiment you cannot do anything. Since you produce the W almost at rest and if it would have a form factor it would not enter here. So it will not be a big effect.

- *LEURER* :

In order to see a missing particle you need to have detectors all around the interaction area. How do you do this without blocking the colliding beams ?

- *RUBBIA* :

We do not know how to measure the missing longitudinal energy at the moment. There are few but highly energetic particles disappearing via the beam pipe and this is killing us. We have full flash calorimetry up to 0.2 degrees which is small but not enough for the energies we are dealing with, so there is room for improvement.

- *BAGGER* :

What signature will you use for detecting Higgs particles ?

- *RUBBIA* :

If the Higgs is there with a mass $\leq$ 40 GeV we should be able to see the Higgs bremsstrahlung according to the following diagram

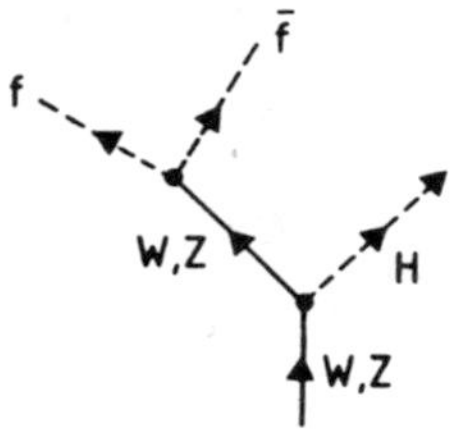

$(W,Z) \to H + f + \bar{f}$ $(W,Z) \to H + f +$

Since in this case the virtual Z^0 will be close to the mass shell the $f\bar{f}$ mass spectrum will show a clear peak and the Higgs should preferentially go to $b\bar{b}$.

- *BALLOCCHI* :

How do you think to get such an incredible increase in luminosity ?

- *RUBBIA* :

Money, just a matter of money.

- *BALLOCCHI* :

What do you hope to see about compositeness at the collider ?

- *RUBBIA* :

The lack of compositeness is becoming one of the most fundamental problems. To see it at a collider you need precise measurements at a high energetic machine (about 1 TeV) with high luminosities. Let me stress the fact that this is a purely experimental question. An experimental physicist should always look for compositeness.

- *MILOTTI* :

Do you expect to be able to measure the width of the Z^o.

- *RUBBIA* :

Yes, it is one of the most important numbers to be measured but we are not able to measure it at this moment with enough accuracy. Presently a limit of 7 neutrino species can be established.

- *MILOTTI* :

What is your opinion about the prospective conversion of LEP II into a $p\bar{p}$ collider ?

- *RUBBIA* :

It is the obvious outcome of the LEP-project. It can easily provide us with a 10 TeV hadron machine and with more sophisticated magnet technology you could reach 20 TeV although I don't know what an increase in energy in this region by a factor of two implies. A high energy and high luminosity machine is a must. It is just a matter of money and organization and there are already several projects, the Tevatron at Fermilab and some other longer term projects.

- *RATOFF* :

Was the event you showed us this morning which had a 20 GeV photon associated with it a Z^o candidate ?

- *RUBBIA* :

We have one event with a significant loss of energy in a shower and UA2 has an even more spectacular event. The probability of these events to occur is down to the percent region and one should put them in a special category and do the physics without them. Although their inclusion in the analysis will not change our result. They are interesting events and it may well be that nature is trying to tell us something here.

- *ZICHICHI* :

What are the masses of the e^+e^- and the $e^+e^- \gamma$ system ?

- *RUBBIA* :

For both events the effective mass of the e^+e^- is about 50 GeV, the mass of the $e\gamma$ is 9 GeV for UA2 and 3 GeV for UA1 and the mass of the $e^+e^- \gamma$ system is close to the Z^o mass.

- *SIMIĆ* :

What is the lower experimental bound on the Higgs mass ?

- *RUBBIA* :

We have no bound on the Higgs mass now. The best bound for the neutral Higgs comes as far as I know from the theory and should be more than 8-9 GeV. In fact, Veltman says that there is absolutely no limit to the Higgs mass and the theory can accommodate masses up to 10, 20, 50, ... TeV. Anyhow, I personally believe that the Higgs mechanism is not the right way and that the mass problem is still an open question.

- *SIMIĆ* :

If the Higgs mass is above a certain value this would imply that the theory must be dynamically broken and we must look at composite structures.

- *RUBBIA* :

You say so. I don't know about it. I don't know what the hell is going on, I'm just a plumber. I'm trying to do my work as well as I can. If there is a leak I'll fix it !

ANALYSIS OF THE HADRONIC FINAL STATES AT THE CERN $\bar{p}p$ COLLIDER

Gösta Ekspong

Institute of Physics, Stockholm University, Sweden
CERN, Geneva, Switzerland

INTRODUCTION

This lecture covers new results on some selected soft processes from the CERN SPS antiproton-proton collider in experiments UA1, UA2 and UA5. The new energy regime of the collider has opened the possibility to further explore phenomena already seen at the CERN ISR or at fixed target accelerators and to find or search for new phenomena.

The first section contains a presentation of the remarkable result that the average transverse momentum in the central region increases with the multiplicity, presents rapidity distributions where a noticeable plateau has developed and, finally, gives the charge particle multiplicity distribution and its moments. The question of KNO scaling will be addressed and evidence for scaling violation presented. In the second section charge particle correlations in rapidity will be presented. The analysis leads to the conclusion that small groups of resonances (clusters) are being produced.

In the last section some results on the production of identified particles will be given, including an estimate of the average particle content of an event.

1. GENERAL FEATURES

1.1. Experimental Details

When comparisons are made with lower energy data, one has generally available proton-proton reactions at ISR and fixed target experiments, whereas the new 540 GeV (c.m.s.) data refer to anti-proton-proton collisions. This does not prevent meaningful comparisons since at 540 GeV the expected differences between $\bar{p}p$ and pp reactions are very small, ($\Delta\sigma/\sigma \simeq 0.001$). Several experiments at ISR have compared $\bar{p}p$ and pp reactions at the same energy and found no large differences[1].

The data refer to the most common type of events, namely inelastic collisions, generally of rather high multiplicity. The average multiplicity of all particles is about 46 out of which 28 are charged particles. The fluctuations are large and there are different kinds of experimental difficulties when the multiplicity is very high (secondary nuclear and γ-conversion interactions in the beam pipe and in the detector material) or is very low (acceptance losses). Each experiment has set up a minimum bias trigger which is based on signals from a set of counters in which hits in coincidence with a beam crossing are required. The larger part of the solid angle is covered by the trigger counters, the smaller will be the necessary corrections. In azimuth the coverage is generally 2π in angle, whereas in polar angle θ (or pseudorapidity $\eta = -\ln \tan \theta/2$) the coverage is less complete although generally large. The situation is summarized in Fig. 1 for the three experiments discussed, all of which have two arms - one downstream from the proton direction, the other downstream from the antiproton direction. The lay-out of UA1, UA2 and UA5 are as in Figs. 2-4. The beam pipe obscures the very forward and backward regions. Other construction features leave dead angles, which all detectors aim to make small. In the case of UA5, the team I

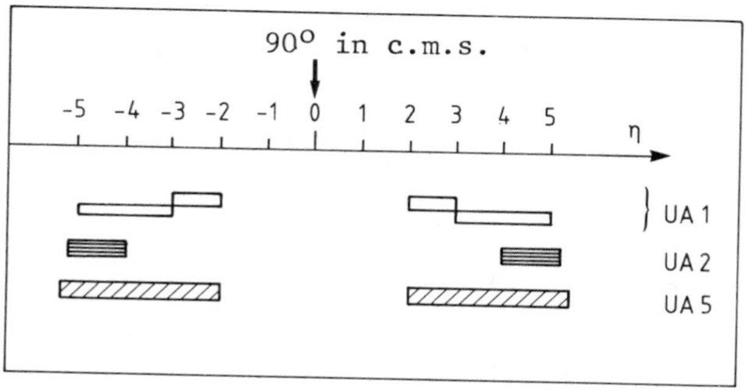

Fig. 1 Coverage in pseudorapidity of minimum bias triggers in the three experiments UA1, UA2 and UA5.

belong to, the two streamer chambers are very long (6 m) and wide (1.25m). The necessary gap between them for the beam pipe is only 9 cm, small enough to allow some strange particles to reach and decay in the sensitive volume. The geometrical acceptance of such a set-up is about 95% for polar angles $\theta > 5.7°$ (or pseudorapidity $|y| < 3$) and falls gradually to zero at $\theta = 0.8°$ (or $|\eta| = 5$). The trigger accepts most of the inelastic events ($\sigma \simeq 53$ mb) and no elastic events, for which according to UA4 and UA1,[2] $\sigma = 13$ mb. There is one class of inelastic events which requires special attention, namely single diffractive events. One of the beam particles then remains in the beam pipe. A trigger which requires at least one hit in each arm will obviously not accept most of the single diffraction events. A single arm trigger with a veto in the other arm is a complement which has been used (caution is needed, since beam-restgas collisions represent a background). From the above it should be clear that corrections are needed to the raw data. Monte Carlo programmes have been developed and tuned to represent the data sufficiently well to serve the purpose of making reliable corrections. For example, the minimum bias 2-arm trigger of UA5 is estimated to accept 96% of the inelastic non-single diffraction events and about 10% of the single diffraction events (which are about 18% of all inelastic events)[3]. The UA1 detector has a huge magnetic field and is equipped with calorimeters for both electromagnetic and hadronic

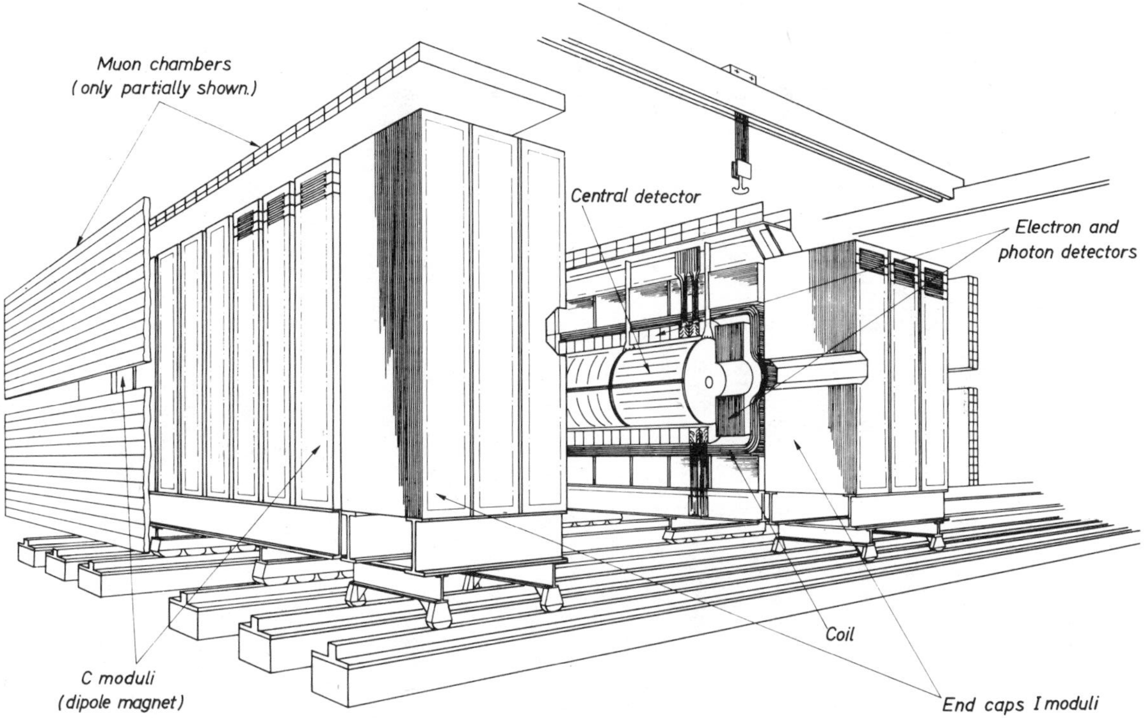

Fig. 2. The UA1 detector with its large dipolemagnet (0.7T), central tracking detector and calorimeters.

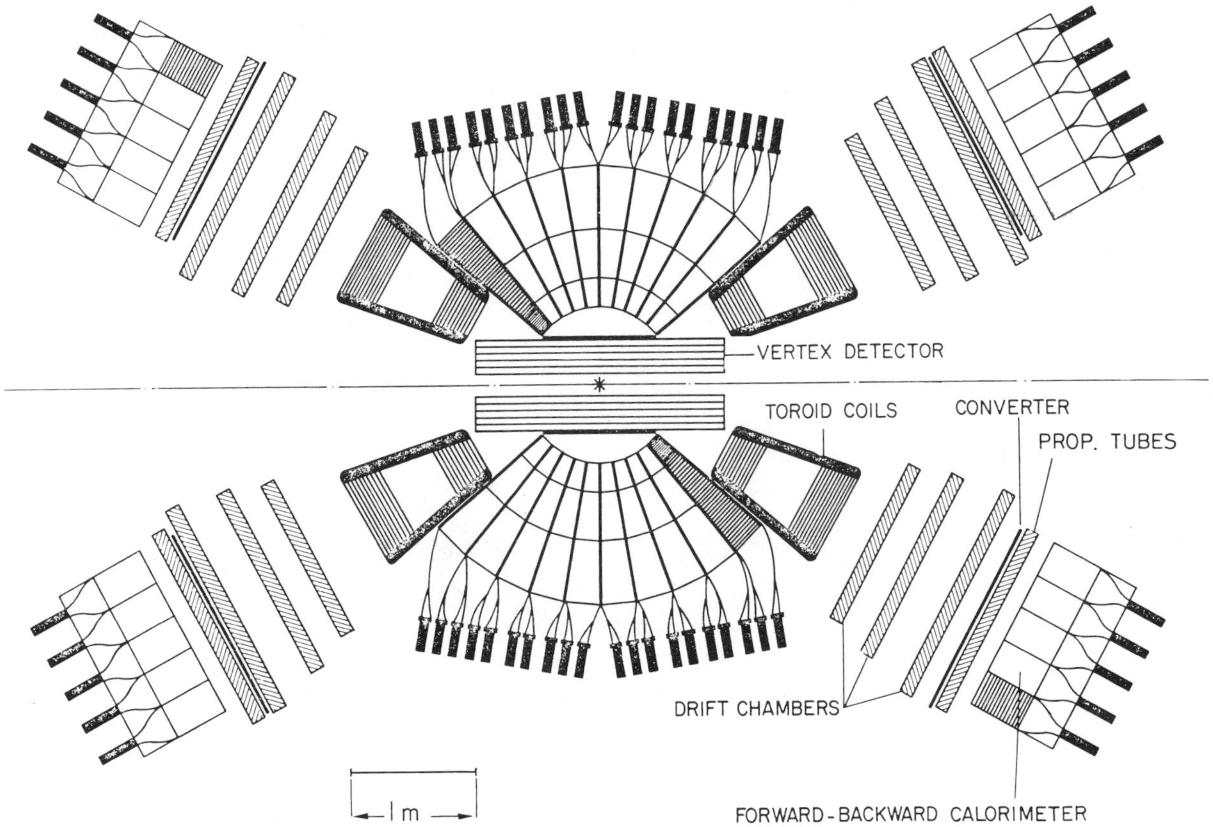

Fig. 3. The UA2 detector with magnetic spectrometers in the forward and backward cones, central tracking detectors and calorimeters.

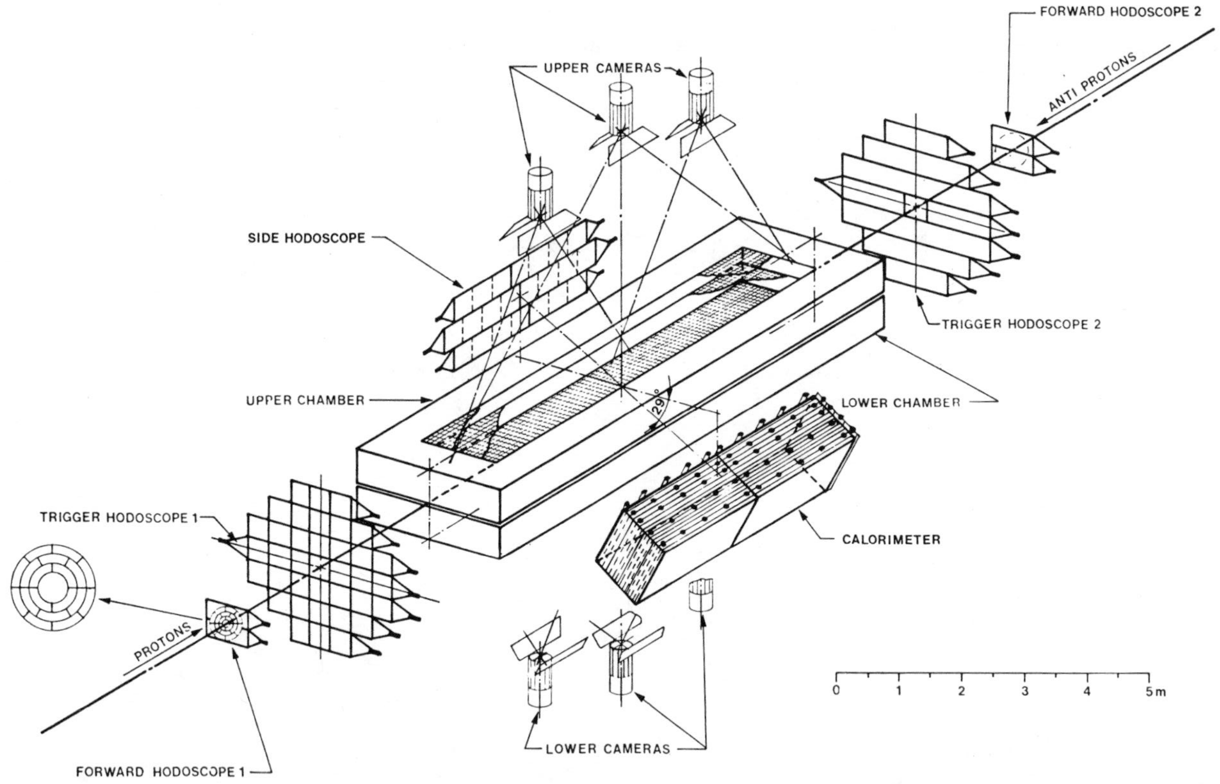

Fig. 4. The UA5 detector with its two large streamer chambers viewed each by three cameras. Counter hodoscopes used for triggering.

showers[4]. The UA2 detector[5] has good electromagnetic calorimeters and a toroidal magnetic field in the end-caps. Both have tracking counters placed before the calorimeters as viewed from the interaction point. The UA5 detector is a visual (photography) tracking device[6] without a magnetic field and calorimeter (later a small calorimeter was added). One can still in certain cases measure or indirectly infer momenta of particles as we will see.

1.2. The Theoretical Framework

As is well known no fundamental theory is at present capable of dealing with soft processes. It is generally assumed that the parton model with quarks and gluons is the appropriate and correct picture. As perturbative QCD is not applicable in the domain of soft processes, one finds various phenomenological approaches in the literature. A physical picture for non-diffractive reactions which was advanced almost ten years ago[7] views the processes in the following way: (i) the gluon sea is excited, (ii) the valence quarks pass through. Hadronization of the gluon sea produces hadrons in the central region. The valence quarks give rise to the leading particles and to hadrons in the fragmentation regions. In the multiperipheral model the central particles come from several strings (chains)[8]. These pictures are given as background for the way one might view these processes. Hadronization as a general phenomena has been emphasized by the group of Zichichi,[9] who compares different processes (pp, $\bar{p}p$, e^+e^-, νN) at the same available energy for particle production, i.e. by removing the leading particles and subtracting their energies. At the SPS $p\bar{p}$ collider no data exist for such a comparison.

1.3. Transverse Momenta of Charged Particles

One of the established experimental facts of high energy physics is the limited transverse momentum (p_T). At the CERN $p\bar{p}$-collider UA1 has found a new effect which is seen in the Figures 5, 6 and 7.

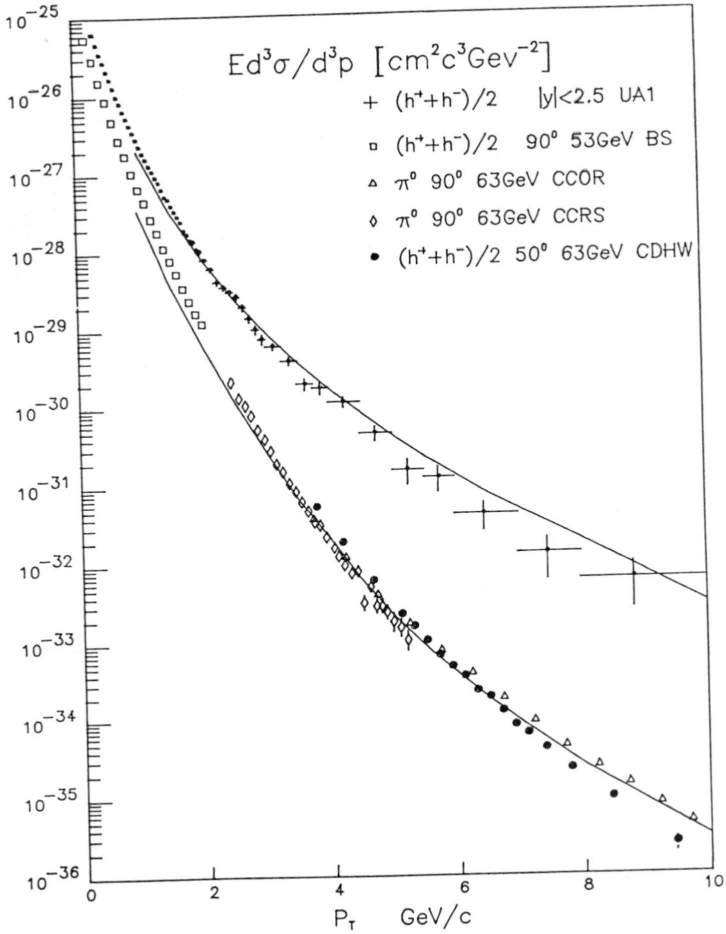

Fig. 5 The distribution in transverse momentum of charged hadrons (UA1). For comparison ISR results (lower curve).

The invariant cross-section for both charged and neutral particle production is much larger at high p_T than at ISR energies and of a different shape. The result[10] can be well described in the observed region of p_T by the form

$$\frac{E\ d^3\sigma}{d^3p} = A\left(\frac{p_T}{p_o} + 1\right)^{-n}$$

with $n \simeq 9$, $p_o = 1.3$ GeV/c. This leads to an average $\langle p_T \rangle = 0.42$ GeV/c.

The new effect is that the spectrum changes shape with particle density in the central region. The curves for the three densities 2.4, 5.7 and 10.2 charged particles per unit of rapidity (the average density is 3.3), are shown in Fig. 6.

The new effect is shown in a more condensed form in Fig. 7, where one sees that the average transverse momentum increases with increasing particle density to a levelling off at $\langle p_T \rangle$ = 0.47 GeV/c. A

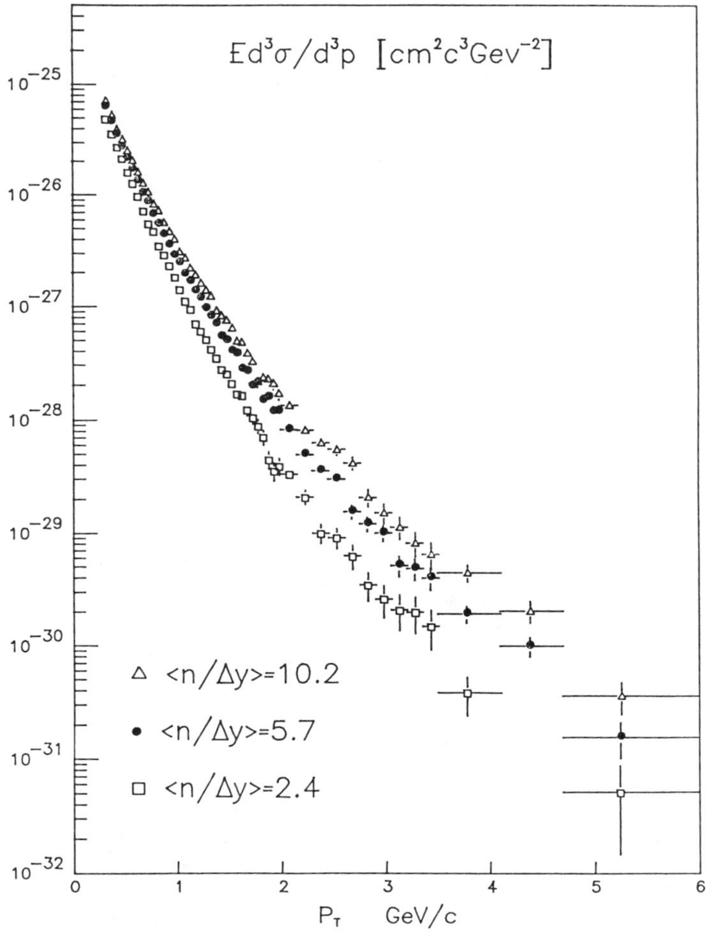

Fig. 6 The distribution in transverse momentum for different multiplicity densitities in the central plateau, (UA1).

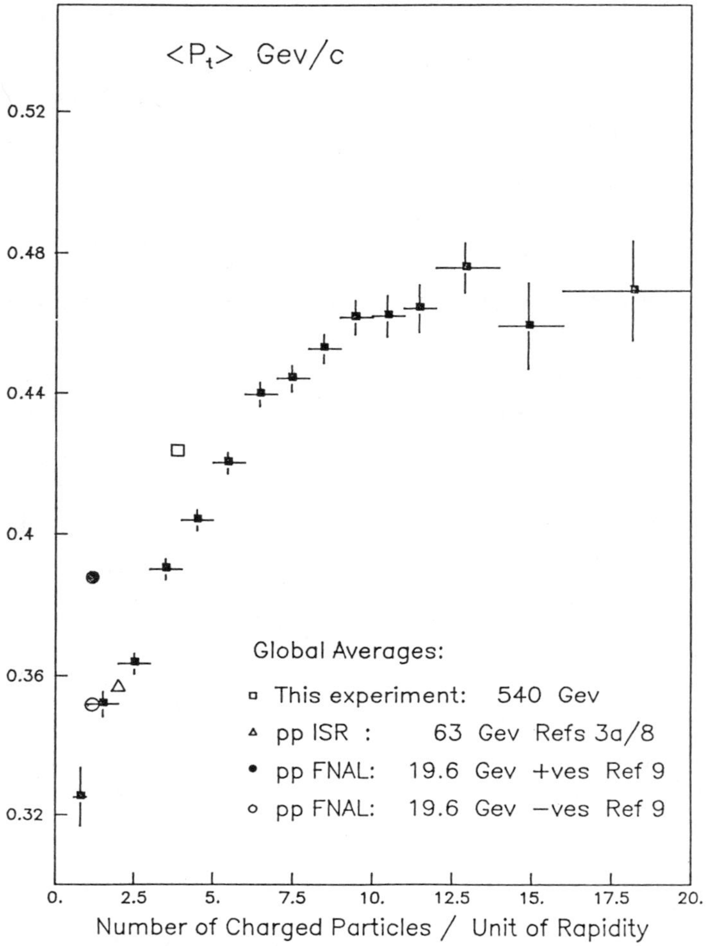

Fig. 7 The average transverse momentum of charged particles vs the multiplicity density, (UA1).

similar behaviour at the ISR, but less pronounced, was reported at the Brighton Conference this summer[11]. This phenomenon has no definite theoretical explanation, but will certainly attract attention and more study. Whether it can be viewed as a signal of the presence of a quark-gluon plasma state of matter was discussed by L. Van Hove[12]. Some theoretical arguments have been advanced that at high energy densities, a phase transition to this kind of plasma should occur when the temperature exceeds a critical value[13].

1.4. Rapidity Distributions

Another new phenomenon is the still increasing particle density (number of particles produced per unit of rapidity). The definition of rapidity is $y = 1/2 \ln (E + p_{\parallel})/(E - p_{\parallel})$, of pseudorapitidy ($\eta$) is $\eta = - \ln \tan \theta/2$, where ($E$, $p_{\parallel}$ and θ) are the energy, longitudinal momentum and polar angle of a produced particle. For massless particles the two concepts are identical and for energetic massive particles ($p >> m$) almost the same. At the ISR the density of charged particles per unit of pseudorapidity (at 90° to the beam axis) is at most about 2 and is increasing slowly over the ISR range of energies[14]. The value at 540 GeV is about 3. Figure 8 shows the pseudorapidity distribution for charged particles as measured by UA1 and UA5, showing good agreement[15,16]. A central plateau is clearly

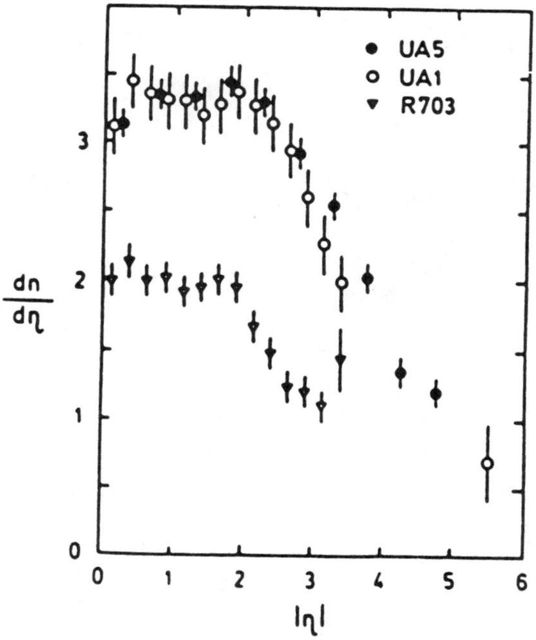

Fig. 8 The distribution in pseudorapidity (folded around $\eta = 0$, i.e. 90° in c.m.s.) for inelastic events with single diffraction excluded. The UA1 and UA5 data are at the SPS $\bar{p}p$ collider (540 GeV), the R703 at the ISR (53 GeV).

developed. The same figure shows data at one ISR-energy (53 GeV)[14]. The above mentioned rise of the central density is approximately linear in log s as is seen in Fig. 9. Since Feynman scaling would require a constant density, scaling is clearly violated. No prediction exists whether this rise should level off or continue.

1.5. The Average Charged Particle Multiplicity.

Since the available rapidity range increases as lns, Feynman scaling predicts the average multiplicity to increase linearly in lns. The observations deviate also from this behaviour in that an

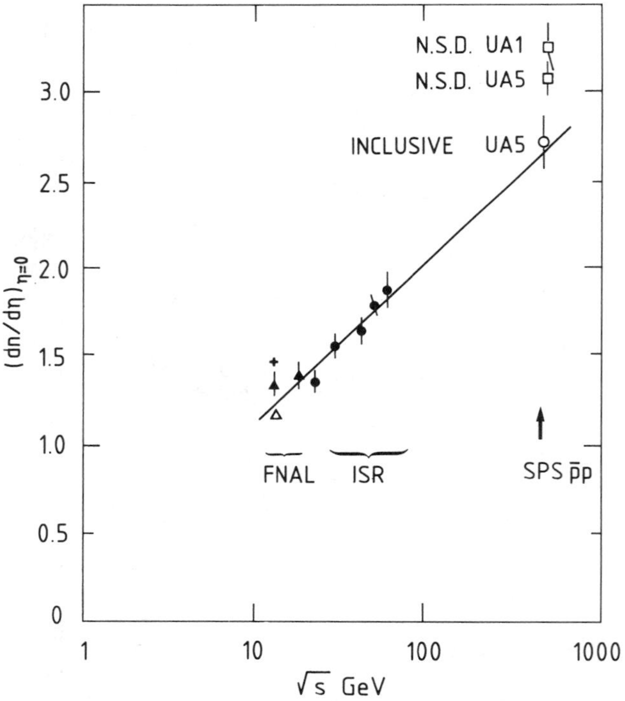

Fig. 9 The central multiplicity density at FNAL, ISR and $\bar{p}p$ Collider energies. The UA1 and UA5 measured points are for non-single diffractive events; the lower energy points are for inclusive data to be compared with the corrected data point, inclusive UA5.

important $(\ln s)^2$-term is needed to describe the energy dependence of the average multiplicity. At ISR energies the coefficients in the expression $\langle n \rangle = a + b \ln s + c (\ln s)^2$ were found to be, a = 0.88, b = 0.44, c = 0.118. The corresponding curve is displayed in Fig. 10 together with UA5 results for the average charge particle multiplicity at 540 GeV. It is to be noted that the curve refers to inelastic events, i.e. including single diffractive events. At the SPS collider UA5 has recently measured the single diffractive component using a single arm trigger. Adding this component to the non single-diffractive one leads to a lowering of the average multiplicity (both data points are given in Fig. 10). The same curve also well describes the new result. It is of interest to note that at 540 GeV the $(\ln s)^2$-term is responsible for most (67%) of the multiplicity. In a recent paper[17] R.V. Gavai and H. Satz have compiled predictions at 540 GeV of four models, all with free parameters fitted to lower energy data. The Fermi/Landau thermodynamical model, in which the

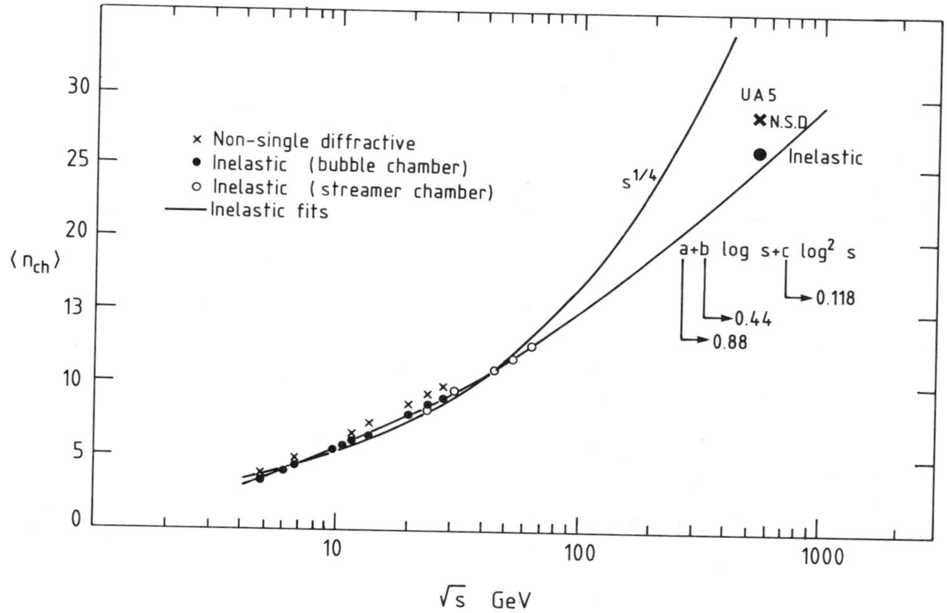

Fig. 10 The energy dependence of the average charged particle multiplicity. The curves are fits to data up to the ISR energy range.

relation $\langle n \rangle = A \cdot s^{1/4}$ should hold, predicts too high a multiplicity, as can be seen in Fig. 11. The uncorrelated jet model of Van Hove 1964, for which Engels and Satz obtained $\langle n \rangle = A \ln s + B + C (\ln s)^{-1}$, leads to too low a value. A model with dominance of soft pion emission leads to a value in good agreement with the inelastic data point of UA5. This model has the energy dependence $\langle n \rangle = a + b \ln s + c \ln^2 s$. The curve $\langle n \rangle = A\, e^{1.7 \sqrt{\ln s/\Lambda^2}}$ suggested by QCD calculations (extrapolated into the soft region) goes somewhat high at 540 GeV but fits well the UA5 non single-diffractive point, i.e. inelastic with single diffraction subtracted. However, since the parameter was fitted to inelastic data at lower energies the agreement is somewhat fortuitous. Models should be tested in many ways, but at

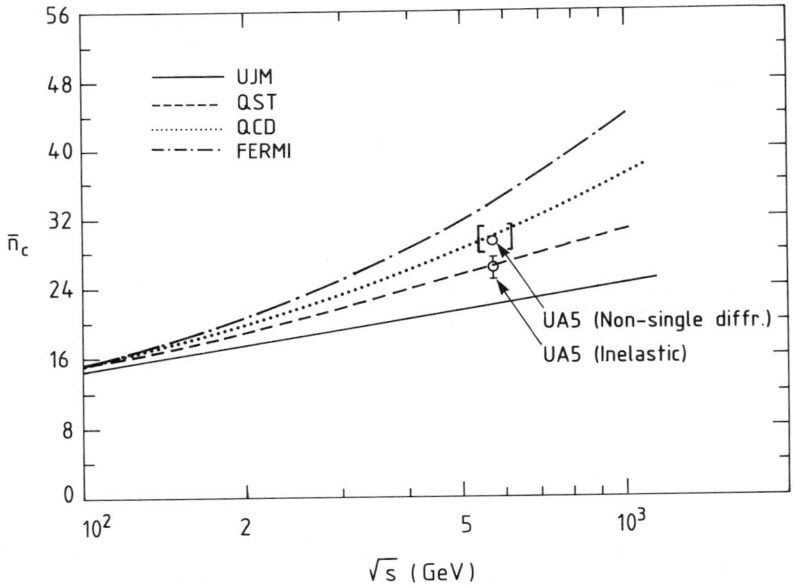

Fig. 11 Extrapolation of four theoretical curves for the average charge particle multiplicity, fitted to inelastic data up to the ISR energies.
UJM = Uncorrelated jet model, QST = Model with Soft pion dominance, QCD = Quantum chromodynamical extrapolation to the soft region, Fermi = Fermi-Landau thermodynamical model.

this stage it seems unavoidable to conclude that the first two models have failed this test.

1.6. The Multiplicity Distribution

The discussed average multiplicity is only the first moment of the multiplicity distribution. Fluctuations in multiplicity are measured by the second moment (D_2^2) and higher moments. As is well established, at lower energies the dispersion D_2 grows linearly with the average multiplicity[18] rather than being proportional to $\langle n\rangle^{1/2}$ as for a Poisson distribution. The Koba, Nielsen, Olesen[19] scaling conjecture implies that all higher moments are proportional to $\langle n\rangle^k$ where k is the order of the moment in question. These authors introduced a scaled variable $Z = n/\langle n\rangle$ and predicted that at very high energies the scaled probability $\langle n\rangle\, P_n$ for n-particles should be equal to a universal, energy-independent function, $\psi(Z)$. Their argument was based on Feynman scaling. Although Feynman scaling is clearly violated, the KNO scaling has been less clearly violated so far. The UA1 group has reported KNO-scaling to hold in the central region even up to 540 GeV[20]. The UA5 group, on the other hand, has found KNO scaling to be violated at 540 GeV for the full phase space region[21].

In the theory KNO scaling is a limiting property valid at asymptotic energies. In practice, experimental data become available at successively higher energies. Approximate KNO scaling was found to hold from about 10 GeV c.m.s. energy in pp collisions[22] up to the then highest available energy, 28 GeV ($= \sqrt{s}$). It was first found by A. Wroblewski[23], that for KNO scaling to hold, one should subtract leading particles. He found that $D_2 = \omega(\bar{n} - n_o)$, where $n_o \sim 1$ is taken to represent the average number of charged leading particles. It has also been argued that single diffraction events should be eliminated. If so, KNO scaling could eventually be expected to hold more accurately for the non single-diffractive component, which at highest energy is the dominating one (about 85% of the inelastic

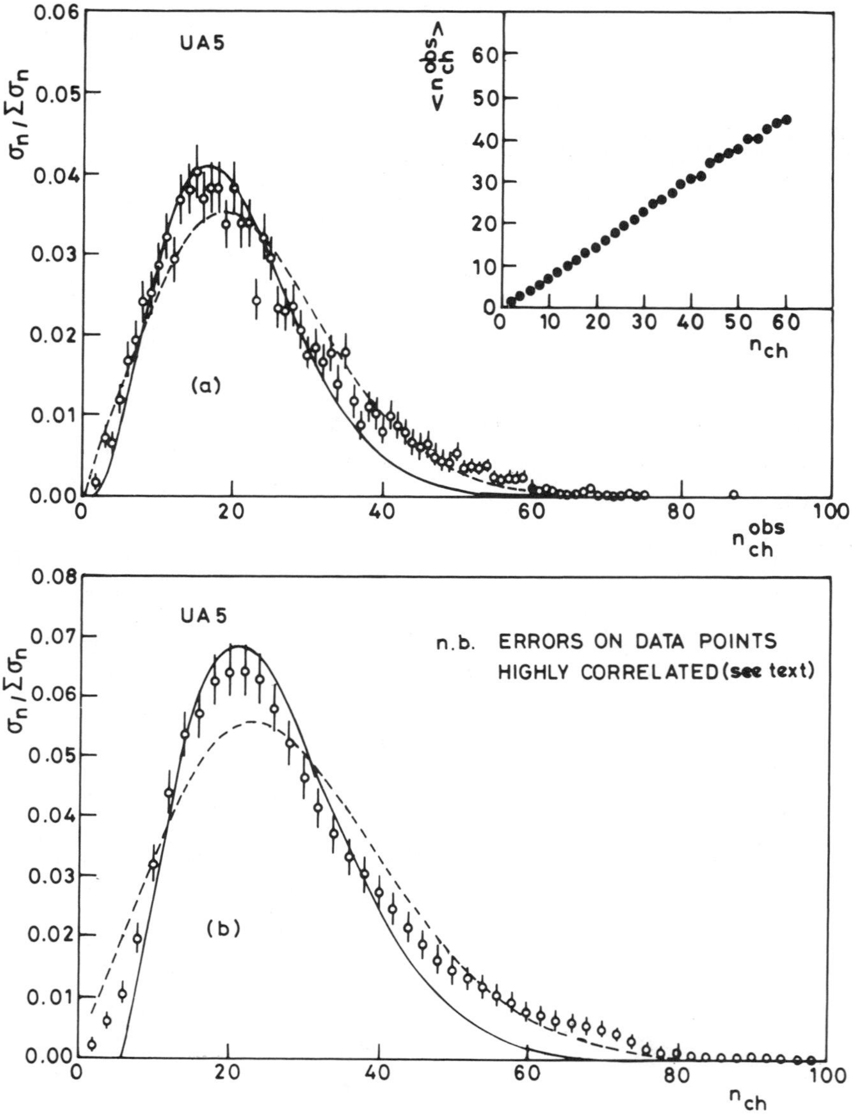

Fig. 12 The distribution of charged multiplicities at 540 GeV c.m.s. energy (UA5).

a) observed; (insert: Monte Carlo results for the relation between true and observed number of charged tracks).
b) fully corrected distribution.

Curves are predictions if KNO scaling holds, dashed line from ref. 22, solid line new fit by UA5 to non-single diffractive data.

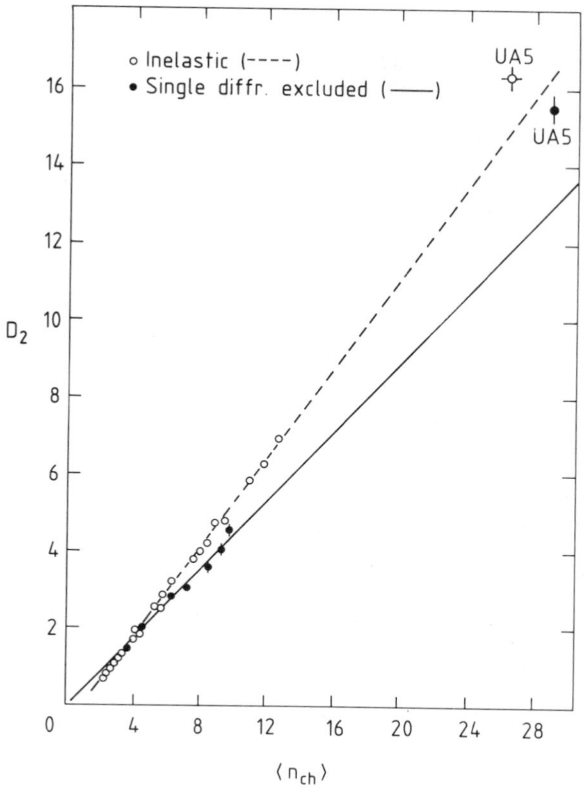

Fig. 13 The relation between the first two moments (D_2 and $<n>$) of the distribution of charged multiplicity. The straight lines are fitted to data at ISR energies and below.

cross-section). New data at several ISR energies were presented recently at the Brighton Conference[24]. Particularly in the non single-diffractive component, one does find energy independent shapes of the scaled multiplicity curves. This means that no KNO scaling violation has been seen up to 60 GeV C.M.S. energy in pp collisions. The CERN $\bar{p}p$ Collider gives an order of magnitude higher energy. The UA5 group has measured the multiplicity distribution with its two large streamer chambers. The result, after corrections for acceptance losses, secondary interactions, trigger biases etc, is given in Fig. 12.

The curves are predictions from fits to data at lower energies based on the assumption that KNO scaling holds. A clear violation of KNO scaling is seen for the non-single diffractive component. The scaling violation is, of course, reflected into the moments of the distribution. The second moment, D_2 vs $\langle n \rangle$, is shown in Fig. 13, where KNO scaling would require all data points to lie on a straight line (in the ideal KNO-scaling the line should pass through the origin). Another way to quantify the violation of scaling is to give the fraction of events in the high multiplicity tail. At the ISR-energies ($<$ 60 GeV c.m.s. energy), where scaling seems to hold, the tail beyond Z = 2 comprises 2.2 % of the events, whereas UA5 finds 6% in this region at 540 GeV c.m.s. energy.

As mentioned, UA1 has reported KNO-scaling to hold for the central region. Results from this detector have been given for the restricted ranges $|\eta| < 1.5$ and $|\eta| < 3.5$ but not for the full phase space. When a comparison is made with ISR data, a similar restricted phase space region is chosen. KNO-scaling holds approximately as is seen in Fig. 14. (NOTE ADDED: The UA5 group has, by restricting the phase space to $|\eta| < 1.5$ in a similar fashion, also found a similar result.)

1.7. KNO Scaling in Models

When KNO scaling seemed to hold at lower energies, several models were constructed to account for this behaviour. Without being complete I list here some of these models together with other models which contain violation of KNO scaling.

I. MODELS WITH KNO SCALING

(a) Unitary uncorrelated cluster model[25]

(b) Geometric-dynamic model[26, 27, 28]

(c) QCD jet calculus[29]

II. MODELS WITH SCALING VIOLATION

(a) Dual parton model[30, 31]

(b) Soft QCD Bremsstrahlung[32]

(c) Three independent "fireballs" model[33]

(d) Quantum Statistical or Chaotic Source model[34]

One should add that some models with KNO scaling are flexible and can be modified to accomodate scaling violation. Several models with KNO scaling violation show the wrong behaviour in that they predict a narrower curve (smaller $D/\langle n\rangle$) at higher energy. Two models show

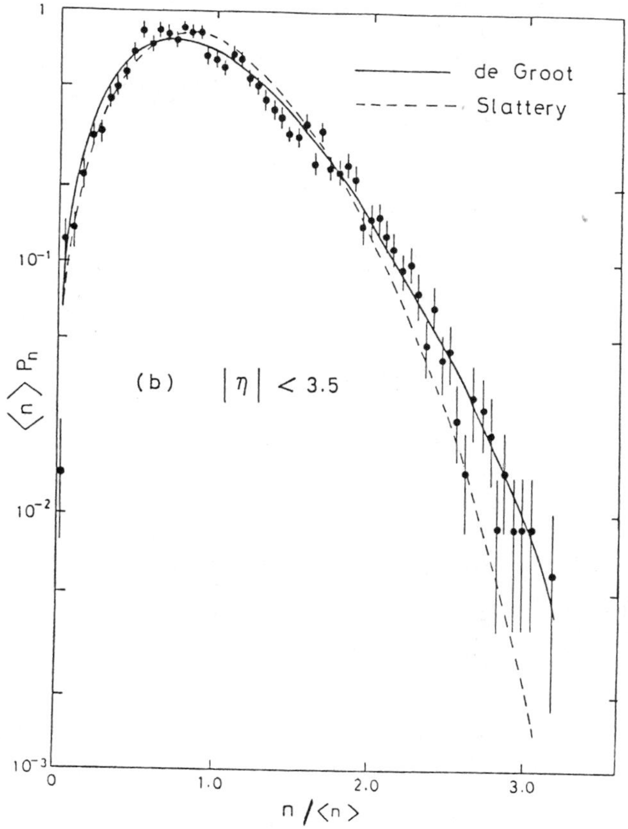

Fig. 14 The KNO scaled multiplicity distribution in the central plateau region (UA1). Curves fit lower energy data well, dashed line FNAL energies, solid line ISR energies.

a scaling violation of the observed type, the dual parton model by Kaidalov and Ter-Martirosyan[30] and also the Three Source Model by Liu and Meng[33]. The Quantum Statistical or Chaotic Model, advanced by Knox[34] in 1974 already and recently by Carruthers et.al.[34], is simple in that it is based on the principle that no detailed dynamics need be specified. The model leads to a discrete multiplicity distribution with two parameters which is known in physics as the Bose-Einstein distribution, in statistical mathematics as the negative binomial (or Pascal's distribution). This model does not have KNO scaling. The mentioned authors, however, interpret the free parameter such as to lead to the wrong behaviour of the scaling violation. As will be shown below there is a connection possible to the Three Source Model[33]. It leads to accidental approximate scaling in the ISR energy range and scaling violation of the correct type at 540 GeV. The model has asymptotic scaling at much higher energies where only one fireball (the central region) is assumed to dominate.

The Bose-Einstein distribution, first derived by Planck (1923)[35], was introduced in particle physics by Knox (1974)[34] and recently revived by Carruthers et al (1982)[34]. It is a discrete distribution in n given by

$$P_n = \binom{n+m-1}{m-1} \left(\frac{\bar{n}/m}{1+\bar{n}/m}\right)^n (1 + \bar{n}/m)^{-m}$$

where there are two parameters; $\bar{n}$ being the mean value of n, which in this model is taken from experiment, and m. The meaning of m is interpreted differently by Knox (number of independent phase space cells) and Carruthers et al (number of independent chaotic sources). Both authors suggest that m increases with energy, leading to KNO scaling violation of the wrong type. At high enough mean values ($\bar{n}$) the distribution can be approximated by a continous one,

$$\bar{n}P_n \to \psi(z) = \frac{m^m}{(m-1)!} Z^{m-1} e^{-mZ}$$

($Z=n/\bar{n}$) as has been pointed out[36]. Obviously asymptotic scaling would require the parameter m to be constant. Fits to data show m to decrease with energy (m = 8, 4.5 and 4.0 at c.m.s. energies 24, 63 and 540 GeV, respectively)[47].

A transition to the Liu and Meng model is obtained if for each source one assumes m = 2, asymptotically represented by $\psi(Z) = 4\,Z\,e^{-2Z}$. This is the input distribution for each of three independent sources - two fragmentation sources and one for the central region. The sources have different average multiplicities $\bar{n}_1 = \bar{n}_2$ (fragmentation) and $\bar{n}_3$ (central).

One can show that if several independent components contribute to the multiplicity, i.e. $n = \Sigma_i n_i$, and each component exhibits KNO scaling, then for the total distribution to KNO scale, it is required that the ratios $\bar{n}_i/\bar{n}$ (i = 1,2---) are energy independent.

Scaling violation is obtained by assuming different energy variations for at least one $\bar{n}_i$. This is what Liu and Meng do. In particular, the fragmentation regions are assumed to stop increasing somewhere in the ISR energy range, when $\bar{n}_1 = \bar{n}_2 = 3.4$ charged particles. At asymptotic energies only one component dominates, the central one, and then the limiting distribution returns to the input distribution, $\psi(Z) = 4Ze^{-2Z}$ which is very wide with $D/\langle n\rangle = 1/\sqrt{2} = 0.71$. Data at ISR has $D/\langle n\rangle = 0.45$ and at the SPS Collider $D/\langle n\rangle = 0.54$ for inelastic events excluding single diffraction. It is of interest to note that the UA1 central ($|\eta| < 1.5$) region distribution (most of the fragmentation regions excluded) has $D/\langle n\rangle = 0.66 \pm 0.02$ not far from the limiting case. Also the form ($4Ze^{-2Z}$) seems to represent the UA1 data well. At intermediate energies the simple three source model is mathematically more complicated (a sum of two exponentials in Z, each multiplied by a different low order polynomial in Z), but with only one energy dependent parameter, namely $\bar{n}_3/\bar{n}$. The mathematical complication disappears if all three sources are of equal strength

$(\bar{n}_1 = \bar{n}_2 = \bar{n}_3)$. In this case the overall distribution is a negative binomial with m = 6 (in agreement with fits to ISR data at $\sqrt{s}$ = 31 GeV), which approximates to $\psi(Z) = 1944/5\ Z^5 e^{-6Z}$.

2. CORRELATIONS

At lower energies it has long been known that the observed charged particles, most of which are pions, are to a large extent produced indirectly as decay products of resonances (ρ, ω, K^* etc). In experiments with many final state particles or in which no momentum measurements and particle identification have been made, one can resort to correlation studies. These are of two types - short-range correlations and long-range correlations. The "range" refers to distances in rapidities or pseudorapidities (η).

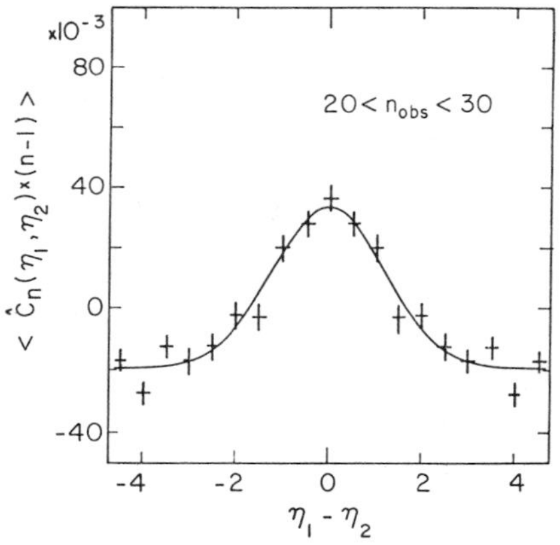

Fig. 15 Two-particle correlation in pseudorapidity showing typical short range correlation of about one unit of pseudorapidity.

2.1. Short Range Correlations

If one particle is produced at η_1, another at η_2, the distance is $(\eta_1-\eta_2)$ between them. It is useful to keep in mind that the decay of a neutral ρ-meson of typical momentum produces two charged pions at a distance of about one unit of rapidity.

One can define a correlation function $C_n(\eta_1, \eta_2)$ which depends on the charged multiplicity (n). By averaging over an interval in n one can increase the statistics. Since theory makes $C_n \propto 1/(n-1)$, one takes $\langle C_n (\eta_1, \eta_2) \cdot (n-1)\rangle$ and averages further over a region in (η_1, η_2) space, keeping the distance $(\eta_1-\eta_2)$ fixed. The result of one such analysis is shown in Fig. 15. Curves for other multiplicities are quite similar to a first approximation.

The conclusion is that one observes clearly a correlation of a range in $\Delta\eta$ typical of small mass resonance particles (ρ, ω, K^*...), namely of about one unit in pseudorapidity. One can further from studies of the height and width find the average size of the cluster of particles that produces this correlation. It is not immediately obvious that not groups (clusters) of particles larger than just resonances are responsible. Such studies are under way within the UA5 group, but here I can state that the sizes are about 2 charged particles.

2.2. Long Range Correlations

If only small sized clusters of small mass were initially produced at first sight, it would seem natural that particles with distances > 2 units in pseudorapidity be uncorrelated. That this is not so was shown in experiments[37] at ISR. A similar and stronger long range correlation has been observed at the SPS collider by the UA5[38] and UA1 groups[39]. In these studies, one observes the number (n_F) of charged particles in a forward interval, say, $0<\eta<4$, and the number

(n_B) of charged particles in a symmetric backward interval, $-4<\eta<0$. If the joint distribution f (n_F,n_B) does not factorize into $f(n_F) \cdot f(n_B)$ then a correlation exists. Experimentally one studies the average of one of the variables (say n_B) at fixed n_F as a function of n_F. If factorization holds, the resulting $\langle n_B(n_F)\rangle$ is independent of n_F. Empirically it turns out that the quantities are linearly related, thus

$$\langle n_B\ (n_F)\rangle = a + bn_F$$

The slope b is taken as a measure of the correlation strength. Factorization (uncorrelated n_F, n_B) implies b = 0 (note, however, that b = 0 does not necessarily imply uncorrelated variables). Another closely related measure is the usual correlation coefficient ρ which is defined as

$$\rho = \frac{\mathrm{cov}(n_F, n_B)}{\sqrt{\mathrm{var}\ n_F \cdot \mathrm{var}\ n_B}}$$

In the case where symmetry holds between the two hemispheres, one has var n_F = var n_B and then ρ = b (if b is determined by a least squares fit where each event is given the same weight). In the ISR energy range b = 0.2-0.3, whereas at the SPS collider b = 0.54±0.01 is reported by UA5. There is agreement between the b-values measured by UA1 and UA5. Fig. 16a shows the result together with the earlier ISR data. The increase of the strength with energy is clearly demonstrated. However, one can suspect that the effect is due to short range correlations. Clusters produced near the centre can emit decay products into both regions. In order to decouple from this effect one opens up a gap between the forward and backward regions. A study shows that a gap of $\Delta\eta$ = 2 units of pseudorapidity is sufficient. The correlation strength (b) is reduced, Fig. 16c, but not to zero. At the SPS collider b = 0.41 ± 0.01 remains, only at the lower range of ISR energies is b ≃ 0. The explanation could be very simple. One does not have to introduce any specific, dynamical long range correlation. It is just unlikely that fluctuations (up or down from the average number of particles) in one region (F) are not accompanied by similar fluctuations in the other region (B). How this suffices to explain the effect will be described below.

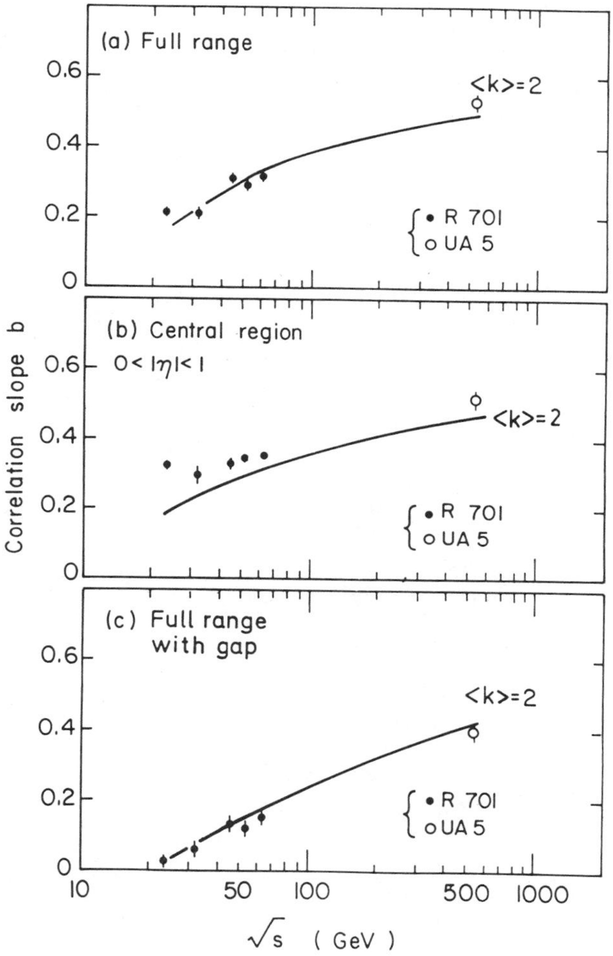

Fig. 16 The long range multiplicity correlation parameter b in dependence of energy

a) forward region $0 < \eta < 4$, backward $-4 < \eta < 0$.

b) central part with $0 < |\eta| < 1$.

c) with a central gap in η; forward region $1 < \eta < 4$, backward $-4 < \eta < -1$.

Data are from the ISR experiment R701 and SPS Collider experiment UA5. The curves are Monte Carlo simulations with an average cluster size of 2 charged particles.

In Fig. 17 is given the two-dimensional empirical distribution (scatterplot) of the variables n_F and n_B. (The definitions of regions F and B are in terms of intervals in pseudorapidity η).

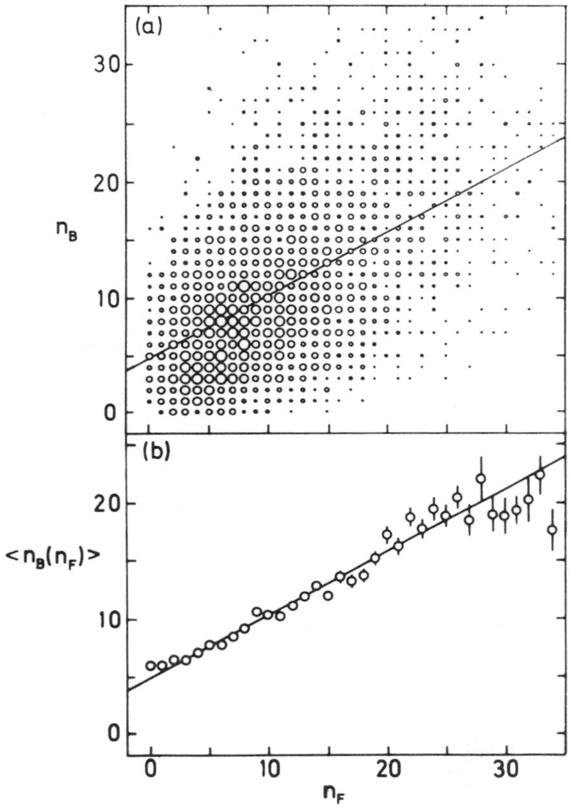

Fig. 17 Forward-backward multiplicity correlation for the case $0<|\eta|<4$

a) the two-dimensional multiplicity distribution with n_F charg particles emitted forward, n_B backwards. The size of the smal areas is proportional to the number of events with a given (n_B n_F).

b) the average n_B for fixed n_F, i.e. $E(n_B/n_F)$ in dependence of n_F.

The figure also shows how remarkably close to a straight line the averages lie. By introducing new coordinates ($n_s = n_F + n_B$ and $\Delta = n_F - n_B = 2n_F - n_s$), one has a joint distribution $f(n_s, n_F)$ and can sum over the variable Δ (or n_F) at fixed n_s to obtain the distribution of multiplicity n_s. The first two moments of this distribution are represented by the mean, $\langle n_s \rangle$, and the dispersion D_s. This distribution is thus obtained by projecting all events onto the 45° diagonal of the scatterplot. At fixed n_s the distribution in n_F $f(n_F|n_s)$ is of interest, since it describes how the n_s particles are shared between the forward and backward regions. The mean of n_F (given n_s) is by symmetry = $1/2\, n_s$. The first interesting moment is the squared dispersion, $d_s^2(n_F)$. One can show that the following identity holds:

$$b = \frac{1/4D_s^2 - \langle d_s^2 \rangle}{1/4D_s^2 + \langle d_s^2 \rangle}$$

where $\langle \quad \rangle$ symbolizes the average over the n_s-distribution. To prove this identity, one uses the fact that

$$b = \frac{\mathrm{cov}(n_F, n_B)}{\mathrm{var}\ n_F} = \frac{\sum_i (n_F(i) - \bar{n}_F)(n_B(i) - \bar{n}_B)}{\sum_i (n_F(i) - \bar{n}_F)^2}$$

(where the sums are over all events (i)) and one carries out the sums in two steps; first for events with $n_F(i) + n_B(i) = n_s$ fixed, second over n_s.

One finds that if the observed particles were emitted (for fixed n_s) with constant probability into F or B the distribution, $f(n_F|n_s)$ would be binomial and $d_s^2(n_F) = 1/4n_s$. In this case

$$b = \frac{D_s^2 - \bar{n}_s}{D_s^2 + \bar{n}_s}$$

A second remark is that if the total n_s-distribution was Poissonian, then $D_s^2 = \bar{n}_s$ and $b = 0$ would follow. However, the n_s-distribution is not a Poisson. It is much broader, and one has approximately $D_s \simeq w \cdot \bar{n}_s$. This goes in the right direction to produce a positive value of the correlation strength parameter b and to give $b \to 1$ as the limit at very high energies. However, inserting the observed values for the case $0 < |\eta| < 4$, which are $D_s = 8.8 \pm 0.1$ and $n_s = 16.0 \pm 0.2$, gives much too high a value for b ($b = 0.66$). The remedy for this discrepancy is to invoke the fact that particles are emitted together in clusters. As a side remark we note that energy-momentum conservation could lead to emission in opposite directions. However, since we have unobserved leading particles and furthermore observe only a restricted range in pseudorapidity, the effect of energy-momentum conservation is not strong. Let us, therefore, return to consider clusters. The relation between the number of clusters, (C), their sizes (k_i) and the observed particles (n), is

$$n = \sum_{i=1}^{C} k_i$$

Generally, both C and k_i fluctuate around their mean values ($\bar{C}$, $\bar{k}$). As an illustration of the way clusters remedy the discrepancy, we ignore the fluctuation in the sizes and take all k_i equal, i.e. $n_s = k \cdot C_s$ where C_s and n_s fluctuate but k is fixed. Then $d^2 = 1/4\, k^2 \cdot C_s = 1/4\, k \cdot n_s$ based on the assumptions that the number of clusters falling into F (or B) is governed by the binomial distribution, and that all k particles from a cluster remain in the same η-interval as the parent cluster. Then

$$b = \frac{D_s^2 - k\bar{n}_s}{D_s^2 + k\bar{n}_s}$$

One finds that $k \simeq 2$ is needed to yield the observed b (=0.41). In a more realistic model the cluster sizes fluctuate. Under the same conditions as above it can be shown[48] that the relation given is changed into

$$b = \frac{D_s^2 - k_{eff}\,\bar{n}_s}{D_s^2 + k_{eff}\,\bar{n}_s}$$

Here $k_{eff} = \bar{k} + d_k^2/\bar{k}$, where ($k$, d_k^2) are the first two moments of

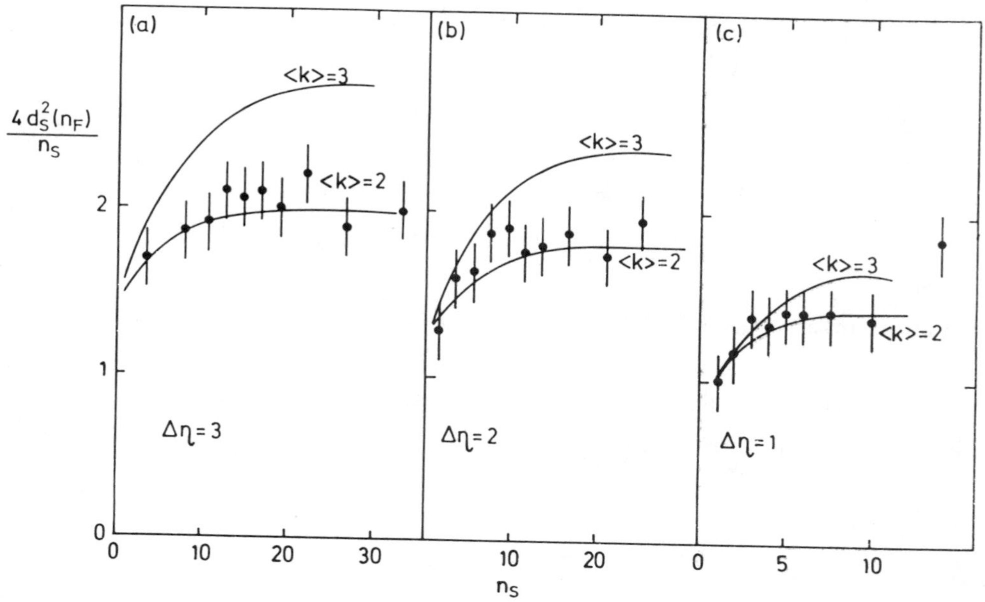

Fig. 18 The variance d^2 of the n_F distribution at fixed total multiplicity n_s as a function of n_s for three selected regions of pseudorapidity η. The curves are obtained from two Monte Carlo simulations with assumed average cluster sizes of 2 and 3 charged particles, respectively.

the cluster decay distribution. There is one more fluctuation to consider. The decay products (k_i particles) of a given cluster may leak out from or into the chosen rapidity interval. Monte Carlo studies show that this effect lowers the effective k in the relation above. In Fig. 18 the quantity $4d_s^2/n_s$ is plotted vs $n_s = (n_F + n_B)$. In the oversimplified case of single particle binomial distribution at fixed n_s, this quantity is expected to be 1, whereas binomially distributed clusters of fixed size k should lead to a value equal to k. Data points are compared to a simple Monto Carlo simulation. In each event with a given number of clusters (C), the C positions in pseudorapidity (η) are chosen at random such as to reproduce the η-distribution after cluster decay (gaussian with a width as observed in the study of short range correlations). The distributions of C is taken so as to reproduce the observed charge particle multiplicities. For this program one has to assume the properties of the cluster decay distribution. It turns out that $\bar{k} \sim 2$ and $d_k^2/\bar{k} \sim 0.6$ (truncated Poisson) gives a good description of the data. The same parameters account for the short range correlation. Fig. 19 shows some examples (of many) of $f(n_F|n_s)$ distributions based on the simple cluster Monte Carlo simulation with cluster sizes fluctuating around an average of $\bar{k} = 2$.[47] The agreement with data is good. Data are not reproduced if particles rather than clusters are binomially distributed, nor if clusters of fixed size are assumed.

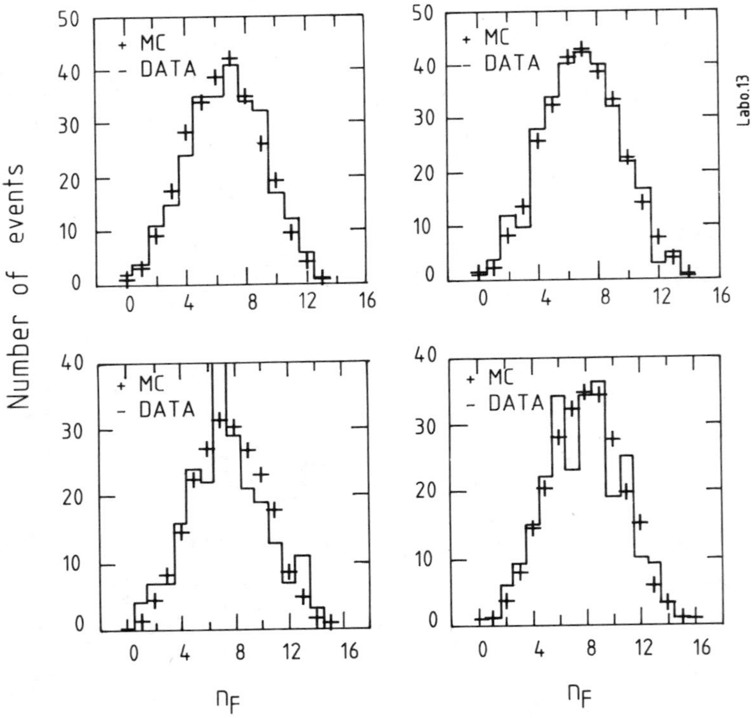

Fig. 19 Distributions of n_F at fixed total multiplicity n_s, i.e. $P(n_F/n_s)$ for n_s = 13, 14, 15 and 16 charged particles respectively. The Monte Carlo simulation carried out with an average cluster size of 2 charged particles and assuming binomial distribution of clusters into each hemisphere (at fixed number of clusters) (UA5 unpublished).

The conclusion is that the clusters are small ($\sim$ 2 charged particles). High multiplicity events would thus correspond to the production of many small clusters rather than a few large clusters. However, the statistics in the extreme high multiplicity tail prevents one making valid conclusions over the full range of multiplicities.

Before leaving the subject of long range correlations, I would like to return to the empirical result as given in Fig. 17. One can ask if there is any good reason why the points of average n_B (at fixed n_F) are so close to a straight line. It must reflect some property of the two-dimensional distribution in the scatterplot. Let us idealize to a scatterplot $f(C_F, C_B)$ of clusters (rather than particles), for which we have reasons to believe that $f(C_B|C)$, i.e. at fixed $C = C_F + C_B$ is a binomial distribution and also assume that the linear relation $\langle C_B \rangle = a' + b' \cdot C_F$ is exact. With these two assumptions one can prove mathematically[40] that the distribution in the sum $C(=C_B + C_F)$ has to be a negative binomial. A corollary is that the projected distributions in C_F and C_B also are negative binomials. One is reminded that this type of discrete distribution (or its asymptotic continous form) plays a central rôle in two of the above mentioned recent models[33,34].

3. RESULTS FOR IDENTIFIED PARTICLES

3.1. Photons, π^0-, η-Mesons

Both the UA1 and UA2 experiments can observe photons in the electromagnetic calorimeters. The UA5 group has observed the tracks of electron-positron pairs from γ-rays converted in the beam pipe or in lead converters inserted in the streamer chamber.

The UA2 group[41] has presented clear evidence for both π^0 and η-meson production. Fig. 20 shows the mass distribution for 2γ's with the two peaks.

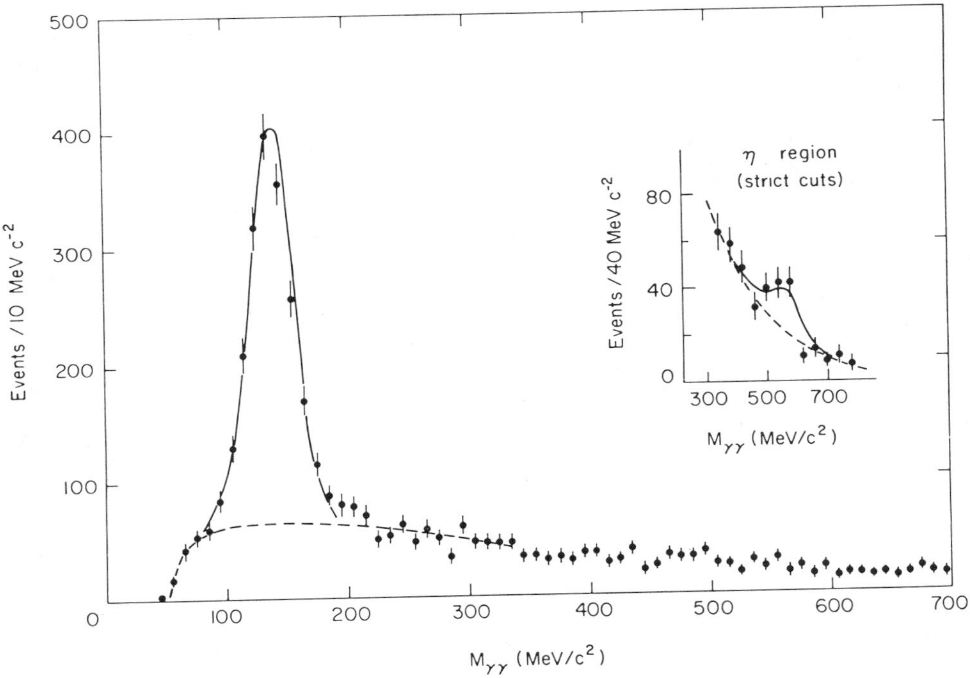

Fig. 20 The π^0 and η peaks in the distribution of invariant 2γ-masses (UA2).

The UA5 experiment[42] has measured the rapidity distribution of photons (γ's), shown in Fig. 21. The same figure shows the density of charged particles. Obviously there is an excess of photons, whereas one would expect equal numbers if all photons came from π^0 decay and if all charged particles were $\pi^{\pm}$. Removing the estimated number of $K^{\pm}$, p, $\bar{p}$ as well as photons from $K_s^{\circ} \rightarrow 2\pi^0 \rightarrow 4\gamma$, one still is left with a sizeable excess of photons as is seen in Fig. 21b. If this is attributed to η-meson production, the UA5 result is $\eta/\pi^0 \simeq 30\%$, whereas UA2 reports $\eta/\pi^0 \simeq 50\%$. Results from the ISR[43] also show a strong η-production with a ratio of 20%.

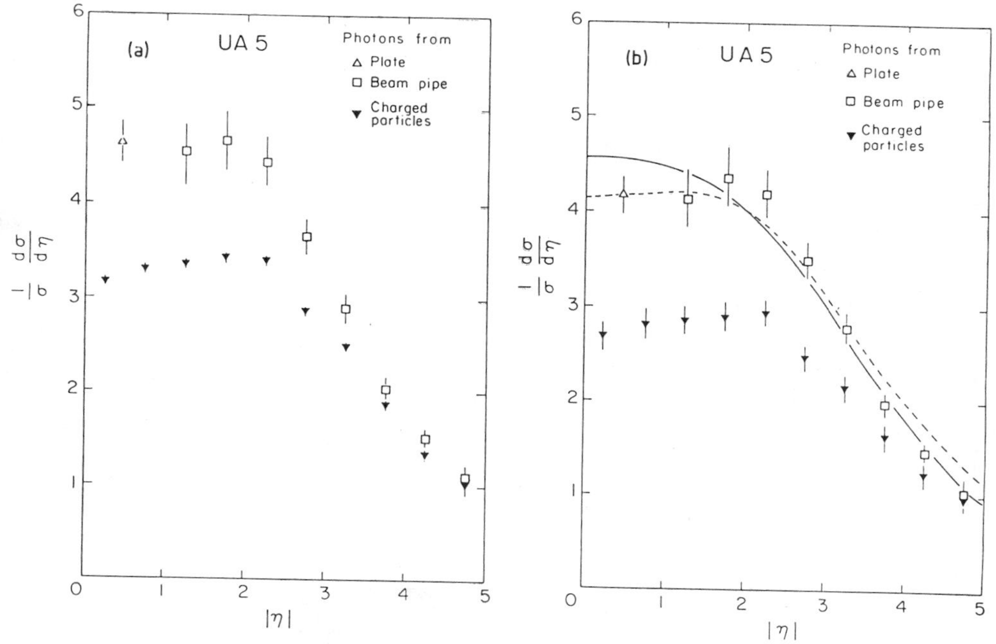

Fig. 21 The excess of γ-rays over charged particles as observed in the UA5 experiment. In (a) all data vs the pseudorapidity; in (b) data corrected for other particles than pions and η-mesons.

3.2. Strange Particles

Results have been reported by the UA2 and UA5 groups. In UA2 the identification of $K^{\pm}$ is made by the time of flight method[44]. In UA5 information on K°_{s} and $\Lambda/\bar{\Lambda}$ is obtained by a statistical analysis of decay angles[45]. It is a good excercise to find out how this is possible and why it also yields the momentum of the particle.

As is seen in Fig. 22 the UA2 group finds the transverse momentum spectrum of charged K-mesons harder (less steep) than that of π-mesons. They also find the particle ratios to be approximately constant when plotted against the "transverse mass", i.e. $M_T = (M^2 + p^2{}_T)^{1/2}$. The ratio $K^{\pm}/\pi^{\pm}$ has increased from that at ISR energies. The UA5 group finds from the measured p_T distribution,

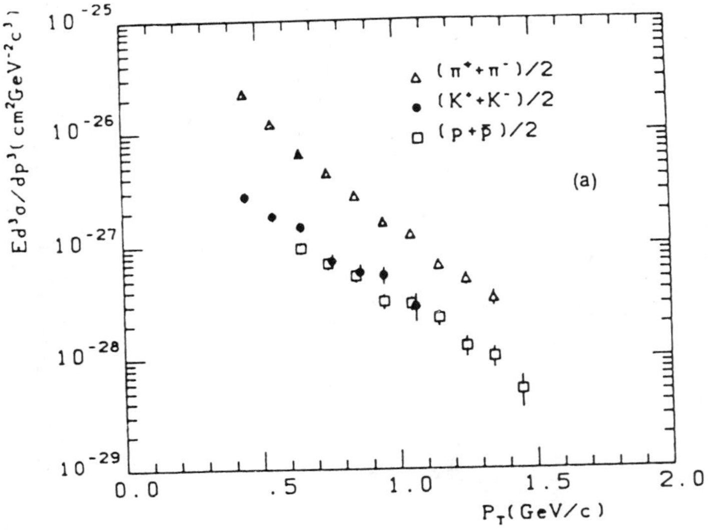

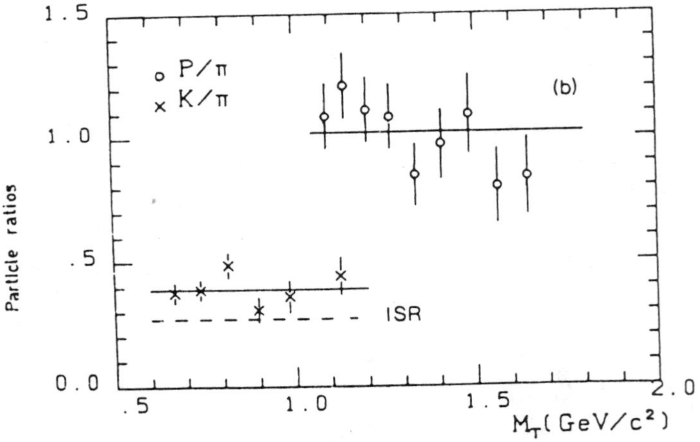

Fig. 22 The distribution of transverse momenta of charged K-mesons in comparison with pions and protons as measured in the UA2 experiment. Also the particle ratios protons/pions and K-mesons/pions as a function of transverse mass.

Fig. 23, that the average p_T is high, being 0.70 GeV/c for K° and 0.65 GeV/c for $\Lambda/\bar{\Lambda}$. Also the production has increased. The ratio $(K° + \bar{K}°)/\pi^{\pm}$ = 0.12 (± 0.03) is reported, to be compared with generally lower values at lower energies, Fig. 24. Also the baryon + antibaryon production has increased from about 5% at ISR to about 9% at SPS Collider. More data are needed on these subjects before we can be sure about these tendencies and relate them to basic processes.

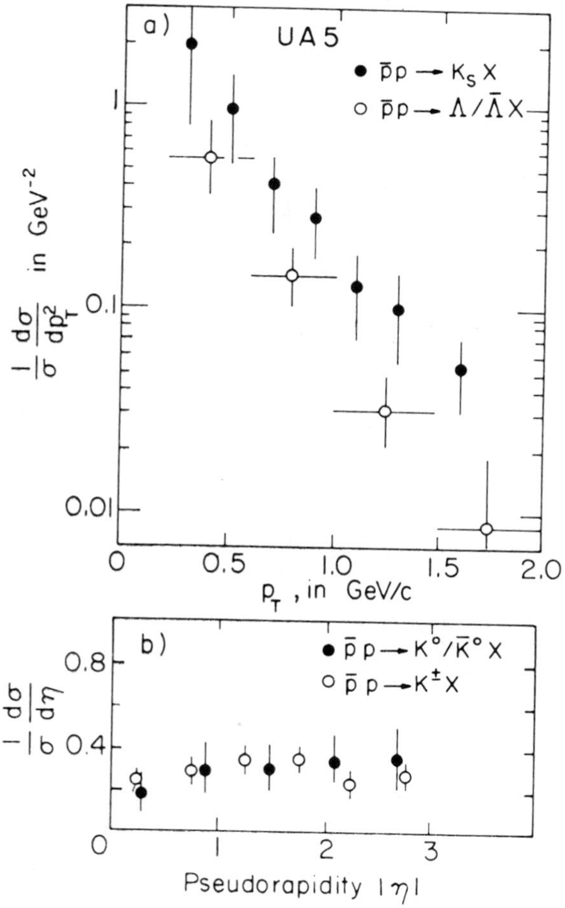

Fig. 23 The distribution of transverse momentum of neutral K_s mesons and $\Lambda/\bar{\Lambda}$ hyperons (UA5 data). Also the pseudorapidity distribution of K-mesons in a limited region of pseudorapidity.

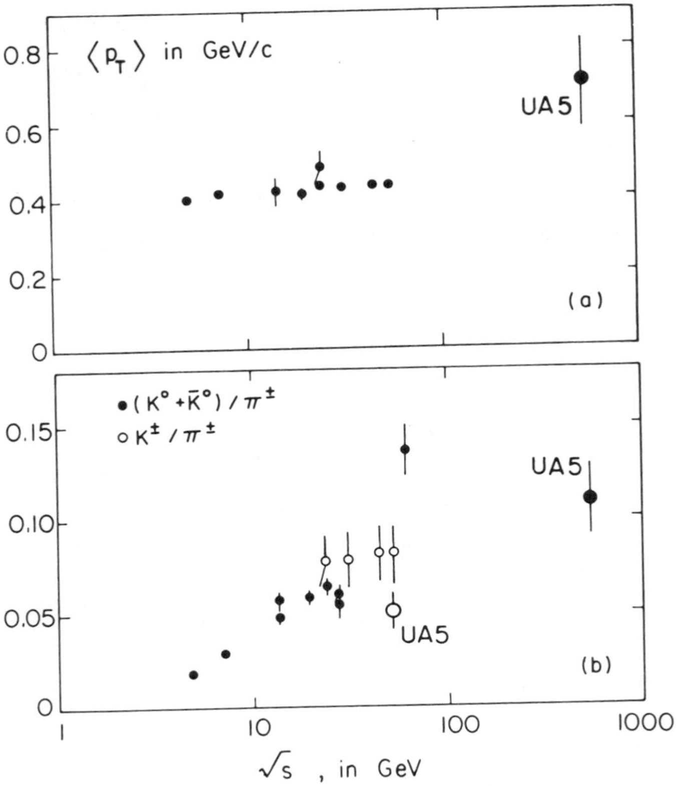

Fig.24 a) The mean transverse momentum of K-mesons.

b) The K/π ratio at various energies including the UA5 result at 540 GeV for ($|\eta| < 3$).

3.3. The Average Event

Compiling what is known about the production of various particles in given rapidity intervals, one tries to extrapolate from $|\eta| < 3$ to $|\eta| < 5$, which is close to full space (same relative increase as for charged particles). In this way $K^\pm$, $(K^\circ + \bar{K}^\circ)$, $(\Lambda + \bar{\Lambda})$ production is found. Furthermore, one can make reasonable extrapolations about particle ratios at lower energies to obtain estimates for $(p + \bar{p})$ and $(n + \bar{n})$ productions based on the observed production of $(\Lambda + \bar{\Lambda})$ hyperons. A proportional amount of Σ's is assumed (∿ 50% of Λ°). The remaining average charged multiplicity is

Table 1

The average particle content of an event at the SPS Collider at 540 GeV as estimated by the UA5 group. Measurements are underlined, inferred values for the pseudorapidity range $-5 < \eta < 5$ are given in the third column (beyond $|\eta| = 5$ one expects very few particles).

Type of particle	540 GeV $\langle n \rangle$ SPS Collider for $|\eta| < 3$	540 GeV $\langle n \rangle$ SPS Collider for $|\eta| < 5$	53 GeV $\langle n \rangle$ ISR
All charged	19.5	26.5	10.1
K^+ and K^-	1.8	2.5	0.75
K^o and $\bar{K}^o$	2.0	2.7	0.7
Λ and $\bar{\Lambda}$	0.35	0.5	0.1
p and $\bar{p}$		1.5	0.3
$\Sigma^\pm$ and $\bar{\Sigma}^\pm$		0.25	0.04
$\pi^\pm$		22.3	9.0
$\bar{n}$ and n		1.5	0.3
all γ		31.5	11.8
η		3.5	1.1
γ from η		11.2	3.4
$\pi^\pm$ from η		2.1	0.6
$\pi^\pm$ (not η)		20.2	8.4
π^0 (not η)		10.1	4.2

assumed to be $\pi^\pm$. Out of a total average of 26.5 charged particles, the charged pions are 22.3 or 84%. The compilation is summarized in the Table (UA5-compilation)[46], which also shows the estimated η-meson production based on the further assumption that $\pi^0/\pi^\pm = 1/2$, after η-meson decays are eliminated.

The total number of charged particles (tracks) is 26.5 on the average and the number of photons is 31.5. Counting identifiable particles (charged and neutral) gives from the Table an average of 42.7 in the interval $|\eta| < 5$. An extrapolation to full phase space (by Monte Carlo) gives about 44 produced particles on the average, not including 2 leading baryons. One should remember that the particles are produced via small groups of resonances (clusters).

REFERENCES

1. K. Alpgård et al., (UA5 Coll.) Phys. Lett. 112B (1982) 183; T. Åkesson et al.,(AFS Coll.) Phys. Lett. 108B (1982) 58; V. Cavasini et al., CERN-EP/83-90 (1983); A. Breakstone et al., CERN-EP/83-101 (1983).
2. G. Arnison et al., (UA1) Phys. Lett. 128B (1983) 336; R. Battiston et al., (UA4) Phys. Lett. 117B (1982), 126.
3. K. Alpgård et al., Phys. Lett. 121B (1983) 209.
4. G. Arnison et al., (UA1) Phys. Lett. 122B (1983) 103.
5. M. Banner et al., (UA2) Phys. Lett. 122B (1983) 476.
6. UA5 Collaboration, Phys. Scripta 23 (1981) 642.
7. S. Pokorski and L. Van Hove, Acta Phys. Pol. B5 (1974) 229.
8. A. Capella and J. Tran Thanh Van, Phys. Lett. 114B (1982) 450; Phys. Lett. 95B (1980) 311.
9. M. Basile et al., Phys. Lett. 99B (1981) 247; Nuovo Cim. 67A (1982) 244; CERN-EP/81-147 (1981).
10. G. Arnison et al., (UA1) Phys. Lett. 118B (1982) 167.
11. A. Breakstone et al., (Ames-Bologna-CERN-Dortmund-Heidelberg-Warshaw Coll.), Contribution to the High Energy Physics Conference, Brighton (UK), 20-27 July 1983.
12. L. Van Hove, Phys. Lett. 118B (1982) 138.
13. R. Hagedorn and J. Rafelski, Phys. Lett. 97B (1980) 136.

14. W. Thomé et al., Nucl. Phys. B 129 (1977) 365.
15. K. Alpgård et al., (UA5) Phys. Lett. 107B (1981) 310 and Contribution to the High Energy Physics Conference, Brighton, July 1983.
16. G. Arnison et al., (UA1) Phys. Lett. 107B (1981) 320
17. R.V. Gavai and H. Satz, Phys. Lett 112B (1982) 413.
18. A. Wroblewski, Acta Phys. Pol. B4 (1974) 857.
19. Z. Koba, H.B. Nielsen, P. Olesen, Nucl. Phys. B40 (1972) 317.
20. G. Arnison et al., (UA1) Phys. Lett. 107B (1981) 320; Phys. Lett. 123B (1983) 108;
M. Calvetti, UA1 Collaboration; Proceedings Third Topical Workshop on Proton-Antiproton Collider Physics, Rome 1983, CERN/83-04, (1983) 10.
21. K. Alpgård et al.,(UA5) Phys. Lett. 121B (1983) 209.
22 P. Slattery, Phys. Rev. D7 (1973) 2073.
23. A. Wroblewski, Acta Phys. Pol. B4 (1973) 857.
24. A. Breakstone et al., Contribution to the High Energy Physics Conference, Brighton (UK) 20-27 July 1983.
25. E.H. de Groot, Phys. Lett. 57B (1975) 159.
26. T.T. Chou and C.N. Yang, Phys. Lett. 116B (1982) 301.
27. C.S. Lam and P.S. Yeung, Phys. Lett. 119B (1982) 445;
K. Goulianos, H. Sticker and S.N. White, Phys. Rev. Lett. 48 (1982) 1454.
28. S. Barshay, Phys. Lett. 42B (1972) 457; Phys. Lett. 116B (1982) 193.
29. D. Amato and G. Veneziano, Phys. Lett. 83B (1979) 87
A. Bassetto, M. Ciafaloni and G. Marchesini, Nucl. Phys. B 163 (1980) 477.
30. A.B. Kaidalov and K.A. Ter-Martirosyan, Phys. Lett. 117 B (1982) 249.
31. A. Capella and J. Tran Thanh Van, Phys. Lett. 93B (1980) 946; Part. & Fields 10 (1981) 249; Phys. Lett. 114B (1982) 450.
P. Aurence, F.W Bopp, and J. Ranft, LAPP-TH-83, SI-83-8.

32. G. Pancheri and Y.N. Srivastava, LNF-82/86 (P); LNF-83-25 (Frascati reports).
G. Pancheri, Proceedings Third Topical Workshop on Proton-Antiproton Collider Physics, Rome 1983, CERN 83-04 (1983) 503.
33. Liu Lian-sou and Meng Ta-chung, Phys. Rev. D27 (1983) 2640; Frei. Univ. Berlin preprint FUB/HEP 12-83.
34. W.J. Knox, Phys. Rev. D10 (1974) 65.
P. Carruthers and C.C. Shih, Los Alamos preprint LA-UR-83-1231 (1983).
35. M. Planck, Sitzungber. Deutsch Akad. Wiss. Berlin 33 (1923) 355.
36. L. Mandel, Proc. Phys. Soc. 74 (1959) 253 and
V. Pessin, Ann. Math. Stat. 32 (1961) 922.
37. S. Uhlig et al., Nucl. Phys. B132 (1978) 15.
38. K. Alpgård et al., (UA5) Phys. Lett. 123B (1983) 108.
39. M. Calvetti, UA1-Coll., Proceedings Third Topical Workshop on Proton-Antiproton Colliders, Rome 1983; CERN 83-04 (1983) 10.
40. B. von Bahr, Stockholm Univ., private communication.
41. M. Banner et al., (UA2) Phys. Lett. 115B (1982) 59.
42. K. Alpgård et al., (UA5) Phys. Lett. 115B (1982) 71.
43. G. Jansco et al., Nucl. Phys. B124 (1977) 365.
44. M. Banner et al., Phys. Lett. 122B (1983) 322.
45. K. Alpgård et al., (UA5) Phys. Lett. 115B (1982) 65.
46. K. Alpgård et al., (UA5) Phys. Lett. 121B (1983) 209.
47. UA5-Collaboration, (unpublished);
B. Åsman, Stockholm Univ., private communication.
48. G. Ekspong, Proceedings Third Topical Workshop on Proton-Antiproton Colliders, Rome 1983; CERN 83-04 (1983) 112.

D I S C U S S I O N

CHAIRMAN : G. EKSPONG

Scientific Secretary : D. Nemeschansky

- *BERNSTEIN* :

Why isn't the slope of the plot $<n_b>$ versus n_f equal to one ?

- *EKSPONG* :

The averages of n_b and n_f are, of course, equal due to $^F/_B$ symmetry. But in a subsample of events with an arbitrary fixed n_f there is no reason for the backward average $<n_b(n_f)>$ to be equal to n_f. (Equality can be approached only in the limit of infinite energy.) In case this conditional n_b - average should be equal to its overall average, independent of n_f, then the slope b is zero and there is no correlation. When $b \neq 0$ one has correlation and one has necessarily $-1 < b < 1$.

- *ROTELLI* :

First I want to make a comment. The KNO scaling is based on Feynman scaling and furthermore the model is no longer in use. The first person to my knowledge to predict KNO scaling was A.M. Polyakov a few years before Koba, Nielsen and Olesen. Polyakov's model is a branching model and does not have Feynman scaling.

My question is, the UA1 results seem to suggest that KNO scaling is valid whereas UA5 results suggest it is not. Can you tell if these two disagree or if it is only due to different η cuts?

- *EKSPONG* :

Yes, probably due to different cuts. The UA5 group has data referring to the full phase space, i.e. no cut, and compares with lower energy data with no cut. The UA1 group has used two sets of data with different cuts. For their data in the range $|\eta| < 1.5$ the comparison is made with data in $|\eta| < 1.3$ at lower energy, whereas for $|\eta| < 3.5$ the comparison is made with no cut data.

- *MARUYAMA* :

You showed a table of composition of identified particles. How do you measure neutrons and antineutrons ?

- *EKSPONG* :

The number of neutrons and antineutrons has not been measured. One assumes that the p and $\bar{p}$ produced in the central region are accompanied with an equal number of neutrons and antineutrons.

- *ARNEODO* :

I would like to know whether there are any other indications on the formation of quark-gluon plasma besides the behaviour of the p_T distribution.

- *EKSPONG* :

As far as I know it is the only one.

- *ARNEODO* :

Can you see the restoration of SU(3) symmetry in π or K yields ?

- *EKSPONG* :

There is an increase in the K meson yield. At SPS it has been estimated to be

$$\text{K mesons/all} \sim 12\% \text{ (UA5)}$$

whereas at ISR it is 9%. The indicated increase may be due to heavy flavour production with decay to strange quarks.

- *BERNSTEIN* :

What is the error on the K yield ?

- *EKSPONG* :

The UA5 estimate is only rough; it will be improved in the future.

- *WEIDBERG* :

Have you checked with the UA1 data if the discrepancy in KNO scaling between UA1 and UA5 is simply due to the smaller acceptance of UA1 ?

- *EKSPONG* :

This has not been done.

- *GASPARANI* :

Has any difference been measured in the x_F and p_T distribution between strange and non-strange particles produced at the collider ?

- *EKSPONG* :

The slope of the p_T distribution of the produced kaons is smaller than of pions. The average p_T is higher. The x_F - region is small ~0.1.

- *ZICHICHI* :

Has anybody measured charged energy/total energy ?

- *EKSPONG* :

No, the only thing I know of is the total transverse energy distribution.

THE PROBLEM OF NEW HEAVY FLAVORS:

TOP AND SUPERBEAUTY

A. Zichichi

Cern
Geneva
Switzerland

1. INTRODUCTION: FROM PAST KNOWLEDGE TO FUTURE POSSIBILITIES

The present status of our knowledge on quarks and leptons may be summarized as follows:

Families:	1st	2nd	3rd
Quarks:	$\begin{pmatrix} u \\ d \end{pmatrix}$	$\begin{pmatrix} c \\ s \end{pmatrix}$	$\begin{pmatrix} ? \\ b \end{pmatrix}$
Leptons	$\begin{pmatrix} \nu_e \\ e \end{pmatrix}$	$\begin{pmatrix} \nu_\mu \\ \mu \end{pmatrix}$	$\begin{pmatrix} \nu_\tau \\ \tau \end{pmatrix}$

There are very good reasons to believe that our knowledge is far from being complete and thus the search for new heavy flavors and the study of their family structure is one of the key problems in Subnuclear Physics.

Let me quote two "theoretical" arguments in favor of further needs of new quarks. The ABJ anomaly cancellation requires that the number of leptons be equal to the number of quarks. This means that a sixth quark is needed. Its natural location would be the up-like member of the 3rd family, i.e. the "top" quark.

According to Supersymmetry, a very heavy quark with a mass in the few $10^2 GeV/c^2$ range is needed in order to produce radiatively (see Figure 1) a gluino with a mass such as to avoid a conflict with existing lower limits[1]. None of the presently known quarks (s, c and b) is heavy enough for this purpose.

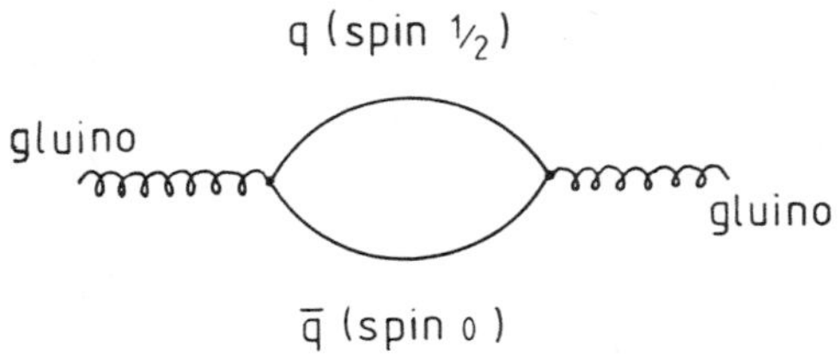

Fig. 1. The diagram illustrates how a gluino can acquire a mass from radiative processes, where a spin ½ quark and a spin 0 antiquark are virtually produced. The quark mass must be in the 10^2 GeV/c^2 range, in order to allow a gluino mass of the order of a few GeV/c^2.

Apart from these theoretical arguments we should not underestimate that Nature has often provided physicists with more regularities than needed (for example the equality between the proton and the electron charges, which took more than three decades to be understood).

We propose to consider the ratio between the masses of the known heavy quarks as a limit for the regularity in their masses. There are good reasons[2] to consider the strange quark heavy enough to be used in our argument.

At present we know that:

i) $(m_c/m_s) \cong (1.8/0.5) \cong 3.5 \cong 4$;

ii) $(m_b/m_s) \cong (5.5/0.5) \cong 11 \cong 10$.

Suppose that (i) and (ii) are of general validity, i.e.:

$$(m_c/m_s) = [m(\text{uplike quark})/m(\text{downlikequark})] = 4 \ , \tag{1}$$

and

$$(m_m/m_s) = [m(\text{family N+1})/m(\text{family N})] = 10 \ . \tag{2}$$

We ignore the 1st family (u, d) because of its very light mass. On the other hand, the ratios (1) and (2) would not be inconsistent with the various models used to derive the quark masses from bound states.

The validity of (1) and (2) would allow to conclude that the "top" mass is in the 20 GeV/c^2 range. This is too light for Supersymmetric models to avoid a gluino mass in conflict with experimental data. On the other hand, Supersymmetry tells us that the maximum number of flavors, n_f, allowed in order to have a consistent theory (for example: the unification limit not above the Planck mass) is $n_f=8$. This means that the maximum number of families is 4. In the theories that ignore Supersymmetry, the asymptotic freedom is lost if $n_f>16$.

What is not forbidden, in Nature, does take place. Thus, the message from Supersymmetry is twofold:

i) four families of quarks are allowed;
ii) quarks heavier than (d, u, s, c, b and t) are needed.

Formulas (1) and (2) tell us that the 4th family would have the heavy masses wanted by Supersymmetry. In fact, using (2), the heavy down-like quark (called, in the following, "superbeauty" or sb) would have a mass in the 50 GeV/c^2 range:

$$m(\text{down-heavy}) = m(\text{"superbeauty"}) \cong 10 \times 5.5 \cong 55 \text{ GeV}/c^2 ,$$

and, using (1), the heavy up-like quark would have a mass in the 200 GeV/c^2 range:

$$m(\text{up-heavy}) = m(\text{"supertruth"}) \cong 55 \times 4 \cong 220 \text{ GeV}/c^2 .$$

The up-like quark of the 4th family should be called "supertruth". In fact, this very heavy mass is wanted by Supersymmetry; moreover, if Supersymmetry is a good theory, the 4th family should really be the last of the quark families, ever to be discovered.

The four families are shown in the next graph, where the main objectives of this first part of my talk are indicated by the dotted circles.

Families:	1st	2nd	3rd	4th
Quarks:	$\begin{pmatrix} u \\ d \end{pmatrix}$	$\begin{pmatrix} c \\ s \end{pmatrix}$	$\begin{pmatrix} t \\ b \end{pmatrix}$	$\begin{pmatrix} u_H \\ d_H \end{pmatrix}$
Leptons:	$\begin{pmatrix} \nu_e \\ e \end{pmatrix}$	$\begin{pmatrix} \nu_\mu \\ \mu \end{pmatrix}$	$\begin{pmatrix} \nu_\tau \\ \tau \end{pmatrix}$	$\begin{pmatrix} \nu_H \\ L_H \end{pmatrix}$

Let us come to a key question: are "top" and "superbeauty" accessible to ($p\bar{p}$) Collider energies? If yes, how can they be detected?

Many methods have been suggested:

i) detection of a hidden state with the study of the invariant mass of the lepton pairs;
ii) detection of an open state with the identification of hadronic decay channels;

iii) study of multilepton events;
iv) study of the inclusive transverse momentum spectrum of the leptons from semileptonic decays;
v) study of the transverse dilepton (l, ν) and jets masses.

All of these methods present, in various degrees, experimental problems related to small production cross sections, low global branching ratios, high background levels, poor experimental resolution of the quantities needed to be measured. Moreover, most of them have big troubles in being able to identify the up-like or down-like nature of the new flavors.

We present here a new method to observe the production of heavy mass states, either up-like ("top") or down-like ("superbeauty") which is based on the "Leading" baryon production mechanism, extended to the heaviest baryon and antibaryon states. In fact, due to this production mechanism, a charge asymmetry of the leptons, originating from these heavy flavors can be observed in a selected region of the phase space. Moreover, this asymmetry will show an energy dependence characteristic of the masses of the decaying states.

More precisely, the "top" baryonic state will decay semileptonically into ℓ^+ and produce a positive asymmetry in the outgoing proton rapidity hemisphere

$$A_p = (\ell^+ - \ell^-)/(\ell^+ + \ell^-) = \text{positive.}$$

The "anti-top" antibaryonic state will produce a negative asymmetry in the outgoing antiproton rapidity hemisphere

$$A_{\bar{p}} = (\ell^+ - \ell^-)/(\ell^+ + \ell^-) = \text{negative.}$$

The signs of these asymmetries will be reversed for the "superbeauty" case.

The energy range where to measure the $\ell^\pm$ asymmetry, and the separation between the maxima, depend on the parent-daughter mass difference in the decay of the two new flavors. Extending the validity of the generalized Cabibbo mixing (GCM) to the 4th family, the mass differences in the superbeauty and top decays would be:

$$\Delta m = m(\text{top}) - m(\text{beauty}) \cong 20\ \text{GeV}/c^2,$$

$$\Delta m = m(\text{superbeauty}) - m(\text{top}) \cong 30\ \text{GeV}/c^2.$$

The $\ell^\pm$ transverse momentum spectra associated with "top" and "superbeauty" will be quite different because:

$m(\text{"top"}) \cong 25\ \text{GeV}/c^2$,
$m(\text{"superbeauty"}) \cong 55\ \text{GeV}/c^2$,

thus the Asymmetry will change sign with increasing lepton energy. Figure 2 shows the main trend of the measurement we propose. Notice that this GCM condition, if not valid, would not spoil our method. It would only shift the energy spectrum of e^- from superbeauty decay to higher values.

As we will see in the following, the amplitude of the effect depends on:

i) production cross sections;
ii) production and decay, angular and momentum distributions;
iii) branching ratios into semileptonic channels;
iv) luminosity;
v) acceptance and rejection power of the experimental set-up designed to observe the leptons produced by these "new" flavors decay.

2. A BRIEF REVIEW OF HEAVY FLAVORS PRODUCTION IN HADRONIC MACHINES

How to look for "top" and "superbeauty" at the ($p\bar{p}$) Collider, is a problem analogous to "charm" and "beauty" in hadronic machines and,

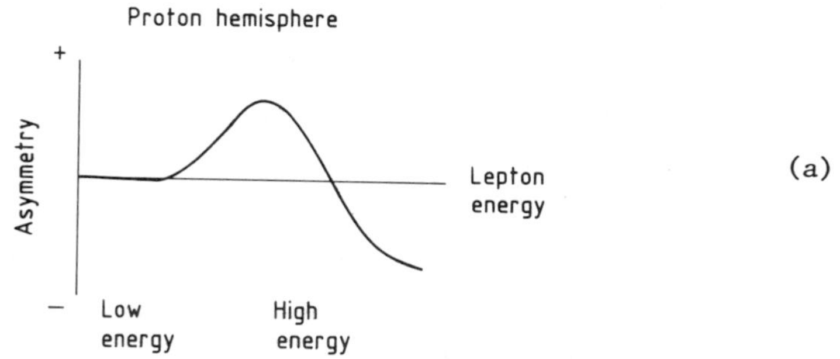

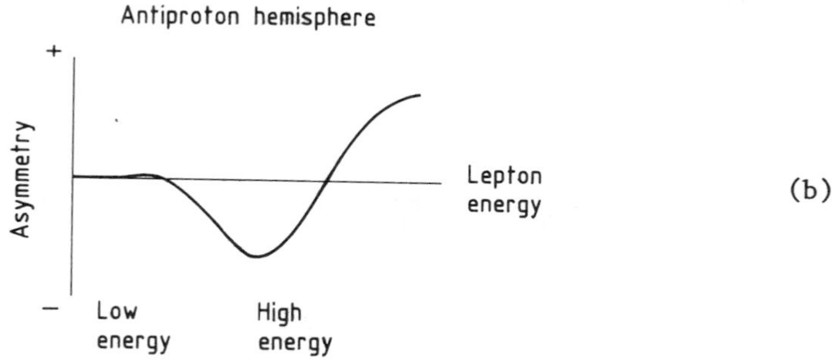

Fig. 2. Main trend of the electron charge Asymmetry in the proton hemisphere (a) and in the antiproton hemisphere (b).

in particular, at the ISR. It is probably instructive to review, very briefly, the main steps in this field.

2.1. Production Cross Sections

The theoretical predictions for the charm production cross section and the experimental findings are shown in Table 1.

If we were to believe in the string model or in the statistical thermodynamical model, or in the first QCD attempts (fusion model), the conclusion should have been that the production and observation of "charm" is out of question in hadronic machines.

Only recently QCD models (flavor excitation) came nearer to experimental findings. In Figure 3 all the experimental data are shown and compared with the various steps in the QCD models. Notice that neither $\ln(s)$ nor $\ln^2(s)$ are compatible with the observed threshold behavior.

2.2. The "Leading" Effect

A result which was theoretically unpredicted is the "Leading" effect which shows up in the production of heavy flavors.

A detailed study of (pp) interactions at the ISR showed that the Λ_c^+ is produced in a "Leading" way[7].

After this experimental result was obtained, a series of theoretical proposals were presented, to account for the "Leading" Λ_c^+ production. The longitudinal momentum distribution[7] for Λ_c^+ was in fact found at the ISR to be:

$$(d\sigma/d|x|) \sim (1 - |x|)^{\alpha} \quad \text{with } \alpha = 0.40 \pm 0.25.$$

Table 1. Charm Cross Section at ISR Energies ($\sigma_\pi = 10^2$mb).

String model	: 10^{-10}	x σ_π
Fermi-Hagedorn	: 10^{-5}	x σ_π
QCD (Fusion)	: 10^{-4}	x σ_π
QCD (Flavor excitation)	: 10^{-2}-10^{-3}	x σ_π
Experimentally	: $= 10^{-2}$	x σ_π

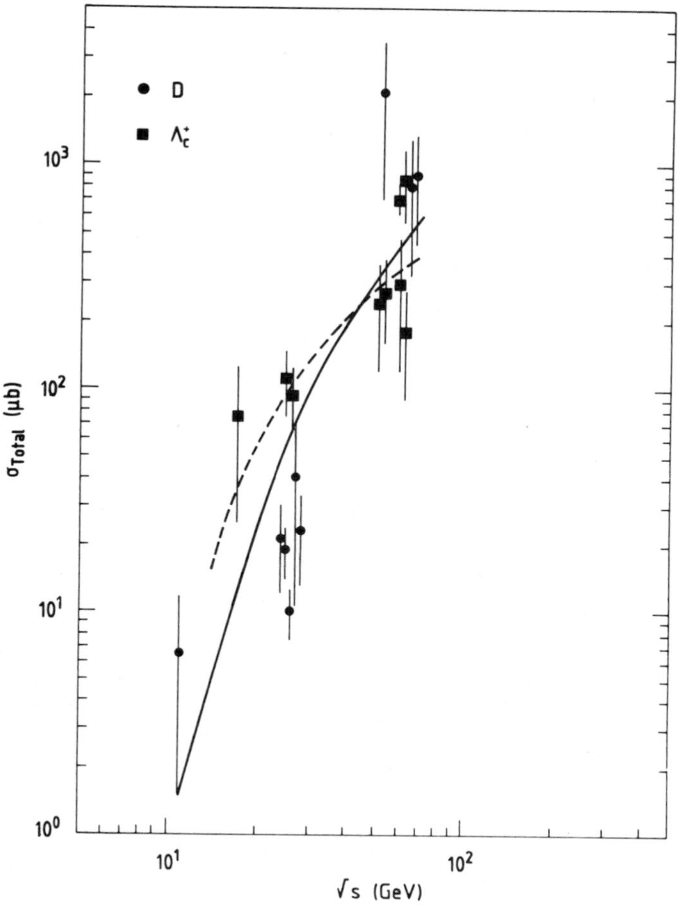

Fig. 3. Cross sections expected on the basis of gluon and quark fusion models ($gg \rightarrow c\bar{c}$ and $q\bar{q} \rightarrow c\bar{c}$ curves). Other QCD models are also shown. The data are taken from 3, 4, 5 and 6.

The results are shown in Figure 4a. The charmed meson production[8] was on the other hand measured to be "non-Leading", i.e.

$$E(d\sigma/d|x|) \simeq (1 - |x|)^{\alpha} \quad \text{with } \alpha \sim 3.$$

This can be "a posteriori" qualitatively understood in terms of the Λ_c^+ obtained by a recombination of the spectator c-quark with a valence (ud) pair in the proton; while the $\bar{D}$ production is given by the recombination of the spectator $\bar{c}$-quark with a most one valence quark[9,10].

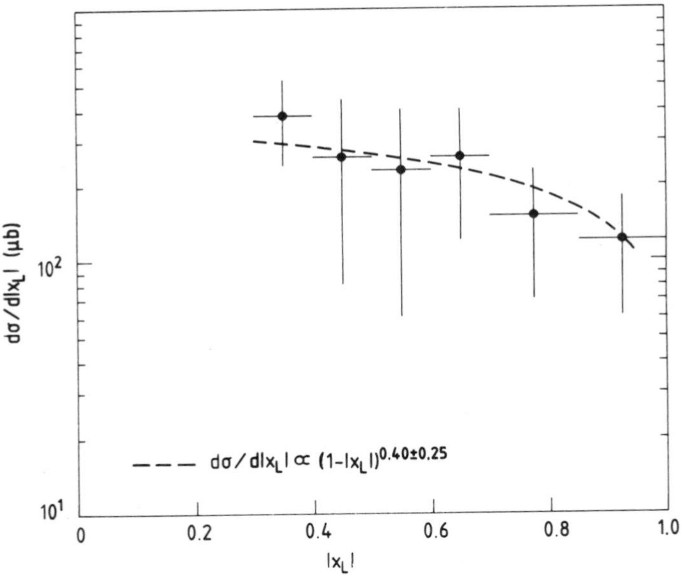

Fig. 4a. Experimental longitudinal momentum distribution of Λ_c^+.

Figure 4b shows the qualitative behavior, as a function of the quark mass, of the quantity (Leading/Total), which will be defined in section 5.4 as the ratio between the inclusive cross-section for producing leading baryons in one hemisphere and the total cross-section for producing $(q\bar{q})$ pairs. This figure shows that, at least at the ISR energies, the "Leading" production increases with increasing quark mass.

A more complete summary of charm production in purely hadronic interactions is reported in the Table 2. There is no model which can fit all measured quantities.

2.3. Further Comments

For those who have strong faith on QCD it could be interesting to extend our review. In fact the photoproduction was considered a simpler case for QCD. Therefore its predictions should have been in agreement with experimental findings.

A summary of QCD problems in photoproduction physics is as follows:

a) large photoproduction cross sections of the Heavy Flavors are impossible to be predicted by perturbative QCD;
b) the p_T dependence of inelastic $(c\bar{c})$, for open and hidden states, cannot be accounted for by QCD;
c) the A-dependence cannot be A^1.

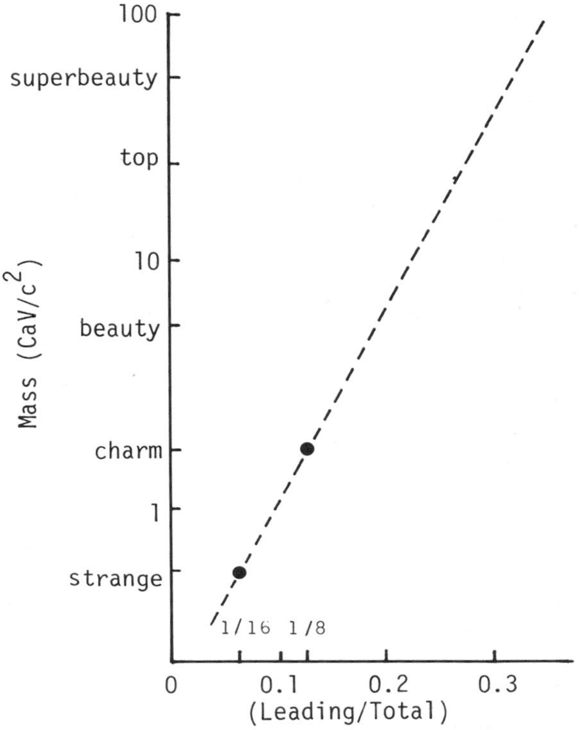

Fig. 4b. Qualitative behavior of the quantity (Leading/Total) as a function of the quark mass, at $\sqrt{s}$ = 62 GeV.

Table 2.

		Models		
	Experiment	Diffractive	Flavor excitation	Fusion
Leading effect	Yes	Yes	Yes	No
Threshold behavior	Steeper than $\ln^2 s$	lns	Steeper than $\ln^2 s$	>> Steeper than $\ln^2 2$
Mass dependence	?	$1/m^2$	Stronger than $1/m^2$	>> Stronger than $1/m^2$
Cross section	Large	Large	Large	Small
A^α dependence	α <2/3*	α = 2/3	α = 1	α = ...

*The p_T dependence is derived from data on strangeness.

2.4. Conclusions

The conclusion of this short review on the "charm" flavor production in (pp) interactions is therefore:

i) the cross section values found are at least an order of magnitude above the "theoretical" predictions of perturbative QCD;

ii) the x-distribution for Λ_c^+, i.e. the "Leading" effect, was theoretically unpredicted;

iii) with "new" models (essentially flavor excitation[11,10] and non-perturbative QCD)[9] both cross sections values and x-distributions can be "theoretically" derived.

All this should be quite a warning for QCD prediction on Heavy Flavors production at extreme energies such as those of the $(p\bar{p})$ CERN Collider.

3.2. EXPECTED NEW HEAVY FLAVORED STATES

The main purpose of this section is to call attention on the enormous number of new states which are expected on the basis of the old and new flavors.

3.1. Examples from Previous Experience with $SU(3)_{uds}$ and $SU(4)_{udsc}$

The following graph illustrates what could indeed happen.

$$
SU(2)_{ud} \longrightarrow
\begin{cases}
s \to SU(3)_{uds} \longrightarrow
\begin{cases}
c \to SU(4)_{udsc} \\
t \to SU(4)_{udst} \\
st \to SU(4)_{uds(st)}
\end{cases} \\[2em]
b \to SU(3)_{udb} \longrightarrow
\begin{cases}
c \to SU(4)_{udbc} \\
t \to SU(4)_{udbt} \\
st \to SU(4)_{udb(st)}
\end{cases} \\[2em]
sb \to SU(3)_{ud(sb)} \to
\begin{cases}
c \to SU(4)_{ud(sb)c} \\
t \to SU(4)_{ud(sb)t} \\
st \to SU(4)_{ud(sb)(st)}
\end{cases}
\end{cases}
$$

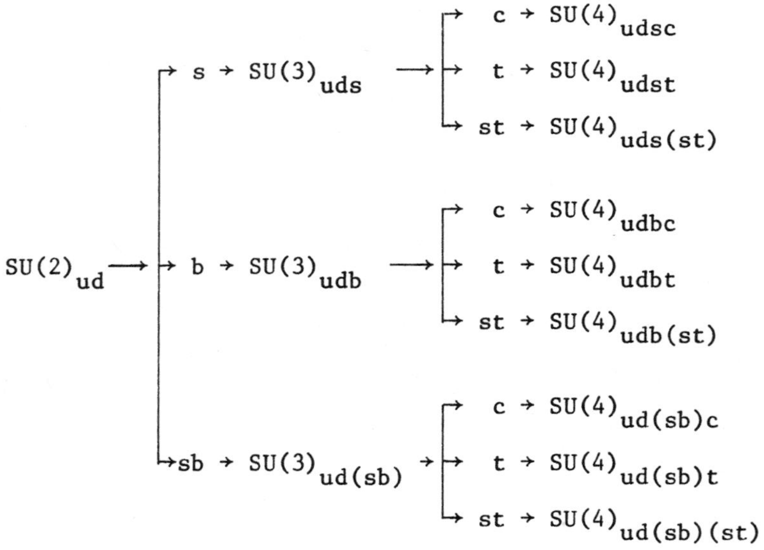

With 3 flavors (u, d and s) the famous $SU(3)_{uds}$ came out. It could be that "beauty" will produce another $SU(3)_{udb}$. The advent of the "charm" with four flavors (udsc) produced $SU(4)_{udsc}$. On the other hand with the "top" there are two possible SU(4): $SU(4)_{udst}$ and $SU(4)_{udbt}$.

Despite the large mass differences among the various flavors, it could be that Nature will provide, as usual, more regularities than wanted. The above global symmetry groups for the structure of the various possible particle states could eventually show up, even if not expected.

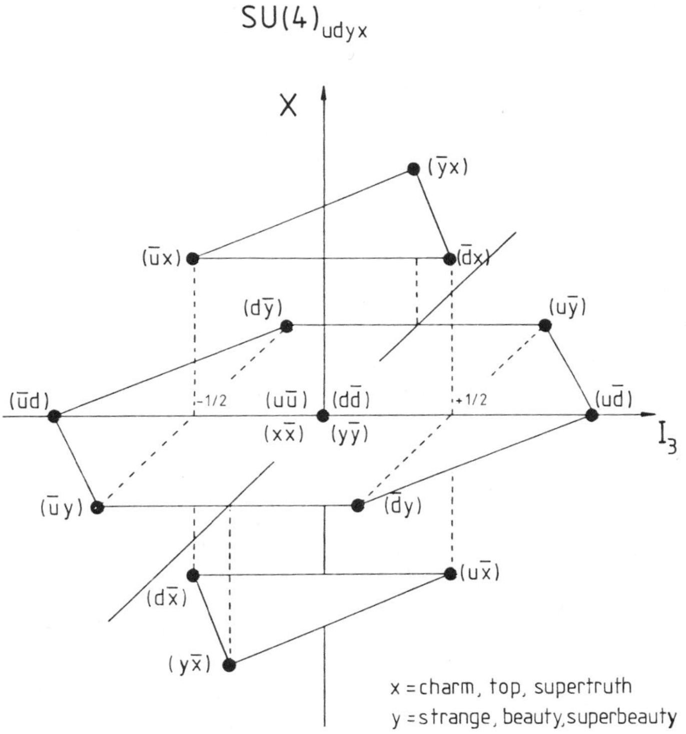

Fig. 5. The $SU(4)_f$ mesonic multiplets for $J^P = 0^-$; the same multiplet structure repeats for $J^P=1^-$. "x" represents the quarks with electric charge + 2/3, "y" the quarks with electric charge - 1/3. Each of the possible udyx combinations should produce an $SU(4)_f$. The quark composition for each state is indicated in parenthesis.

If "superbeauty" was there we would have even a larger set of SU(3) and of SU(4) states, in addition to all combination of purely singlets like (s, c, b, t and sb).

Examples of SU(4) structures using (u, d, s, c, b, sb and super-truth) are presented in Figures 5 to 7. They include mesonic and baryonic states.

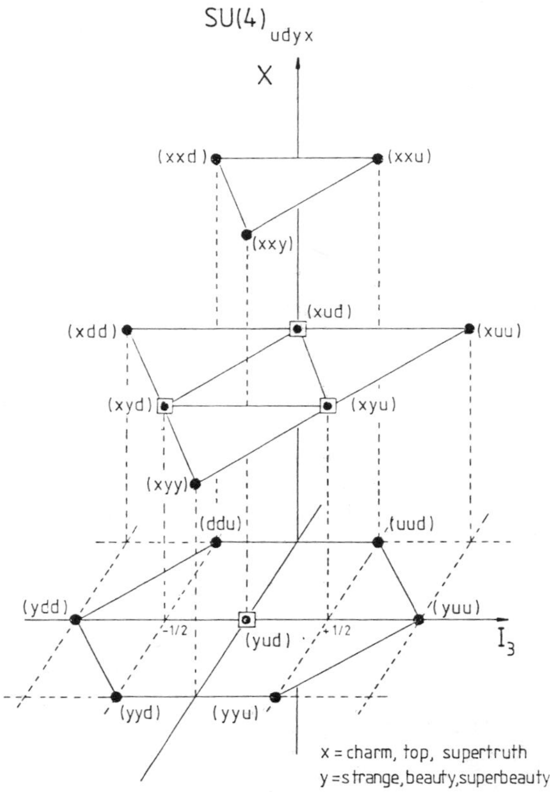

Fig. 6. Showing the structure of the expected baryonic $J^P=1/2^+$ $SU(4)_f$ 20-plets.

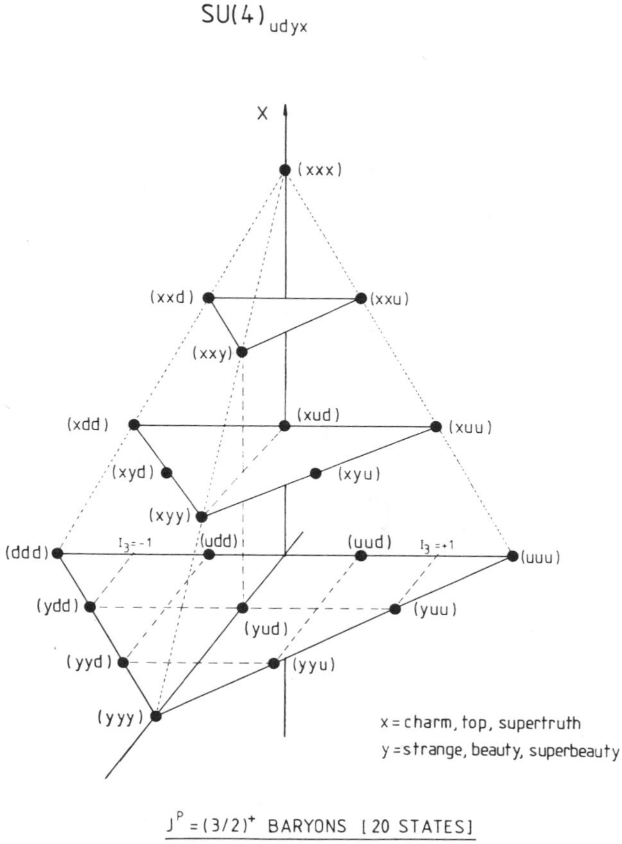

Fig. 7. Showing the structure of the expected baryonic $J^P = 3/2^+$ $SU(4)_f$ 20-plets.

If we enlarge the symmetries to the intrinsic spins the multitude of states increases further. These are indicated, for the first 3 flavors (u, d and s), in the Tables 3 and 4, respectively for the mesons and for the baryons in SU(6).

An example of how the multitude of the states goes with the mass is shown in Figures 8a and 8b, where the masses of the particles run from few GeV/c^2 up to 60 GeV/c^2.

Table 3. SU(6) Mesonic Supermultiplets

SU(6)	$SU(3)_f$	J^{PC}	Particle states	Number of states
$[(35 \oplus 1) \otimes 1]$; (L=0)	$8 \oplus 1$	0^{-+}	π, K, η, η'	36
	$8 \oplus 1$	1^{--}	ρ, K*, ω, ϕ	
$[(35 \oplus 1) \otimes 3]$; (L=1)	$8 \oplus 1$	1^{+-}	B, $Q_{1,2,\ldots,?}$	108
	$8 \oplus 1$	0^{++}	S, χ, S*, ε	
	$8 \oplus 1$	1^{++}	A_1, $Q_{1,2}$, D, E	
	$8 \oplus 1$	2^{++}	A_2, K**, f, f'	

Table 4. Baryons in SU(6) Supermultiplets

$[Su(6), L^P]$	$SU(3)_f$	J^P	Standard names of particle states
$(56,0^+)$	8	$1/2^+$	N, Λ, Σ, Ξ^-
	10	$3/2^+$	N*, Σ*, Ξ*, Ω^-
$(70,1^-)$	1	$1/2^-$	Repeat singlet
	8	$1/2^-$	Repeat octet
	10	$1/2^-$	Repeat decuplet
	1	$3/2^-$	Repeat singlet
	8	$3/2^-$	Repeat octet
	10	$3/2^-$	Repeat decuplet
	8	$1/2^-$	Repeat octet
	8	$3/2^-$	Repeat octet
	8	$5/2^-$	Repeat octet

Repeat means that the quantum numbers (isospin and strangeness) of the states are identical to the "octet" and "decuplet" already known for the 56-case.

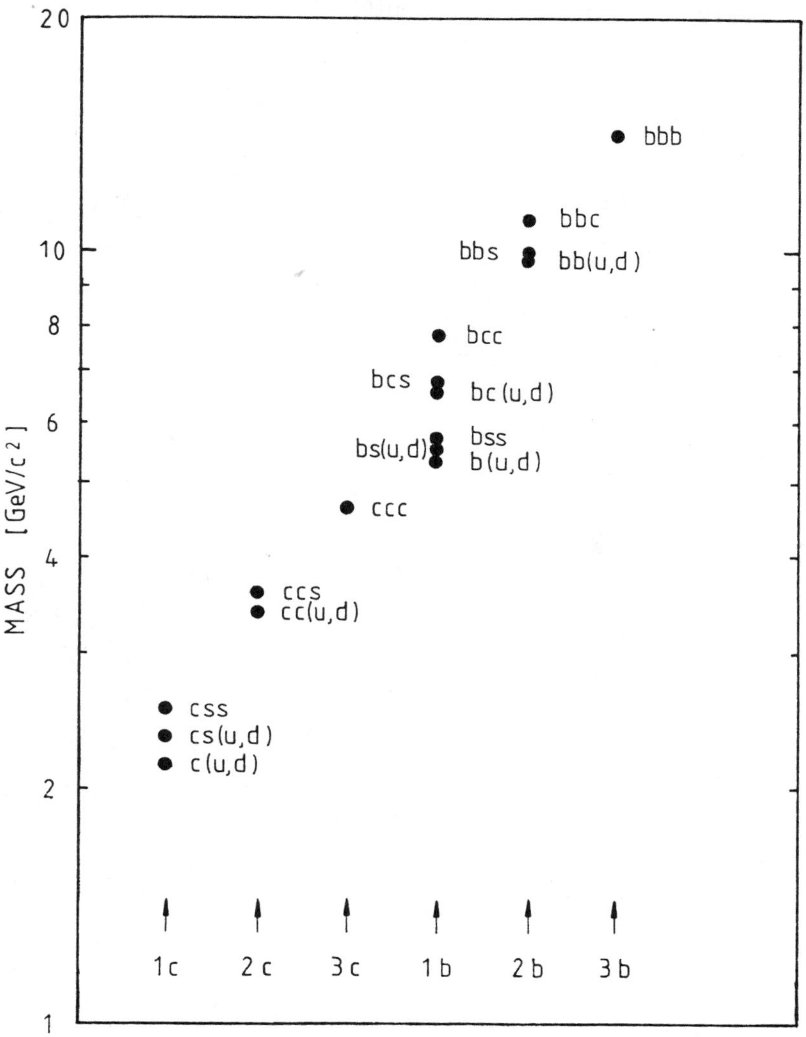

Fig. 8a. The figure shows the mass ranges of the baryon states with: (a) from 1 to 3 charm or beauty quarks; (b) from 1 to 3 top quarks.

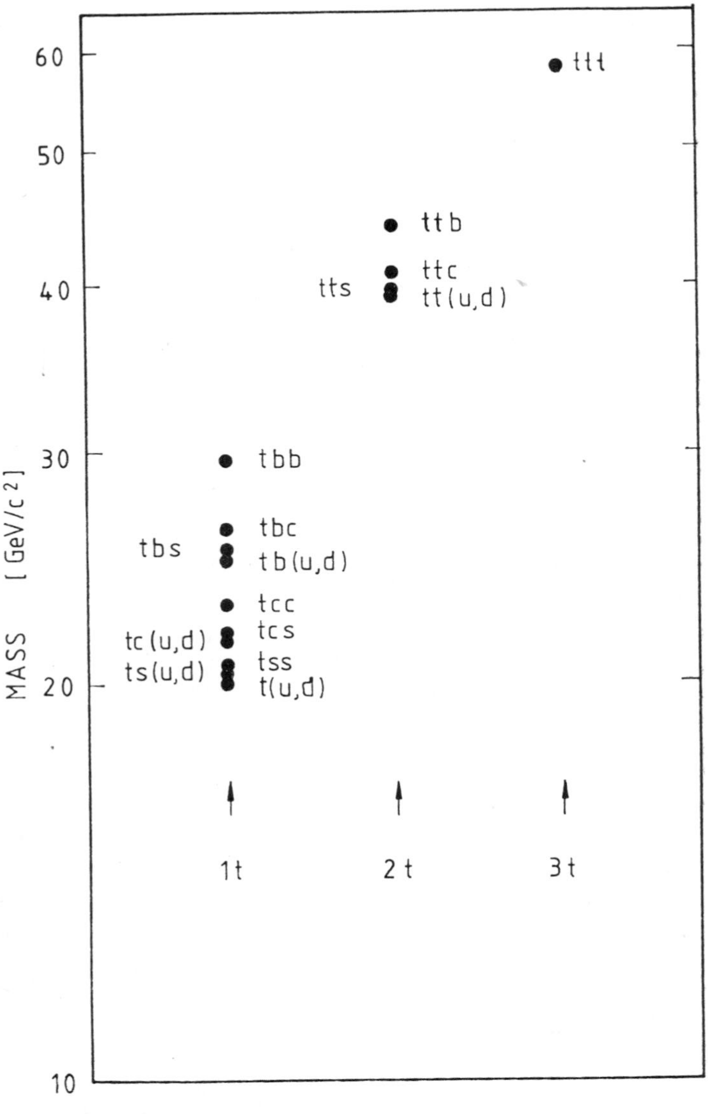

Fig. 8b. The figure shows the mass ranges of the baryon states with: (a) from 1 to 3 charm or beauty quarks; (b) from 1 to 3 top quarks.

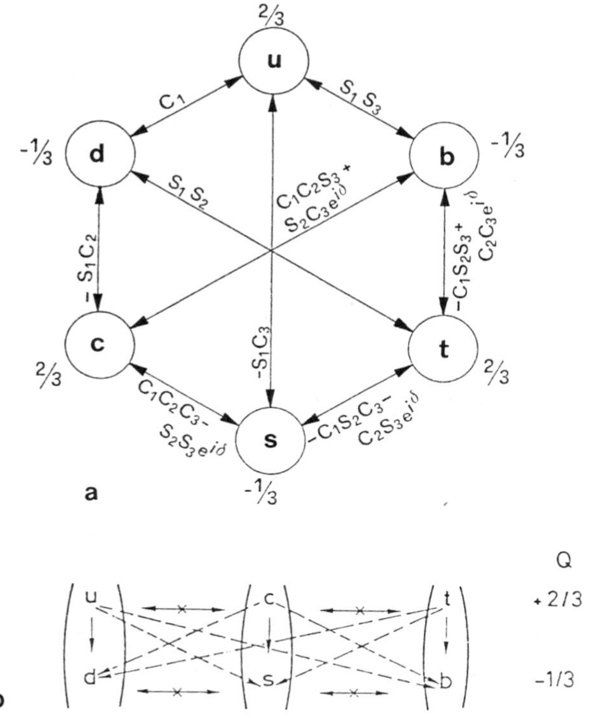

Fig. 9. (a) The six quark mixing with CP violation. $S_i = \sin\theta_i$, $C_i = \cos\theta_i$. (b) Transitions among the various states. The Cabibbo mixing opens the dashed channels. The horizontal transitions are forbidden for any value of the mixing angle. Allowed neutral currents are: $u\bar{u}$, $c\bar{c}$, $t\bar{t}$, $d\bar{d}$, $s\bar{s}$ and $b\bar{b}$.

3.2. Note on the Semi-Leptonic Decay Modes: Generalized Cabibbo Mixing

A fact of Nature is that the matrix which relates the down-like "weak" flavors "Cabibbo mixed"

$$\begin{pmatrix} d_c \\ s_c \\ b_c \end{pmatrix}$$

to the "strong" flavors

$$\begin{pmatrix} d \\ s \\ b \end{pmatrix}$$

is approximately a unit matrix

$$\begin{pmatrix} d_c \\ s_c \\ b_c \end{pmatrix} \cong \begin{pmatrix} 1 & 0 & 0 \\ 0 & 1 & 0 \\ 0 & 0 & 1 \end{pmatrix} \begin{pmatrix} d \\ s \\ b \end{pmatrix}$$

as shown in Figures 9a and 9b.

In order to extend the generalized Cabibbo mixing to the 4th Family, we make the following extrapolations:

i) all the generalized Cabbibbo angles, even those coming from the existence of the 4th family are small;
ii) the flavor-changing neutral currents are forbidden to any order of family;
ii) the amplitude for the transition from family N to family $N \pm \alpha$ has a coefficient

$$\prod_{i=1,\alpha} (\sin\theta_i) \text{ .}$$

As a consequence, the Cabibbo-favored decay chains of flavors c, b, t and sb are:

$c \to s$

$b \to c \to s$

$t \to b \to c \to s$

$sb \to t \to b \to c \to s.$

All we need to know is the charge sign of the lepton in a transition from an "up-like" to a "down-like" flavor and viceversa. This can be summarized as follows:

$$\begin{pmatrix} u \\ d \end{pmatrix} \begin{pmatrix} c \\ s \end{pmatrix} \begin{pmatrix} t \\ b \end{pmatrix} \begin{pmatrix} st \\ sb \end{pmatrix}$$

$$\Downarrow \quad \Downarrow \quad \Downarrow \quad \Downarrow$$

$$\begin{pmatrix} 1 \\ 2 \end{pmatrix} \begin{pmatrix} 3 \\ 4 \end{pmatrix} \begin{pmatrix} 5 \\ 6 \end{pmatrix} \begin{pmatrix} 7 \\ 8 \end{pmatrix} \quad \begin{matrix} \leftarrow \text{ODD (= UP-LIKE) QUARKS} \\ \leftarrow \text{EVEN (= DOWN-LIKE) QUARKS} \end{matrix}$$

with the charge formula written as:

$$Q = (1/3 + f_i)/2$$

with:

$f_i = +1$ for odd quarks (i = 1, 3, 5, 7)

$f_i = -1$ for even quarks (i = 2, 4, 6, 8).

From this follows the electric charge sign of the lepton in the semileptonic decay of the flavor:

ODD $\longrightarrow$ EVEN (UP-LIKE $\longrightarrow$ DOWN-LIKE)

TRANSITION $\Longrightarrow$ POSITIVE LEPTON

EVEN $\longrightarrow$ ODD (DOWN-LIKE UP-LIKE) TRANSITION $\Longrightarrow$

$\Longrightarrow$ NEGATIVE LEPTON

As will be seen in the Figure 15, a sequence $t \to b \to c \to s$ will be accompanied by the semileptonic series giving rise to $e^+ \to e^- \to e^+$. For the antiquark sequence $\bar{t} \to \bar{b} \to \bar{c} \to \bar{s}$, the charges will be reversed ($e^- \to e^+ \to e^-$). These results are straightforward consequences of the previous table.

4. CROSS SECTION ESTIMATES: HOW TO GO FROM STRANGENESS TO CHARM, BEAUTY, TOP AND SUPERBEAUTY

Now comes a key question: once we know the "strange" and "charm" cross sections, is it possible to predict the heavier flavors (c, b, t or sb) cross sections in hadronic collisions?

Simple arguments bring to the conclusion that:

$$\sigma(m) \sim (1/m^2) \times f(s/m^2). \qquad (3)$$

In fact, the only quantities which enter in the problem of producing a $(q\bar{q})$ pair, having at disposal the total energy $\sqrt{s}$, are the quark mass, m, and the total energy, $\sqrt{s}$.

Formula (3) is based on dimensional and scaling arguments:

- Dimension says that: $\sigma \sim (1/m^2)$;
- Scaling says that the two quantities m^2 and s are such that nothing changes if their ratio (s/m^2) is kept constant; the ratio (s/m^2) is the dimensionless quantity needed if no other scale should remain in the game.

The basic formula is therefore:

$$\sigma_i[(\sqrt{s})_{pp} = E_i] = (m_j/m_i)^2 \times \sigma_j[(\sqrt{s})_{pp} = E_j = (m_j/m_i)E_i] \quad (4)$$

where:

σ_i, σ_j are the production cross sections,

m_i, m_j are the masses,

E_i, E_j are the energies,

at which flavors f_i and f_j are produced.

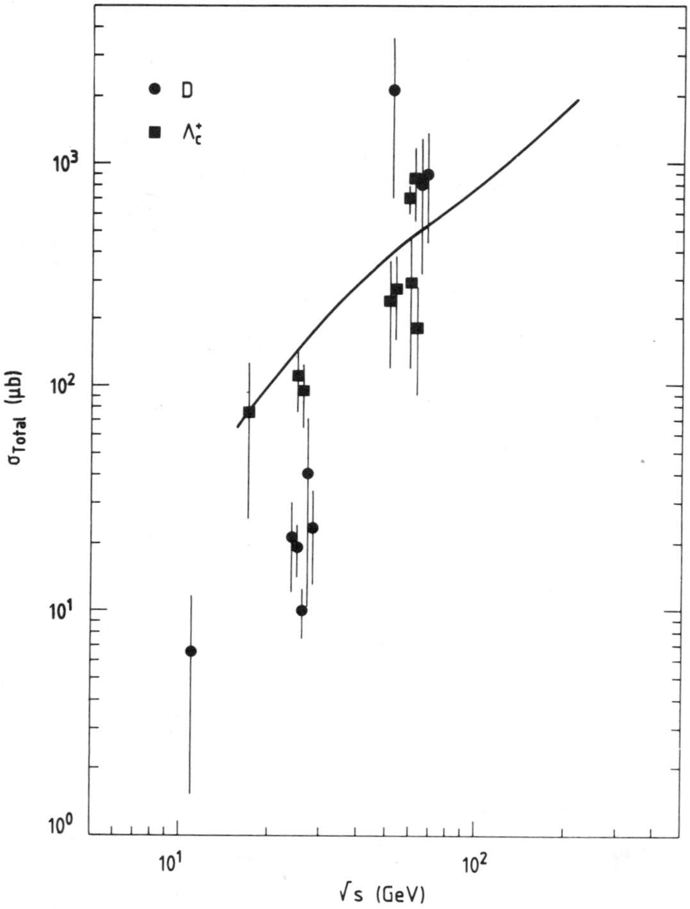

Fig. 10. Charm cross section derived from strange cross section following formula (4).

The results are shown in Figures 10-13, where we have used:

i) the strangeness data to predict c, b, t and sb;
ii) the "charm" data to predict b, t and sb;
iii) the "beauty" data to predict t and sb.

Finally, for completeness, we also report the most recent QCD predictions of references 10 and 9 (Figure 14).

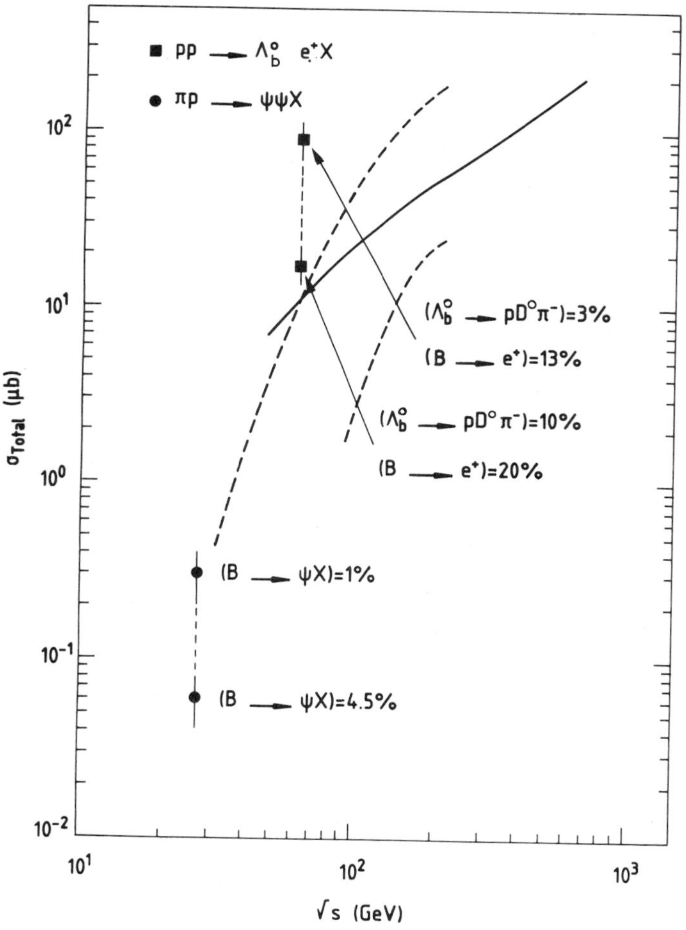

Fig. 11. Beauty cross section derived from strange (full line) and charm (dashed lines - notice the width due to the experimental uncertainties) cross sections following formula (4). The data are taken from references 12 and 13.

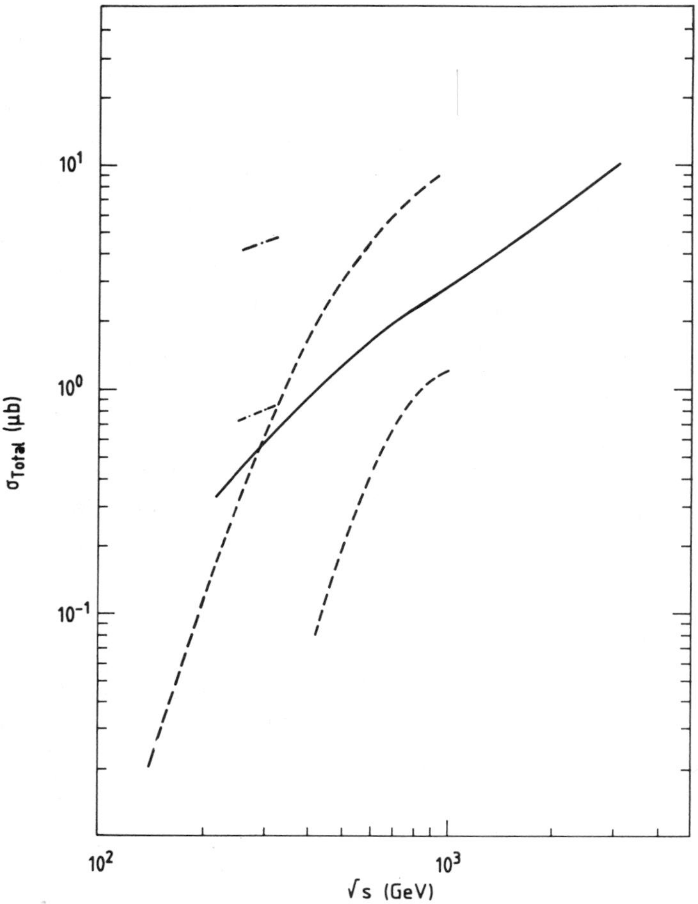

Fig. 12. Top cross section derived from strange (full line), charm (dashed lines) and beauty (dash/point lines) cross sections following formula (4).

5. THE STUDY OF THE LEPTON CHARGE ASYMMETRY AND ITS ENERGY DEPENDENCE AS A WAY TO DETECT THE HEAVIEST FLAVORED STATES (BARYONIC AND ANTIBARYONIC) AT THE ($p\bar{p}$) COLLIDER

The leptonic decay chains, following the generalized Cabibbo dominance, for the various flavors c, b, t, sb, are shown in Figure 15.

Once a particle-antiparticle pair has been produced, on the average the number of positive and negative leptons from its decay is equal. However we will discuss under which conditions an asymmetry in the number of positive and negative leptons can be observed, due

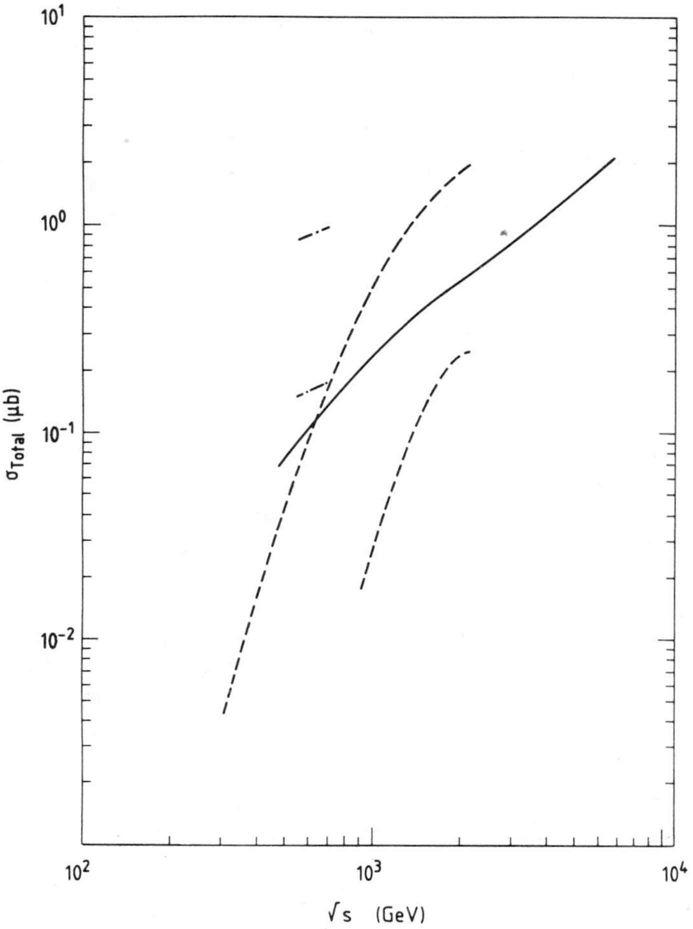

Fig. 13. Superbeauty cross section derived from strange (full line), charm (dashed line) and beauty (dash/point lines) cross sections following formula (4).

to the different longitudinal momentum production distribution for baryons and mesons, and to the dependence of the lepton p_T spectra from the product particle mass.

Let us define the Asymmetry parameters as

$$A^\circ(p_T, \theta_{cut}) = \frac{N(\ell^+) - N(\ell^-)}{N(\ell^+) + N(\ell^-)} \tag{5}$$

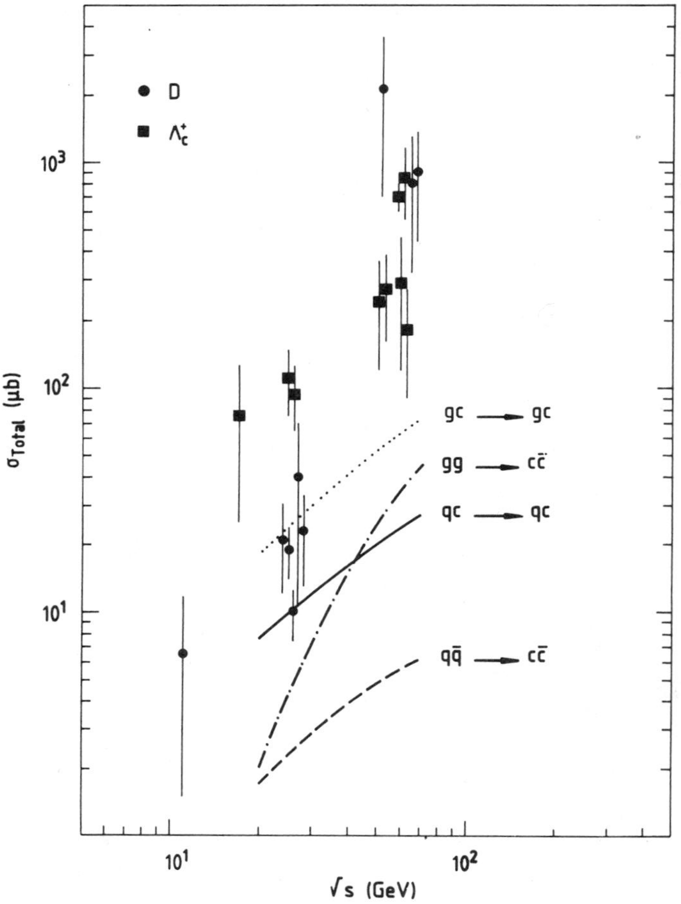

Fig. 14. Non perturbative QCD predictions (dashed line) and flavor excitation perturbative QCD predictions (full line) for charm hadroproduction.

where $N(\ell^+) \equiv N(\ell^+; p_T, \theta_{cut})$ and $N(\ell^-) \equiv N(\ell^-; p_T, \theta_{cut})$ are the number of positive and negative leptons produced in the angular range $0° < \theta < \theta_{cut}$ and with transverse momentum p_T.

The number of positive leptons ℓ^+ is expressed by

$$N(\ell^+) = L[n_{sb}(\ell^+) + n_t(\ell^+) + n_b(\ell^+) + n_c(\ell^+)]$$

where L is the total integrated luminosity and $n_f(\ell^+)$ (with f = sb, t, b, c) is the contribution from the direct production of sb, t, b, c states.

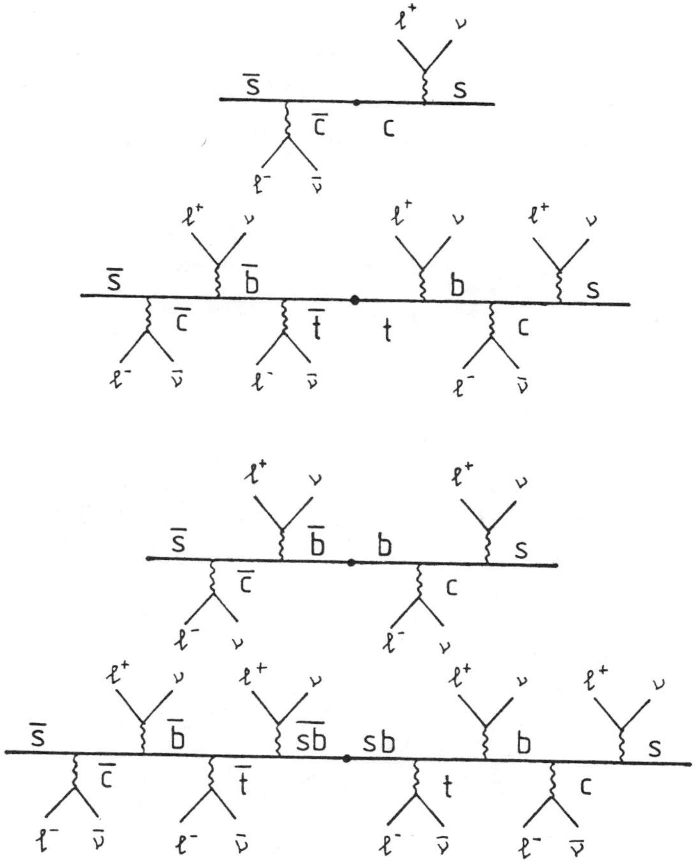

Fig. 15. Diagram illustrating all the possible electric charge signs of the electrons originating from the semileptonic decay of the quarks c, b, t and sb.

Analogously the number of negative leptons is given by

$$N(\ell^-) = L[(n_{sb}(\ell^-) + n_t(\ell^-) + n_b(\ell^-) + n_c(\ell^-)].$$

The leptons originated by the decay of the various flavors and antiflavors are summarized in Tables 4 and 5.

In order to write down explicity $n_f(\ell^+)$ let us define:

i) $\sigma_f^T \equiv$ total cross section for the production of open $(f, \bar{f})$ pairs;

ii) ρ_{Mf}, $\rho_{\bar{M}\bar{f}}$, ρ_{Bf}, $\rho_{\bar{B}\bar{f}}$ ≡ ratio between, the inclusive cross-section for producing (M = meson, $\bar{M}$ = antimeson, B = baryon, $\bar{B}$ = antibaryon) states, and the total cross section σ_f^T;

iii) $BR_{Mf'}$, $BR_{M\bar{f}'}$, $BR_{Bf'}$, $BR_{B\bar{f}'}$ ≡ semileptonic branching ratio of the various states with flavor f' (f' = c, b, t, sb);

iv) $\varepsilon_{Mf}(\ell^{\pm}_{f'})$, $\varepsilon_{\bar{M}\bar{f}}(\ell^{\pm}_{f'})$, $\varepsilon_{Bf}(\ell^{\pm}_{f'})$, $\varepsilon_{\bar{B}\bar{f}}(\ell^{\pm}_{f'})$ ≡ acceptance for $\ell^{\pm}$ from the leptonic decay of the flavor f' produced in the decay chain of the state with flavor f. This acceptance is a function of the lepton p_T and of the cut in $\theta < \theta_{cut}$ applied to the lepton polar angle.

Accordingly we have, for the case of "superbeauty":

$$n_{sb}(\ell^+) = \sigma^T_{sb} \left\{ \sigma_{Msb} \left[BR_{Mt}\varepsilon_{Msb}(\ell^+_t) + BR_{Mc}\varepsilon_{Msb}(\ell^+_c) \right] + \right.$$
$$+ \rho_{\bar{M}\bar{s}\bar{b}} \left[BR_{\bar{M}\bar{s}\bar{b}}\varepsilon_{\bar{M}\bar{s}\bar{b}}(\ell^+_{\bar{s}\bar{b}}) + BR_{\bar{M}\bar{b}}\varepsilon_{\bar{M}\bar{s}\bar{b}}(\ell^+_b) \right] +$$
$$+ \rho_{Bsb} \left[BR_{Bt}\varepsilon_{Bsb}(\ell^+_t) + BR_{Bc}\varepsilon_{Bsb}(\ell^+_c) \right] +$$
$$\left. + \rho_{\bar{B}\bar{s}\bar{b}} \left[BR_{\bar{B}\bar{s}\bar{b}}\varepsilon_{\bar{B}\bar{s}\bar{b}}(\ell^+_{\bar{s}\bar{b}}) + BR_{\bar{B}\bar{b}}\varepsilon_{\bar{B}\bar{s}\bar{b}}(\ell^+_b) \right] \right\}$$

and

$$n_{sb}(\ell^-) = \sigma^T_{sb} \left\{ \rho_{\bar{M}\bar{s}\bar{b}} \left[BR_{\bar{M}\bar{t}}\varepsilon_{\bar{M}\bar{s}\bar{b}}(\ell^-_{\bar{t}}) + BR_{\bar{M}\bar{c}}\varepsilon_{\bar{M}\bar{s}\bar{b}}(\ell^-_{\bar{c}}) \right] + \right.$$
$$+ \rho_{Msb} \left[BR_{Msb}\varepsilon_{Msb}(\ell^-_{sb}) + BR_{Mb}\varepsilon_{Msb}(\ell^-_b) \right] +$$
$$+ \rho_{\bar{B}\bar{s}\bar{b}} \left[BR_{\bar{B}\bar{t}}\varepsilon_{\bar{B}\bar{s}\bar{b}}(\ell^-_{\bar{t}}) + BR_{\bar{B}\bar{c}}\varepsilon_{\bar{B}\bar{s}\bar{b}}(\ell^-_{\bar{c}}) \right] +$$
$$\left. + \rho_{Bsb} \left[BR_{Bsb}\varepsilon_{Bsb}(\ell^-_{sb}) + BR_{Bb}\varepsilon_{Bsb}(\ell^-_b) \right] \right\} .$$

Table 4

Flavor	Decays producing ℓ^+	Decays producing ℓ^-
sb	$sb \to t \to b + \ell^+$ $sb \to t \to b \to c \to s \to + \ell^+$	$sb \to t + \ell^-$ $sb \to t \to b \to c + \ell^-$
t	$t \to b + \ell^+$ $t \to b \to c \to s + \ell^+$	$t \to b \to c + \ell^-$
b	$b \to c \to s + \ell^+$	$b \to c + \ell^-$
c	$c \to s + \ell^+$	

Table 5

Antiflavor	Decays producing ℓ^+	Decays producing ℓ^-
$\overline{sb}$	$\overline{sb} \to \bar{t} + \ell^+$	$\overline{sb} \to \bar{t} \to \bar{b} + \ell^-$
	$\overline{sb} \to \bar{t} \to \bar{b} \to \bar{c} + \ell^+$	$\overline{sb} \to \bar{t} \to \bar{b} \to \bar{c} \to \bar{s} + \ell^-$
$\bar{t}$	$\bar{t} \to \bar{b} \to \bar{c} + \ell^+$	$\bar{t} \to \bar{b} + \ell^-$
		$\bar{t} \to \bar{b} \to \bar{c} \to \bar{s} + \ell^-$
$\bar{b}$	$\bar{b} \to \bar{c} + \ell^+$	$\bar{b} \to \bar{c} \to \bar{s} + \ell^-$
$\bar{c}$		$\bar{c} \to \bar{s} + \ell^-$

The analogous expressions for $n_t(\ell^\pm)$, $n_b(\ell^\pm)$, $n_c(\ell^\pm)$ can be easily derived and are not reported here.

From the above formulas it can be seen that in order to evaluate the Asymmetry parameter A° one needs to know:

i) the total cross section $\Rightarrow \sigma^T$;
ii) the decay branching ratios $\Rightarrow$ BR;
iii) the production distribution of the baryons or mesons states $\Rightarrow \varepsilon$;
iv) the relative fraction of baryons and mesons $\Rightarrow \rho$;
v) the lepton distribution in the decays $\Rightarrow \varepsilon$.

We will now discuss in some detail the assumptions we made for these quantities.

5.1. The Total Cross Sections

We will extrapolate the total cross sections for the heavy flavors at the $(p\bar{p})$ Collider energy, $\sqrt{s}$ = 540 GeV, using formula (4) and starting from the strangeness cross section. Using the known masses for "charm" and "beauty" baryons and mesons, and the values:

$$m_t = 25 \text{ GeV}/c^2, \quad m_{sb} = 55 \text{ GeV}/c^2,$$

for the "top" and "superbeauty" particles, one obtains:

$$\sigma_c \simeq 2000 \ \mu b, \quad (6a)$$
$$\sigma_b \simeq 140 \ \mu b, \quad (6b)$$
$$\sigma_t \simeq 1.5 \ \mu b, \quad (6c)$$
$$\sigma_{sb} \simeq 0.15 \ \mu b. \quad (6d)$$

Other estimates, from perturbative QCD, will however be taken into account when discussing the results. It will be shown that, under some conditions, even these very low cross sections ($\sigma_b \simeq 10\mu b$ and $\sigma_t \simeq 0.1\ \mu b$) [14] give rise to a measurable Asymmetry.

5.2. The Decay Branching Ratios

Recent data from CLEO[15] give for the semileptonic branching ratio of the "beauty" mesons:

$$(M_b \to \ell^{\pm})/(M_b \to \text{all}) \simeq 0.13.$$

In our Monte Carlo we assume the known semileptonic branching ratios for "charm":

$$(D \to \ell^{\pm})/(D \to \text{all}) \sim 0.085,$$

$$(\Lambda_c^+ \to \ell^{\pm})/(\Lambda_c^+ \to \text{all}) \sim 0.045.$$

and the conservative value of 0.1 for all other heavier particles.

5.3. The Production Distributions of Baryon and Meson States

The study of the reactions:

$$pp \to D + e^- + \text{anything},$$
$$pp \to \Lambda_c^+ + e^- + \text{anything},$$
$$pp \to \Lambda_b^o + e^+ + \text{anything},$$

at the ISR, indicate that in baryon-baryon collisions the heavy flavored baryons are produced according to a rather flat x-distribution:

$$(d\sigma/dx) \sim \text{const.} ,$$

while the heavy flavored mesons are produced with softer x-distribution of the type:

$$E(d\sigma/d|x|) \sim (1 - |x|)^3 .$$

These distributions will be assumed all along the following discussion, together with the p_T dependence:

$$(d\sigma/dp_T) \sim p_T \exp(- 2.5\ p_T)$$

observed at the ISR in the production of heavy flavors[16,17].

5.4. The Relative Yield of Mesons and Baryons

From the data on strangeness production at the ISR, it can be assumed that, in (pp) collisions the following reactions dominate

$$pp \to \bar{M}_{\mathrm{Central}} + B_{\mathrm{Leading}} + \text{anything} \equiv \{\bar{M}_C;\ B_L\}\ ,$$

$$pp \to \bar{M}_{\mathrm{Central}} + M_{\mathrm{central}} + \text{anything} \equiv \{\bar{M}_C;\ M_C\}\ .$$

In ($p\bar{p}$) collisions, due to the anti-baryonic nature of the p hemisphere, the following reactions can take place:

$$p\bar{p} \to \bar{M}_{\mathrm{Central}} + B_{\mathrm{Leading}} + \text{anything} \equiv \{\bar{M}_C;\ B_L\}\ ,$$

$$p\bar{p} \to M_{\mathrm{Central}} + \bar{B}_{\mathrm{Leading}} + \text{anything} \equiv \{M_C;\ \bar{B}_L\}\ ,$$

$$p\bar{p} \to \bar{M}_{\mathrm{Central}} + M_{\mathrm{Central}} + \text{anything} \equiv \{\bar{M}_C;\ M_C\}\ ,$$

$$p\bar{p} \to \bar{B}_{\mathrm{Leading}} + B_{\mathrm{Leading}} + \text{anything} \equiv \{\bar{B}_L;\ B_L\}\ ,$$

In this case, the ratio of the inclusive cross sections for producing the four classes of particles $\bar{M}_C$, M_C, $\bar{B}_L$ and B_L and the total cross sections are:

$$\rho_M = \left[\sigma\{M_C;\ \bar{B}_L\} + \sigma\{\bar{M}_C;\ M_C\}\right]/\sigma^T\ ,$$

$$\rho_{\bar{M}} = \left[\sigma\{\bar{M}_C;\ B_L\} + \sigma\{\bar{M}_C;\ M_C\}\right]/\sigma^T\ ,$$

$$\rho_B = \left[\sigma\{\bar{M}_C;\ B_L\} + \sigma\{\bar{B}_L;\ B_L\}\right]/\sigma^T\ ,$$

$$\rho_{\bar{B}} = \left[\sigma\{M_C;\ \bar{B}_L\} + \sigma\{\bar{B}_L;\ B_L\}\right]/\sigma^T\ ,$$

with

$$\sigma^T = \sigma\{\bar{M}_C;\ B_L\} + \sigma\{M_C;\ \bar{B}_L\} + \sigma\{\bar{M}_C;\ M_C\} + \sigma\{\bar{B}_L;\ B_L\}\ .$$

Defining the ratio:

$$\text{Leading/Total} = \rho_B$$

the four inclusive cross sections can be written as:

$$\rho_B = \rho_{\bar{B}} = (\text{Leading/Total})\ ,$$

$$\rho_M = \rho_{\bar{M}} = [1 - (\text{Leading/Total})]\ .$$

At ISR, in each hemisphere, the ratio (Leading/Total) is $\sim$ 1/16 for strangeness and $\sim$ 1/8 for "charm". In our discussion we will study the behavior of A° as a function of (Leading/Total).

5.5. The Lepton Decay Distributions

The data from CLEO[18] show that in the semileptonic decay of "beauty" mesons, M_b, the magnitude of the mass recoiling with respect to the leptons is very near to the D mass:

$$M_b \to Xe\nu, \quad \text{with } M_X \sim M_D \approx 2.0 \text{ GeV}/c^2.$$

Moreover, the mean charged multiplicity of the decay is 3.5, where the D contributes with 2.5 charged particles on average. We can conclude that, even at values of the mass as high as the mass of the M_b, the semileptonic decay proceeds via a 3-body decay. On the contrary, the mean charged multiplicity in the hadronic decays of the M_b mesons is 6.3, i.e. the hadronic decay of the M_b produces, on average, one D plus four charged particles plus two neutral particles:

$$M_b \to D + 6\text{-bodies}.$$

In the following we will consider two possibilities:

i) the total multiplicity of all decays is 3: this is the worst case for the Asymmetry A°;

ii) the total multiplicity of all semileptonic decay is 3 and the decay is $K_{\ell 3}$-like for mesons and phase-space for baryons, while the total multiplicity of all the hadronic decays have the known values for "charm" and "beauty" ($\sim$3 for "charm", $\sim$ charm + 6 for "beauty"), and, for "top" and "superbeauty", the same multiplicity as "beauty". It should be noted that this is already a conservative hypothesis, since the hadronic decays of "top" and "superbeauty" can be expected to produce more particles than "beauty". Higher values of multiplicity would produce higher values for the asymmetry.

5.6. Estimates of the Asymmetry A°

In order to estimate the Asymmetry A°, we have restricted our study to the case of electrons and positrons.

The detection acceptances ε have been evaluated by means of a Monte Carlo simulation, with the conditions set in the previous section and for 5 values of θ_{cut} (θ_{cut} = 10°, 20°, 30°, 40° and 90°).

Figures from 16a to 16t show the $e^{\pm}$ acceptances for θ_{cut} = 30° and model (ii) of section 5.5 for baryons, mesons and antimesons. The acceptances for e^{-} originated by the decay of antibaryons are, of course, negligible.

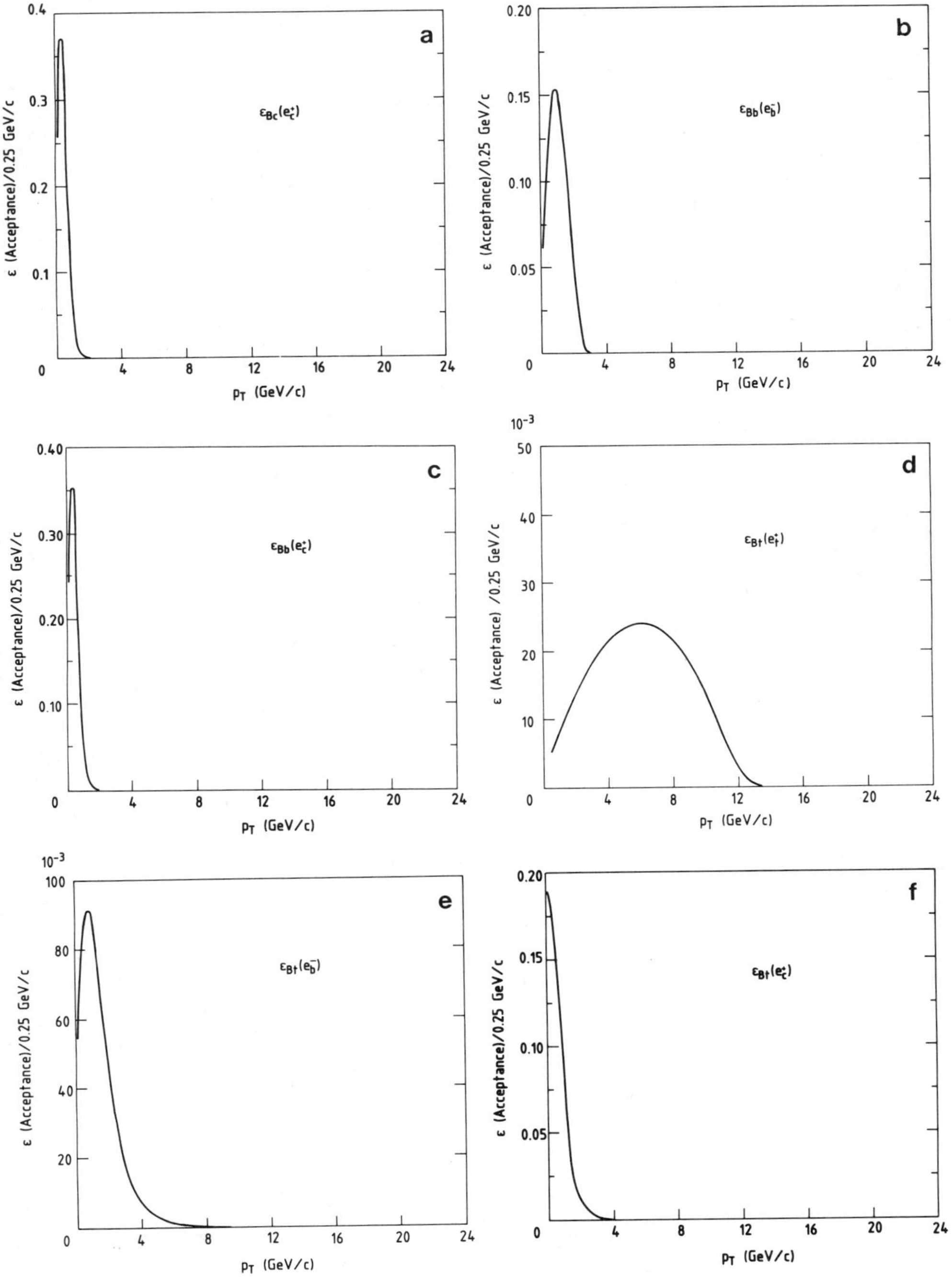

Fig. 16. (a) to (f): Acceptances ε for the various states and decay chains.

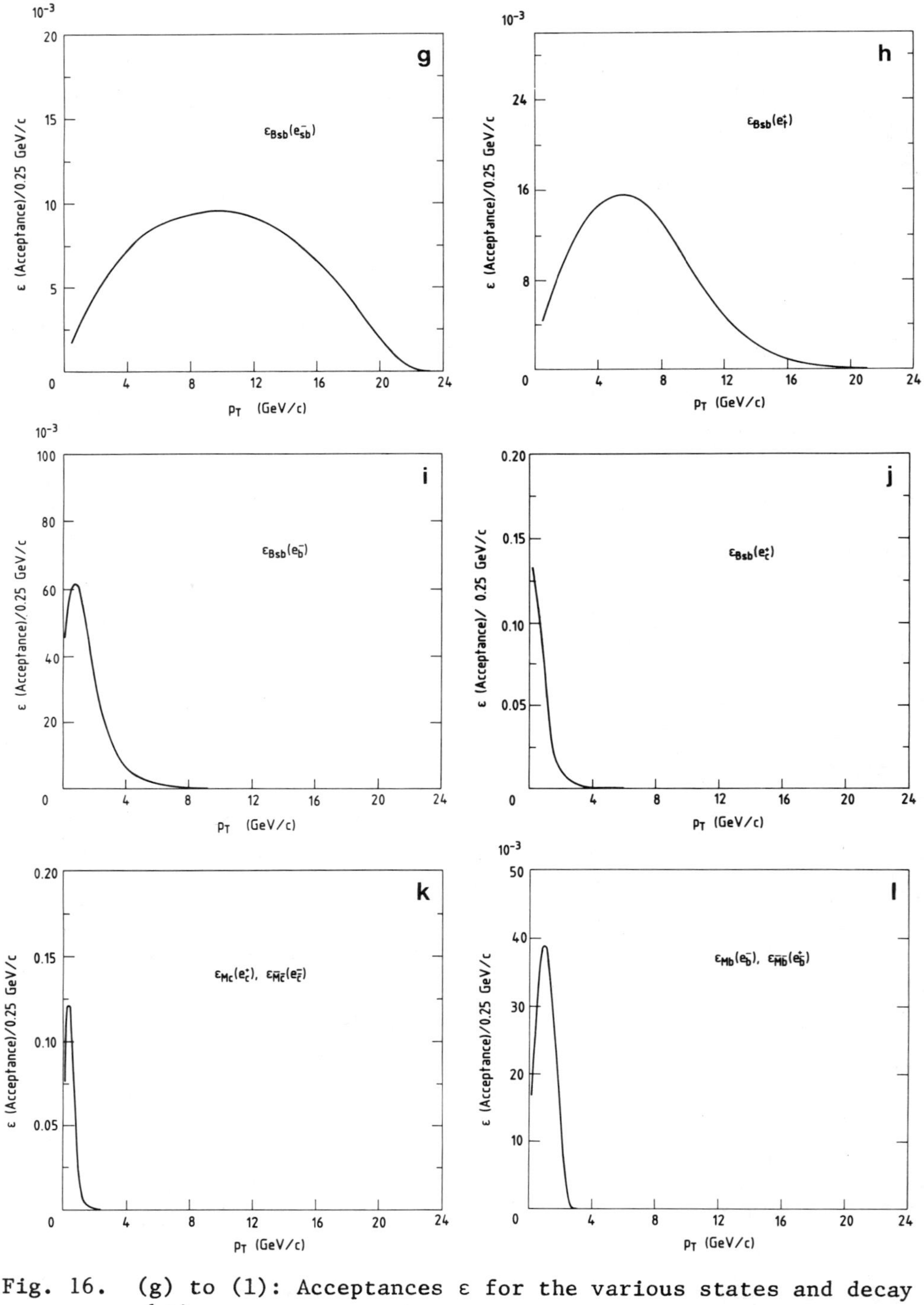

Fig. 16. (g) to (l): Acceptances ε for the various states and decay chains.

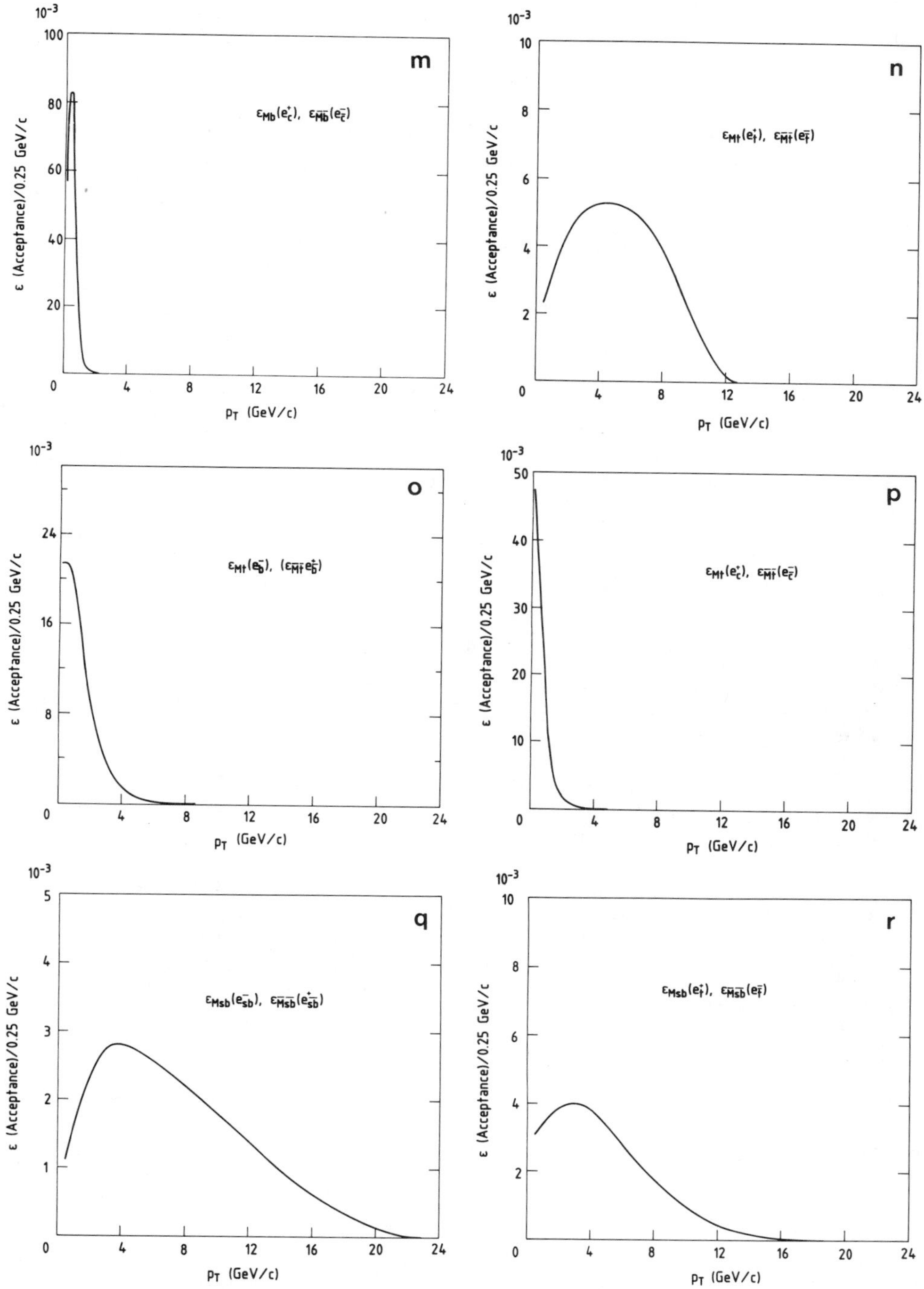

Fig. 16. (m) to (r): Acceptances ε for the various states and decay chains.

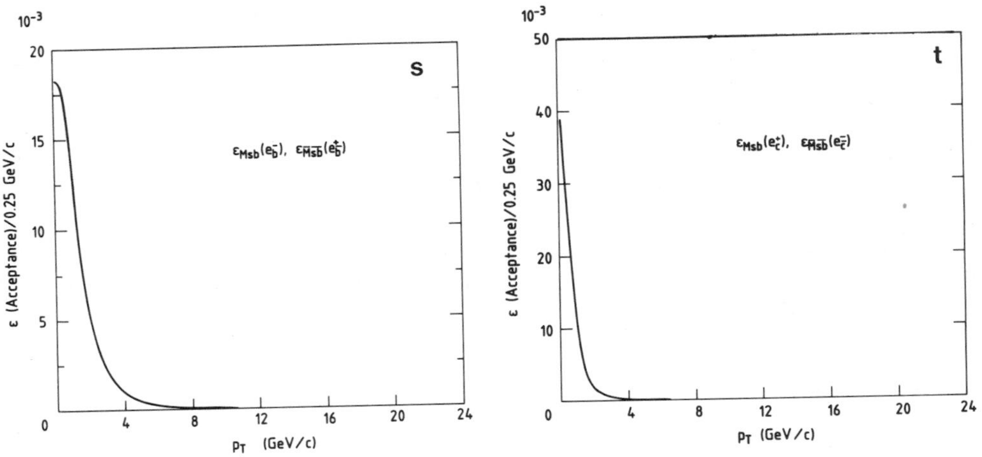

Fig. 16. (s) to (t): Acceptances ε for the various states and decay chains.

Figure 17 shows the behavior of A° (p_T, 30°), for (Leading/Total) = 0.25 and model (ii) of section 5.5. There are two main peaks, one positive around p_T = 10 GeV/c, due to the "top" baryon decay into e^+, and one negative around p_T = 19 GeV/c, due to "superbeauty" baryon decay into e^-.

It is interesting to note that the separation between the two peaks depends only on the mass differences in the semileptonic decays of "superbeauty" and "top" states.

In fact in the 3-body semileptonic decay the transverse momentum spectrum of the electrons or positrons scales with $p_T/\Delta m$ where Δm is the difference between the parent mass and mass of the hadronic particle produced in the decay. This is shown in Figure 18 where the normalized $p_T/\Delta m$ spectra of the electrons and positrons produced in the decays:

i) $\Lambda^o_{sb} \rightarrow \Lambda^+_t e^- \bar{\nu}$;

ii) $\Lambda^+_t \rightarrow \Lambda^o_b e^+ \nu$;

iii) $\Lambda^o_b \rightarrow \Lambda^+_c e^- \bar{\nu}$;

iv) $\Lambda^+_c \rightarrow \Lambda^o_s e^+ \nu$;

are reported.

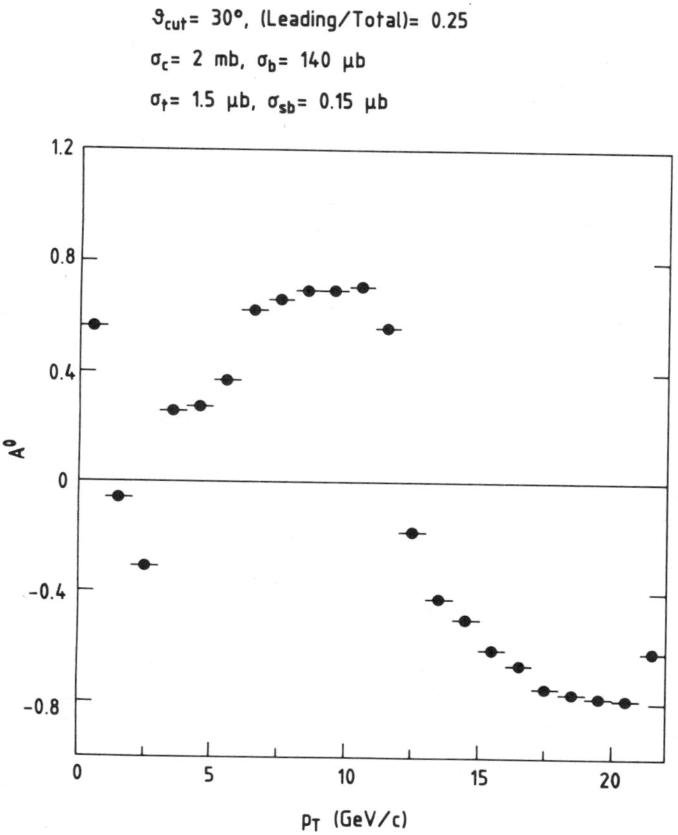

Fig. 17. Plot of $A°(p_T, \theta_{cut} = 30°)$ as a function of p_T. The values assumed for the cross sections and for (Leading/Total) are indicated in the figure.

The mass differences Δm have the following values:

i) $\Delta m = m(\Lambda^o_{sb}) - m(\Lambda^+_t) \approx 30$ GeV/c² for the "sb" baryon decay;

ii) $\Delta m = m(\Lambda^+_t) - m(\Lambda^o_b) \approx 19.5$ GeV/c² for the "t" baryon decay;

iii) $\Delta m = m(\Lambda^o_b) - m(\Lambda^+_c) \approx 3.2$ GeV/c² for the "b" baryon decay;

iv) $\Delta m = m(\Lambda^+_c) - m(\Lambda^o_s) \simeq 1.2$ GeV/c² for the "c" baryon decay;

Figures 19a, b and 20a, b show the amplitude of the two peaks as a function of θ_{cut} and (Leading/Total) for: a) model (i) of section 5.5 (all 3-body decays) and: b) model (ii) of section 5.5 (greater multiplicity for the hadronic decays).

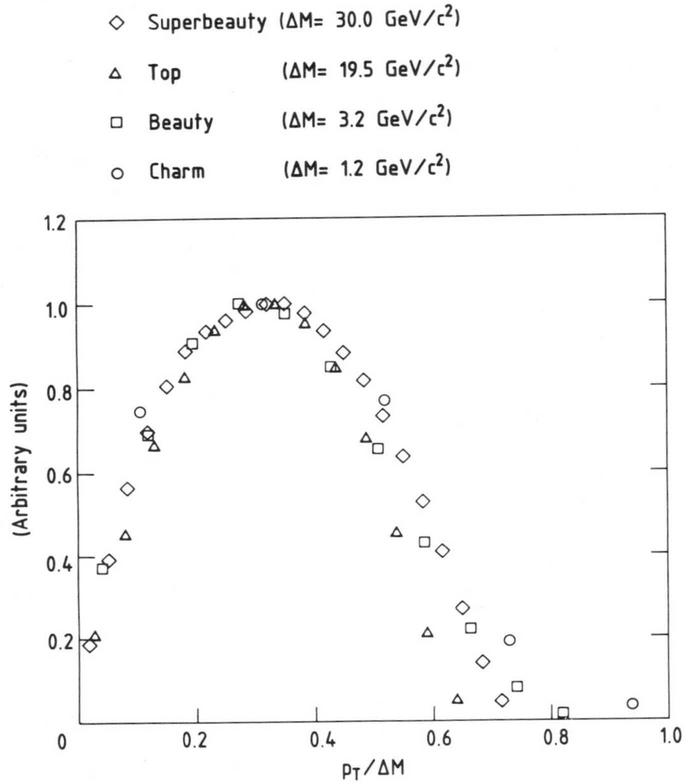

Fig. 18. Normalized ($p_T/\Delta m$) spectra of the electrons from the decays: $\Lambda^\circ_{sb} \to \Lambda^+_t e^- \bar{\nu}$; $\Lambda^+_t \to \Lambda^\circ_b e^+ \nu$; $\Lambda^\circ_o \to \Lambda^+_c e^- \bar{\nu}$; $\Lambda^+_c \to \Lambda^\circ_s e^+ \nu$. The Δm values relative to the four decays are indicated in the figure.

5.7. Background Evaluation

In what has been described so far, the background contamination in the sample of prompt e^+ and e^- has not been considered. It is mainly due to:

i) the misidentification of charged and neutral hadrons in the experimental apparatus;

ii) the prompt e^+ or e^- production from sources other than open heavy flavor states.

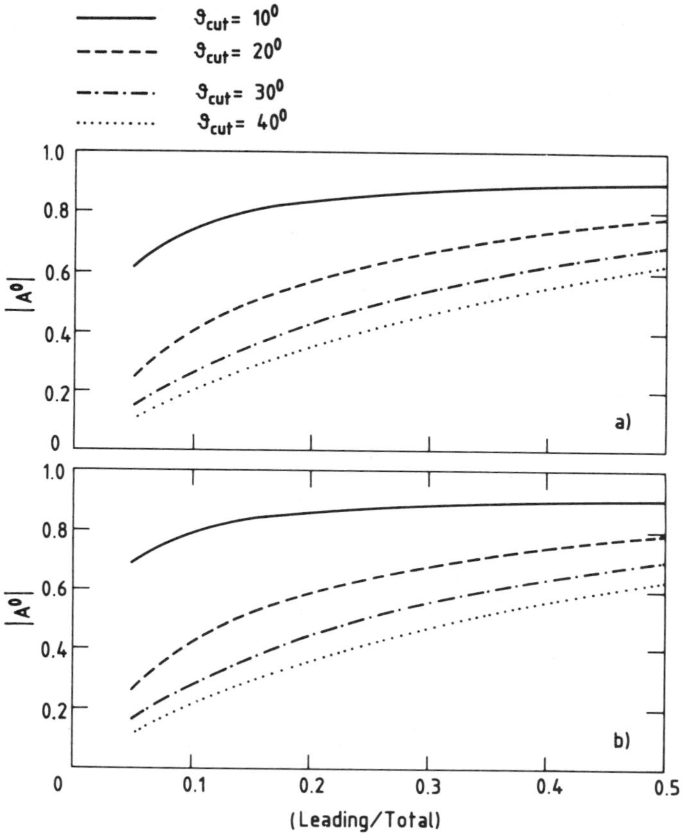

Fig. 19. Plot of $A°(p_T = 10$ GeV/c, θ_{cut})("top" peak) as a function of (Leading/Total) for different values of θ_{cut}, and using: a) model (i) of section 5.5, and b) model (ii) of section 5.5.

The contribution (i) can be derived by extrapolating, above $p_T \cong 10$ GeV/c the inclusive pion cross section as measured by the UA1 experiment[19], using the fit to their data:

$$E(d^3\sigma/dp^3) = A \times p_o^n/(p_o + p_T)^n \tag{7}$$

with $A = 0.46 \pm 0.10$ $mb^2c^2GeV^{-2}$, $p_o = 1.3 \pm 0.18$ $GeVc^{-1}$ and $n = 9.14 \pm 0.77$.

Formula (7) is relative only to charged pions (averaged over the two charges) and is given in unit of rapidity. We have assumed the contribution to the background due to the neutral pions to ~ 0.2 of

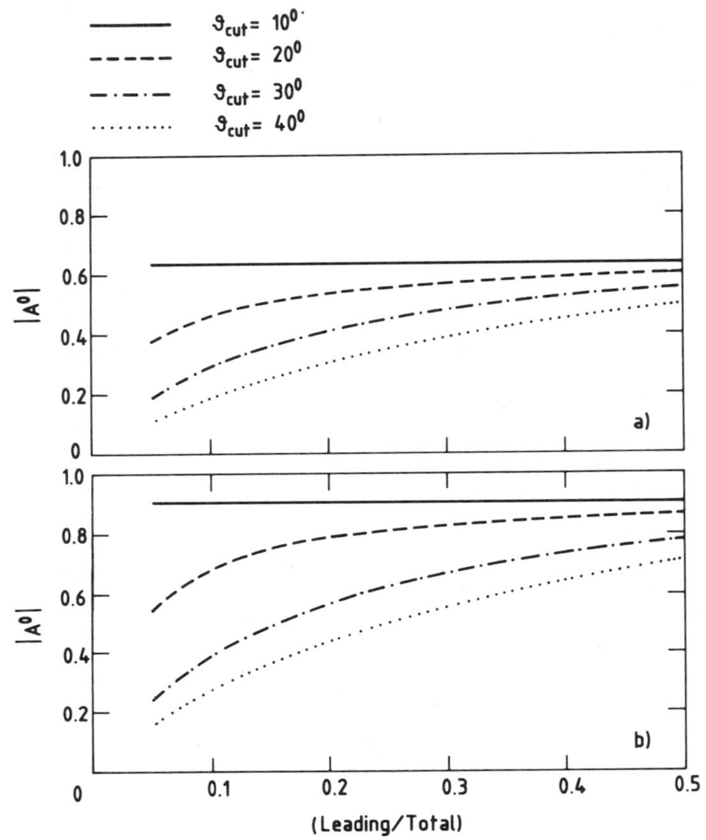

Fig. 20. Plot of $A°(p_T = 19$ GeV/c, θ_{cut})("superbeauty" peak) as a function of (Leading/Total) for different values of θ_{cut}, and using: a)model (i) of section 5.5, and b) model (ii) of section 5.5.

the cross section (7). The rapidity interval over which we integrated the background depends on the θ_{cut}:

$$\theta_{cut} = 90° \Rightarrow \Delta y \simeq 4.0 \ ,$$
$$\theta_{cut} = 40° \Rightarrow \Delta y \approx 3.0 \ ,$$
$$\theta_{cut} = 30° \Rightarrow \Delta y \approx 2.7 \ ,$$
$$\theta_{cut} = 20° \Rightarrow \Delta y \approx 2.3 \ ,$$
$$\theta_{cut} = 10° \Rightarrow \Delta y \quad 1.5 \ .$$

The extrapolated background rates should be multiplied by a reduction factor representing the rejection of the background from source (i) in the experimental apparatus.

Concerning the prompt electron background (ii), we assume, as a first approximation, that it would be negligible when compared with the contribution (i).

5.8. Estimate of the Asymmetry Parameter Inclusive of Background

Due to the background sources described in the previous section, the experimental Asymmetry parameter is given by:

$$A^{exp}(p_T, \theta_{cut}) = \frac{[N(e^+)+N_{bg}(e^+)] - [N(e^-)+N_{bg}(e^-)]}{[N(e^+)+N_{bg}(e^+)] + [N(e^-)+N_{bg}(e^-)]}$$

where $N_{bg}(e^{\pm})$ is the number of background positrons or electrons. A^{exp} can then be expressed as a function of $A°$ and of the Signal-to-Background ratio defined as:

$$\frac{\text{Signal}}{\text{Background}} \equiv \frac{N(e^{\pm})}{N_{bg}(e^{\pm})}$$

by assuming $N_{bg}(e^+) = N_{bg}(e^-) = N_{bg}(e)$ and by substituting, in turns, in the above equation of A_{exp}, the expression of $N(e^-)$ and $N(e^+)$ derived from the definition of $A°$. The result is:

$$A^{exp} = \frac{A°}{1 + \dfrac{1 + A°}{N(e^+)/N_{bg}(e)}}, \quad \text{for } A°>0 ;$$

$$A^{exp} = \frac{A°}{1 + \dfrac{1 - A°}{N(e^-)/N_{bg}(e)}}, \quad \text{for } A°<0 ;$$

Figure 21 shows the quantity A^{exp} plotted as a function of the Signal/Background ratio for various values of $A°$. The results can be expressed with curves of constant A^{exp} as function of (rejection power) versus (cross section for "top" and "superbeauty" production), for different θ_{cut}, (Leading/Total) and decay models (Figures 22 to 31).

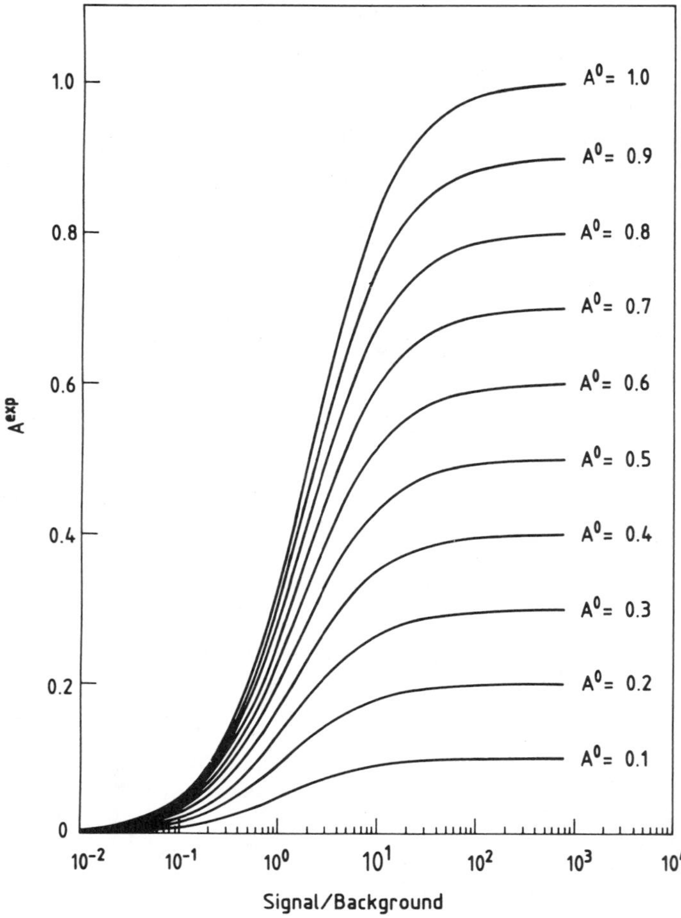

Fig. 21. The curves give the behavior of A^{exp} as a function of Signal/Background, for different values of A°.

The Table 6 summarizes results given in Figures 22 to 31, for two values of the total cross section. They show the rejection powers needed to obtain $A^{exp}>0.3$ or $A^{exp}< - 0.3$, i.e. a reasonably high value for the Asymmetry, at the "top" and "superbeauty" peaks respectively. It can be seen that, with a rejection power of the order of 10^{-3}, a large range of θ_{cut}, (Leading/Total) and cross section values are accessible.

Table 6

θ_{cut}	(Leading/ Total)	$\sigma_t = 1.5\ \mu b$, $\sigma_{sb} = 0.15\ \mu b$ (formula (9)) Model (i)	Model (ii)	$\sigma_t = 0.1\ \mu b$, $\sigma_{sb} = 0.01\ \mu b$ (perturbative QCD) Model (i)	Model (ii)
P_T = 10 GeV					
10°	0.1	3.0	3.5	0.2	0.3
	0.25	10.0	12.0	0.6	0.7
	0.5	15.0	18.0	1.0	1.3
20°	0.1	1.3	1.5	0.1	0.1
	0.25	8.0	9.0	0.5	0.6
	0.5	20.0	23.0	1.0	1.2
30°	0.1	-	-	-	-
	0.25	6.0	7.0	0.4	0.4
	0.5	19.0	22.0	1.0	1.2
40°	0.1	-	-	-	-
	0.25	3.5	4.8	0.3	0.3
	0.5	17.0	21.0	1.0	1.1
P_T = 19 GeV/c					
10°	0.1	0.4	1.0	0.02	0.1
	0.25	0.9	2.5	0.1	0.2
	0.5	1.7	4.5	0.1	0.3
20°	0.1	2.5	5.5	0.2	0.4
	0.25	10.0	20.0	0.5	1.0
	0.5	25.0	50.0	1.3	2.2
30°	0.1	-	2.2	-	0.2
	0.25	9.0	25.0	0.6	1.1
	0.5	23.0	45.0	1.6	2.0
40°	0.1	-	-	-	-
	0.25	4.0	19.0	0.3	1.0
	0.5	20.0	40.0	1.5	2.0

Rejection power in units of 10^{-3}.

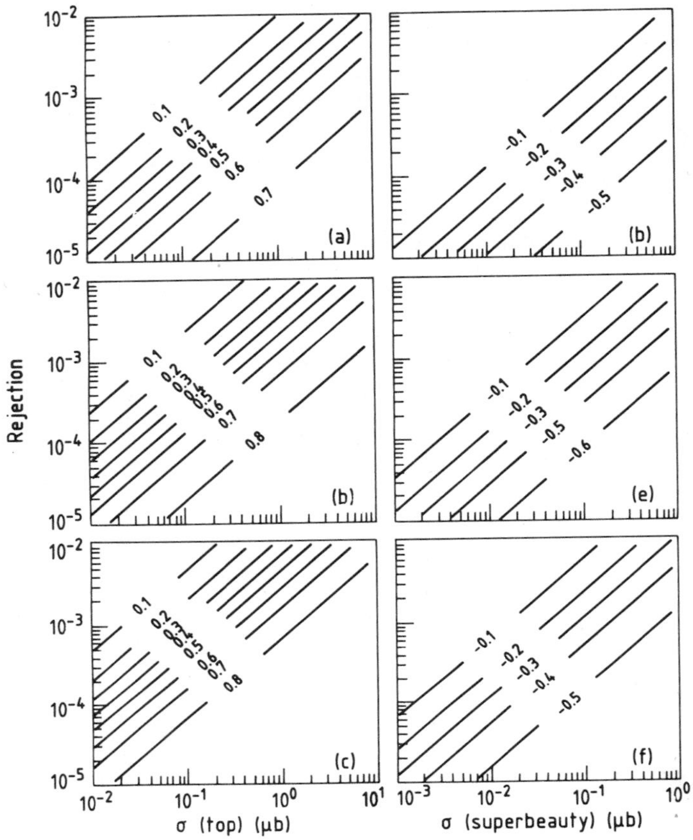

Fig. 22. Curves of constant A^{exp}, in the plot (rejection power) versus (cross section for "top" or "superbeauty") for θ_{cut} = 10° and for model (i). Plots (a), (b) and (c) refers to p_T = 10 GeV/c ("top" peak), plots (d), (e) and (f) refers to p_T = 19 GeV/c ("superbeauty" peak). Plots (a) and (d) are obtained with (Leading/Total) = 0.1, plots (b) and (e) with (Leading/Total) = 0.25, plots (c) and (f) with (Leading/Total) = 0.5.

5.9. A Detailed Case

As an example of what can be obtained experimentally in terms of the Asymmetry A^{exp}, let us fix in a reasonable way some of the parameters. We take (as in Figures 29b and e):

i) (Leading/Total) = 0.25;
ii) θ_{cut} = 30°;
iii) decay model (ii).

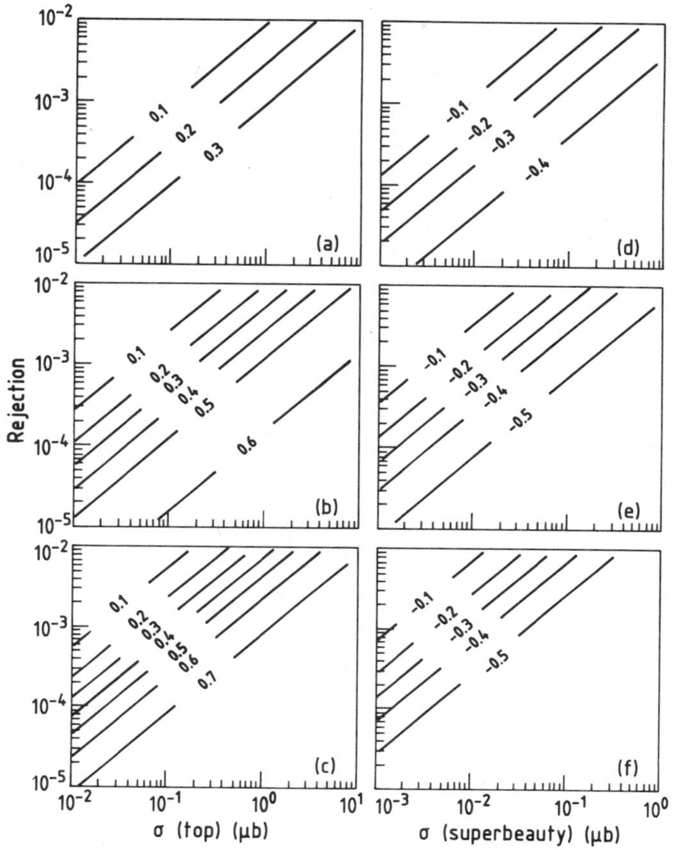

Fig. 23. As Figure 22 but for θ_{cut} = 20° and model (i).

Moreover, in order to have an estimate for the experimental errors on A^{exp}, we assume a total integrated luminosity L = 300 nb^{-1} (foreseen for 1983 at the CERN ($p\bar{p}$) Collider).

Figures 32a and 32b show the plot of A^{exp} as a function of p_T, for a rejection power of 10^{-3} and for the two cross section estimates: as in section 5.1 (formula (6)) and as from perturbative QCD ($\sigma_t:\sigma_{sb}$ = 10:1), respectively. The errors are purely statistic.

Figures 33a and 33b show the expected number of produced electrons as a function of p_T, with the same two assumptions for the total cross sections. Superimposed is the expected background, as defined in section 5.7, and with a rejection factor of 10^{-3}.

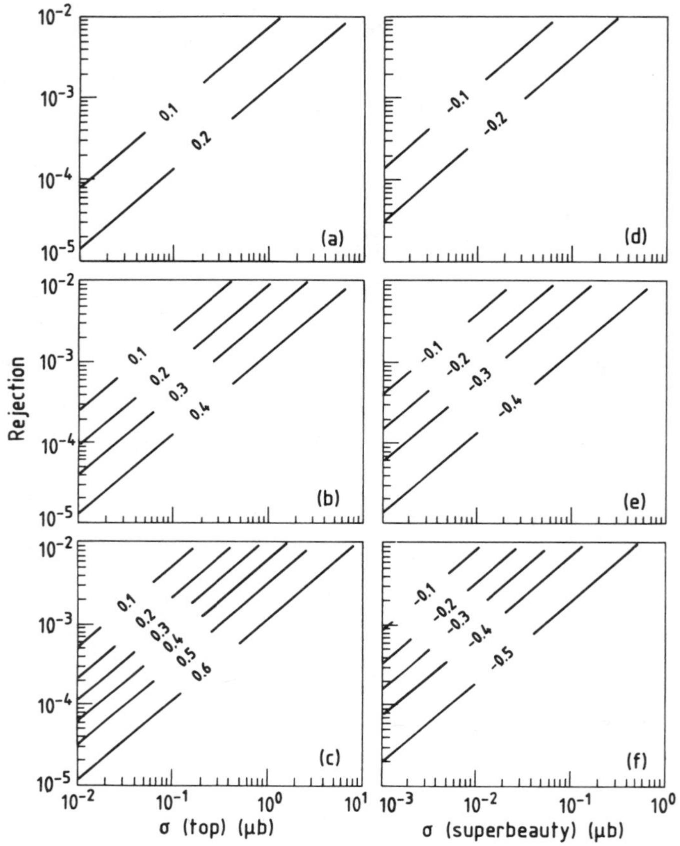

Fig. 24. As Figure 22 but for θ_{cut} = 30° and model (i).

It could be that the rejection factor needed is much less than 10^{-3}. In fact, a very heavy state (such as Λ_t^+ or Λ_{sb}^0) decaying semi-leptonically, may have the hadronic "jet" recoiling against the lepton pair. The study of the hadronic pattern associated with the $(e^{\pm})$ could be of such an help in the selection of good events, that a rejection power much below 10^{-3} could be sufficient. For example, an order of magnitude improvement would mean that the data which, at present, are quoted with a rejection of order 10^{-3} (Figures 22-34), would reach the level of 10^{-4}. In this case, all our expectations would be scaled by this factor.

For completeness, let us mention that, at present, the (e/π) ratio in the p_T range above ~ 20 GeV is not known at the $(p\bar{p})$ Collider.

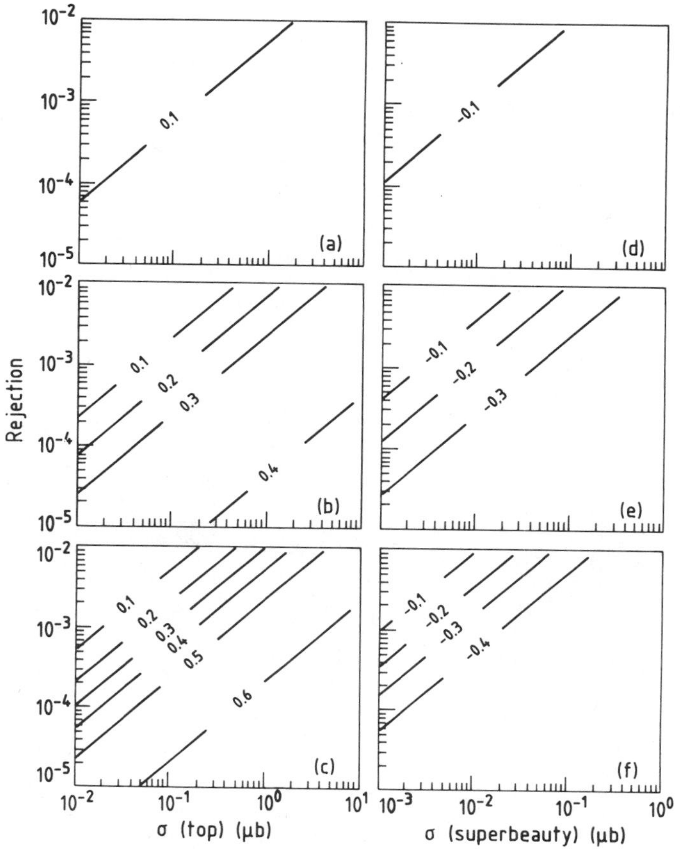

Fig. 25. As Figure 22 but for $\theta_{cut} = 40°$ and model (i).

In order to compute the statistical significance of the observed effect, it is convenient to integrate A^{exp} over the p_T ranges:

i) $7 \leq p_T \leq 12$ GeV/c, corresponding to the "top" region;
ii) $14 \leq p_T \leq 23$ GeV/c, corresponding to the "superbeauty" region.

This is equivalent to the experimental procedure of fitting the data to reduce the statistical errors on the single points.

Figures 34a and 34b show the number of standard deviations that can be obtained in the measurement of A^{exp}, respectively in the "top" and in the "superbeauty" regions, as function of the total cross sections σ_t and σ_{sb}, in the same conditions specified above. The 90% confidence level in the measurement is also shown. These results show that, especially in the "top" case, a high statistical significance can be reached with a moderate rejection power, even if the total cross section for "top" production is as low as 0.1 μb, i.e. the value predicted by QCD.

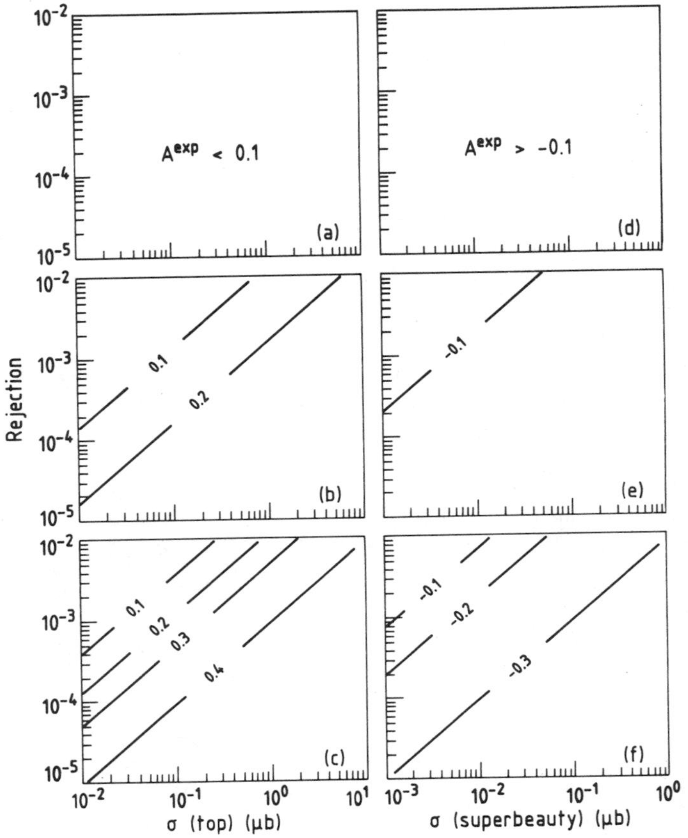

Fig. 26. As Figure 22 but for $\theta_{cut} = 90°$ and model (i).

In Figure 35, the behavior of the ratio (Signal/Background) as a function of the total cross section for "top" and "superbeauty" and for different values of the rejection power, is shown.

6. COMPARISON WITH PRELIMINARY DATA FROM CERN ($p\bar{p}$) COLLIDER

As a first step in the study of the heavy-quark physics using the ($p\bar{p}$) Collider, we propose to compare our predictions with the data already available from the ($p\bar{p}$) Collider.

In their search for electron candidates, the UA1 Collaboration finds 16 events with an isolated electron[20]. Five of these events are attributed to $W^{\pm}$ decay, whilst the remaining 11 are characterized by a "jet activity" in the azimuthal region opposite the isolated

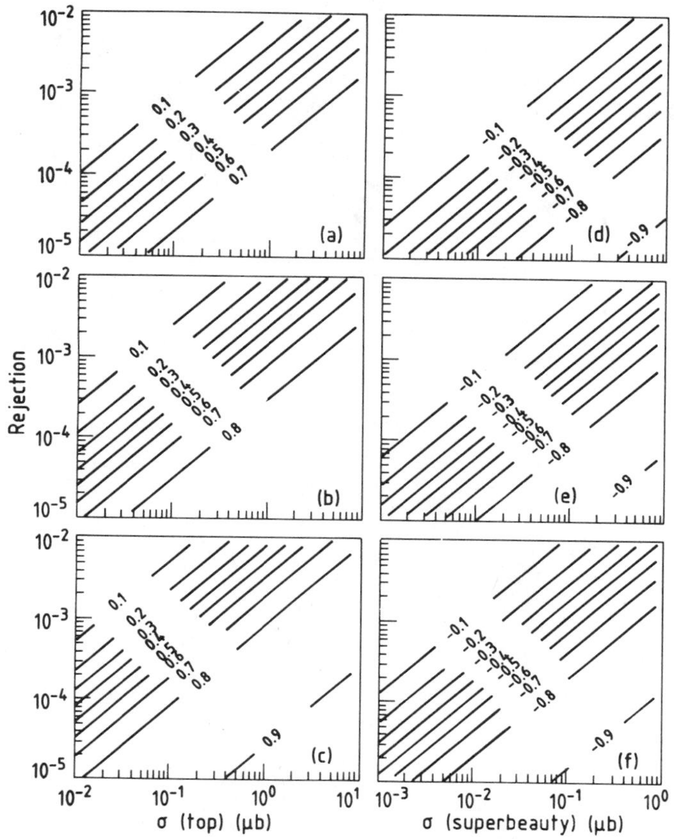

Fig. 27. As Figure 22 but for $\theta_{cut} = 10°$ and model (ii).

electron. The events correspond to a total integrated luminosity of $L = 20\ nb^{-1}$.

According to the Monte Carlo discussed in detail in the previous section, the acceptance for electrons originating from "superbeauty" decays in the phase-space region defined by the UA1 data ($p_T > 15$ GeV/c and, for the polar angle, $25° < \theta < 155°$) is, for baryon and antibaryon decays,

$$\varepsilon_B(e^{\pm}) = 0.08 ,$$

and the meson and antimeson decays

$$\varepsilon_M(e^{\pm}) = 0.12 .$$

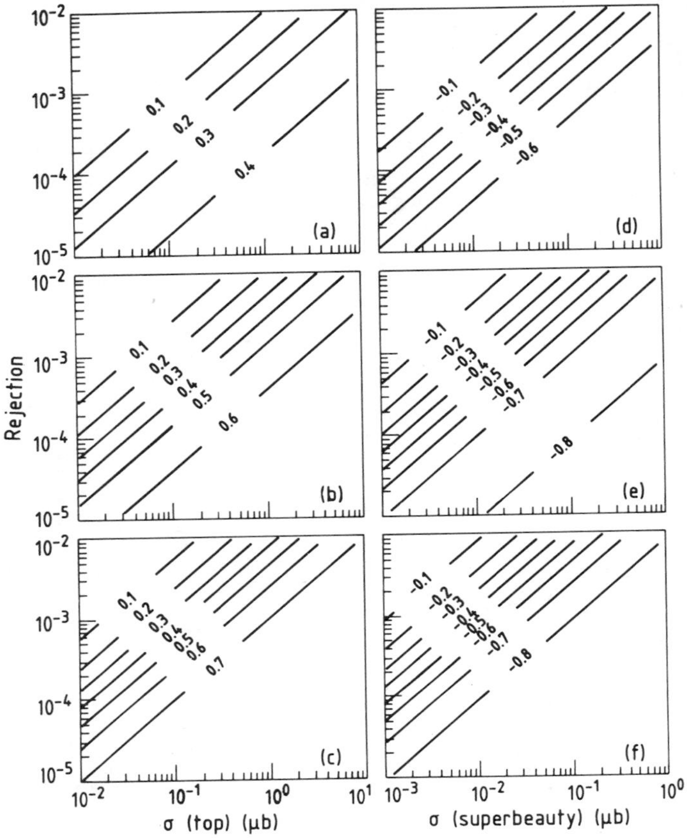

Fig. 28. As Figure 22 but for $\theta_{cut} = 20°$ and model (ii).

We can therefore use the approximation

$$\varepsilon(e^{\pm}) = \varepsilon_B(e^{\pm}) = \varepsilon_M(e^{\pm}) = 0.1 \ .$$

Since the heavy flavors are produced in pairs, the total efficiency for seeing at least one electron from the leptonic decay of "superbeauty" is

$$\varepsilon_T(e^{\pm}) = 2 \times \varepsilon(e^{\pm}) = 0.2 \ ,$$

where we have assumed equal semileptonic branching ratios for baryon and meson decays.

The request for "jet activity" opposite in azimuth to the electron gives rise to another acceptance factor, ε_T(jet), which we

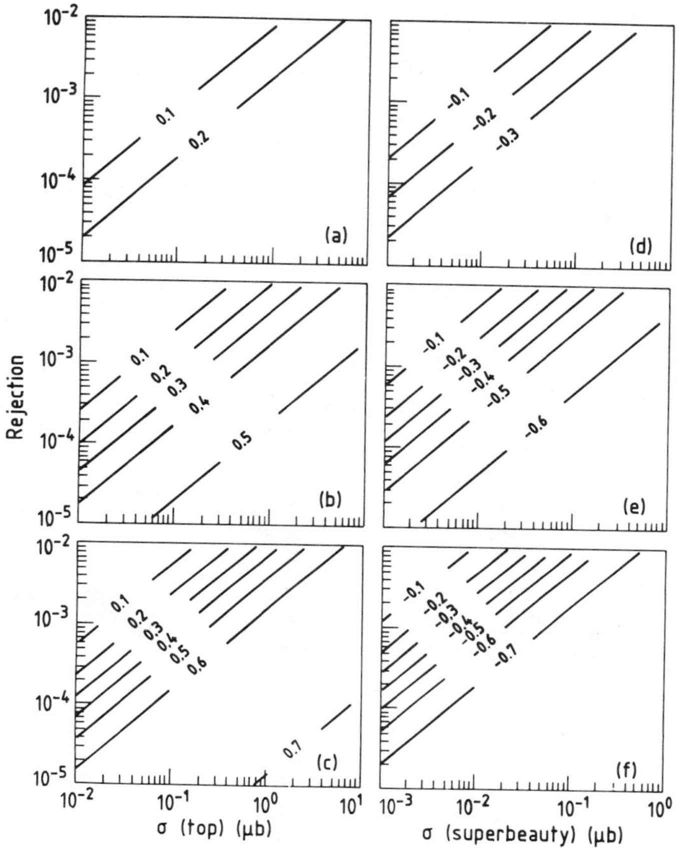

Fig. 29. As Figure 22 but for θ_{cut} = 30° and model (ii).

can derive by analyzing the hadronic pattern of the "superbeauty" semileptonic decays predicted by our Monte Carlo, according to the "jet" definition outlined by the UA1 Collaboration, i.e.

i) all particles with $p_T > 2.5$ GeV/c are associated with a jet if their separation in phase-space is

$$\Delta R = \sqrt{(\Delta\phi)^2 + (\Delta\eta)^2} < 1 ,$$

with $\Delta\phi$ in radians and η(= pseudorapidity) = $-\ln(\mathrm{tg}\theta/2)$;

ii) all other particles are associated with the jet defined as in point (i), if they satisfy the conditions

p_T relative to the jet < 1 GeV/c,
θ relative to the jet < 45°;

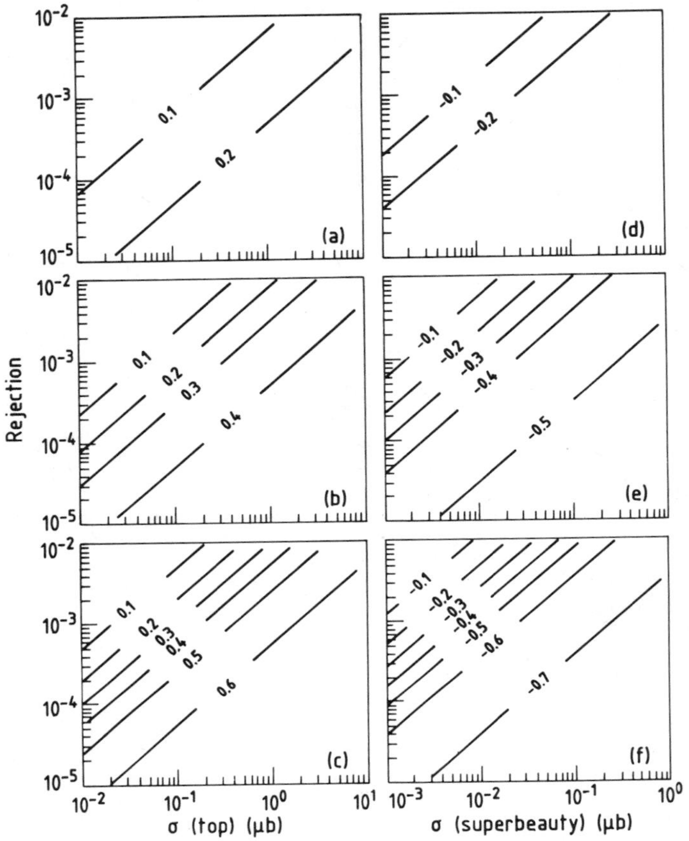

Fig. 30. As Figure 22 but for $\theta_{cut} = 40°$ and model (ii).

iii) the total transverse energy of the jet must be greater than 10 GeV.

The result is

$$\varepsilon_T(\text{jet}) \cong 70\%.$$

On the other hand, as shown in Figure 36, the condition that the jet is opposite in azimuth to the electron within $\Delta\phi = 30°$ is nearly always satisfied.

The number of electrons from "superbeauty" semileptonic decays in the UA1 electron sample is therefore given by

$$N(e^{\pm}) = \sigma_{sb} \times BR \times L \times \varepsilon_T(e^{\pm}) \times \varepsilon_T(\text{jet}) \ , \tag{8}$$

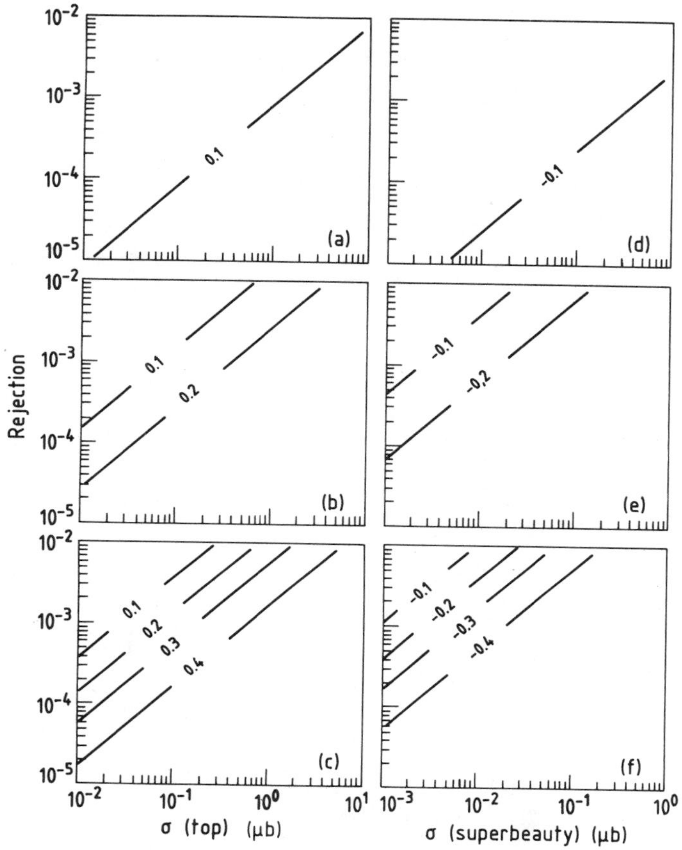

Fig. 31. As Figure 22 but for θ_{cut} = 90° and model (ii).

where BR is the semileptonic branching ratio of the "superbeauty" state, taken to be BR ≅ 0.15, and σ_{sb} is the cross section for the production of "superbeauty" particle states at the $(p\bar{p})$ Collider; the other symbols have already been defined. The UA1 results show that $N(e^{\pm})$ = 11±3).

Note that in Equation (8) the efficiencies for the electron trigger have not been taken into account. They are as follows:

i) The efficiency for detecting "isolated" electrons, i.e. with no other particles with p_T > 2 GeV/c in a 20° cone around the electron direction. Here the risk is in the random vetoing, otherwise the efficiency for genuine events is very high. From UA1 data[20] it is possible to deduce the upper limit for random vetoing: it must be below 75%, and it could, in fact, be almost zero.

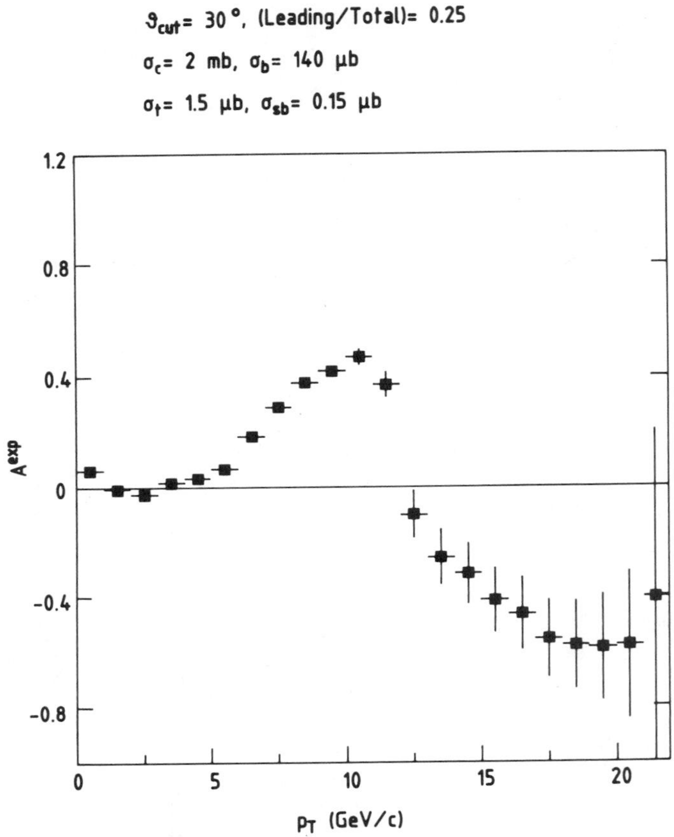

Fig. 32a. Plot of A^{exp} as a function of p_T, for a total Luminosity $L = 300\ nb^{-1}$, a rejection power of 10^{-3}, $\theta_{cut} = 30°$ and (Leading/Total) = 0.25, using the cross section estimates. (a) from formula (6); (b) from perturbative QCD. The errors are statistical.

ii) The efficiency of the energy cut $E_T > 15$ GeV/c owing to the finite resolution of the electromagnetic shower detectors (EMSDs). This efficiency can be evaluated to be $\geq 95\%$, using the quoted EMSDs energy resolution ($\Delta E/E = 0.15/\sqrt{E}$).

Figure 37a shows the cross-section corresponding to the (11±3) events observed. Notice that the "experimental" finding of UA1 falls in a remarkable range of agreement with the crude extrapolation from "charm" and QCD.

Going further, we have compared the p_T distributions of the $(e^{\pm})$ from the UA1 with that from our Monte Carlo simulation. This is

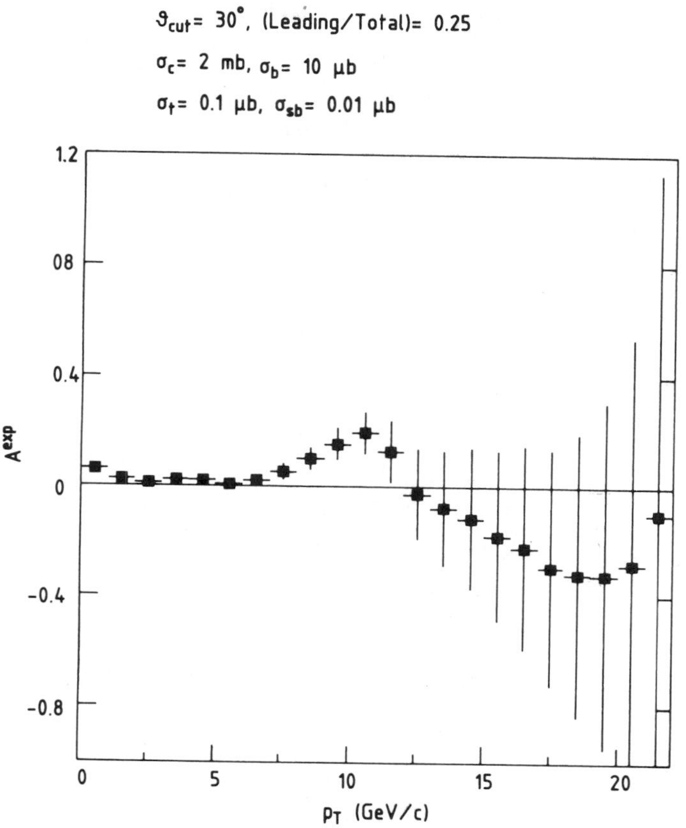

Fig. 32b. Plot of A^{exp} as a function of p_T, for a total Luminosity $L = 300\ nb^{-1}$, a rejection power of 10^{-3}, $\theta_{cut} = 30°$ and (Leading/Total) = 0.25, using the cross section estimates. (a) from formula (6); (b) from perturbative QCD. The errors are statistical.

shown in Figure 38, where the Monte Carlo expectations are obtained using different values for the parameter b in the "superbeauty" production process

$$d\sigma/dp_T \propto p_T \exp(-bp_T).$$

Data and Monte Carlo distributions are normalized to the total number of events with $p_T > 15$ GeV/c.

A value of $b \cong 0.20\ GeV^{-1}c$, i.e. a mean value for the production average transverse momentum $\langle p_T \rangle$ of the order of 10 GeV/c, fits the UA1 data quite well. This value for b is much smaller than the value (b = 2.5) which we found in our study of "charm" production at the

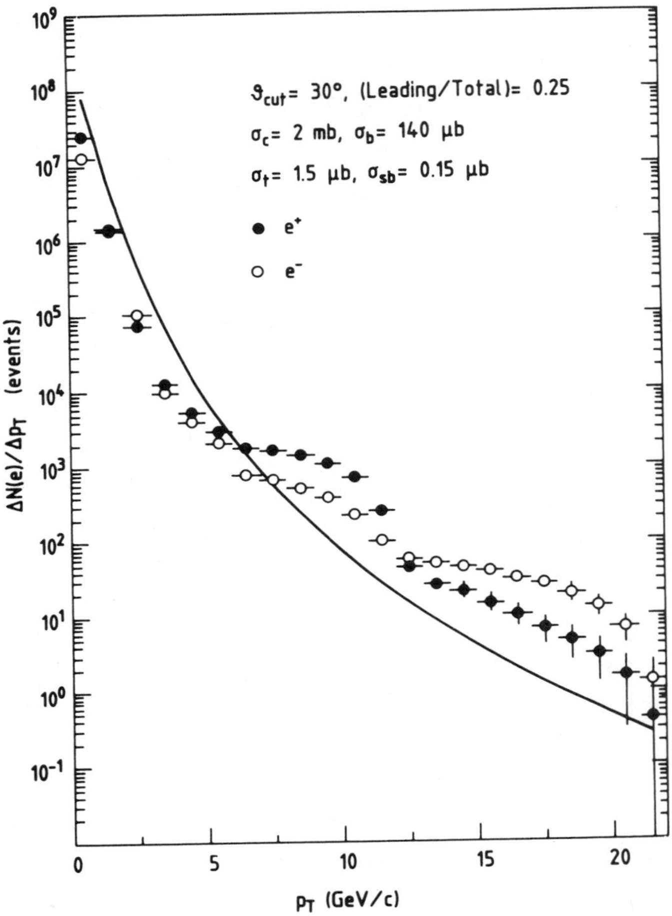

Fig. 33a. Expected number of produced electrons as a function of p with the same assumptions as in Figure 29, and the two cross section estimates: (a) from formula (6); (b) from perturbative QCD. The errors are statistical.

ISR[4,16]. However, it should be noticed that here we are dealing with the production of a flavor much heavier than "charm". The value b = 0.2 is in good agreement with the prescription $\langle p_T{}^2 \rangle \cong m^2/4$ (m is the quark mass) used by Odorico[10] to compute the "charm" production properties from flavor excitation.

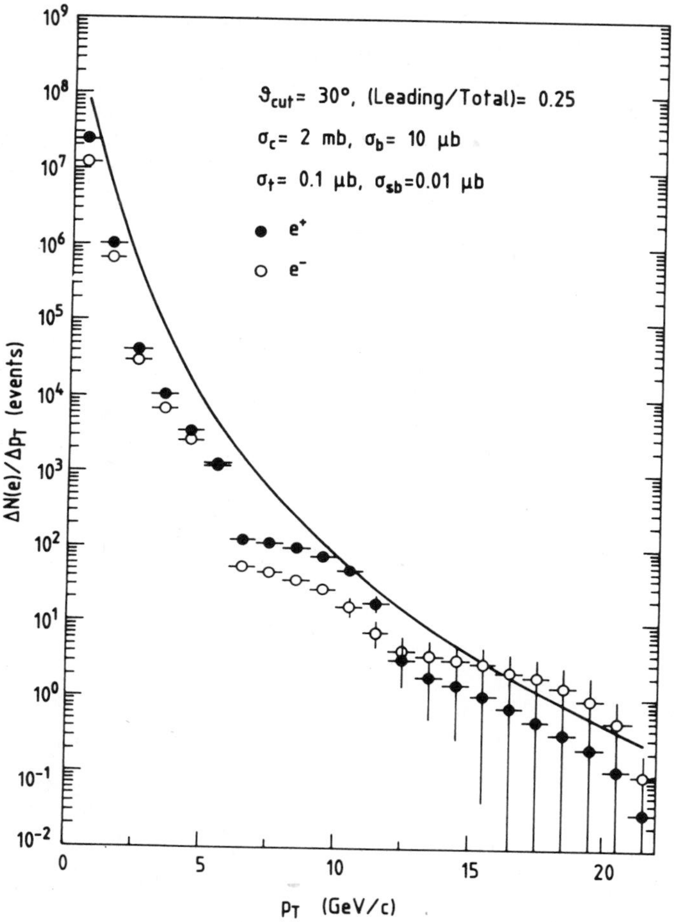

Fig. 33b. Expected number of produced electrons as a function of p with the same assumptions as in Figure 29, and the two cross section estimates: (a) from formula (6); (b) from perturbative QCD. The errors are statistical.

The efficiency for detecting electrons with p_T > 15 GeV/c in the Monte Carlo simulation does not change very much for b > 0.2, as shown in Figure 39. Thus the total "superbeauty" cross-section derived by Equation (8) holds, within ± 30%, even with this very low but expected value of b.

As mentioned above, the "down-like" nature of the observed (11±3) events cannot be established in a direct way. It is based on a chain of self-consistent arguments. Let us give up the mass ratio

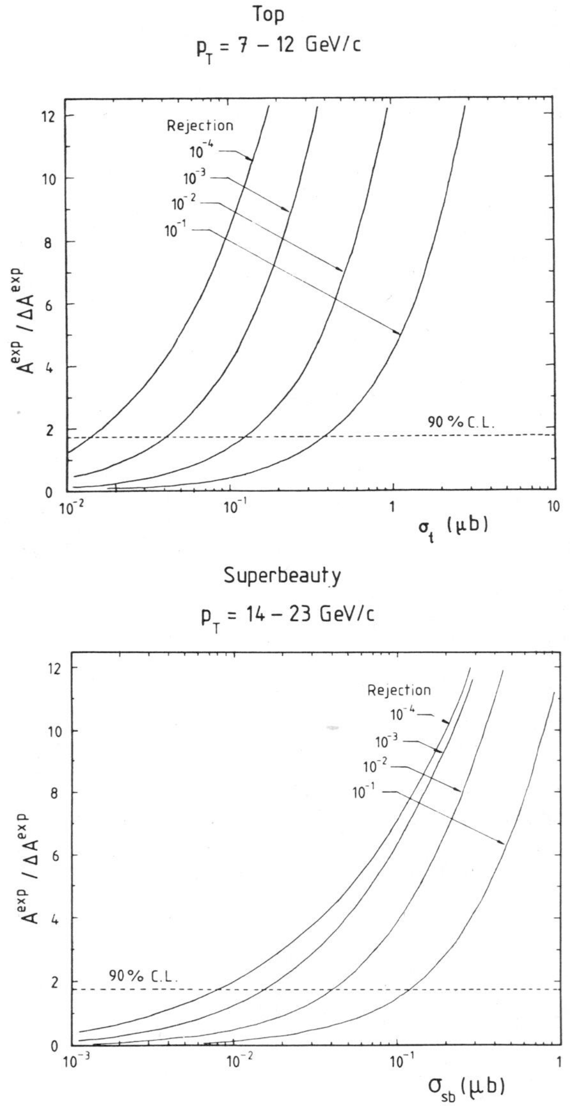

(a)

(b)

Fig. 34. The statistical significance of the measurement of A^{exp} ($A^{exp}/\Delta A^{exp}$), is shown, for a total luminosity of L = 300 nb^{-1}. θ_{cut} = 30° and (Leading/Total) = 0.25, and for the two p_T ranges: (a) p_T = 7-12 GeV/c; (b) p_T = 14-23 GeV/c.

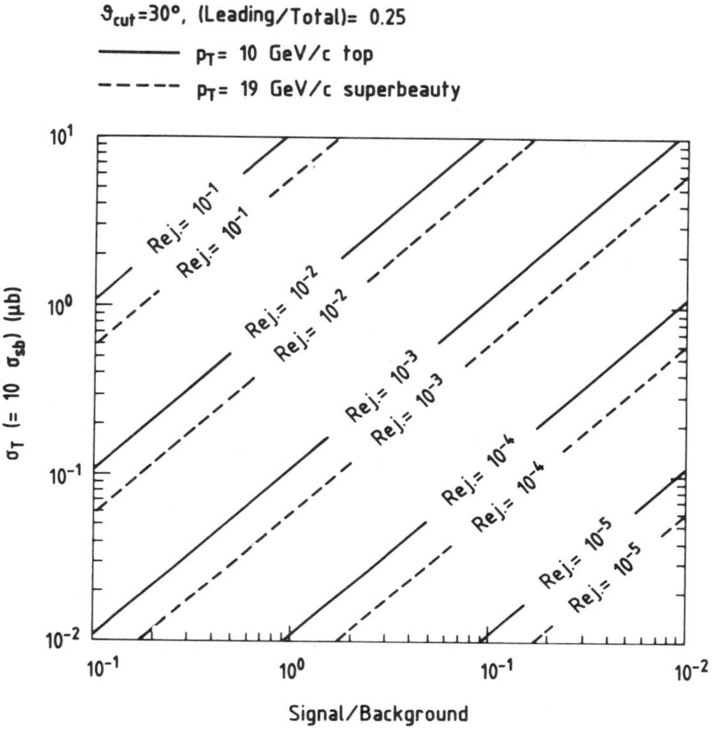

Fig. 35. Correspondence between rejection power and signal-to-background ratio (Signal/Background) as function of total cross section for "top" and "superbeauty" production, and at the two p_T values: 10 GeV/c and 19 GeV/c.

(1) which binds the "top" flavor to be in the 25 GeV/c² range. If we repeat the analysis without this constraint, the (11±3) events can be reinterpreted as the "up-like" signature with the "top" flavor mass 30 GeV/c² above the beauty flavor. The value of the cross-section would in this case be as shown in Figure 37b.

A sequence of arguments based on known facts and on simple hypotheses, extrapolated to the CERN ($p\bar{p}$) Collider energies, allow us to conclude that the (11±3) events observed by the UA1 Collaboration and consisting each of a single ($e^{\pm}$) accompanied by a jet activity in the opposite hemisphere, correspond to a value of the cross-section expected for the production of a very heavy flavored state, in the 55 GeV/c² mass range. Moreover, the observed transverse momentum spectrum of the ($e^{\pm}$) follows the expectations for the semileptonic decay of a very massive state, again in the 55 GeV/c² mass range.

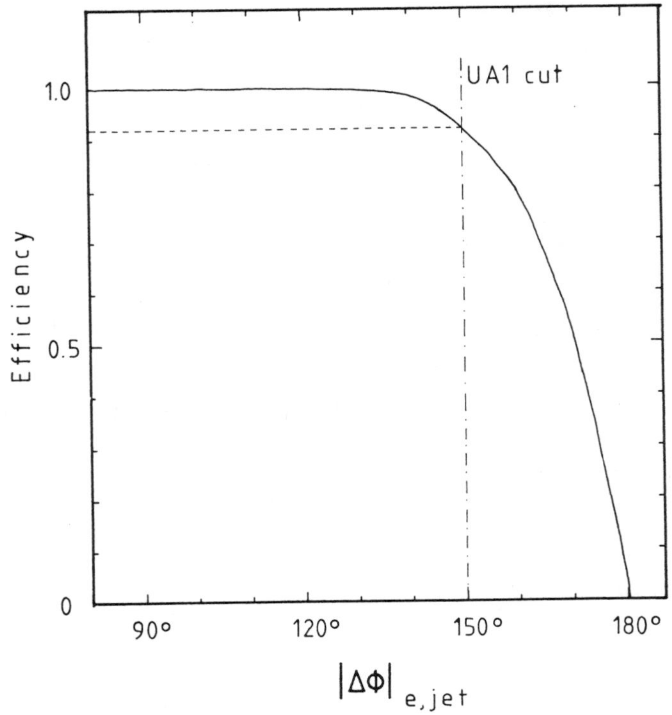

Fig. 36. Efficiency versus the cut value in the difference of azimuth $|\Delta\phi|$ between the electron and the hadronic jet in "superbeauty" decay, as derived from the Monte Carlo simulation.

It should, however, be noticed that the identification of the "down-like" nature of the heavy-flavored state is based on a series of hypotheses which produce the correct $\Delta m \cong 30$ GeV/c^2 for the $(e^{\pm})$ transverse momentum spectrum, and the correct magnitude for the cross-section. If we were to ignore the cross-section and the quark mass ratios which allow us to predict the masses of the 4th family, the only parameter left to fit the observed $(e^{\pm})$ transverse momentum spectrum would be the value $\Delta m \cong 30$ GeV/c^2. In this case the conservative interpretation of the UA1 results would be a "top" with a mass 30 GeV/^{2} above the "beauty".

This shows the importance of our proposal to study in detail the production of new heavy flavors at the CERN $(p\bar{p})$ Collider by measuring the (e^+/e^-) asymmetry. In fact, the sign of the asymmetry allows the identification of the "up-like" or "down-like" nature of the heavy-flavored state in a direct and unambiguous way. Moreover, the

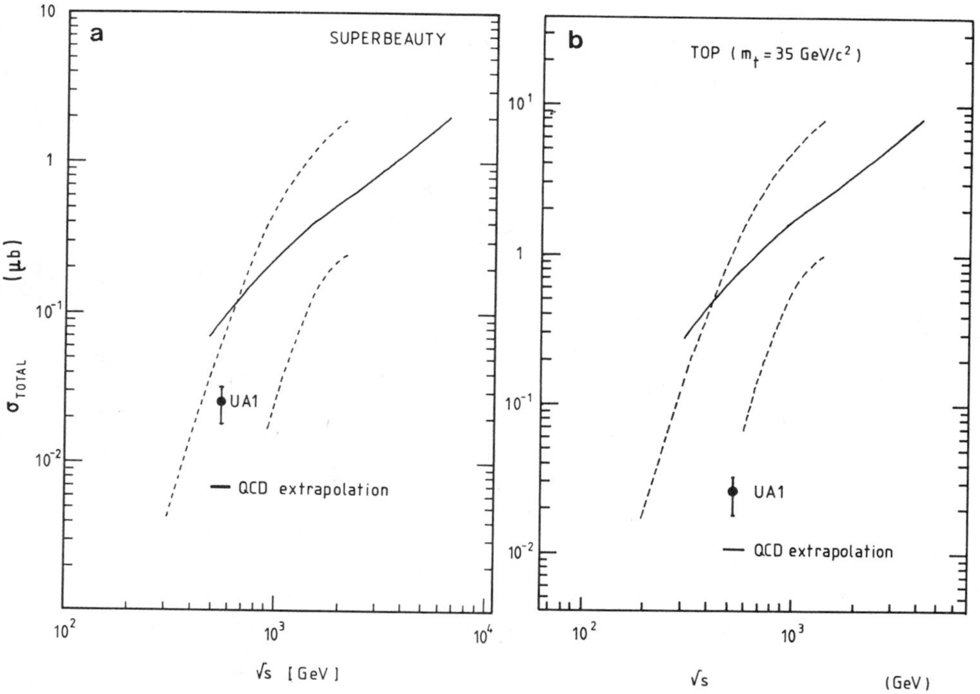

Fig. 37. Comparison between the UA1 results and the cross-sections for: (a) superbeauty (m_{sb} = 55 GeV/c^2; (b) top (m_t = 35 GeV/c^2); derived from strange (full line) and charm (dashed lines) cross-sections following formula (4).

(e^+/e^-) energy dependence of the asymmetry enables us to establish the correct sequence of

"down-like" ↔ "up-like"

decay chains for the new heavy flavors, and their mass difference.

7. CONCLUSIONS

The following conclusions are in order:

i) Past experience says: do not take too seriously the "theoretical" QCD predictions; many things still do not fit between theory and experiments. In particular, neither the large "charm" cross sections, nor the "Leading" effect were predicted.

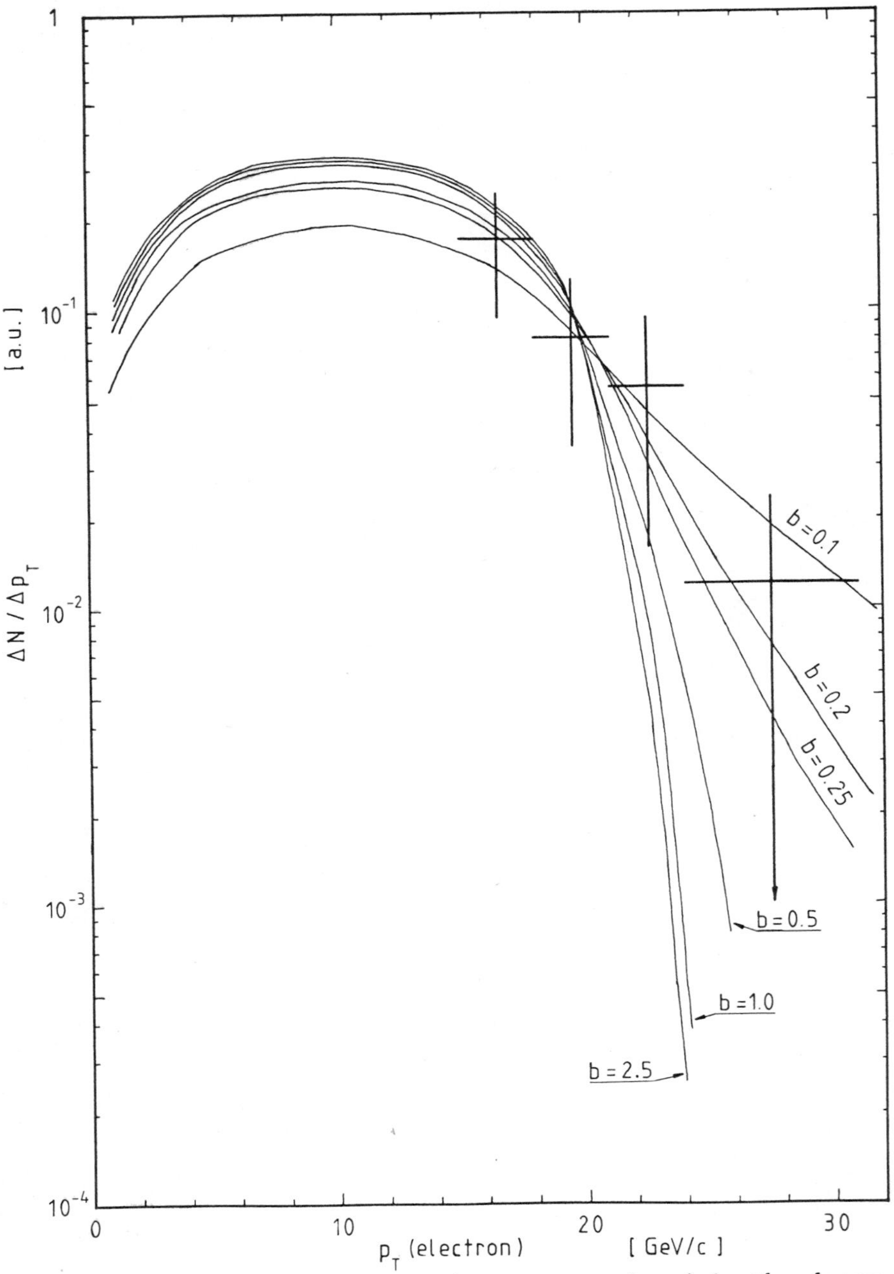

Fig. 38. p_T spectra of the electrons produced in the decay (sb) $\rightarrow$ teν, for different $(d\sigma/dp_T) \propto p_T \exp(-bp_T)$ production distributions of the parent (sb) particle, and comparison with UA1 data. The normalization is for $p_T > 15$ GeV/c.

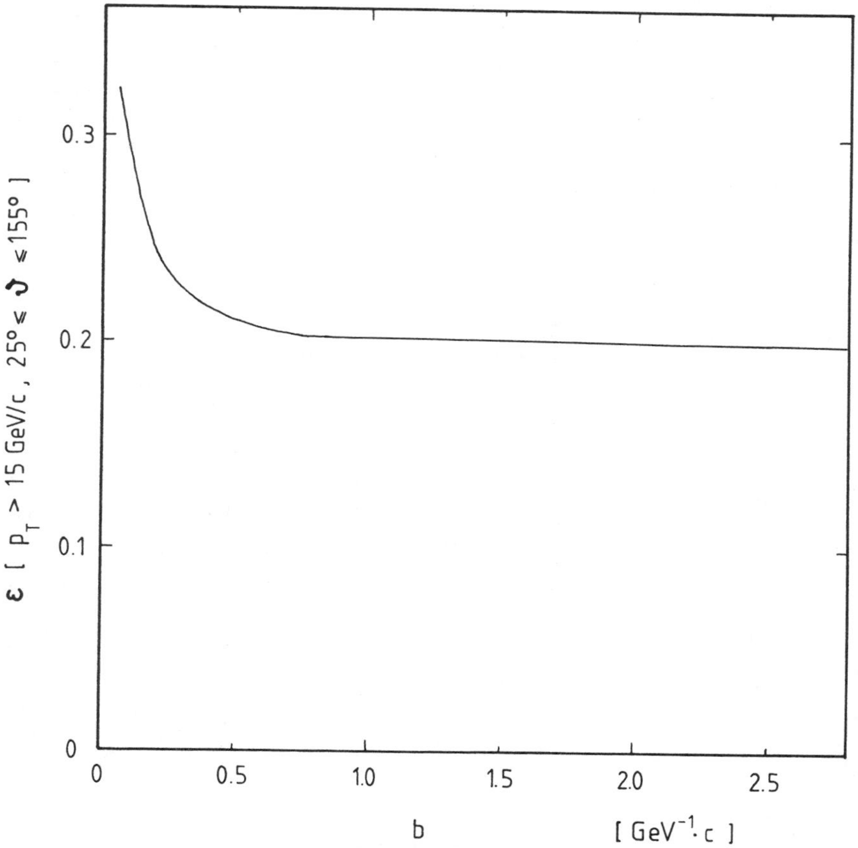

Fig. 39. Efficiency of the Monte Carlo simulation for the UA1 electron selection, as a function of the exponent parameter b in the parent (sb) p_T production distribution.

ii) A detailed study of the production mechanism of heavy flavors at the ISR is important in order to make reasonable extrapolations to the ($p\bar{p}$) Collider energy.

iii) The number of "new" states with heavy flavors is very large. If their production cross sections follow the simple extrapolation proposed by us, the CERN ($p\bar{p}$) Collider would be a quasi-factory for these new states. The problem is to have the instrumentation able to detect their existence.

iv) The study of the electron-positron asymmetry and of its energy dependence is of the utmost importance at the ($p\bar{p}$) Collider. If the "Leading" effect follows the same trend as "charm" at the ISR, this asymmetry is expected to be detectable, even if the production cross sections of the heavy flavored states would follow the QCD predictions.

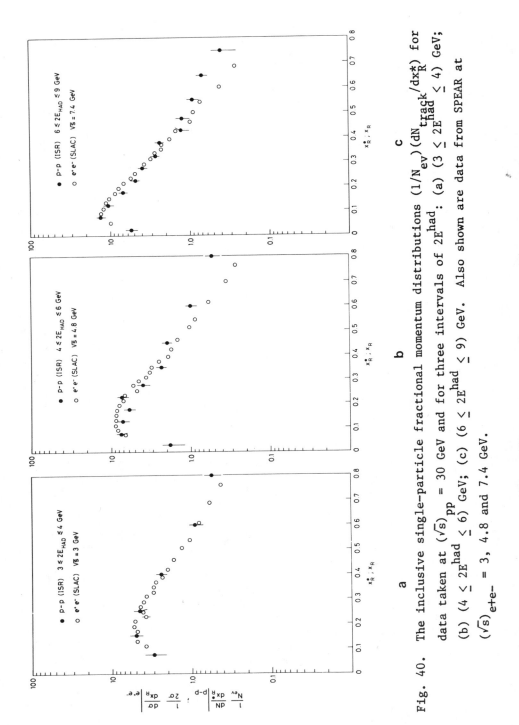

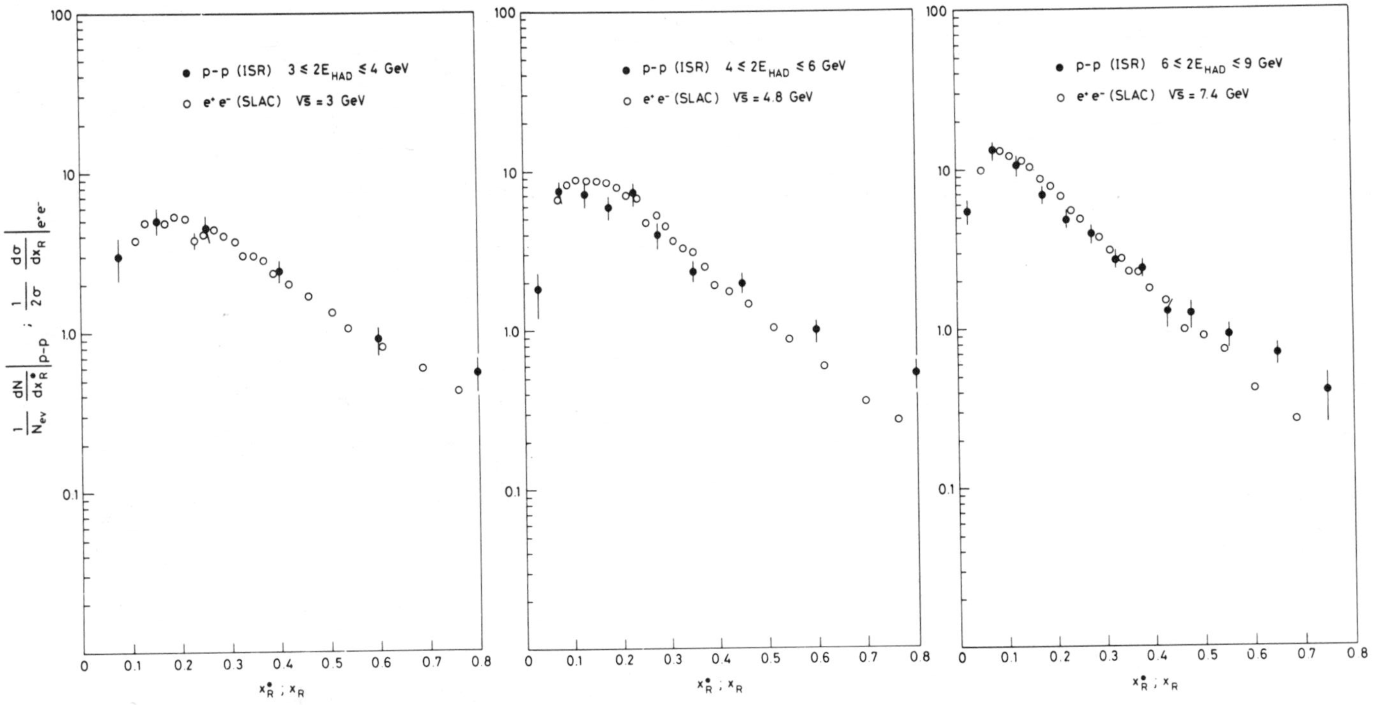

Fig. 40. The inclusive single-particle fractional momentum distributions $(1/N_{ev})(dN_{track}/dx_R^*)$ for data taken at $(\sqrt{s})_{pp} = 30$ GeV and for three intervals of $2E^{had}$: (a) $(3 \leq 2E^{had} \leq 4)$ GeV; (b) $(4 \leq 2E^{had} \leq 6)$ GeV; (c) $(6 \leq 2E^{had} \leq 9)$ GeV. Also shown are data from SPEAR at $(\sqrt{s})_{e+e-} = 3$, 4.8 and 7.4 GeV.

The study of the electron asymmetry and of its energy dependence is not less important than the searches for the Z° and the $W^{\pm}$. Finding the Z° and the $W^{\pm}$ would tell us nothing about one of the most crucial problems of Subnuclear Physics: the families problem.

A detailed study of the electron asymmetry allows to investigate the presence, in a mass range so far unaccessible to any other machines, of the new flavors up-like ("top") and down-like ("superbeauty"), and to determine their mass relation.

REFERENCES

1. S. Ferrara, private communication.
2. A. Martin, The masses of the Heavy Flavoured Hadrons, CERN, Preprint TH-3314 (1982).
3. M. Basile et al., Nuovo Cimento, 63A:230 (1981).
4. M. Basile et al., Nuovo Cimento, 65A:457 (1981).
5. M. Basile et al., Nuovo Cimento, 67A:40 (1982).
6. D. Drijard et al., Phys.Lett., 85B:452 (1979); K. Giboni et al., Phys.Lett., 85B:437 (1979); W. Lockman et al., Phys.Lett., 85B:443 (1979); J. Eickmeyer et al., XX ICHEP (Madison, 1981); D. Drijard et al., Phys.Lett., 81B:250 (1979); P. F. Jacques et al., Phys.Rev., D21:1206 (1980); P. Coteus et al., Phys.Rev.Lett., 42:1438 (1979); A. Soukas et al., Phys.Rev. Lett., 44:564 (1980); A. E. Asratyan et al., Phys.Lett., 74B:497 (1978); P. Alibran et al., Phys.Lett., 74B:134 (1978); T. Hansl et al., Phys.Lett., 74B:139 (1978); A. Bosetti et al., Phys.Lett., 74B:143 (1978); M. Fritze et al., Phys.Lett., 96B:427 (1980); D. Jonker et al., Phys.Lett., 96B:435 (1980); H. Abramowicz et al., CERN Preprint CERN-EP/82-17 (1982); M. Aguilar-Benitez et al., CERN Preprint CERN-EP/81-131 (1981); T. Aziz et al., Nuclear Phys., B199:424 (1982); J. Sandweiss et al., Phys.Rev.Lett., 44:1104 (1980).
7. M. Basile et al., Lett.Nuovo Cimento, 30:487 (1981).
8. M. Basile et al., Lett.Nuovo Cimento, 33:33 (1982).
9. F. Halzen, W. Y. Keung and D. M. Scott, Madison Report MAD/PH/63 (1982).
10. R. Odorico, Phys.Lett., 107B:231 (1981).
11. V. Barger, F. Halzen and W. Y. Keung, Phys.Rev., D24:1428 (1981).
12. M. Basile et al., Nuovo Cimento, 65A:391 (1981).
13. J. Badier et al., XXI Intern. Conf. on High Energy Physics, Paris, 1982.
14. F. Halzen, Rapp. Talk XXI Intern. Conf. on High Energy Physics, Paris, 1982.
15. K. Chadwick et al., Preprint CLNS/82/546 (1982).
16. M. Basile et al., Lett.Nuovo Cimento, 30:481 (1981).
17. M. Basile et al., Lett.Nuovo Cimento, 33:17 (1982).

18. L. Olsen, Monriond Workshop on New Flavours, Les Arcs, France, 1982.
19. UA2 Collaboration, Phys.Lett., B118:203 (1982).
20. G. Arnison et al., UA1 Collaboration, Preprint CERN-EP/83-13 (1983).

DISCUSSION

CHAIRMAN: A. ZICHICHI

Scientific Secretary: J. Bagger

- *OHTA:*

Although your proposal to measure charge asymmetry is simple and interesting, is this method viable even when top and superbeauty are almost degenerate in mass?

- *ZICHICHI:*

It is true that the charge asymmetry will vanish if up-like and down-like quarks are degenerate, but this would be an accident of nature. We should look for evidence of the asymmetry, and not expect any accidental cancellation.

- *NEMESCHANSKY:*

You mentioned that we need heavy quarks to get phenomenologically acceptable gluino masses. How model-dependent are these results?

- *ZICHICHI:*

These results are terribly model-dependent.

- *FERRARA:*

Many people have investigated this question, including Barbieri, Masiero and Girardello. Heavy fermions give one-loop fermion masses of a few GeV.

- *ZICHICHI:*

This argument should not be taken as the main reason for performing the experiment. We should not be too biased by our theoretical friends. We should measure this asymmetry at the CERN $p\bar{p}$ Collider, and any result would be interesting, even zero.

- *NEMESCHANSKY:*

How do your results depend on the heavy quark masses?

- *ZICHICHI:*

We have investigated various possibilities, and we have demonstrated that there are good reasons to expect the asymmetry. Playing with the parameters gives different results, but the real interest is in making the measurements.

- *HOU:*

Has charm been found at the ISR? If so, in what experiment and in what channel? I understand that searching for charm has always been a problem at the ISR.

- *ZICHICHI:*

This figure shows the world data on charm production in pp collisions. It is very unlikely that all these experiments are wrong. It is too pessimistic to say that charm has not been established in hadronic reactions. For more information I refer you to my report at the 1981 EPS Conference in Lisbon.

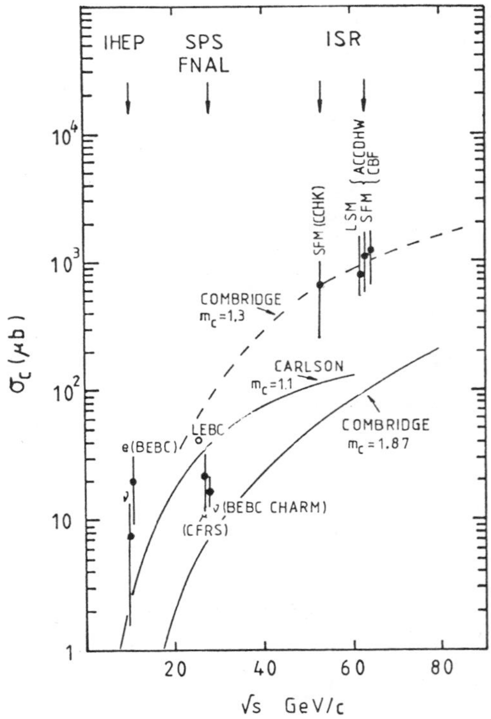

- *HOU:*

Do these experiments see a Λ_c mass peak?

- *ZICHICHI:*

Not only that -- they also demonstrate leading-particle production.

- *HOU:*

Has the charm asymmetry been measured for charm?

- *ZICHICHI:*

The effect is difficult to see because the Λ_c mass is too small and the p_T of the e^+ is too low to be clearly distinguished from the background.

- *SALATI:*

Can you detect the supersymmetric partners of the quarks? If so, how would you tell the difference between a heavy quark and the supersymmetric partners of a quark?

- *ZICHICHI:*

It would be fantastic to find the supersymmetric partners of quarks when searching for heavy flavours.

- *MARUYAMA:*

What is the possibility of finding the fourth generation of leptons at the $p\bar{p}$ Collider? Are leptons harder to detect that quarks?

- *ZICHICHI:*

Production cross-sections for heavy lepton pairs at the $p\bar{p}$ Collider are much smaller than those for heavy quarks. This is because the heavy leptons must be created by virtual photons. Very massive time-like photons are damped by the inelastic strong form factor (see diagram).

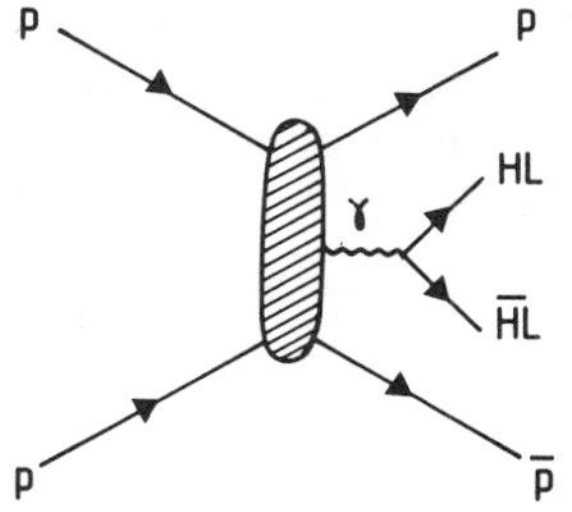

- *BAGGER:*

What is the theoretical reason for expecting significant production of heavy-flavoured baryons in a leading manner?

- *ZICHICHI:*

I will address this question in my lecture on universality features in pp, e^+e^-, and DIS processes. For the moment, let me say that the theoretical situation is not yet clear. However, experiments tell us that the final state of a hadronic process with two hadrons in the initial state do not produce an equal energy sharing among all particles of the final state. On the average, 50% of the total energy is taken by the two hadrons which propagate in the final state. This is why leading hadrons appear in the final state. These leading hadrons are formed from the incoming hadrons.

- *BENCHOUK:*

I would just like to comment that the leading proton effect has been obserbed by the EMC in deep-inelastic muon scattering.

- *ZICHICHI:*

I am glad that you confirm our DIS (pp) results.

- *MANA:*

You stated that supersymmetry requires a heavy quark of order 100 GeV in order to avoid a low-mass gluino. The contribution of quark and lepton loops usually goes as ln (q^2/m^2). How can a heavy quark have any effect on the gluino mass?

- *FERRARA:*

The mass of the heavy fermion is related to the mass splitting of its two scalar partners. The diagrams with the two scalar partners sum in such a way that the gluino mass is proportional to the mass of the heavy fermion. This is why you need a heavy fermion. This is true only in an SU(3) × SU(2) × U(1) theory without unification and with vanishing tree-level gaugino masses. In N = 1 supergravity the gaugino mass is an arbitrary input parameter. Also, in a supersymmetric grand unified theory, the heavy fermionic partners of the heavy gauge bosons contribute to the gaugino mass.

- *MANA:*

If the heavy quark is *not* a top quark, then supersymmetry requires a fourth family. Can a neutrino counting experiment rule out supersymmetry if the number of families turns out to be three?

- *ZICHICHI:*

Your proposal is very strange. You want neutrino counting to produce three families, and a new quark which is not the missing member of the third family. This has nothing to do with supersymmetry.

- *TAVANI:*

The search for heavy particles is related to one of the most important parameters of grand unified theories -- the number of families. It is difficult to determine this number in the laboratory. In your lecture you mentioned an astrophysical limit on the number of families. Can you elaborate on this?

- *ZICHICHI:*

This would involve a long explanation. Astrophysical arguments consider the He and D production in the early Universe. With more than four families, it would be difficult to explain the presently known He and D abundances.

- *HWANG:*

One motivation for the top quark was cancellation of the ABJ anomaly. Can you explain that cancellation?

- *ZICHICHI:*

Cancellation of the SU(2) × U(1) anomaly requires

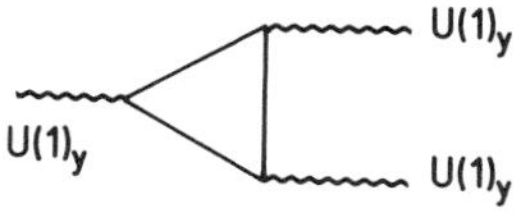

$$\sum_{\substack{\text{quarks}\\ \text{leptons}}} e_i^3 = 0$$

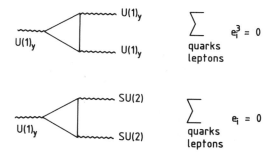

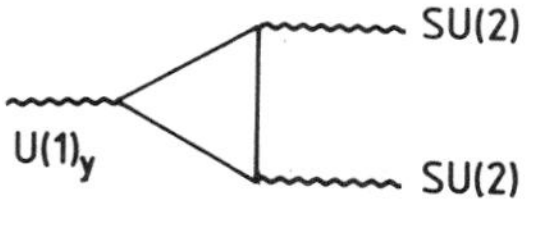

$$\sum_{\substack{\text{quarks}\\ \text{leptons}}} e_i = 0$$

where e_i is the hypercharge. The anomaly cancels if the top quark has hypercharge $^1/_6$ (electric charge $^2/_3$).

- *BERNSTEIN:*

What new apparatus do you need for running your experiment?

- *ZICHICHI:*

We need a powerful electron detector with good pion rejection. Neither UA1 nor UA2 would do.

- *BERNSTEIN:*

What are the systematic problems?

- *ZICHICHI:*

The systematics are no problem. When one measures an effect which changes sign with energy and geometry, the systematics are under control, if present.

- *BENCHOUK:*

My question is connected with the one of Bernstein. It is not clear to me how you treat the background from other sources of leptons. For example, in which hemisphere do you put leptons from central meson decays?

- *ZICHICHI:*

We have taken this into account in our Monte Carlo simulation, with central meson production falling like $(1 - |x|^3)$.

- *OLEJNÍK:*

What makes you believe in the regularity of the masses of up-like and down-like quarks (the factors of 4 and 10 you mentioned in your lecture)? Is there a theoretical argument?

- *ZICHICHI:*

There is no argument except simplicity. Professor Fritzsch, however, has told me that the factor of 4 within one generation can be explained within subconstituent models. From this point of view, the factor arises as the ratio of the squared up-like to down-like charges: $(\text{up-like})^2/(\text{down-like})^2 = (4/9)/(1/9) = 4$. We will hear more about this in his lecture. The ratio of masses of quarks in succeeding generations cannot be explained in this way. In any case, it is interesting to explore such simple hints without relying too much on theory. I would like to see whether the top quark mass is really in the 25 GeV range.

QCD AT THE COLLIDER

G. Altarelli

Cern, Geneva, Switzerland
Università di Roma I, Roma
INFN - Sezione di Roma

INTRODUCTION

Although QCD[1] imposes itself as the only theory of strong interactions within reach of the weapons of conventional quantum field theory, yet it is still the least established sector of the standard model. Testing QCD is, in fact, more difficult than testing the electroweak sector. In the latter domain the theory is more explicit because perturbation theory can always be applied. Besides that, the leptons and the weak gauge bosons are at the same time the fields in the Lagrangian and the particles in our detectors. On the other hand, QCD is the theory of quarks and gluons while only hadrons are observable; also, perturbation theory can only be applied in that particular domain of the strong interactions where approximate freedom, which is only asymptotic, can be reached.

This difficulty of testing QCD is the reason why substantial evidence in support of the theory has only been gathered during the last few years from a collection of converging indications from many different processes, while no single feasible test is by itself decisive.

We can actually identify three broad classes of QCD tests:

a) Purely topological tests Asymptotic freedom implies that a perturbative expansion is valid in the deep inelastic region. This fact induces a hierarchy of levels that one tries to expose by a suitable analysis. For example, at zero order in e^+e^- annihilation the naïve parton model is obtained. The final state consists of two jets and shows a collinear topology. At order α_s a fraction of three-jet events is also produced with a planar topology.

The transverse momentum increases in a plane, while $(p_T)_{out}$, the transverse momentum out of the event plane, stays fixed. In a three-jet event the slim jet is the same as the jets in a two-jet event (at a somewhat smaller energy). At order α_s^2 four jets appear, $(p_T)_{out}$ and the acoplanarity start increasing, the slim jet becomes fatter and so on. This group of experiments has led to spectacular successes for QCD, not only in e^+e^-, but also in deep inelastic scattering (DIS) (where the final state analysis is essential as a cross examination of the QCD explanation of scaling violations).

b) Correct estimates of rates and cross-sections For many processes, definite predictions are obtained at the leading approximation in α_s, which cannot be made completely quantitative, because of ignorance, either of next-to-leading corrections, or of some non-perturbative aspects. This kind of prediction is still significant because the answer often changes by orders of magnitude or is qualitatively different in alternative approaches. For example: jets in γ-γ processes, large p_T photons, charm production in DIS and photoproduction, slopes of p_T versus $\sqrt{S}$ (at fixed scaling variables), etc.

c) Fully quantitative tests The typical exponents of this class are $R_{e^+e^-}$, the scaling violations in DIS, the γ structure functions and, to some extent also, the three-jet distributions in e^+e^-, the widths of quarkonium decays, the Drell-Yan cross-section, etc. These are predictions that also include the complete evaluation of next-to-leading corrections, so that every quantity which appears in the leading term is well defined. In principle one can use these results in order to measure α_s, to fix the gluon spin, and to test the gauge nature of the strong interactions. For example, we report in Fig. 1 a number of measurements of α_s, in the $\overline{MS}$ definition. We see a plot of $\alpha_s(Q^2)$ for different values of $\Lambda_{\overline{MS}}$ (with next-to-leading accuracy). We observe that even a rough determination of α_s at low Q^2 is transformed into a rather restrictive prediction at large Q^2. Thus, the mere observation of precocious scaling, which is interpreted to imply $\alpha_s \lesssim 1$ at $Q^2 \sim 1$ GeV^2, leads to a strong constraint on the value of α_s at the largest available Q^2, if QCD is right. This prediction is in fact in agreement with the existing measurements in spite of many uncertainties and ambiguities in the extraction of α_s from the data. Note that the sensitivity of α_s on $\Lambda_{\overline{MS}}$ decreases with Q^2, so that at $Q^2 \sim 10^3 \div 10^4$GeV an order of magnitude in Λ corresponds to less than a factor 1.5 on α_s.

The importance for QCD of experiments at the SPS collider rests on the possibility they offer of testing parton dynamics in a new and highly non-trivial configuration. For example, hadron-hadron interactions in the deep inelastic, large p_T, region are non-linear in parton densities. The relevant predictions cannot be derived

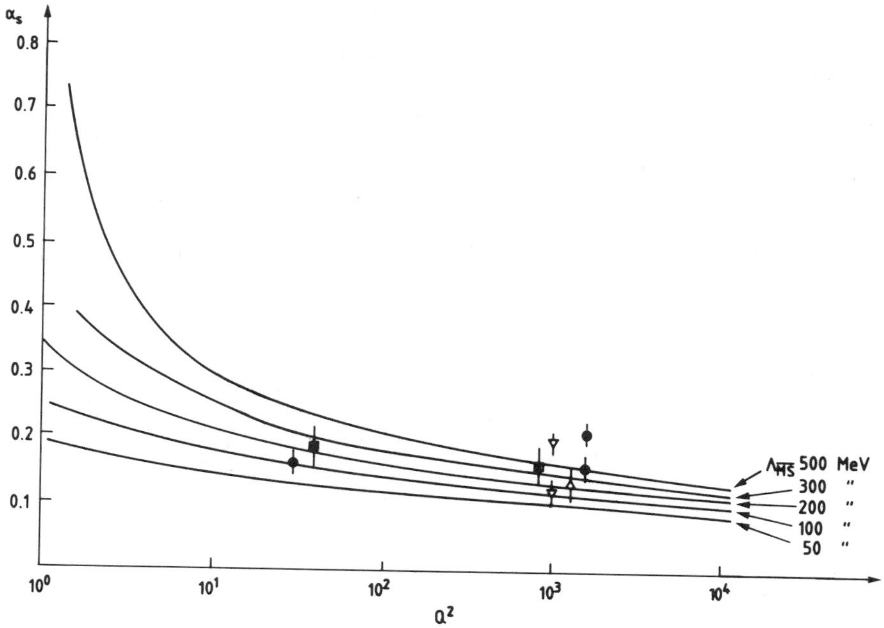

Fig. 1 α_s as a function of $\Lambda_{\overline{MS}}$ to next-to-leading accuracy $\alpha_s = \alpha_0[1 - b'\alpha_0 \ln\ln Q^2/\Lambda^2_{\overline{MS}}]$ with $\alpha_0 = 1/(b \ln Q^2/\Lambda^2_{\overline{MS}})$, $b = (33 - 2f)/12\pi$, $b' = (153 - 19f)/[2\pi(33 - 2f)]$, for $f = 4$. The experimental points are from: ● Υ decays, ■ CDHS (deep inelastic scattering), ◉ TASSO (e^+e^-), △ MARK J (e^+e^-), ▽ CELLO (e^+e^-), ⊠ JADE (e^+e^-).

by less committed formulations than the explicit QCD improved parton model, such as, for example, light cone dominance and operator expansion. This complexity, which is important for providing qualitatively new testing grounds is, however, paid for by a loss of precision in predictive power. Thus experiments at the SPS collider give a very valuable contribution to classes a) and b) of QCD tests as listed above.

In addition to that, $p\bar{p}$ collisions are also important as jet sources with an energy scale comparable to that of an e^+e^- ring with beam energy of about 50 GeV and more. As partons are studied through their jets and there are many kinds of partons, one needs different jet factories in order to be able to disentangle quark flavours and gluons. In particular, the problem of distinguishing gluon jets is of special relevance. As we shall see, most of the jets produced at large p_T and high energies in $p\bar{p}$ collisions are predicted to be gluon jets. This fact adds further interest to the jet studies that will be carried out in the future at the SPS collider.

INCLUSIVE JET YIELD

The clear observation of large p_T jets of the SPS collider[2] is in itself a distinct success of the QCD parton picture of deep inelastic processes. Once more, the early apparent failure of a parton prediction has been demonstrated to be merely due to lack of sufficient energy. Not only that, but the observed cross-section for inclusive jet production, as a function of p_T and $\sqrt{S}$, agrees remarkably well with the QCD estimates[3-6] (see Fig. 2). This is particularly significant in that the very steep dependence from energy and transverse momentum greatly facilitates the appearance of large factors, such as, for example, in the comparison[3] of jet yields at ISR[7] and at SPS energies.

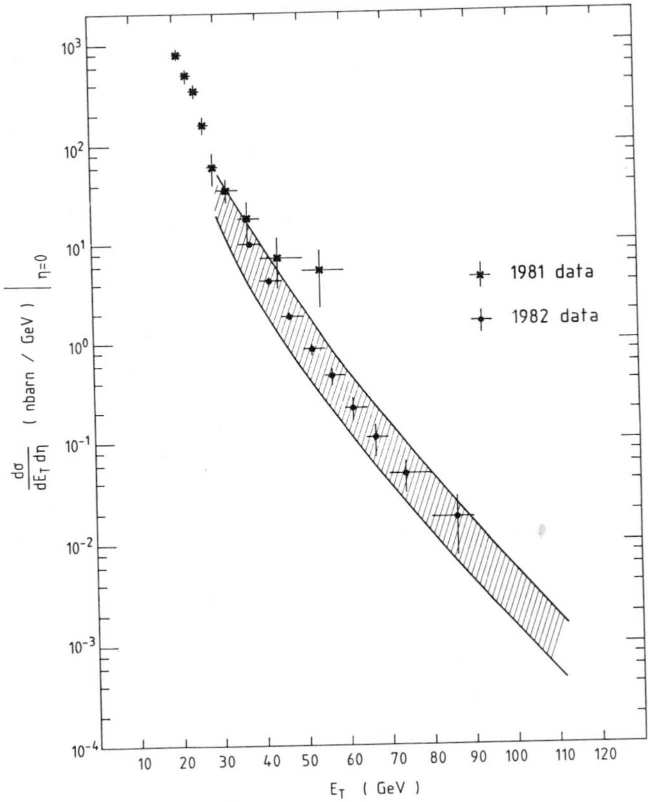

Fig. 2 The jet p_T distribution (as measured by UA2) compared with the QCD predictions[3-6]. The shaded band is obtained from different forms of parton densities and different values of Λ as explained in Ref. 6.

The leading term of the inclusive jet yield is computed by a sum of convolutions of products of two given parton densities with the corresponding cross-section into any two final partons, evaluated in lowest order QCD:

$$\frac{Ed\sigma}{d^3p} = \sum_{A,B} \int dx_1 dx_2 p_A(x_1,p_T^2) p_B(x_2,p_T^2) \delta(s+t+u) \alpha_s^2(p_T^2) \sum_f \frac{|M|^2_{AB\to f}}{s} \tag{1a}$$

where s,t and u are the Mandelstam variables for the (massless) parton subprocesses and $p_{A,B}$ are the parton densities in the colliding hadrons. In the Table we report the relevant parton processes and their associated $|M|^2$. Also shown is a figure of merit F_M, namely the value of $|M|^2$ at $\theta = 90^o$ in the parton centre of mass. In this frame,

$$s = 4\varepsilon^2; \quad t,u = -2\varepsilon^2(1 \pm \cos\theta) \tag{1b}$$

so that at $\theta = \pi/2$, s:t:u = 4:-2:-2. The importance of qg and gg subprocesses at large angles is immediately seen.

Table

Parton cross-sections: $d\sigma/dt = (\pi\alpha_s^2/s^2)|M|^2$ [averaged (summed) over initial (final) colours and spins]. s,t,u refer to the parton processes. F_M is the value of $|M|^2$ at c.m. angle $\theta = 90^o$ (s:t:u = = 4:-2:-2).

PARTON PROCESS	$\|M\|^2$	F_M
$qq' \to qq'$ $q\bar{q}' \to q\bar{q}'$	$\frac{4}{9}\frac{s^2+u^2}{t^2}$	2.22
$qq \to qq$	$\frac{4}{9}(\frac{s^2+u^2}{t^2} + \frac{s^2+t^2}{u^2}) - \frac{8}{27}\frac{u^2}{st}$	3.26
$q\bar{q} \to q'\bar{q}'$	$\frac{4}{9}\frac{t^2+u^2}{s^2}$	0.22
$q\bar{q} \to q\bar{q}$	$\frac{4}{9}(\frac{s^2+u^2}{t^2} + \frac{t^2+u^2}{s^2}) - \frac{8}{27}\frac{u^2}{st}$	2.59
$q\bar{q} \to gg$	$\frac{32}{27}\frac{u^2+t^2}{ut} - \frac{8}{3}\frac{u^2+t^2}{s^2}$	1.04
$gg \to q\bar{q}$	$\frac{1}{6}\frac{u^2+t^2}{ut} - \frac{3}{8}\frac{u^2+t^2}{s^2}$	0.15
$qg \to qg$	$-\frac{4}{9}\frac{u^2+s^2}{us} + \frac{u^2+s^2}{t^2}$	6.11
$gg \to gg$	$\frac{9}{2}(3 - \frac{ut}{s^2} - \frac{us}{t^2} - \frac{st}{u^2})$	30.4

Precise trigger requirements are demanded in order for Eq. (1a) to apply. An inclusive trigger is necessary for us to be able to identify the $\vec{p}_T$ of each final parton with the measured $\vec{p}_T$ of the corresponding jet (defined as the sum of the $\vec{p}_T$ of the hadrons in the jet). An inclusive trigger in principle allows us to do without fragmentation functions, thus simplifying the expression of the cross-section. On the other hand, the presence of jets must be defined by demanding energy clustering in a limited region of phase space. The allowed cones must be sufficiently small to define precisely the jet momentum and sufficiently large to include most of the jet fragments.

For central production, i.e., for almost zero rapidity, which is the dominant configuration, each incoming parton has an energy $\varepsilon = xE$ where E is the P or $\bar{P}'$ c.m. energy. That is, the energy fraction of the two incoming partons are nearly the same, $x_1 \simeq \simeq x_2 \simeq x$. Then the average energy per parton is given by $\bar{E} = = (\sqrt{S}/2)\bar{x}$, where $\sqrt{S} = 540$ GeV at the collider and $\bar{x}$ is the average x of a given kind of partons in the nucleon. For valence $\bar{x} \simeq \simeq 0.25 \div 0.30$, which corresponds to $\bar{\varepsilon} \simeq 70 \div 80$ GeV. The final parton transverse momentum is $p_T = \varepsilon \sin\theta$. The perturbative region is at large angle θ, where p_T is almost as large as $\bar{\varepsilon}$, i.e., $\bar{x}_T \simeq 2\bar{p}_T/\sqrt{S} \lesssim \bar{x}$. At the collider, this means $p_T \gtrsim 10 \div 20$ GeV. At such small values of x_T, the production of gluon jets is seen to be dominant, both because there are many gluons in the nucleon at small x and because the colour factors are favourable. At fixed x_T, $d\sigma/dp_T^2$ is predicted to fall as p_T^{-4}, apart from logarithms:

$$\frac{d\sigma}{dp_T^2} = \frac{f(x_T, \alpha(p_T^2))}{p_T^4} \tag{2}$$

At fixed S a change in p_T implies a proportional variation of x_T. The steeper dependence on p_T which is observed in this case is well reproduced by the decrease of parton densities as x_T is increased.

There are a number of ambiguities and uncertainties that affect both the predicted cross-section and the data. First the measurements are affected by systematic errors, which are quoted to be of 60% for UA1 and of 40% for UA2 (in addition to the errors shown in the plots such as that in Fig. 1). From the point of view of theory there is, first of all, a scale ambiguity, as in all leading logarithmic calculations. In Eq. (1a), the parton densities $p(x,Q^2)$ and the coupling $\alpha_s(Q^2)$ all depend on the scale Q^2. It is only by a computation of next-to-leading terms that in principle one can distinguish a priori equivalent possibilities, such as $Q^2 = p_T^2$, $\frac{1}{2}p_T^2$, $2stu/(s^2+t^2+u^2)$, etc. Note that the QCD evolution must be applied to the parton densities as measured in DIS at $Q_0^2 \sim 5\div20$ GeV^2, to boost them up to $Q^2 \simeq 10^3\div10^4$ GeV^2. It makes a quite sizeable difference if one stops at $Q^2 = p_T^2$ or $\frac{1}{2}p_T^2$ or $2p_T^2$. Scaling violations influence the shape of the p_T distribution at fixed energy by a very significant amount, which is somewhat different for different choices of the scale. Also note that the rate is proportional to $\alpha_s^2(Q^2)$ which doubles its sensitivity to first-order variations of α_s. A related ambiguity is the choice of Λ. For α_s this is exactly the same as the scale ambiguity, because α_s only depends on the ratio Q^2/Λ^2. For the evolution equations the situation is different, because the solutions only depend on Q^2/Λ^2 provided the initial conditions are also scaled in terms of a fixed Q_0^2/Λ^2 ratio.

On the other hand, the calculation of non-leading terms is quite difficult because there are many subprocesses. The only existing[8] fragment of this calculation refers to $q_i + q_j \rightarrow q_k + x$ where i,j,k indicate different flavours. In this case the "correct" scale (the one which minimizes the next-to-leading corrections) was found to be $Q^2 = \frac{1}{2}p_T^2$. Also in Ref. 4) the large π^2 terms in the corrections to all subprocesses were computed.

The calculation of next-to-leading terms is not only very involved, but also it is not clear that it would be completely worthwhile in view of the fact that the gluon density in the nucleon is poorly known. In fact, the exact definition of the gluon (or any other) density is to be specified in terms of some measurement. Different definitions change $g(x,Q^2)$ by terms of order α_s. In DIS this is not very important because the gluons only enter at order α_s in the structure functions, so that a variation of order α_s in $g(x,Q^2)$ only affects the structure functions at order α_s^2. But in the case of jet production in $p\bar{p}$, the gluons already enter in lowest order, so that, if the next-to-leading corrections to the production rate are to be meaningfully treated, one must know the gluon density to a resolution which is comparable to the terms of order α_s. This is certainly not the case because the gluon density is only roughly measured from the scaling violations in DIS, which are themselves a small effect (order α_s). Fortunately, the sensitivity of the different empirical forms of the gluon density appears[5] to be incidentally reduced in the interval of x values involved in the calculation of the jet yield at the relevant energies. In fact, in this particular range of x values the various determinations of the gluon density show a limited dispersion.

These ambiguities do not allow a greater accuracy than a factor of two or so. In spite of that the observed agreement with the data is still highly significant because of the very steep dependence of the jet yield on $\sqrt{S}$ and p_T.

At small angles, i.e., for $\Lambda^2 << p_T^2 << \varepsilon^2$, the situation becomes more complicated. Terms of order $\log \varepsilon^2/p_T^2 \sim \log \theta^2$ are no longer negligible in comparison to $\log p_T^2/\Lambda^2$. There are two scales in the parton processes that make all the scale ambiguities more acute. In addition to that, it also becomes more difficult to disentangle spectator jet effects. Further work is needed in order to extend the theory at small angles (as well as to different triggering conditions).

ANGULAR DISTRIBUTION OF JETS

The study of the angular distribution of jets offers a valuable double check of the underlying QCD dynamics by testing the vector nature of gluons. From the table, for each subprocess, the angular distributions in $\cos\theta$ are simply derived (averaged over $\cos\theta \rightarrow -\cos\theta$ because one cannot tell which jet is which), θ being the scattering angle of the subprocess in the parton centre of mass. This angle can be identified with the jet-beam angle in the large p_T jets rest frame, provided that one neglects the errors introduced by the intrinsic momentum spread of partons in the nucleon and by fragmentation effects. For example, for $gg \rightarrow gg$ one finds:

$$\frac{dn}{d\cos\theta} \sim 3 - \frac{\sin^2\theta}{4} + \frac{4(1+3\cos^2\theta)}{\sin^4\theta} \qquad (3)$$

The Coulombic singularity $\sin^{-4}\theta$ is typical of gauge vector boson exchange. In the subprocesses $qg \rightarrow qg$ and $gg \rightarrow gg$, it arises from the three-gluon vertex, while in $q\bar{q} \rightarrow q\bar{q}$ it would be present in the Abelian case as well. It is thus of particular interest to study experimentally the jet angular distribution near the forward direction. A word of caution is, however, necessary in that we already remarked that other contributions than the Born parton diagrams become important at small angles. The latter presumably are important at angles such that $\alpha_s/\pi \ln \theta^2 \sim 1$, so that one expects that the much stronger Coulombic singularity is still unaffected at not too small θ values, where it should be already clearly visible.

The comparison with the data is shown in Fig. 3. The larger acceptance at small angles of UA1 is crucial in order to probe the interesting angular region. The results are encouraging but more statistics at small angles are needed.

GLUON JETS VERSUS QUARK JETS

A lot of the jets produced at large P_T at the collider are predicted to be gluon jets. It would be important to find an objective difference between the jets produced in $p\bar{p}$ collisions (mostly gluons) and in e^+e^- (mostly quarks). Up to now, the difference did not show up, although maybe the matter has not yet

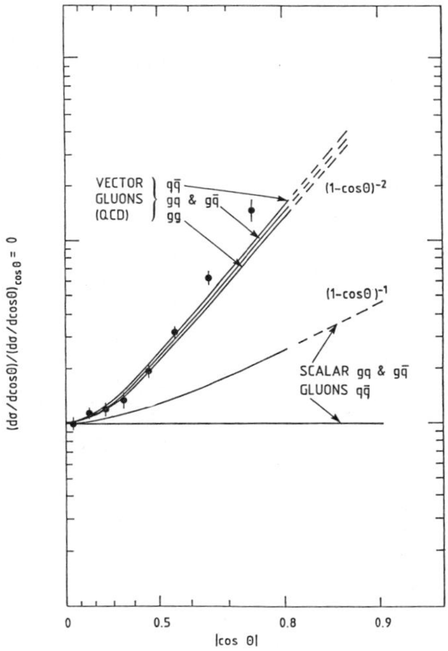

Fig. 3 The jet angular distribution (data from UA1)

been attacked with adequate means. The longitudinal momentum distributions of particles in jets of comparable energies look the same at $p\bar{p}$ and in e^+e^-. The same is true for the p_T distributions of jet fragments (transverse with respect to the jet axis in this case). We recall that some indications have actually been reported by the JADE collaboration[9] at PETRA that in a three-jet event in e^+e^- the p_T spread of fragments is more pronounced for the less energetic jet (which has the largest probability of being a gluon). I also recall that there are hints from several sources[9] (DIS, quarkonium decay final states versus continuum, etc.) that baryons are more abundant in gluon than in quark jets. Perhaps the most interesting analysis is the attempt[10] at comparing the multiplicity in the jets at $\bar{p}p$ and e^+e^-. The idea is that the multiplicity should be larger in a gluon jet than in a quark jet. In fact, at the level of parton multiplicities (i.e., as if the partons were directly observable) the ratio of 9 to 4 is asymptotically predicted[11] for gluon and quark jets respectively. However, some assumptions on fragmentation are needed in order to translate the parton multiplicity into a hadron multiplicity. The main experimental ambiguity which up to now makes the result still inconclusive, has to do with the criterion for attributing to one jet a given particle in the central region of $p\bar{p}$ where the four jets merge.

In conclusion, there is no evidence as yet of a difference between gluon jets and quark jets. However, the collider appears to be in a good position for adding much to this issue in the near future.

THREE JETS AT LARGE p_T

The next level in topological complexity is given by events with three jets at large p_T. In principle, all the experience gained from the analysis of three-jet events in e^+e^- can be carried to $p\bar{p}$. On the theoretical side, the very simple, well-known formula for the three-jet distribution[1] in e^+e^- is replaced in this case by a much more complicated expression, which is obtained by considering the whole plethora of $2 \rightarrow 3$ parton subprocesses. This calculation has been completed[12] as far as the distribution away from the collinear and soft singularities is concerned, while the more involved one-loop corrections to the $2 \rightarrow 2$ subprocesses have not been yet attacked (the addition of the virtual corrections is needed to compute the next-to-leading corrections to the two-jet production rate).

The result can be cast in a relatively simple form by using a method by Berends et al.[13] which is a generalization of the well-known factorization of infra-red singularities valid in the soft limit. They argue that in the massless theory, when no two four momenta are parallel, the $2 \rightarrow 3$ cross-section can be factorized as:

$$\sigma(1+2 \rightarrow 3+4 + g) = |M|^2 I \qquad (4)$$

where $|M|^2$ and I in the limit $k \rightarrow 0$ (k being the g four momentum) approach the non-radiative cross-section and the soft infra-red factor respectively. The correct form of the two factors, checked by comparison with the exact result, has been obtained for the relevant subprocesses.

These calculations allow us to derive a whole set of testable predictions to be compared with the data. In particular, a $2 \rightarrow 2$ process is planar, while a $2 \rightarrow 3$ process is not. Thus, an acoplanarity and a momentum component p_{out} (out of the "event plane") can be defined and their distributions studied. A preliminary comparison of the p_{out} distribution with the data shows a remarkable agreement (see Fig. 4).

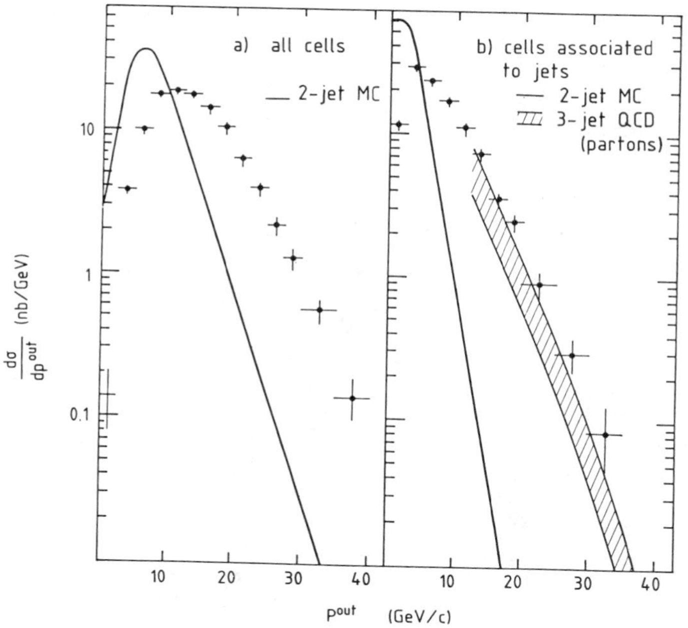

Fig. 4 The distribution in p_{out}, the momentum perpendicular to the plane defined by the trigger jet and the beam momentum.

MULTIPLICITIES

A wealth of very interesting data on multiplicity distributions have been collected at the SPS collider[14]. Experiment has shown that $dn/dy|_{y=0}$, i.e., the total charged multiplicity per unit rapidity y at y = 0, increases from ISR to collider energy at a rate compatible with $dn/dy|_{y=0} \sim \ell n\ S$. This implies that Feynman scaling is violated (which is no news) and suggests that the total average multiplicity $\bar{n}$ increases with energy with a law compatible with $\bar{n} \sim \ell n^2 S$, as confirmed by direct measurements of $\bar{n}$. The bulk of the multiplicity is determined by soft processes which lie far away from the domain of perturbative QCD. This is why we shall be very brief here. On the other hand, QCD leads to important predictions for the scale dependence of multiplicities within hard jets, such as, for example, those produced at large p_T^2 in $p\bar{p}$ or in e^+e^- annihilation at large Q^2. In hard jets, the parton multiplicity (i.e., as if partons could be directly observed) is predicted[15] to evolve in Q^2 with the law:

$$\bar{n}(Q^2) \simeq C(\ell n \frac{Q^2}{\Lambda^2})^{k_1} \exp(k_0\sqrt{\ell n\ Q^2/\Lambda^2}) \qquad (5)$$

where

$$k_0 = \sqrt{\frac{72}{33-2f}} \sim 1.7 \qquad (6)$$

$$k_1 = (\frac{111}{4} - \frac{11}{18}f)/(33-2f) \sim 1 \qquad (7)$$

(f being the number of excited flavours). The non-leading term with exponent k_1 has been recently derived[16]. Thus if one assumes that the hadronization mechanism does not significantly depend on Q^2, the same behaviour can be expected to hold for the measurable hadronic multiplicity in the jet. Extending this prediction to the total multiplicity in $p\bar{p}$ collisions is not justified, because the mechanisms at work in soft processes might be different. However, it is interesting to observe that $\bar{n} \sim \ell n^2 S$ and $\bar{n} \sim \exp\{k_0\sqrt{\ell n\ S/S_0}\}$ are both compatible [17] with the available data.

Another interesting experimental result is the approximate validity of Koba, Nielsen, Olesen (KNO) scaling. As is well known, Feynman scaling implies KNO scaling. Since the former is violated the question is whether there are arguments for the latter to be respected. The question can be discussed in QCD. Again one must distinguish the case of multiplicity distributions in hard jets and in soft processes. In the first case, one can study the distribution of colour singlet clusters of gluons in a jet[15,18]. This distribution shows KNO scaling broken by logarithms. Probably this conclusion is likely to be true in general. On the other hand, the shape of the KNO scaling function appears to be more model-dependent.

W and Z PRODUCTION

The discovery[19] of $W^{\pm}$ and Z^0 at the expected masses is so important at a very fundamental level that the QCD aspects of W and Z physics are relatively secondary. They are still very interesting, though. Indeed, the production mechanism is a Drell-Yan type process with $\sqrt{S} \simeq 540$ GeV and $Q \simeq 90$ GeV. All the very non-trivial QCD improved parton model predictions for this sort of process[1] can be confronted with experiment in a domain of energies far deeper inside the deep inelastic region than previously attained. In the context of W and Z physics, one interesting challenge to QCD theory from experiment is the measured shape of the transverse momentum distribution, (in addition to the determination of the production cross-section). The QCD theoretical analysis of p_T distributions in Drell-Yan processes not only allows a very severe and demanding test of the validity of our views on the underlying, quite complex parton dynamics, but is also interesting in itself because it involves many new issues not present in the calculation of the total production cross-section, such as, for example, the presence at intermediate p_T of two different scales and the related appearance of double logs.

The p_T distribution is built up from three components. The first component is the wave function spread of partons in hadrons, the so-called intrinsic p_T. This is a bound state effect, and is not within reach of theory at present. Thus, it is a model-dependent term in the distribution. However, we know empirically that the intrinsic p_T is small, roughly of the order of some units of Λ, as expected. Thus we expect that, as $(p_T)_{MAX}$ increases, this uncertainty will at most affect a thin slice of the p_T spectrum at small p_T. The second component is the "hard" component that determines the distribution at $p_T \sim Q$. At large values of p_T, $\alpha_s(p_T)$ becomes small and the lowest order contribution, where the heavy lepton pair is produced in association with a massless parton, is dominant. This term is completely under control. Finally, there is a "soft" component arising at $\Lambda << p_T << Q$ from multiple gluon emission which is enhanced by logs, in the sense that the effective coupling becomes $\alpha(p_T)\ell n\, Q^2/p_T^2$. One is able to take into account these logarithmically enhanced terms by resumming them to all orders in perturbation theory[20].

It is important to stress that while a complete determination of the p_T distribution requires the knowledge of all three sources of p_T as listed above, on the other hand, some particular quantities can be more directly obtained. The most important example is perhaps the slope of the average p_T (p_T^n) versus $\sqrt{S}(S^{n/2})$ at fixed $\tau = Q^2/S$:

$$\langle p_T \rangle = \alpha_s(Q^2)[1 + O(\alpha_s(Q^2))]\sqrt{S}\, f(\tau,\ell n\, Q^2) + \text{const.} \qquad (8)$$

While the constant terms (apart from logs) also contain non-perturbative contributions, the leading term is completely determined by the perturbative result. Actually, the experimental verification of the size with energy of p_T at fixed τ and the rough check of the rise of the slope in $\sqrt{S}$ is one of the striking semi-quantitative successes of QCD. The uncertainties which are left over by a lowest order evaluation have to do once more with a scale ambiguity, with the poor knowledge of the gluon density and with the presence of large corrective terms in the total cross-section (K factor[21]). In the quark-antiquark channel a calculation of the next-to-leading terms in the p_T distribution was completed[22]. This allows us to remove some important uncertainties, while those connected with the rather large size of the resulting corrections are still present.

Going back to the general problem of a calculation of the whole p_T distribution, the existing[23] theoretical p_T distributions for W and Z are still not completely satisfactory in that not all the available information is used, and, in some cases, useless and

and questionable assumptions were made. What it is important to stress is that a reasonable precise prediction of the p_T distribution is indeed possible, and it will be extremely important to test it in all its complicated features both in Drell-Yan processes at moderate energies, where statistics are large, and in W and Z production (or in Drell-Yan processes at collider energies in the future), where the energy is much larger.

REFERENCES

1. For a recent review, see, for example
   G. Altarelli, Phys. Reports 81:1 (1982).
2. UA1 Collaboration: G. Arnison et al., Phys. Lett. 123B:115 (1983);
   UA2 Collaboration: M. Banner et al., Phys. Lett. 118B:203 (1982).
3. B.L. Combridge, J. Kripfganz and J. Ranft, Phys. Lett. 70B:234 (1977);
   R. Horgan and M. Jacob, Nucl. Phys. B179:441 (1981).
4. N.G. Antoniou et al., Phys. Lett. 128B:257 (1983).
5. W. Furmanski and H. Kowalski, Nucl. Phys. B244:523 (1983).
6. Z. Kunszt and E. Pietarinen, CERN preprint TH.3584 (1983).
7. T. Akesson et al., Phys. Lett. 118B:185 (1982); 121B:133 (1983); 128B:354 (1983);
   A.L.S. De Angelis et al., Phys. Lett. 126B:132 (1983).
8. R.K. Ellis, M.A. Furman, H.E. Haber and I. Hinchliffe, Nucl. Phys. B173:397 (1980);
   M.A. Furman, Columbia Univ. preprint CU-TP-182 (1982).
9. W. Bartel, Proceedings of the EPS Conference, Brighton, 1983.
10. UA2 Collaboration: P. Bagnaia et al., CERN preprint EP/83-94 (1983).
11. See, for example,
    K. Konishi, A. Ukawa and G. Veneziano, Nucl. Phys. B157:45 (1979).
12. J. Kripfganz and A. Schiller, Phys. Lett. 79B:317 (1978);
    A. Schiller, J. Phys. G5:1329 (1979);
    C.J. Maxwell, Nucl. Phys. B149:61 (1979);
    T. Gottschalk and D. Sivers, Phys. Rev. D21:102 (1980);
    Z. Kunszt and E. Pietarinen, Nucl. Phys. B164:45 (1980).
13. F.A. Berends et al., Phys. Lett. 103B:124 (1981).
14. UA5 Collaboration: K. Alpgard et al., Phys. Lett. 115B:71 (1982).
15. A.H. Mueller, Phys. Lett. 104B:616 (1981);
    A. Bassetto et al., Univ. of Florence preprint 82/11 (1982), see also
    A. Bassetto et al., Nucl. Phys. B163:477 (1980);
    W. Furmanski et al., Nucl. Phys. B155:253 (1979).

16. A.H. Mueller, Columbia Univ. preprint CU-TP-247 (1982); see also
M. Ciafaloni, Proceedings of the EPS Conference, Brighton, 1983.
17. R.V. Gavai and H. Satz, Phys. Lett. 112B:413 (1982).
18. F. Hayot and G. Sterman, Stony Brook preprint ITP-SB-82-60 (1982);
G. Pancheri and Y. Srivastava, Phys. Lett. 128B:433 (1983).
19. UA1 Collaboration: G. Arnison et al., Phys. Lett. 122B:103 (1983), Phys. Lett. 126B:398 (1983);
UA2 Collaboration: M. Banner et al., Phys. Lett. 122B:476 (1983), Phys. Lett. 129B:130 (1983).
20. G. Parisi and R. Petronzio, Nucl. Phys. B154:427 (1979) [see also Y.L. Dokshitzer, D.L. D'yakonov and S.I. Troyan, Phys. Lett. 78B:290 (1978), Phys. Reports 58:269 (1980)];
J.C. Collins and D.E. Soper, Oregon Univ. preprint OITS-155 (1981).
21. G. Altarelli, R.K. Ellis and G. Martinelli, Nucl. Phys. B143:521 (1978), (E) B146:544 (1978); Nucl. Phys. B157:461 (1979);
J. Kubar-André and F. Paige, Phys. Rev. D19:221 (1979).
22. R.K. Ellis, G. Martinelli and R. Petronzio, Nucl. Phys. B211:106 (1983).
23. P. Chiappetta and M. Greco, Nucl. Phys. B199:77 (1982);
P. Aurenche and R. Kinnunen, LAPP preprint TH-78 (1983);
G. Pancheri and Y.N. Srivastava, Phys. Lett. 128B:235 (1983);
F. Halzen, A.D. Martin and D.M. Scott, Phys. Rev. D25:754 (1982).

D I S C U S S I O N

CHAIRMAN : G. ALTARELLI

Scientific Secretaries : G. D'Ambrosio and C. Ogilvie

- *KLEVANSKY* :

You mentioned that in a jet event with large p_T, color has to be rearranged in a plane, whereas in a $W \to e_\nu$ event color has to be rearranged on a line. Could you explain this, please ?

- *ALTARELLI* :

In the case of W production, experimentalists have observed that spectator jets are slimmer than in the case of large p_T jets. The W is observed as a very energetic electron, plus two jets in the forward and backward directions. When one sees a four-jet event with two back-to-back jets at large p_T and two spectator jets, the spectator jets are fatter in this case. The subprocess is, for example, two gluons going into two gluons, and there is a color mismatch. Hadrons must be born in this plane between any two contiguous jets to neutralize the color, which makes the forward jets fatter. In the case of the W, the underlying process is a q and $\bar{q}$ giving a W, and color rearrangement is a colinear process, since the W is colorless.

- *HOU* :

A question on quark and gluon jets : what are the possibilities of distinguishing them, using things like baryon abundance, charge multiplicity, et cetera ?

- *ALTARELLI* :

There are no sharp theoretical predictions that necessarily

must be respected, or the theory is wrong. On the other hand, since the quark and gluon are so different, there is a feeling that the difference should be observable. I feel, for example that the famous predictions of different multiplicities in gluon and quark jets are completely unreliable, because they are obtained as if partons were directly observable, which is not the case. It appears that the most promising effect is baryon abundance, as observed by the abundance of Λ's and p's at the upsilon, on and off resonance and also in the final state of deep inelastic scattering.

- *HOU* :

Toponium will certainly have jet-like decays. Would two- and three-gluon jets be at all distinguishable in these decays, for example at UA1 ?

- *ALTARELLI* :

Certainly the chance seems minimal of finding top through toponium production at the collider.

- *ZICHICHI* :

Open top is easier to observe because the cross-section drops as $(\text{mass})^4$ for hidden flavor.

- *EREDITATO* :

What is the state of the art in understanding the Fermi motion of partons inside hadrons ?

- *ALTARELLI* :

One must distinguish between the case of Fermi motion in heavy flavor decay and of nucleons in nuclei. One cannot compute these things precisely for hadrons. You ask for the state of the art; I'm not sure there is such an art, at least for hadrons.

- *MOUNT* :

Is there any possibility of finding convincing experimental evidence for the three gluon coupling, for instance, in toponium decays at an e^+e^- machine ?

- *ALTARELLI* :

It will be very difficult to find experimental confirmation of the three gluon coupling. Toponium decay doesn't have much to do with the three gluon coupling; it is mainly radiation of gluons from quark lines.

- *OGILVIE* :

People are starting to look at things like matrix elements in lattice QCD. Can we get information from this that would be useful, or would it be too crude ?

- *ALTARELLI* :

I am aware of some research in this area, and I do think useful information can eventually be obtained.

- *BENCHOUK* :

You mentioned that the p_T distribution of hadrons within a jet from $p\bar{p}$ collisions, where one is presumably dealing mostly with gluon jets, and in e^+e^- annihilation, which is mostly quark jets, look experimentally the same. On the other hand, you also showed results from e^+e^- where the third jet, presumably a gluon jet, has a broader p_T distribution. Aren't these results inconsistent ?

- *ALTARELLI* :

The results from e^+e^- are not so sure yet. Even if there is an effect it is so small that the precision from colliders is not sufficient to find it.

- *POSCHMANN* :

Can you measure α_s at the $p\bar{p}$ collider by counting the number of three-jet versus two-jet events ?

- *ALTARELLI* :

You can never measure α_s that way. The number of three-jet events is not defined. You must define what you mean by a three-jet event by making suitable cuts. Next, you must do a precise calculation up to next-to-leading order of the same quantity. However, this is not the simplest possibility. A much better way is to compute the subleading corrections to the rate, i.e. the total yield of large p_T jets, which is of order α_s^2. It is useless, however, to produce such a calculation if one cannot get a handle on the gluon distribution to a comparable precision.

- *ROTELLI* :

I have a question about the multiplicity prediction. The original (leading) prediction of $\exp\sqrt{\ln(s)}$ was claimed to be a good success of the planar quark model. Does the result of Mueller ruin this ?

- *ALTARELLI* :

All these fits depend on where you think a particular parametrization should start to be valid. The most significant thing is that the famous prediction of Feynman scaling of a ln(s) increase in multiplicity is now ruled out by experiment.

- *TAVANI* :

In W decay color must be rearranged on a line. In about 10% of the W events, there is a jet balancing a W which has significant P_T. Can we understand this qualitatively ?

- *ALTARELLI* :

In W production, the transverse momentum of the W is not necessarily small, because there can be a hadron jet balancing the transverse momentum. In canonical QCD, this can be interpreted as the radiation of a gluon in the production mechanism. It is an interesting question to measure the distribution of these events. I do not think the computations of this have been done well yet, as I explained this morning.

- *GASPARINI* :

Experimental low energy fits of Λ which gave $\Lambda \simeq .5$ GeV are not consistent with the highest energy results. How do you explain this from a theoretical point of view ?

- *ALTARELLI* :

The extraction of Λ is a delicate matter. Essentially, the different values come from a different way of analysing data. Also, the earlier experiments included lower energy data so that pre-asymptotic effects in the low q^2 region can affect the results. It should be remarked that the relevant physical quantity is α_s, not Λ whose value doesn't depend very much at high q^2 on the precise value of Λ.

TESTS OF QCD AT PETRA

Min Chen
Massachusetts Institute of Technology
Cambridge, Massachusetts 02139

INTRODUCTION

Quantum Chromodynamics (QCD)[1] is a gauge theory based on the group SU(3) of color, which is necessary in order to explain deep inelastic scattering, the π^0 lifetime, and the total hadronic cross section measured in e^+e^- annihilation.[2]

The key ingredients and predictions of the QCD theory are as follows:

(a) The carriers of the force are eight doubly colored gluons with spin = 1.

(b) Quarks and gluons are not directly observable. Their properties are inferred from a study of their fragmentation products, the hadrons.

(c) Gluons, unlike the photons, can couple to other gluons via the triple gluon vertex.

(d) A natural consequence of (c) is that the jets from gluons should be broader than the quark jets.

(e) The interactions of quarks and gluons are characterized by a unique parameter, the strong fine structure constant α_s.

(f) α_s decreases at shorter distances or larger momentum transfer q^2.

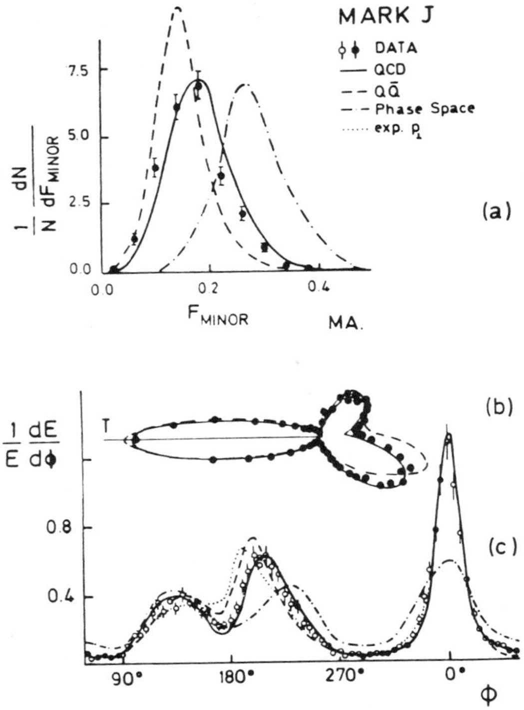

Fig. 1 (a) The distribution $N^{-1}\ dN/dF_{minor}$ in the fraction of the visible energy flow of the entire event which is projected along the minor axis (perpendicular to the event plane).

(b) Comparing the Mark J data with QCD and $q\bar{q}$ models, using energy flow diagrams in the thrust major event plane for events with $O_b > 0.3$, $T_n > 0.98$ or $\theta_{minor} > 60^o$.

(c) Same as Fig. 1b but with $T_n < 0.98$ and $\theta_{minor} < 60^o$.

(d) The unfolded energy flow diagram of Fig. 1c compared with the models of QCD, $q\bar{q}$, phase space, and a $q\bar{q}$ model with an $\exp(-P_T/650\ \text{MeV})$ fragmentation distribution.

All of the above features have been investigated extensively by the five detectors (CELLO, JADE, MARK-J, PLUTO AND TASSO) at PETRA,[3] the currently highest energy e^+e^- collider. Owing to limited time, in this talk I will cite only a few examples to illustrate the major progress made in our understanding of QCD in the past few years. The list is by no means complete.[8]

I. DISCOVERY OF THREE JETS

Studies of hadronic event topology at SPEAR[4] and DORIS[5] have shown that in the process $e^+e^- \to$ hadrons, the hadrons are produced with limited transverse momentum (P_T) with respect to the original quark and antiquark directions. As the e^+e^- center of mass energy increases, the Lorentz boost due to the high velocities of the fragmenting quarks results in two back-to-back jets which are more and more collimated.

PETRA began with a center of mass energy around 12 GeV and has gradually increased its energy to 42 GeV. The low energy (12 to 20 GeV) data could also be described by a two-jet model ($e^+e^- \to \bar{q}q$) with an average quark transverse momentum of $\sim$ 300 MeV/c with respect to the jet axis. However, as the energy increased to the level $\sim$ 30 GeV, the data could no longer be described so simply. A detailed analysis of the event topology showed[6] that events often had a "narrow" jet, in which $\langle P_T \rangle \sim 300$ MeV/c, and a "broad" jet in which $\langle P_T \rangle \sim 450$ MeV/c in a certain plane and $\langle P_T \rangle \sim 300$ MeV/c out of that plane. An analysis of the energy distribution in the event plane of a sample of events with large $\langle P_T \rangle$ in the plane revealed that most of the energy is concentrated in three collimated regions, leading to the discovery of three jet events[7] which are shown in Fig. 1. We see that the rate and distribution of these three jet events agree well with QCD predictions but not with other types of models such as phase space and $\bar{q}q$ models with Gaussian or exponential distributions. For other evidences of gluon jets, one should refer to more detailed review works[8] on QCD.

II. SPIN OF THE GLUON

The discovery of three jet events establishes the mechanism of hard

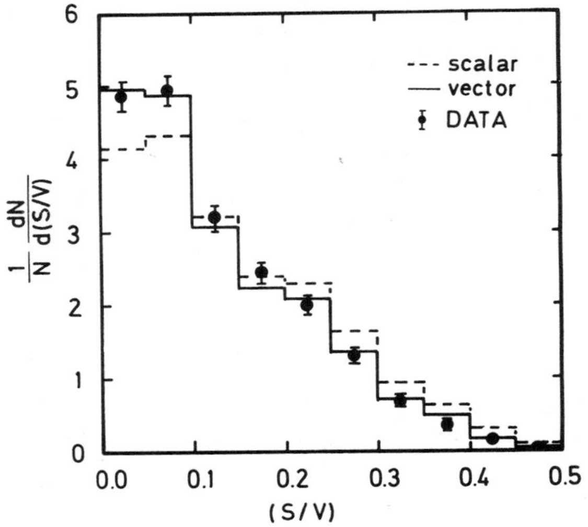

Fig. 2 The distribution of the ratio of spin 0 to spin 1 matrix element $X_3^2/(x_1^2 + x_2^2)$ for Mark J data (black points), and predictions assuming vector gluon (solid histogram) and scalar gluon (dashed histogram).

gluon bremsstrahlung. In lowest order this process is described as

$$e^+e^- \to q\bar{q}g.$$

Denoting by x_1 and x_2 the fractional energies of the quarks, with $x_i = 2E_i/\sqrt{s}$, the cross section for gluon emission is given by

$$\frac{d\sigma(q\bar{q}g)}{dx_1 dx_2} = \frac{2\alpha_s}{3\pi} \sigma_o \frac{x_1^2 + x_2^2}{(1-x_1)(1-x_2)}$$

for a gluon with spin = 1 and

$$\frac{d^2\sigma}{dx_1 dx_2} = \frac{\alpha_s}{3\pi} \sigma_o \frac{x_3^2}{(1-x_1)(1-x_2)}$$

for a gluon with spin = 0, where $\sigma_o = 3\sigma_{\mu\mu}\Sigma e_q^2$. Let θ_i be the angle opposite to the ith jet in a three jet event. The parton energies x_i are related to the angles θ_i by kinematics:

$$x_i = \frac{2\sin\theta_i}{\sin\theta_1 + \sin\theta_2 + \sin\theta_3}$$

The x_i are ordered such that $x_1 > x_2 > x_3$. The distribution of $x_3^2/(x_1^2+x_2^2)$, which is the ratio of spin 0 and spin 1 gluon matrix elements, is thus a sensitive test of the spin of the gluon. Fig. 2 shows the MARK-J data[9], which favor the vector gluon hypothesis over the scalar gluon hypothesis. Independently, one can define a quantity

$$\cos\tilde{\theta} = \frac{x_2 - x_3}{x_1} .$$

In the rest frame of the second and third partons, for massless partons $\tilde{\theta}$ is the angle between the first parton and the axis of the second and third partons.

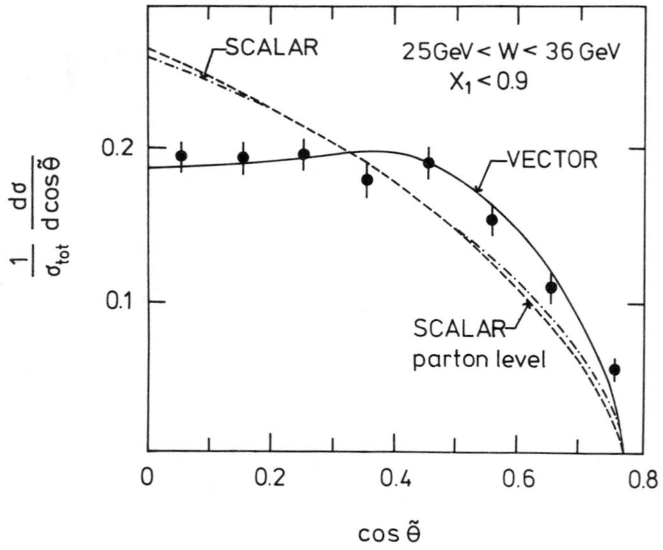

Fig. 3 The cos $\tilde{\theta}$ distribution for the TASSO data and predictions assuming vector (solid) and scalar (dashed) gluons.

Fig. 3 shows the distribution of $\cos\tilde{\theta}$ measured by the TASSO group[10] together with the QCD prediction (vector gluon) and the prediction with a scalar gluon. The data agree with the vector hypothesis and rule out a scalar gluon.

III. IS THE GLUON JET BROADER THAN THE QUARK JET?

An important step in testing QCD is to identify the gluon jet and to analyze its structure. This is predicted to be different from that of quark jets of the same energy, as a result of the larger color charge of the gluon and of the 3-gluon coupling. The analysis carried out by the JADE group[11] compares jets of similar energy but, according to perturbative QCD, with different gluon content. The models used are essentially limited to the first order QCD processes $e^+e^- \rightarrow q\bar{q}$ and $q\bar{q}g$, with which one studies the question of whether a model with identical quark and gluon fragmentation can describe the data and how sensitive the observed hadron distributions are to different schemes of gluon fragmentation. Global properties of jets such as transverse and parallel momentum distributions, particle and energy flow, and average particle multiplicity have been investigated. Two methods were used in the analysis:

(1)

The lowest energy jet, which is most likely produced by a primary gluon, is compared with the other two jets within 3-jet events with $\sqrt{s}$ = 33 GeV. In the energy range of 6 to 10 GeV the third jet has about 50% probability to be from the gluon, while the second jet has a probability of about 25%. Fig. 4 shows the average transverse momentum of charged and neutral particles within a jet relative to the jet axis for the three jets for (a) the experimental data, (b) for a model assuming identical gluon and quark jets, (c) for a model with broader gluon jets (500 MeV/c), and (d) for the color string Lund model, which will be described in section IV in detail. Here we see that the average P_T is larger for the third jet than for the second jet, contrary to the prediction of the g = q model, but that it is in agreement with the broader gluon hypothesis and also with the Lund model.

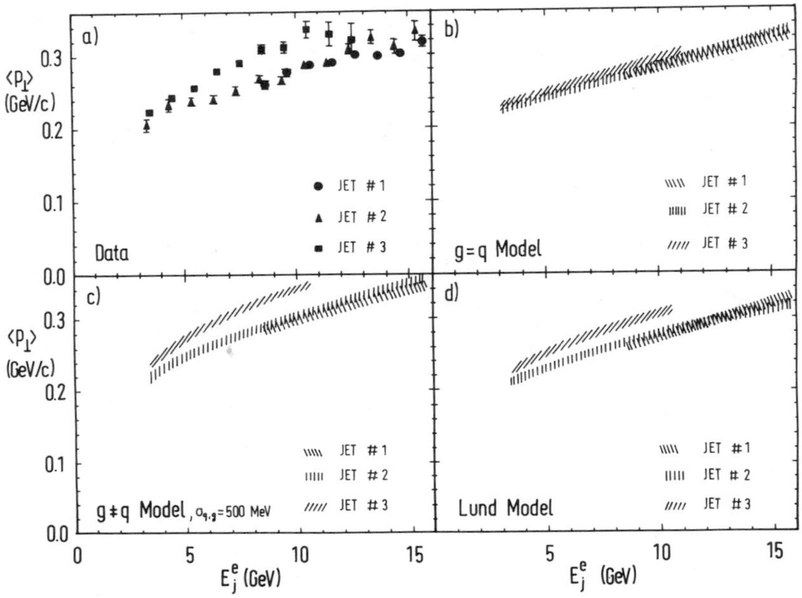

Fig. 4 The distributions of the average transverse momentum of charged and neutral particles within a jet relative to the jet axis for the three jets as a function of the jet energy for (a) the experimental data, (b) for a model assuming identical gluon and quark jets, (c) for a model with broader gluon jets (500 MeV/c), and (d) for the color string Lund model.

(2) The lowest energy jet from 3-jet events at $\sqrt{s} = 33$ GeV is compared with jets from 2-jet events at $\sqrt{s} = 14$ GeV. Fig. 5 shows the differential P_T distribution of jet number 3 and that of jets from 2-jet events at $\sqrt{s} = 14$ GeV together with the same distributions predicted by the g = q model. We see that the third jet at $\sqrt{s} = 33$ GeV (which half of the time is the gluon) is broader than the second jet at $\sqrt{s} = 14$ GeV, which is dominated by quark jets, in contradiction with the prediction of the g = q model.

To describe these observations by standard models, the gluon fragmentation has to differ from the quark fragmentation. For independent parton fragmentation parametrized according to Field and Feynman, one needs a mean transverse momentum $\sigma_q \simeq 300$ MeV for quark jets, and $\sigma_q \simeq 500$ MeV for gluon jets to describe the data. The Lund model, where by definition the gluon fragments differently, also provides a good description of the particles within jets.

IV. DETERMINATION OF α_s TO COMPLETE SECOND ORDER QCD

As described earlier, the discovery of three-jet events at PETRA has been interpreted in terms of the emission of energetic gluons. The amount of 3 jet events has been used to determine α_s to the first order of QCD to be[12] around ~ 0.2. Since this number is quite large it is necessary to go to higher order to obtain more reliable values of this strong coupling constant. To second order the process $e^+e^- \rightarrow$ hadrons can be visualized via the following steps:

(1) Shortly (~ 10^{-24} seconds) after the electron-postron pair annihilates into a quark-antiquark pair the quarks radiate one or two gluons or an additional quark-antiquark pair as shown in Fig. 6. The coupling strength between quark and gluon is proportional to α_s.

(2) The partons fragment into colorless hadrons in a way that is not yet calculable perturbatively.

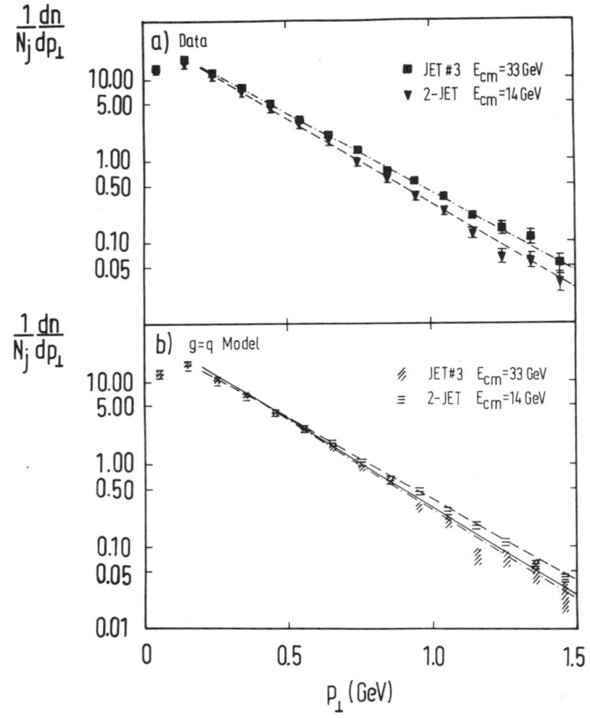

Fig. 5 (a) The differential P_T distribution of jet number 3 and that of jets from 2-jet events at $\sqrt{s}$ = 14 GeV; (b) the same distributions predicted by the g = q model.

Therefore, a fragmentation (hadronization) model must be used to connect the perturbative calculation of the parton distribution with the measured hadron distribution. Furthermore, since perturbative QCD makes no clear prediction about the parton distribution in infrared (low energy) regions, we must make sure that the method of determining α_s does not depend upon the cuts which are used to eliminate the infrared divergences.

The first attempt of measuring α_s to second order QCD was made by the JADE group(13) fitting the thrust distribution of the three jet cross section to an analytic formula by FSSK(14). They obtained $\alpha_s = 0.16 \pm 0.015 \pm 0.03$. However, as we will show in the following sections, the result using thrust distribution is dependent on the cut values. Also terms of the order $0(\varepsilon)$ have been omitted in the analytic formulas.

The determination of α_s by the MARK-J group(15) uses a complete second order QCD calculation, and employs two hadronization models in order to investigate the fragmentation model dependence of α_s. The cut dependence of the measured values of α_s for several independent experimental quantities are also investigated and their relative merits demonstrated.

We measure the rate of flat events, in which energetic gluons are emitted, by two methods. In the first method we attempt to identify these events by using the event hemisphere containing the broad jet which most likely contains the gluon jet and its parent (anti) quark jet. This we achieve by determining the event axis (thrust) and the event plane on an event-by-event basis. All event parameters described earlier are determined for the event. Our studies indicate, however, that events having large oblateness for the broad jet ($O_b > 0.3$) are preponderantly those where one or more energetic gluons have been emitted, and thus the rate of such events is a measure of α_s. In the second procedure, we need not identify jets at all; rather, we use the energy flow in all of the hadron events to measure the gluon emission. This method uses the correlations in energy flow.[16] This energy correlation function is written as

$$\frac{1}{\sigma}\frac{d\Sigma}{d\cos\chi} = \frac{1}{N}\sum_{\text{event}}^{N} \sum_{i,j} \frac{E_iE_j}{E_{vis}^2} \delta\ (\cos\chi_{ij}-\cos\chi)$$

where the sum is over all hadronic events, E_i is the energy measured in the solid angle element i, χ_{ij} is the opening angle between E_i and E_j, and E_{vis} is the measured total event energy. This is a particularly elegant and unbiased analysis requiring no selection of hadronic events with special topology and it is infrared finite. The non-zero correlation at $\chi \neq 0$ or π arises from the presence of large angle energetic gluons ($e^+e^- \rightarrow q\bar{q}g$ or $e^+e^- \rightarrow q\bar{q}gg$) and the components of transverse momentum which the visible hadrons receive in the fragmentation process. To isolate the effect arising from gluon emission we use the asymmetry in $\cos\chi$

$$A\ (\cos\chi) = \frac{1}{\sigma}\left(\frac{d\Sigma}{\cos\chi}\ (\pi-\chi) - \frac{d\Sigma}{d\cos\chi}\ (\chi) \right)$$

The Monte Carlo studies have shown that the region $|\cos\theta| < 0.72$ receives only a small contribution from the two jet events, thus allowing a comparison with a QCD calculation.

To calculate the hadronic cross section to second order in α_s we use the standard three level calculation[17] of three- and four-parton events and complete the calculation by doing a Monte Carlo integration of the second order virtual contributions computed by Ellis, Ross, and Terrano.[18] We have attempted to ensure that our calculation is independent of the cutoffs employed by using two distinct procedures.[19] In one, we impose a minimum energy fraction on each of the partons and a minimum angle between parton pairs, in the spirit of the Sterman-Weinberg

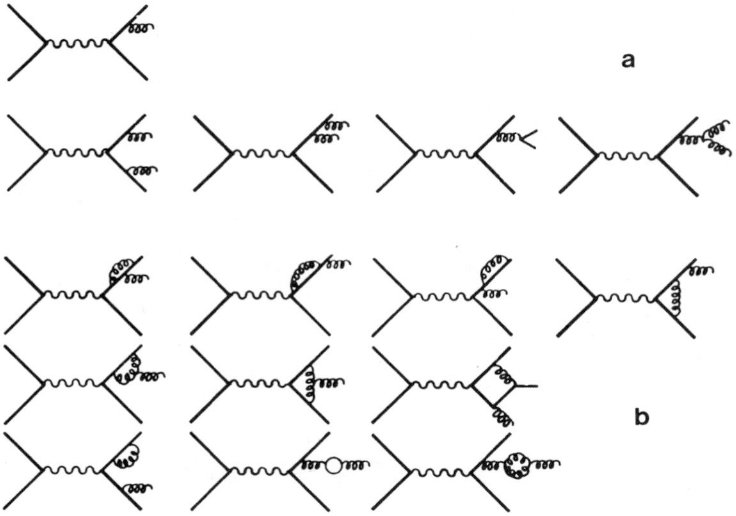

Fig. 6. a) Tree level diagrams for 3- and 4-parton final states.

b) Second order virtual parton diagrams for 3-parton final state.

prescription[20], i.e., the so-called ε and δ cuts. In the other we require that the invariant mass, m, of all pairs of partons in an event exceeds a given fraction of the available energy. The matrix element calculated by ERT[18] is exact, corresponding to the case of infinite resolution. It is therefore suitable for generating Monte Carlo events. The problem of finite resolution is solved by applying either of the above two cuts to the events generated.

In Fig. 7 we display the results of this calculation for the rate of low thrust events and the rate of large O_{b_2} events versus the mass cutoff, expressed as the fraction $x = m^2/s$. We also display the integrated asymmetry

$$\text{Asy} = \int_0^{0.75} A(\cos\chi)\ d\cos\chi$$

for several values of the minimum parton energy fraction, using a fixed minimum parton energy fraction, using a fixed minimum parton-parton angle of 12.6°. The curves show the Monte Carlo results of these three quantities at the parton level as functions of the cuts, while the points are Monte Carlo results after fragmentation and detector simulation. Similar dependence on the cut parameter persists with and without fragmentation. We note that the integrated asymmetry yields a result which is stable for the full range of the parameters.

To compare this calculation with the data, which is a measurement on real hadrons and not partons, we must use a model of fragmentation of the partons into hadrons (hadronization). Two different models were used to estimate the effect via a Monte Carlo simulation. In the first model, implemented by Ali et al.[17] the three or four partons fragment independently of one another so that the jets are formed close to the direction of each parton according to the Field-Feynman model[21]. However, when massless partons are converted into massive hadron jets, energy and momentum cannot be conserved between individual partons and the corresponding jet. It can be conserved only when the whole system of

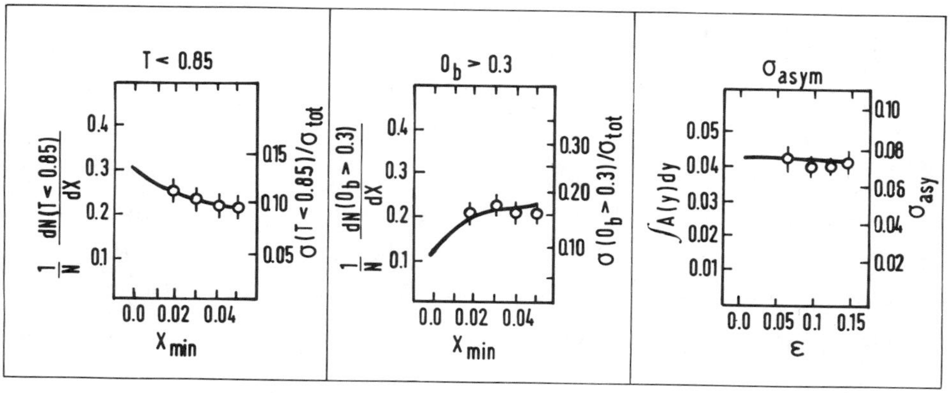

Fig. 7 Monte Carlo results with $\alpha_s = 0.17$ of the fraction of events with

a) Thrust less than 0.85,

b) O_b greater than 0.3 as functions of the mass cutoff $x = m^2/s$, and

c) the integrated asymmetry, Asy, as a function of the parton energy fraction cutoff ε.

The curves, corresponding to the right-hand scales, are the results of a parton-level calculation; the points, corresponding to the left-hand scale, correspond to a calculation of these quantities as measured in our detector, including QED radiative corrections and fragmentation.

partons and hadrons is taken into account. Thus, as the result of balancing momentum, there is a net Lorentz boost in this model, with mean velocity $\beta \simeq P/E \sim 0.05$, randomly distributed around the direction of the gluon. In other words, owing to energy momentum conservation, jets in the so-called independent jet models are not completely independent. In the second, the Lund model,[22] color strings connect quark to gluon and antiquark to gluon. The actual fragmentation follows the Field-Feynman prescription, but it occurs along the direction of the color strings rather than along each parton direction. The two models are rather extreme with regard to the question of the effect of color strings; one assumes no additional effect beyond the constraint of energy-momentum conservation while the other assumes that the two color strings in a three-jet event are preserved throughout the entire fragmentation process. For the case of three energetic partons (i.e., $q\bar{q}g$), the final state hadrons are significantly closer to the original parton distributions in the independent jet model. Thus, the α_s determined to the first order of QCD is quite different in these two models. However, for the case of four energetic partons (e.g., $q\bar{q}gg$) the distributions are similar in the two models.

The data were collected at a center of mass energy $\sqrt{s}$ = 34.7 GeV. A selected sample of 21,000 hadron events from the one photon annihilation process was used in this study. The energy-energy correlation is shown in Fig. 8. Data are compared to a second order QCD Monte Carlo prediction. The Monte Carlo represents the data well using either of the fragmentation models, which are indistinguishable in the figure. Fig. 9 shows the asymmetry data as black points with error bars. The smooth curve is a Monte Carlo prediction at the parton level assuming α_s = 0.13 and including detector simulation and radiative effects[23]. The histogram is a complete simulation of the data including fragmentation into hadrons. The two fragmentation models are indistinguishable on the scale of the figure. We can immediately see that the fragmentation effects are small within these two models for $|\cos\chi| < 0.72$.

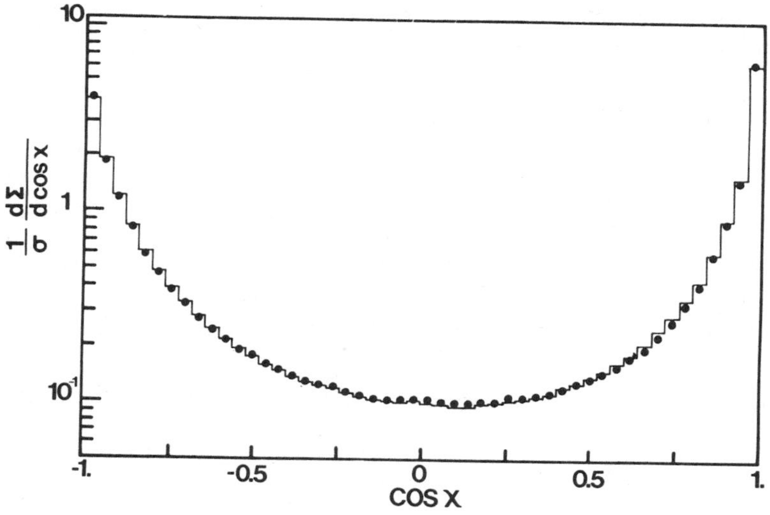

Fig. 8 Comparison of energy correlation data with Monte Carlo prediction, with $\alpha_s = 0.13$, for both fragmentation models, which are indistinguishable.

In Table I we display the values of α_s resulting from the second order QCD fits for the energy correlation asymmetry and the rate of large oblateness events ($O_b > 0.3$). We observe that in second order, α_s is only mildly dependent upon the fragmentation model used. For completeness we also show the first order results, although the corresponding distributions (in thrust, major, minor, etc.) do not fit well due to the lack of 4-jet events.

TABLE I

Result for α_s to second order calculated from the asymmetry of the energy correlations. For comparison, results from fitting the 3-jet event rate via O_b, as well as results of a first order calculation, are given.

Models	2nd Order		1st Order	
	Asymmetry	O_b	Asymmetry	O_b
"String"	0.14 ± 0.01	0.16 ± 0.01	0.20 ± 0.01	0.25 ± 0.01
Independent Jet (Ali)	0.12 ± 0.01	0.14 ± 0.01	0.15 ± 0.01	0.21 ± 0.01

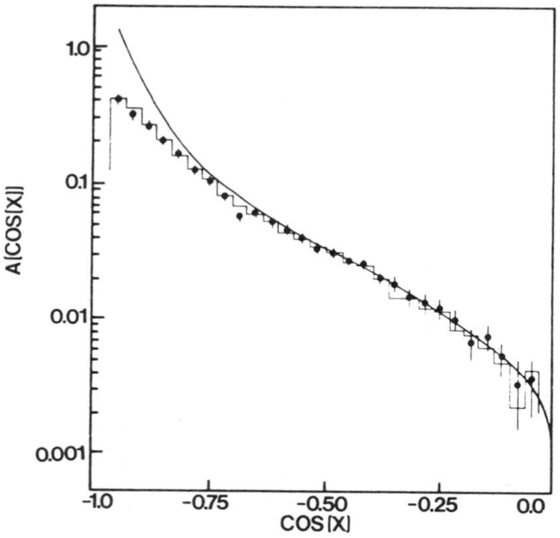

Fig. 9 Asymmetry data compared with predictions at parton level (curve) for $\alpha_s = 0.13$ and predictions for the two fragmentation models (Lund, Ali; histogram) for the best fit values of α_s. These two histograms are indistinguishable.

TABLE II

Systematic errors in the determination of α_s to second order

Sources of errors	$\lvert \Delta\alpha_s \rvert$
Fit over $\lvert \cos\chi \rvert < 0.85$ to 0.60	< 0.01
Use $\varepsilon = 0.07$ to 0.15	< 0.01
or $x = m^2/s = 0.02$ to 0.05	< 0.01
Mean transverse momentum in fragmentation	
250 to 350 MeV for indep. jet	< 0.01
350 to 480 MeV for Lund	
in the gluon jet, 300 to 500 MeV	< 0.01
Detector effect	< 0.01
Background subtraction ($\tau\tau$, and 2γ process)	< 0.01
Fragmentation model dependence	0.01
TOTAL	$< \pm 0.02$

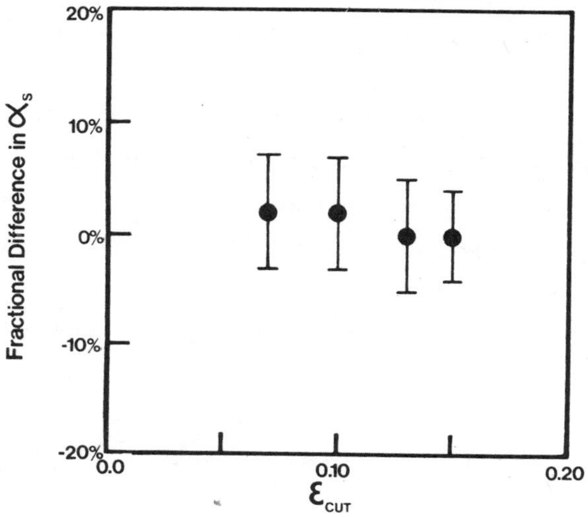

Fig. 10 Percentage change of α_s determined in second order as a function of the ε cut.

The α_s values obtained in second order do not change when specific cut-off parameters are varied over a wide range of values,

$$0.07 \leq \varepsilon \leq 0.15 \text{ or}$$
$$0.02 \leq x < 0.05$$

as shown in Fig. 10. The systematic errors are listed in Table II.

The MARK-J group has used the asymmetry in the energy correlation distribution to determine that $\alpha_s = 0.13 \pm 0.01 \pm 0.02$ in second order. They chose the energy correlation method to derive the primary result because it is less sensitive to variations in the cuts, and all hadron events may be used, without dependence upon event shapes for the selection of an appropriate sample of events. The systematic error is due primarily to the uncertainty arising from alternative descriptions of the fragmentation process. The value of α_s can be converted into a value of $\Lambda_{(\overline{MS})}$, using the second order formula[21]: $55 < \Lambda_{(\overline{MS})} < 350$ MeV. The MARK-J result is comparable with the measurement made by the JADE group[13] using the thrust distribution of the three jet cross section.

V. DOES THE COUPLING CONSTANT DECREASE WITH INCREASING q^2?

The asymptotic freedom of QCD leads to a dependence of the strong coupling constant α_s on q^2. The relation between α_s and q^2 can be expressed as[24]

$$\alpha_s = 2\pi / \left(b_0 \ell n\, (q^2/\Lambda^2) + \frac{b_1}{b_0}\, \ell n \,\left|\, \ell n\, (q^2/\Lambda^2) \right| \right)$$

where $b_0 = (33 - 2 N_F)/6$ and
$b_1 = (153 - 19 N_F)/6$ with
N_F being the number of flavors and
Λ being the QCD scaling parameter.

Since the predicted variation of α_s with q^2 is quite small, to answer the question of whether α_s decreases with increasing q^2 one must measure

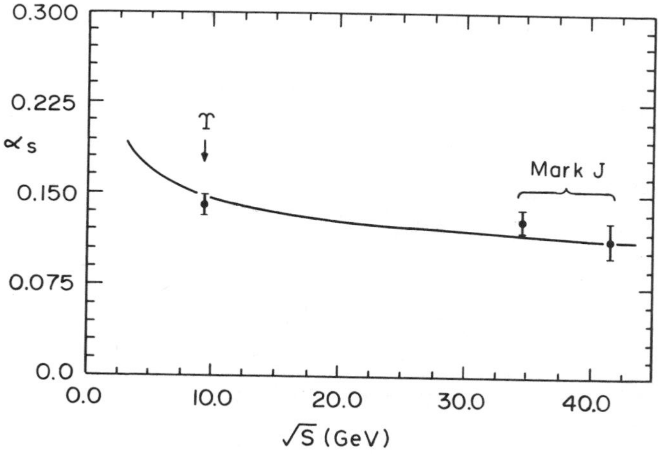

Fig. 11 The measured values of α_s at Υ and at $\sqrt{s}$ = 34.7 GeV and 41 GeV at PETRA, together with the best fit QCD curve as function of $\sqrt{s}$.

α_s over a wide range of q^2. In addition to the determination of α_s at PETRA at $q^2 = 1300$ GeV^2 described above, it has also been suggested that one particularly good measurement of α_s is provided by Υ decay[25]. The three gluon decay rate can be determined by the difference of the total hadronic decay rate and radiative and photon induced contributions, which can be deduced from off-resonance measurements and extrapolation from charmonia decays. The prediction of QCD, including next to leading effects is

$$\frac{\Gamma(\Upsilon \to 3g)}{\Gamma(\Upsilon \to \mu^+\mu^-)} = \frac{10}{81 e_b^2} \; \frac{\pi^2-9}{\pi} \; \frac{\alpha_s^3(M_\Upsilon)}{\alpha_{em}^2} \; \frac{\left(1 + (3.8 \pm 0.5)\,\frac{\alpha_s}{\pi}\right)}{\left(1 - \frac{16}{3}\,\frac{\alpha_s}{\alpha}\right)} .$$

Since the ratio in the above equation depends on the third power of α_s, it is a sensitive measure of the strong coupling. The conclusion of Mackenzie and Lepage is

$$\alpha_s(M_\Upsilon)_{\overline{MS}} = 0.141 \begin{array}{l} +.009 \\ -.008 \end{array}$$

or

$$\alpha_s(0.48\ M_\Upsilon) = 0.158 \begin{array}{l} +0.012 \\ -0.10 \end{array}$$

when the next to leading order corrections are absorbed into the leading α_s^3 term. This yields $\Lambda = 100^{+34}_{-25}$ MeV. If electromagnetic corrections[26] are included, α_s (0.48 M_Υ) would become 0.163 or $\Lambda = 115^{+48}_{-37}$ MeV. The measured values of α_s at Υ and at $q^2 = 1300$ GeV^2, as shown in Fig. 11, are used to fit for values of Λ using the above QCD formula and the data is consistent with the prediction. Since PETRA has already accumulated more than 10 pb^{-1} at $\sqrt{s} \sim 41$ GeV, a precise determination of α_s at $q^2 \sim 1700$ GeV^2 wouldd further help clarify the issue.

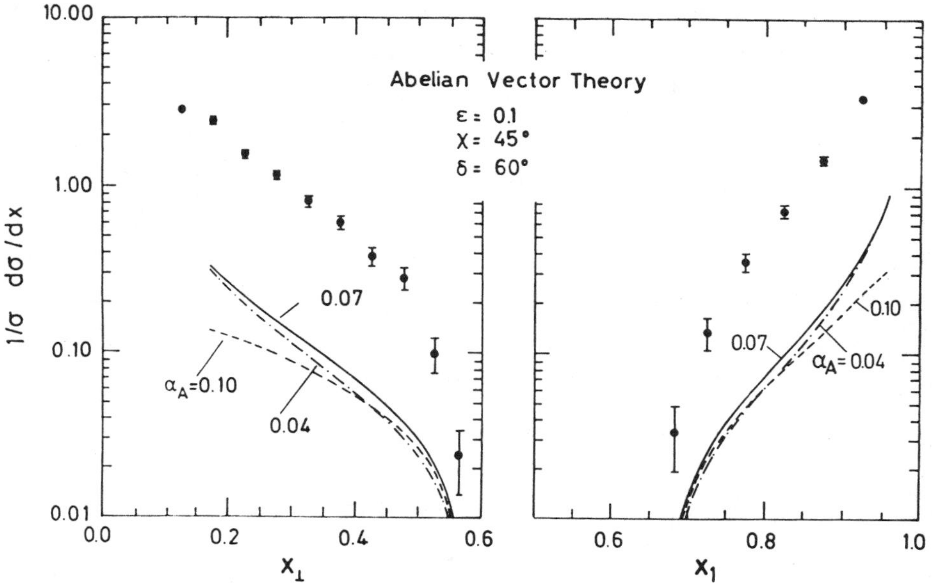

Fig. 12 The corrected x_1 and x_T-distributions for three-jet cross-section compared to the second order predicitons of the Abelian theory for three different coupling constants α_A. The one yielding results closest to the data corresponds to the full line. The dotted line shows the predictions for a coupling constant smaller than the "best fit" and the broken line for a bigger one. The ε, χ jet definitions have been used for the selection of three-jet events with $x_1 > x_2 > x_3$ and x_T is the transverse component of x_2 and x_3 with respect to x_1.

VI. TESTS ON THE TRIPLE GLUON COUPLING

To test the hypothesis of gluon self-coupling as prescribed by QCD, one can artifically use a QED type of Abelian vector theory by removing all processes containing the triple gluon vertex in the process shown in Fig. 6. Since this term contributes only to second order, the first order prediction of the Abelian theory is the same as the QCD after the quark-gluon coupling constant is renormalized to be $\alpha_A = 4/3\ \alpha_s$. When second order effects are included the triple gluon term, although it contains four final state partons, contributes significantly to the three jet cross-section when the energy of one of the partons is below the ε cut or when the opening angle of any pair of partons is smaller than the δ cut with the $\varepsilon + \delta$ defined previously in Section IV. The omission of this term results in a much smaller three jet crosssection than both the first results and the second order QCD prediction for all values of the renormalized quark-gluon coupling strength α_A.

The resultant 3-jet thrust distribution as obtained by the JADE group[27] is shown in Fig. 12. The theoretical curves are obtained with α_A defined in the $\overline{MS}$ scheme in analogy to QCD. The larger number of quarks caused by the color degree of freedom in QCD is simulated by tripling the number of flavors. In this framework the Abelian vector theory gives reasonable results for the total cross-section,

$$R = R_o \left(1 + \frac{3}{4}\ \frac{\alpha_A}{\pi}\right)^2 .$$

But the differential 3-jet cross-section is nearly an order of magnitude below the data and the first order results. Thus one concludes that the second order Abelian vector theory fails to describe the data and the triple gluon vertex is indispensable. This is a direct test of the original Yang-Mills theory.[1]

Acknowledgement

I would like to thank Professors P. Soeding and S. C. C. Ting for valuable discussions and Drs. L. L. Chau, A. Ali, R. Rau, G. Swider, B. Adeva, R. Mount, Z. Y. Zhu, and D. Barber for many useful comments.

REFERENCES

1. C. N. Yang and R. L. Mills, Phys. Rev. 96, 191 (1954);
   H. Fritzch et al., Phys. Lett. 47B, 365 (1973);
   D. J. Gross and F. Wilczek, Phys. Rev. Lett. 30, 1343 (1973);
   H. D. Politzer, Phys. Rev. Lett. 30, 1346 (1973).
2. E. D. Bloom et al. Phys. Rev. Lett. 23, 930 (1969);
   S. L. Adler, Lectures on Elementary Particles and Quantum Field Theory, Brandeis Summer Institute, 1970, MIT Press (1971);
   D. P. Barber et al., MARK-J Collab., Phys. Rep. 63, 337 (1980).
3. PETRA proposal, DESY, Hamburg (February, 1976).
4. G. Hanson et al., Phys. Rev. Lett. 35, 1230 (1975).
5. Ch. Berger et al., Phys. Lett. 78B, 176 (1978).
6. TASSO Coll., R. Brandelik et al., Phys. Lett. 86B, 243 (1979);
   MARK-J. Coll., D. P. Barber et al., Phys. Rev. Lett. 43, 830 (1979);
   PLUTO Coll., Ch. Berger et al., Phys. Lett. 86B, 418 (1979);
   JADE Coll., W. Bartel et al., Phys. Lett. 88B, 171 (1979).
7. D. P. Barber et al., Phys. Rev. Lett. 43, 830 (1979);
   D. P. Barber et al., Phys. Lett. 108B, 63 (1982).
   H.J. Behrend et al., Phys. Lett. 113B (1982) 427.
8. References could be found in the following review papers, e.g.
   P. Soding and G. Wolf, Am. Rev. Nucl. Sci. 31 (1981) 231.
   G. Wolf, DESY 82-077 (1982).
   D.P. Barber et al., MARK J Collaboration, Phys. Rep. 63 (1980).
9. R. Clare, Ph.D. Thesis, M.I.T. (1982).

10. R. Brandelik et al., Phys. Lett. 87B, 453 (1980).

11. W. Bartel et al, DESY 82-086 (1982).

12. D.P. Barber et al., Phys. Lett. 89B, 139 (1979) and 108B, 63 (1982).

W. Bartel et al., Phys. Lett. 91B, 142 (1980).

R. Brandelik et al., Phys. Lett. 94B, 437 (1980).

Ch. Berger et al., Phys. Lett. 97B, 459 (1980).

H.J. Behrend et al., Phys. Lett. 110B, 329 (1982), and Nucl. Phys. B218, 269 (1983).

13. W. Bartel et al., Phhys. Lett. 119B, 239 (1982)

14. K. Fabricius et al., Phys. Lett. 97B, 431 (1981).

G. Schierholz and G. Kramer, private communication.

T. Gottschalk, Phys. Lett. 109B, 331 (1982).

15. B. Adeva et al., Phys. Rev. Lett. 50, 2051 (1983).

16. C. L. Basham et al., Phys. Rev. Lett. 41, 1583 (1978) and Phys. Rev. D19, 2018 (1979).

17. A. Ali et al., Phys. Lett. 93B, 155 (1980);

A. Ali et al., Nucl. Phys. B168, 490 (1980).

18. R. K. Ellis et al., Nucl. Phys. B178, 421 (1981);

J. Vermaseren et al., Nucl. Phys. B187, 30 (1981).

19. A. Ali and F. Barreiro, DESY Report 82-033 (1982).

20. G. Sterman and S. Weinberg, Phys. Rev. Lett. 39, 1436 (1977).

21. R. D. Field and R. P. Feynman, Nucl. Phys. B136, 1 (1978).

22. B. Anderson, G. Gustafson, and T. Sjostrand, Zeitschr. f. Phys. C6, 235 (1980);

B. Anderson, G. Gustafson, and T. Sjostrand, Nucl. Phys. B197, 45 (1982).

23. F. B. Berends et al., Nucl. Phys. B68, 541 (1974).

24. W. Caswell, Phys. Rev. Lett. 33, 244 (1974).

D.T.R. Jones, Nucl. Phys. B75, 531 (1974).

A. Ali and F. Barreiro, Phys. Lett. 118, 155 (1982).

25. P. B. Mackenzie and G. P. Lepage, Phys. Rev. Lett. 47, 1244 (1981).

26. F.J. Yndurain, Phys. Rev. Lett. 48, 897 (1982).

27. W. Bartel et al., DESY 82-086 (1982).

DISCUSSION

CHAIRMAN : M. CHEN

Scientific Secretaries: J. Berdugo and C. Mana

- *EREDITATO* :

Would you please clarify why a bump in the jets thrust distribution is an evidence for the presence of an extra quark ?

- *CHEN* :

When the C.M. energy is much larger than the mass of the produced quarks, the quarks have large velocities. The value of the thrust is close to one. However, if the C.M. energy is just above the threshold of the quarks, they are produced almost at rest. The decay products are almost isotropic and you expect then a bump in the low thrust region ($T \leqslant .7$).

- *HOU* :

Is there any way to distinguish quark and gluon jets ? For example, will you see any difference between the three gluon jets from toponium decay and the $g\bar{g}g$ jets from gluon bremsstrahlung ?

- *CHEN* :

The best place to look for gluon jets is the three gluon decay of quarkonium. Since the toponium has not been found, the best source we have is the T. However, it's not massive enough (~ 10 GeV) so the particles from each gluon are not very jet-like. You cannot identify three and four jet events. The decay distribution is not very different from what one expects from phase space. Even at PETRA, we do not see any clear signal of gluon bremsstrahlung below 25 GeV.

- *HOU* :

Did you say that quark and gluon jets are different ?

- *CHEN* :

I have shown that there is evidence that quark and gluon jets are different in $<P_t>$ because the gluon can fragment into two gluons.

- *BAGGER* :

You said that the effect of the three gluon vertex has been seen. Is there any hope at all to see the four gluon vertex ?

- *CHEN* :

The four gluon vertex is an even higher order effect which is too small to detect at present.

- *ALTARELLI* :

I don't believe that you have seen the three gluon vertex. Can you explain the argument again ?

- *CHEN* :

The complete second order quantum Abelian dynamics (by QAD I mean that you drop all Feynman diagrams with three gluon vertex and reduce the color factor to 1) cannot reproduce the data and thus is ruled out.

- *ALTARELLI* :

Why is it ruled out ?

- *CHEN* :

I described in the lecture that QAD fails to describe the observed $x_{\parallel}$ and $x_{\perp}$ distributions of the three jet cross-section. The four parton final state goes to two when the cut parameter goes to 0. And the virtual diagrams interfering with those three parton final states give $- \infty$. When we take out the three gluon vertex diagram, the three jet cross-section, which includes the soft contribution from four parton final states, becomes significantly reduced independently of the used α_A value. Thus you find that QCD reproduces the data and QAD not. The only remaining question is how big are the third order corrections for QAD, which are not calculated.

- *ALTARELLI* :

The difference between keeping or dropping the three gluon vertex is something which has to do with a change in the radiative corrections. I do not believe that such a radiative correction can produce such a marked difference because, on top of everything, we know that there is a dramatic dependence of the measurements on the way you apply the radiative corrections and on how you do the hadronization process.

- *CHEN* :

I did not believe the results at the beginning either, until I and Dr. Ali from DESY checked through the computation of the JADE group.

- *ALTARELLI* :

In discussing the determination of α_s at PETRA, the spread of the values from the four experiments is much larger than quoted errors. Why should I believe one or the other ?

- *CHEN* :

If you keep track of which experiments are using ERT scheme and which ones use FSSK scheme, you will find that all the measurements are consistent with each other. Let me start with the transparency Walsh showed at the Brighton conference a week ago. Among all the measurements presented I will leave out JADE (which is consistent with all the others) and MAC (because they use a method which is not guage invariant, their numbers are quoted without errors and I suspect energy-momentum is not conserved in their case). Then, we have three measurements from TASSO (which have data as good or even better than ours), CELLO and MARK-J. Let me quote the numbers from Walsh's transparencies, to make the following table. The values of α_s as presented at Brighton are (with energy-momentum conservation and $\cos\chi < 0.72$ for asymmetry).

Table

	ERT		FSSK		$\alpha_s(SM)$ / $\alpha_s(IJ)$
	IJ	SM	IJ	SM	
TASSO			.16 ± .015	.21 ± .015	1.3 ± .1
CELLO			.15 ± .02	.18 ± .03	1.2 ± .2
MARK-J	.12 ± .01	.14 ± .01			1.2 ± .1

From this table one first concludes that the ratios of α_s(string) / α_s(IJ) agree well among three experiments. The next important step is that MARK-J has used both ERT and FSSK to determine α_s and found that the $\alpha_s(\text{FSSK}) \simeq 1.25\ \alpha_s(\text{ERT})$. Once you use this conversion factor you find α_s measured by all three experiments agree very well. The discrepancy you were talking about is mainly due to misinterpretation. However, I would never quote a measurement with ~ 8% error while the method (FSSK) itself is accurate to $O(\varepsilon \sim 0.2)$ and I am not authorized to change other people's data. Therefore, I can only quote the sole available measurement with ERT: $\alpha_s = 0.13 \pm 0.01 \pm 0.02$ from MARK-J. I am afraid if I present all the numbers as Walsh did, people would again confuse α_s(FSSK) with α_s(ERT).

- *ALTARELLI* :

Both calculations are approximate in the sense that at the end you measure hadrons and not partons. All these procedures seem reasonable to me and the final spread of the values which is found is a systematics that you cannot avoid.

- *CHEN* :

The calculations are similar in principle, but FSSK made some mathematical approximation, thus it is accurate only to $O(\varepsilon, \delta)$. In the lecture I described the three steps to carry out the calculation: 1) use mathematically rigorous ERT calculation to generate parton final state; 2) make (ε, δ) cuts to avoid double counting of the hadronization effect and 3) use experimentally determined parameters to describe hadronization. The systematics you described about hadronization has already been included in the systematic errors.

- *RATOFF* :

Does the MARK-J limit on the quark radius take into account the QCD predicted dependence of the hadronic cross-section on the c.m. energy ? Please, comment on the limit of the quark radius reported by the UA1 collaboration.

- *CHEN* :

Yes, both QCD and electroweak corrections to the hadronic cross-section are included. The UA1 limit recently reported is $R_q < 5.10^{-17}$cm. This is done by comparing the measured data with QCD predictions as a function of P_t for high P_t jets. The error is very big and in addition there is a ± 40% systematic error.

They get this limit of R_q by assuming a form factor which is divergent when $\Lambda = \sqrt{g^2}$. This formula is valid when $g \ll \Lambda$ so, if $q \approx \Lambda$ you should use a different form factor e.g. Gaussian. Then they cannot set any limit.

- *PEPE* :

Can you explain why gluon jets are broader than quark jets ?

- *CHEN* :

In the fragmentation procedure a la Feynman-Field, for $q\bar{q}$ pair you have

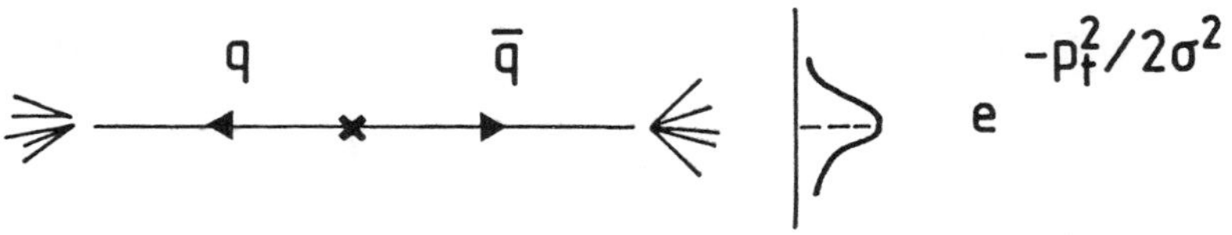

When we have a gluon we expect it to be broader due to the larger color charge of the gluon and the gluon self-coupling.

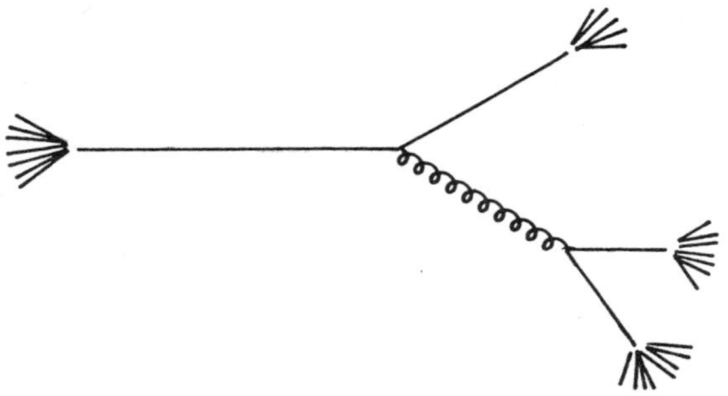

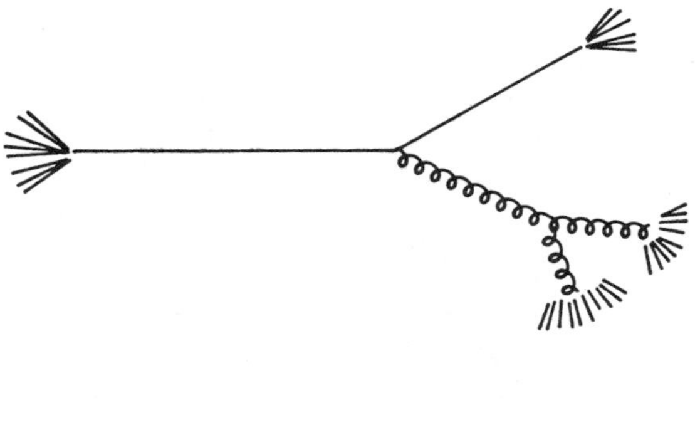

In the string model the hadronization goes through color strings so the gluon is converted into a $q\bar{q}$ pair and each of them is connected with the initial $q(\bar{q})$ to form a color singlet. For every color singlet a Lorentz boost from the c.m. system to the lab. system is performed

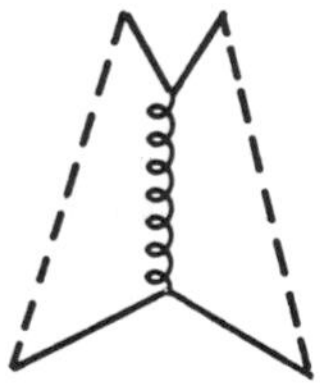

and after that you get a larger $\langle P_t \rangle$ for the g jets than for the quark jets, because the gluon has gone through two Lorentz boosts.

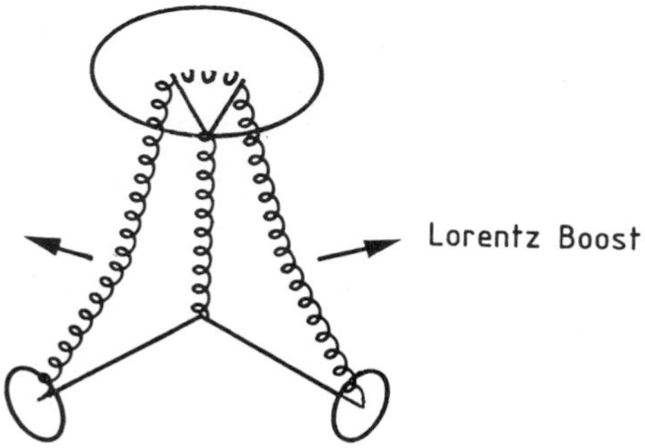

- *KLEVANSKY* :

Would you explain why in your plot of α_s versus q^2 the curve is disjoint and what value of Λ would fit your curves ?

- *CHEN* :

The value is $\Lambda \simeq 100 \pm 30$ MeV and is preliminary. The discontinuity is because below the T there are 4 flavors and above there are 5 quark flavors.

- *VAN DEN DOEL* :

Up to which order has QCD been confirmed ?

- *CHEN* :

Right now the second order QCD is favoured over first order QCD. A third order calculation is desirable.

- *BERNSTEIN* :

How do you extract the value of α_s from your asymmetry plot ?

- *CHEN* :

You have to compare the data with the M.C. simulation which includes detector effects and hadronization. Each time you change α_s you have to repeat the whole procedure. The corrections you have to apply to the pure parton curve are about 30%.

- *DEMARTEAU* :

You showed us your result for the fragmentation function of the c-quark, $f_c(z)$. Is your result comparable with $f_c(z)$ measured with D^* ?

- *CHEN* :

There are results presented by CLEO, TASSO and MARK II for D^*. There is an apparent discrepancy between theirs and our result. We can see a shift of approximately 10%. The average D^* z-value they quote is ≈ 0.6 and ours is $<z> \approx 0.46$ for D meson. However, more than half of this difference is due to the mass difference between the D^* and D, which is of ~ 10%. After this correction the two data agree well.

- *WEIDBERG* :

Why do you regard the string and independent jet fragmentation models as extreme and why don't you use other models ?

- *CHEN* :

The string model implies maximum correlation and IJ minimal correlation between jets. There are many other models available, most of them do not agree with data.

- *JOHNSON* :

Why do the first and second jet not show a similar difference in $<P_t>$?

- *CHEN* :

It is due to the fact that the first and second jets come from different kinematical regions, i.e. for $6 < E_{jet} < 12$ GeV the mean energy of the first jet is at the high energy end while the second jet is around 9 GeV.

- *ALTARELLI* :

There is a region where you cannot follow perturbation theory anyway and this is within some cones (ε, δ) where the infrared sensitivity goes into the confinement and hadronization mechanism. I think that the fact that there is a sensitivity to how you treat the region inside these cones (small ε, δ) is a signal that there is a lot of importance from non-perturbative effects and it is out of the capabilities of the present theory to find a precise value of α_s. A sensitivity to ε, δ means a sensitivity to non-perturbative phenomena. If FSSK and ERT would coincide for a large region of ε, δ everybody would be much happier because that could mean a little sensitivity to how you treat the forbidden region where you cannot enter by perturbation theory because there is confinement and hadronization. This sensitivity warns you that whatever you do is model dependent.

- *CHEN* :

There is certainly a lot to be improved. However ERT and FSSK are really in good agreement, since the difference is only 25% and FSSK is accurate only to ~ θ(20%) according to the authors. Furthermore it is within the present theoretical ability to calculate these correction terms. The ε, δ cut has been made and has been described above already.

- *BATTISTON* :

Can you explain what is the meaning of the fit labelled exp (P_t) to the 3 jets data of MARK J ?

- *CHEN* :

We have tried different models to explain the data without assuming gluon bremsstrahlung. For instance, $q\bar{q}$ with gaussian P_t distributions and exponential distributions. None of them fit the energy flow distribution.

STATUS OF THE GLUEBALLS†

S.J. Lindenbaum

Brookhaven National Laboratory, Upton, New York 11973

City College of New York, New York, New York 10031

INTRODUCTION

QCD[1] was invented to explain hadronic states and interactions. The most crucial and elegant part of QCD is $SU(3)_{color}$ with local gauge invariance.

The most characteristic feature of this non-abelian local gauge theory is the self interaction of the gauge bosons (gluons) which becomes stronger as their relative energy decreases (i.e. asymptotic freedom - infrared slavery).

Thus in a hadronic theory based on $SU(3)_c$ (i.e. a pure Yang Mills theory[2]) and confinement, all the hadrons in the world would be glueballs[3] (i.e. multi-gluon resonant states). The quarks at present are in the category addressed by Rabi's question "who ordered that? However when they are added to the theory we face the anamoly that the hadrons were found to consist of quark-built states, and until recently there was no good evidence for glueballs. Therefore the discovery of glueballs - the missing link in QCD - is (in my opinion) a very necessary condition for

† This research was supported by the U.S. Department of Energy under Contract Nos. DE-AC02-76CH00016 (BNL) and DE-AC02-79ER10550A (CCNY).

establishing QCD and locally gauge invariant $SU(3)_c$. It is also very important if one wishes to maintain the viability of unification or partial unification theories including $SU(3)_c$.[39]

Due to the recent successes of the Electro-Weak Group,[4] $SU(2)_L \times U(1)$, if $SU_3(c)$ controls hadronic interactions, we would have the situation where local gauge invariant groups explain the dynamics of strong as well as electromagnetic and weak physical phenomena at least at energies ~ those presently attainable.

In this lecture I will show that the BNL/CCNY $g_T(2120)$, $g_T'(2220)$ and $g_T''(2360)$ all with $I^G J^{PC} = 0^+2^{++}$ observed[5] in $\pi^- p \to \phi\phi n$ are produced by glueball(s):

1. If QCD is correct;

2. If the OZI rule is universal for weakly coupled glue in Zweig disconnected diagrams where the disconnection is due to the creation or annihilation of new quark flavors.

Since the above axioms merely represent modern QCD practice and agree with experiments, I consider this the discovery of glueball(s).[5–6,20–21]

It is obvious that every conclusion of a "discovery" depends on input axioms - implied or explicitly stated (which I prefer). The simpler, more fundamental and the better their justification by experiment the axioms are, the more justified is the conclusion of a discovery. I will also briefly discuss the status of some other glueball candidates, two in particular, those arising from the J/ψ radiative decay.

In my Erice Lecture last summer[6] I presented the results of the BNL/CCNY glueball search in the reaction $\pi^- p \to \phi\phi n$.[15,19–20] At that time we had ≈ 1203 events. These events clearly demonstrated a complete breakdown of the Zweig (or OZI) suppression[7–9] in the Zweig disconnected diagram $\pi^- p \to \phi\phi n$ (see Fig. 1) which we also had observed in the first experiment[10] to ever observe double ϕ production.

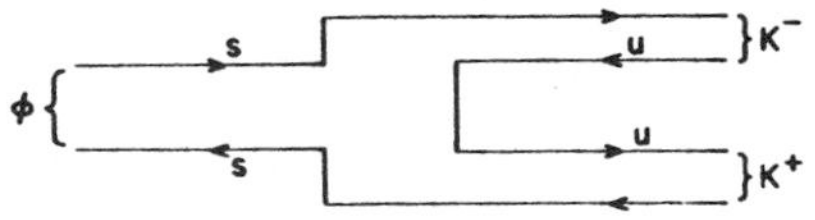

$\phi \rightarrow K^+K^-$

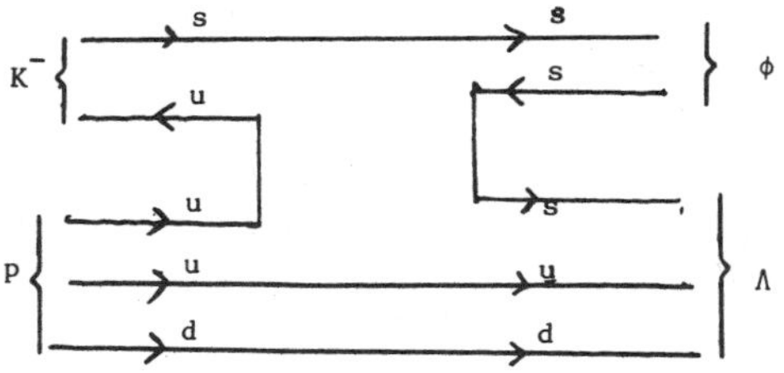

$K^-p \rightarrow \phi\Lambda$

Figure 1a: Zweig connected (allowed reaction) diagrams for the u,d,s quark system.

Except for these experiments the Zweig rule appears to be universally followed in disconnected diagrams in hadronic interactions where the disconnection is due to creation or annihilation of new flavor(s) of quark(s). This is shown clearly in Fig. 1 for the u,d,s quark system where the matrix element for the Zweig connected diagram is two orders of magnitude larger than for the corresponding Zweig disconnected diagram.[8,11–12]

Figures 1a and 1b show that this occurs both in the decay and production processes. Figures 2a and 2b show that the J/ψ system exhibits even much greater Zweig suppression factors for Zweig disconnected diagrams. It should be noted in Fig. 2b that in addition to the well-known and striking Zweig suppression which

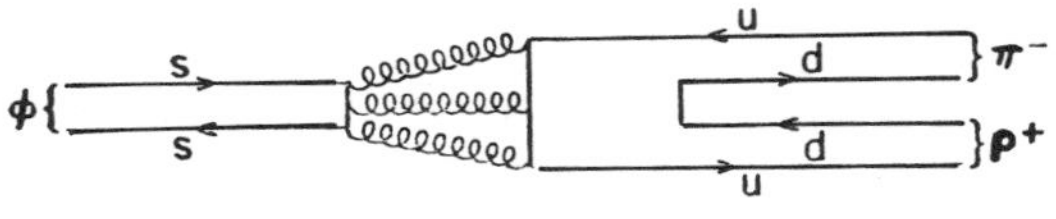

$\phi \to \pi^- \rho^+$

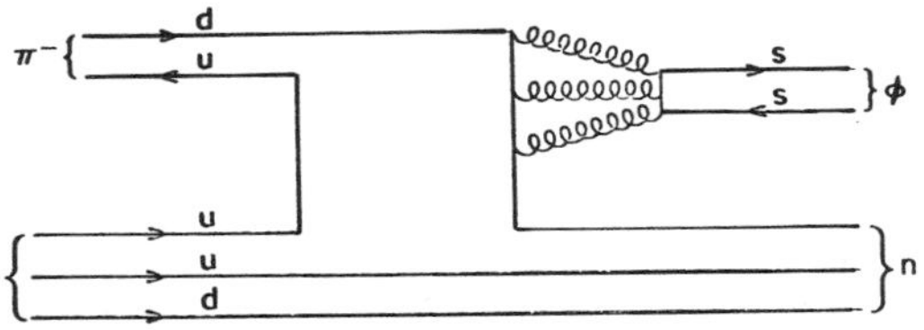

$\pi^- p \to \phi n$

Figure 1b: Zweig disconnected diagrams (suppressed reaction) for the u,d,s, quark system. The helixes represent gluons bridging the disconnection.

occurs when the $c\bar{c}$ quarks annihilate there is a huge suppression in the Zweig disconnected diagram where $\psi(3685) \to J/\psi\ (3100 + 2\pi)$ which results in a width of the $\psi(3685) = 250 \pm 40$ kev even though the $\pi^+\pi^-$ case occurs in $(33 \pm 2)\%$ of the cases and the $\pi^0\pi^0$ case occurs in $(17 \pm 2)\%$ of the cases.

Figure 3 shows a similar and even more striking situation existing in the upsilon system since the $\Upsilon'(10,020) \to \Upsilon(9460)\pi\pi$ $(30 \pm 6)\%$ of the time with the $\Gamma_{\Upsilon(10,020)} = (30 \pm 10)$ kev whereas the $\Gamma_{\Upsilon(9460)} = 42 \pm 15$ kev. Thus the suppression in the first Zweig disconnection is strong enough to maintain the width of the Υ' consistent within errors with the width of the Υ. The

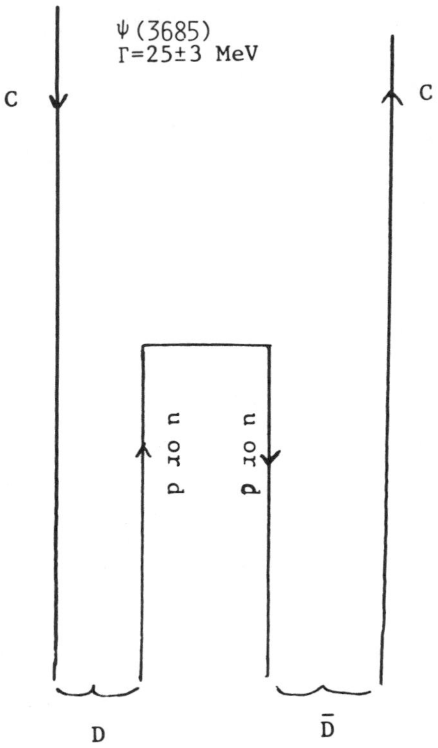

Figure 2a: A Zweig connected diagram for the ψ(3685) decay.

same striking phenomena occurs in the process ϒ"(10,020) → ϒ(9460) + 2π which although it occurs ~ 10% of the time results in a width of the ϒ" which is consistent with the width of the ϒ. Thus it is experimentally clear from the ψ and ϒ systems that what I will call a double hairpin type of disconnection in a Zweig diagram is strongly suppressed.

Lipkin has argued[13] that what I call a double hairpin type of disconnected Zweig diagrams such as Fig. 6, $\pi^- p \rightarrow \phi\phi n$ (which is the process we are observing) should not be Zweig suppressed (or only suppressed by a very small factor) since it is related by crossing to $\phi + n \rightarrow \phi + \pi^- + p$. He refers to this as a crossed pomeron diagram which is just elastic ϕ-nucleon scattering with additional pion production states and there is no reason to

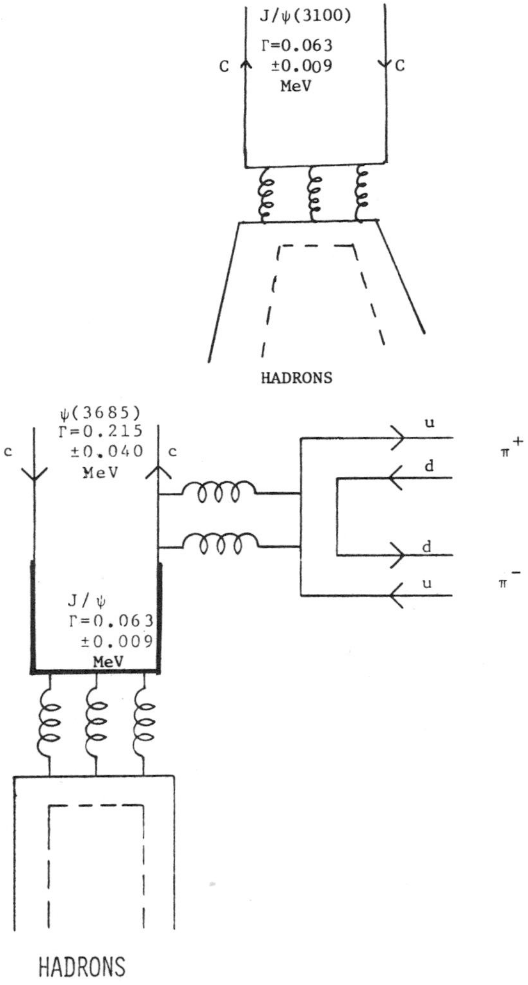

Figure 2b: Zweig disconnected diagrams in the J/ψ and excited ψ states.[40]

believe this process is forbidden. Reference (13) has overlooked the fact that when you cross in that manner you get into different kinematic and physical regions and that you cannot simply relate the two reactions.[14] For example considering the kinematics only the crossed reaction (e.g. $\phi + n \to \phi\pi^- + p$) corresponds to very high momentum transfers and a very high mass for the $\pi^- + p$ system. Diffraction dissociation at very high momentum transfers and very high masses would be expected to be negligibly small and

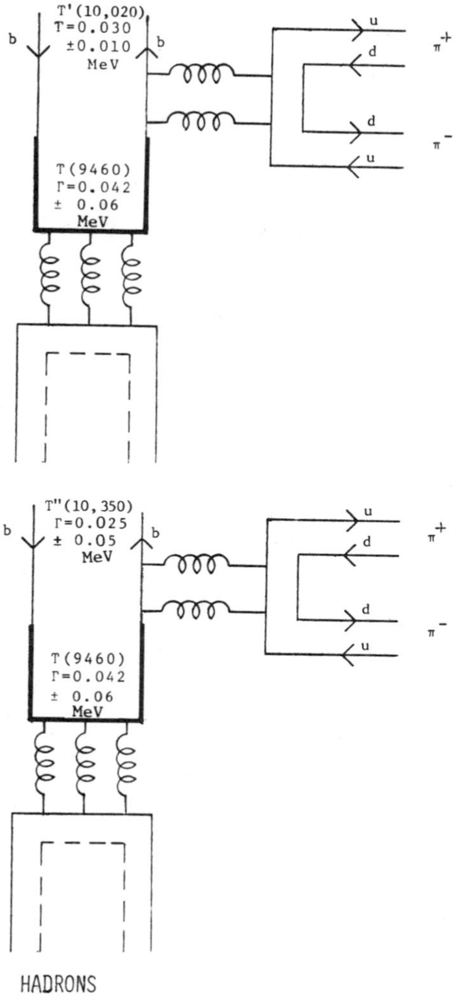

Figure 3: Zweig disconnected diagrams in the Υ system.[40–41]

thus these processes would be expected to be suppressed much more than the Zweig suppression factors we are dealing with. The fact that $\sigma(\pi^- p) \to \phi\phi n \approx 20$ nanobarns whereas diffraction dissociation which Ref. 13 says is large (~ 10 mb) differ by a factor of 10^6 emphasizes that it is not justified to relate the two processes in the naive way Ref. 13 has.

There are other erroneous statements in Ref. 13 that were addressed in Ref. 14. In particular one should note that Ref. 13 concludes the reaction $\psi(3685) \to J/\psi(3100) + 2\pi$ is Zweig allowed since it is also a crossed Pomeron diagram. Ref. 13 ignores the fact that the full width of $\psi(3685)$ is only $\approx$ 215 kev and thus this Zweig disconnected diagram (our Fig. 2b) is strikingly suppressed.

$T(T'') \rightarrow T + 2\pi$ decays also impressively show that so-called crossed Pomeron diagrams (in the notation of Ref. 13) also exhibit very strong suppressions and thus this line of reasoning is obviously fallacious for the reasons I have already mentioned.

The reason why the $\psi' \rightarrow J/\psi + 2\pi$, and $T'(T'') \rightarrow T + 2\pi$ have large branching ratios is probably at least partly due to the fact that these transitions can proceed by two relatively softer gluons compared to the direct three-gluon decays of the ψ', T' and T'', and also the kinematics of the decay favor the 2π channel, whereas there are many channels which compete for the three gluon partial decay width. Thus if we assume the OZI rule is universal for weakly coupled glue in Zweig disconnected diagrams where the disconnection is due to the creation or annihilation of new flavors of quarks, then the breakdown of the OZI suppression that we observe in $\pi^- p \rightarrow \phi\phi n$ must be due to strongly coupled glue. A glueball being a multi-gluon resonance would like in all hadronic resonance phenomena correspond to effectively strong coupling and thus the OZI suppression which in QCD is viewed as due to weakly coupled multi-glue intermediate states would be broken down by a glueball. Thus in the reaction $\pi^- p \rightarrow \phi\phi n$ the multi-gluon system in the intermediate state which forms the $\phi\phi$ system would in the absence of glueballs lead to only Zweig suppressed $\phi\phi$ production. However the $\phi\phi$ system has a variable mass and all the possible glueball quantum numbers for C = +. Thus at those masses where the multigluon intermediate state forms a glueball with C = + the Zweig suppression will be broken down and the $\phi\phi$ system will contain the glueball resonance parameters and quantum numbers. Thus the $\phi\phi$ system in the reaction $\pi^- p \rightarrow \phi\phi n$ will act as a filter passing glueball states and rejecting the other $q\bar{q}$ states.

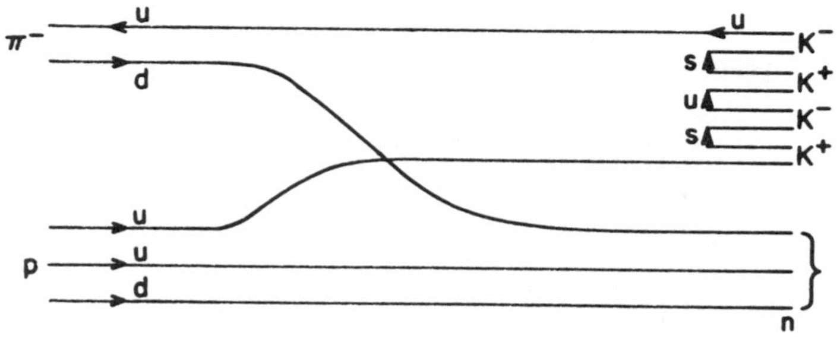

Figure 4: The Zweig quark line diagram for the reaction $\pi^- p \to K^+ K^- K^+ K^- n$, which is connected and OZI allowed.

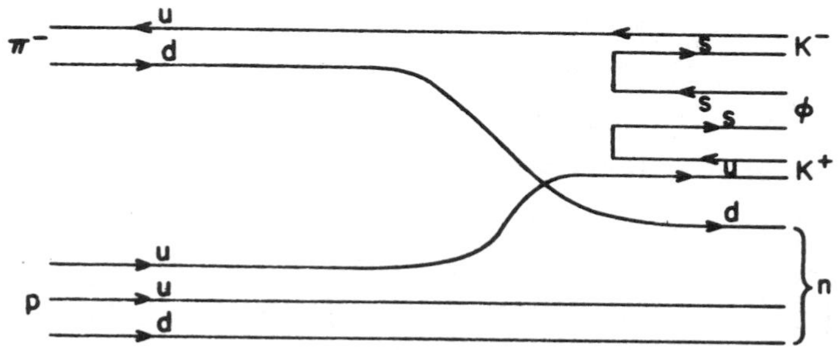

Figure 5: The Zweig quark line diagram for the reaction $\pi^- p \to \phi K^+ K^- n$, which is connected and Zweig allowed.

Other alternatives such as the possibility of more complicated hadronic states will be discussed later. Figures 4-6 show the three reactions we have observed.

During last years ERICE lecture[6] I discussed the 1203 $\pi^- p \to \phi\phi n$ events we had then. This spring we finished a run which raised the statistics to ≈ 4,000 $\pi^- p \to \phi\phi n$ events[5] and in this lecture I will discuss these results and their analysis and interpretation.

Figure 7 demonstrates the dramatic breakdown of the OZI (or Zweig) suppression we saw in the earlier data[10,15,16,19] also occurs in the new sample. We see the general ≈ uniform background

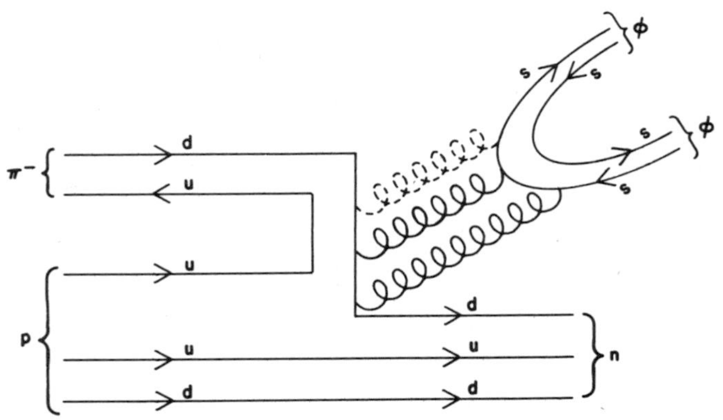

Figure 6: The Zweig quark line diagram for the reaction $\pi^-p \rightarrow \phi\phi n$ which is disconnected (i.e. a double hairpin diagram) and is OZI forbidden. Two or three gluons are shown connecting the disconnected parts of the diagram depending upon the quantum numbers of the $\phi\phi$ system.

from the reactions a) $\pi^-p \rightarrow K^+K^-K^+K^-n$ which is OZI (or Zweig) allowed and the two ϕ bands representing b) $\pi^-p \rightarrow \phi K^+K^-n$ which is also Zweig allowed. Where the two ϕ bands cross we have the Zweig forbidden reaction $\pi^-p \rightarrow \phi\phi n$. The black spot shows an obvious more-or-less complete breakdown of the Zweig suppression. This has been quantitatively shown[16] to be so in these reactions, and also by comparing K^- induced ϕ and $\phi\phi$ production.[17–18] The black spot when corrected for double counting and resolution ≈ 1,000 times the density of reaction (a) and ≈ 50 times the density of reaction (b). If one projects out the ϕ bands as shown in Fig. 8, there is a huge $\phi\phi$ signal which is ≈ 10 times greater than the background from reaction (b) even with rather wide cuts. The recoil neutron signal is shown in Fig. 9 and is also very clean ≈ 97% neutron.

The acceptance corrected $\phi\phi$ mass spectrum in the ten mass bins which were used for the partial wave analysis is shown in Fig. 10. All waves with J = 0 – 4, L = 0 – 3, P ± and η (exchange

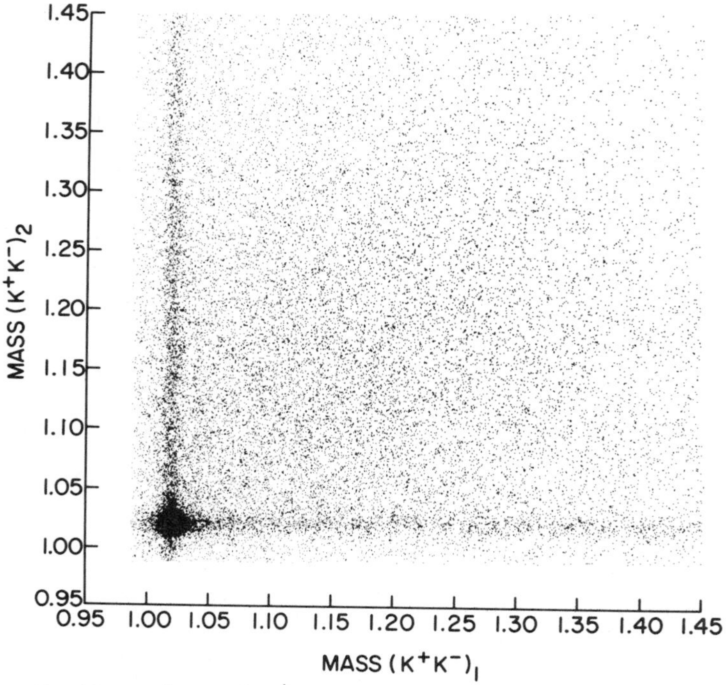

Figure 7: Scatter plot of K^+K^- effective mass for each pair of K^+K^- masses. Clear bands of $\phi(1020)$ are seen with an enormous enhancement (black spot) where they overlap (i.e. $\phi\phi$) showing essentially complete breakdown of OZI suppression.

naturality) = ± were allowed in the partial wave analysis. Thus 52 waves were considered. The incident π^- lab momentum vector and the lab momentum vectors of the four kaons completely specified an event. The Gottfried-Jackson frame angles β(polar) and γ(azimuthal) are shown in Fig. 11. These and the polar angles (θ_1, θ_2) of the K^+ decay in the ϕ rest systems relative to the ϕ direction and the azimuthal angles α_1 and α_2 of the K^+ decay direction in the ϕ_1, ϕ_2 rest systems (see Fig. 12) were also used to specify an event.

The MPS II at BNL (see Fig. 13) was used in the same experimental arrangement as described earlier.[10,19] The results of

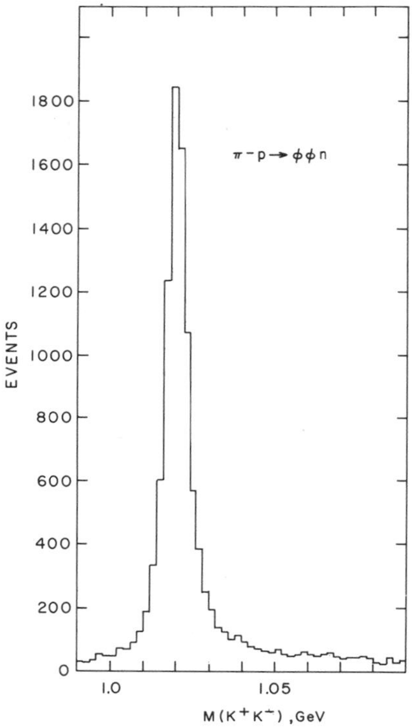

Fig. 8

The effective mass of each K^+K^- pair for which the other pair was in the ϕ mass band.

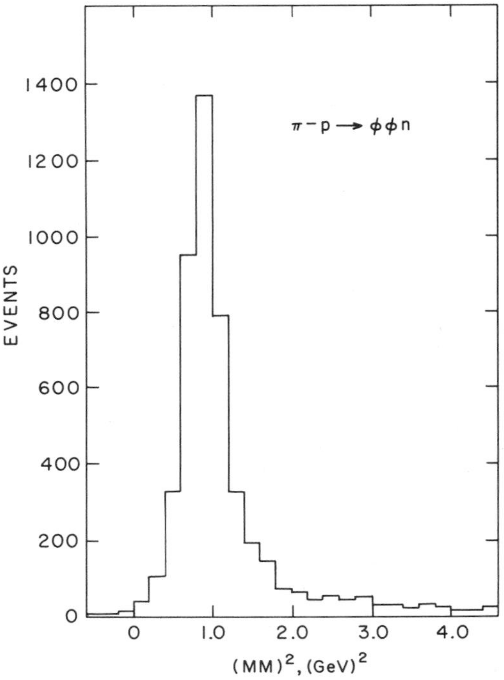

Fig. 9

The missing mass squared for the neutral recoiling system for the $\phi\phi$.

the mass independent partial wave analysis are shown in Figs. 14 and 15. In the analysis of 1200 events performed last year we had determined that our data contained two $J^{PC} = 2^{++}$ waves.[6,19] The predominant one being an S-wave with spin 2 peaked in the lower mass region and the other being a D-wave with spin 2 peaked at higher masses.

In this analysis of ≈ 4,000 events,[5] these two waves were again selected with a very high statistical precision >> 10σ. However the fit was totally unacceptable and required a third D-wave with spin 0 as shown in Fig. 14. The relative phase motion

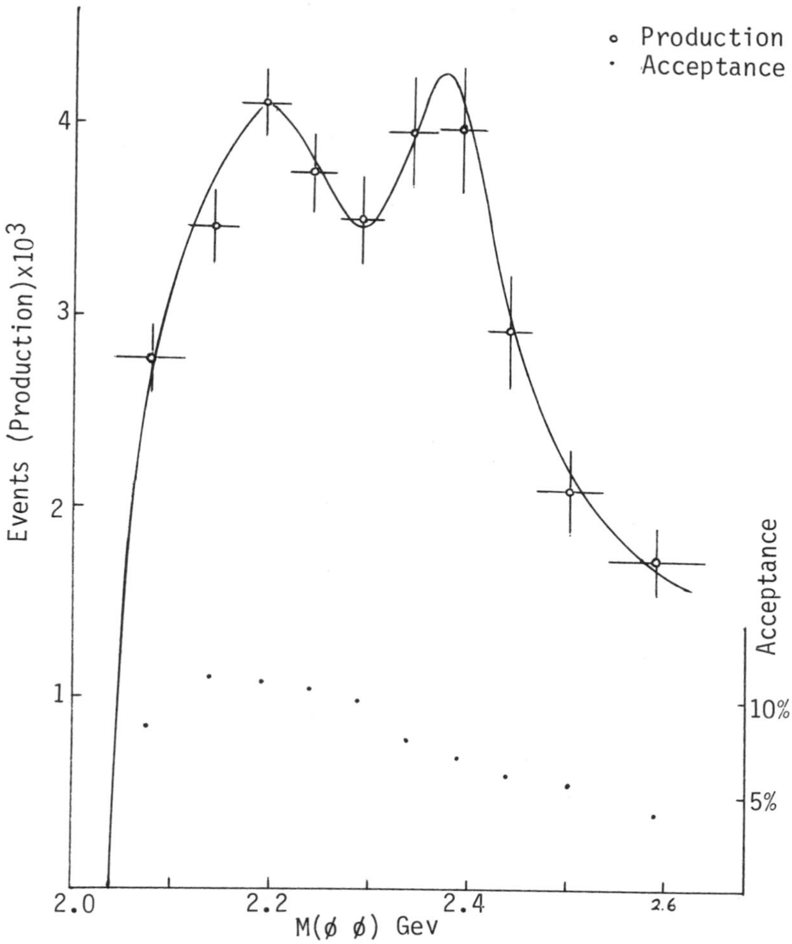

Figure 10: The $\phi\phi$ mass spectrum corrected for acceptance. The solid line is the fit to the data with the three resonant states to be described later. The points at the bottom of the diagram are the acceptance for each mass bin to be read with the scale at the right.

of the D waves using the S wave as a reference is shown in Fig. 15. The statistical significance of this third wave was $\approx 25\sigma$. Although there was an indication for this third wave in the earlier 1200 event sample, it could not be considered statistically

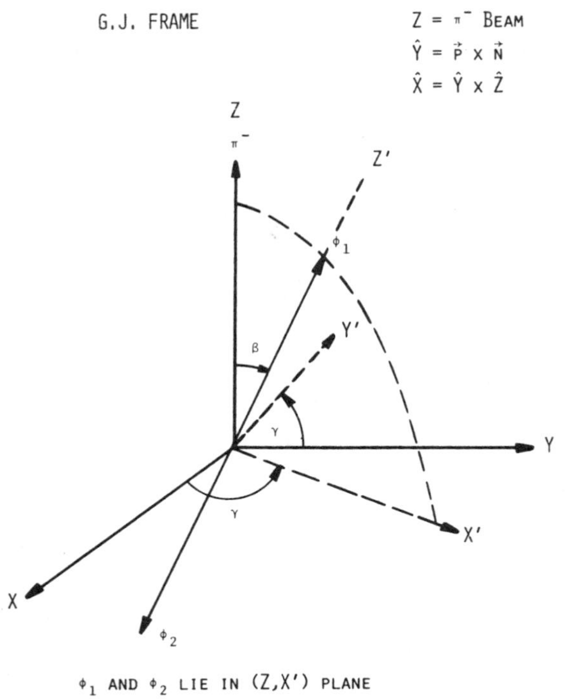

Figure 11: The Gottfried-Jackson frame with polar angle β and azimuthal angle γ.

significant at that time. It should be noted that the 1200 event data sample and the new ≈ 4,000 event data sample agree very well with each other within statistical errors. One should note that the results of the partial wave analysis are quite insensitive to the acceptance and the detailed shape of the mass spectrum. We also found that for $|t'| < 0.3$ GeV^2, the t' distribution is consistent with $e^{(9.4 \pm 0.7)t'}$. If one looks at the quark structure of Fig. 6, one essentially has a pion exchange radiating several gluons (thought to represent a glueball) and thus one would expect a peripheral production mechanism, which is what we observe.

One might ask at this point why are we so incredibly selective - picking 3 waves out of 52 with the statistical significance of the third wave ≈ 25σ. The answer is that the background is small enough and incoherent and thus does not have a significant

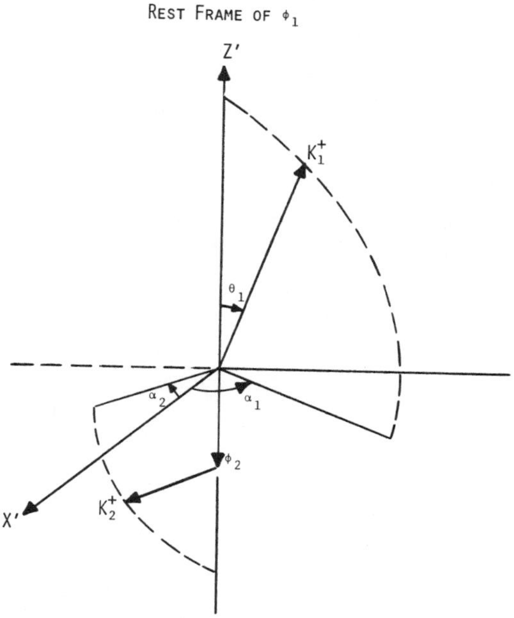

Figure 12: The ϕ_1 rest frame with the polar angle θ_1 of the decay K_1^+ (relative to ϕ direction) and the azimuthal angle α_1 of the decay K_1^+.

effect on the $\phi\phi$ system individual wave signals. The $\phi\phi$ system wave signals are shown (roughly to scale) in Fig. 16 for M = 0 waves. The PWA clearly demonstrated that only M = 0 waves were significant in the fit, thus these are the most relevant. It is clear from Figs. 16a and 16b, that every wave has its own characteristic signature and thus the $\phi\phi$ system is an unusually selective wave content analyzer. This is in large measure due to the fact that each ϕ has spin 1 and thus the six angular variables and their correlations have large characteristic signatures which are very sensitive to the exact quantum numbers of each wave. Furthermore our very low incoherent background allows us to see the characteristics of the $\phi\phi$ system clearly.

Figures 17a–17c show the comparison of Monte Carlo generated events to the observed angular variables and their characteristic

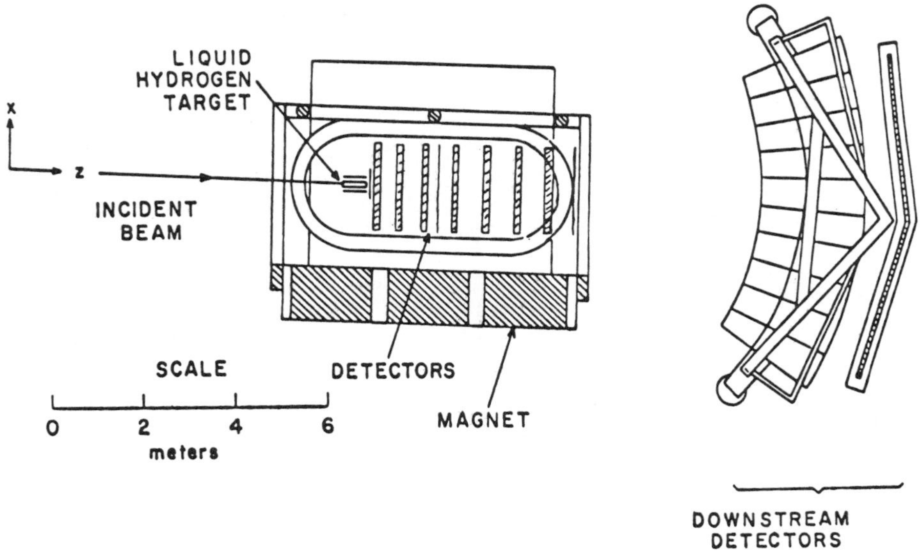

Figure 13: The MPS II and the experimental arrangement (see Refs. 10 and 19 for further details).

combination. It is clear the agreement is very good, and this is true for all ten mass bins. The amplitudes and phase motion (see Figs. 14 and 15) of the waves relative to the S-wave clearly reveals resonance or Breit-Wigner behavior. The S-wave had to be used as a reference due to the fact that the background is both small and incoherent. It is important to note that the appropriate phase motion is the most sensitive test of resonant behavior, and we have clearly demonstrated that it occurs in just the required manner. In the analysis we actually employed the K-matrix method[22] which is approximately equivalent to but a somewhat more realistic approach to fitting the relativistic Breit-Wigner's. Nevertheless in this case either method would give results consistent with each other since the effects of other channels (taken into account in the K-matrix) are small.

Three resonant states (or K-matrix poles) were required to obtain an acceptable fit. Attempts to fit the results with two resonant states (or K-matrix poles) in which the three required

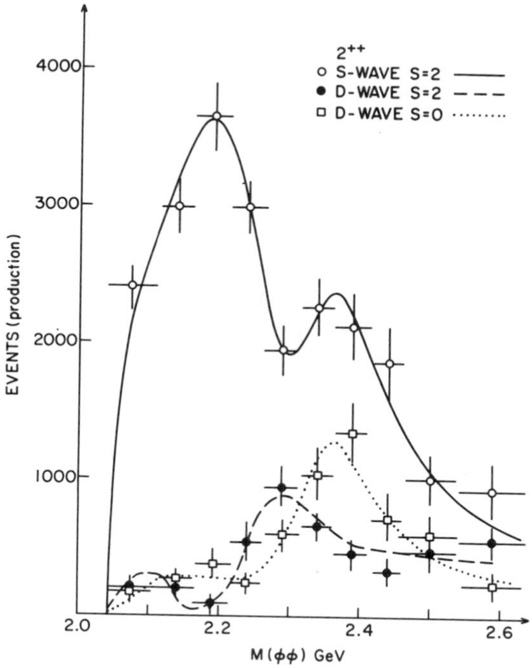

Figure 14: The three 2^{++} partial waves at production in 50 MeV bins (except ends). The smooth curves are derived from a K-matrix fit.

waves were used were rejected by 13σ, whereas the three resonance fit was quite good. The deduced Breit-Wigner parameters, quantum numbers and estimated content of the individual waves for the three states and the estimated errors are shown in Table I. The Argand plot deduced from the K-matrix fit is shown in Fig. 18, and it clearly shows the characteristics expected of resonance behavior. By increasing the statistics from ≈ 1200 events to ≈ 4,000 events the upper of the two resonant $I^G J^{PC} = 0^+ 2^{++}$ states was resolved into two states with the same quantum numbers.

It should be noted that the mixing of waves is substantial in these three $J^{PC} = 2^{++}$ states and the exact wave content of each resonance or K-matrix pole is therefore sensitive to details and somewhat uncertain. However from the glueball physics point of

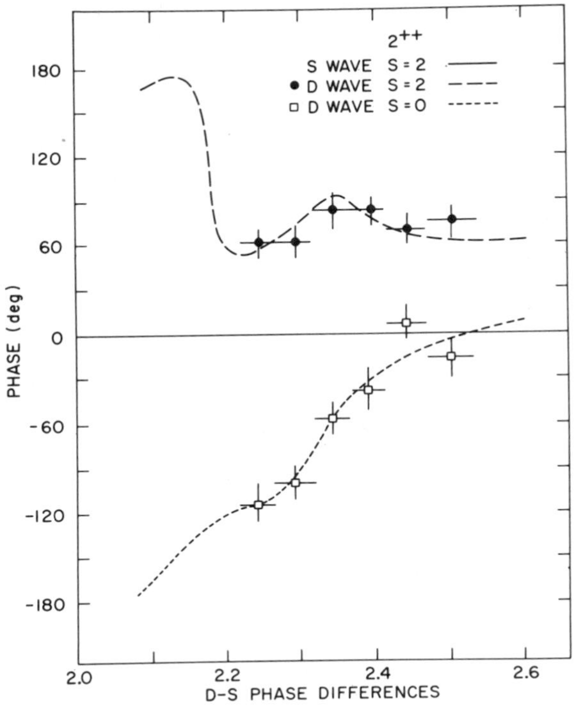

Figure 15: D-S phase difference from the partial wave analysis. The smooth curves are derived from a K-matrix fit.

view we are at present mostly interested in the quantum numbers and parameters of the resonant states and not very concerned about their exact wave contents.

If one assumes as input axioms:

1. QCD is correct;

2. The OZI rule is universal for weakly coupled glue in Zweig disconnected diagrams where the disconnection is due to the introduction of new flavors of quarks, then the states we observe must represent the discovery of 1-3 glueballs.[20,21,5]

Note that axiom (2) allows only resonating glue (i.e. glueballs) to break the Zweig suppression. One primary glueball could break down the Zweig suppression and possibly mix with two quark or other possible states.

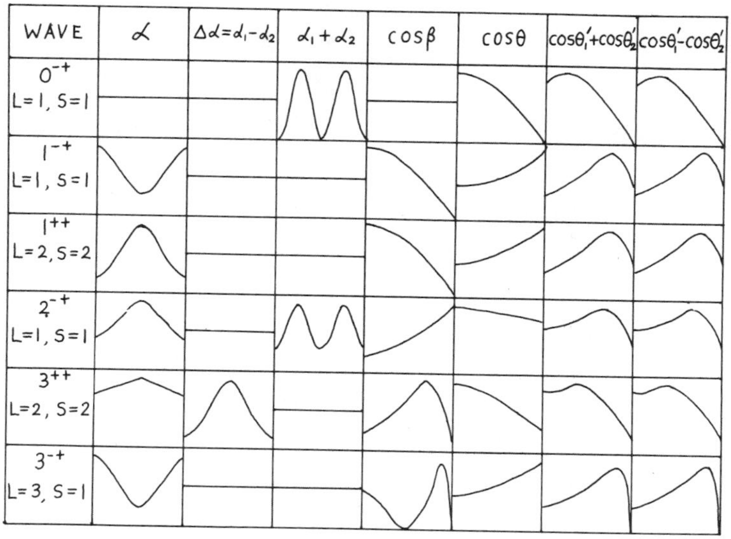

Figure 16a: Various pure waves from $J^{PC} = 0^{++}$ to $J^{PC} = 4^{++}$ with M = 0.

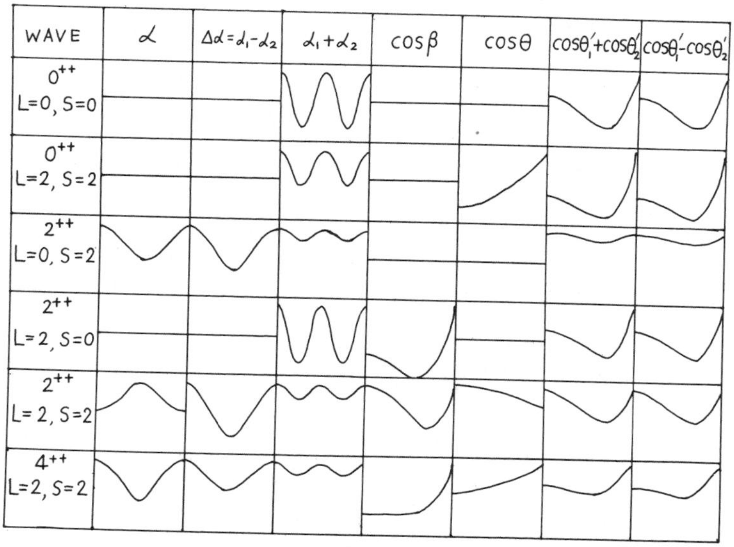

Figure 16b: Various pure waves from $J^{PC} = 0^{-+}$ to $J^{PC} = 3^{-+}$ with M = 0.

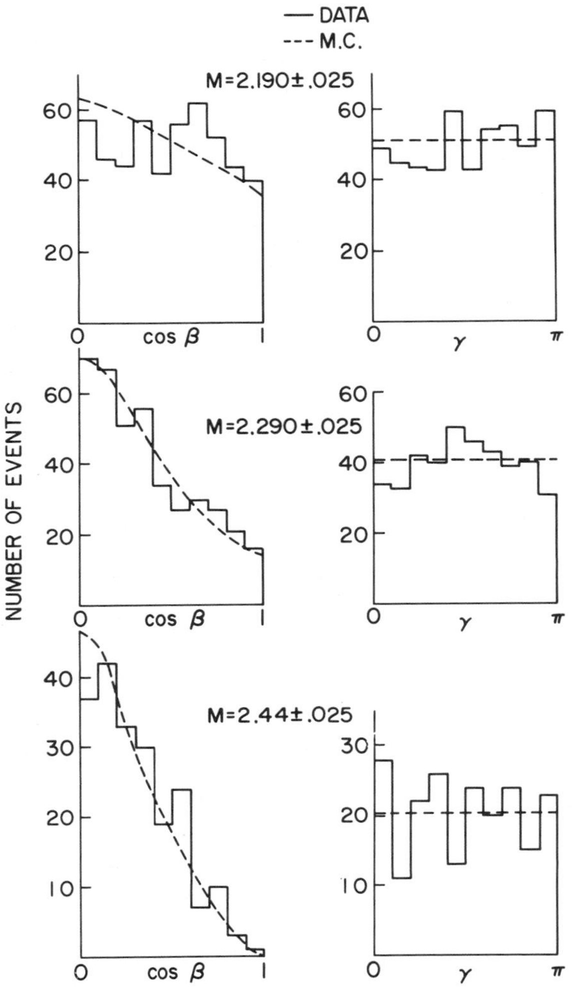

Figure 17a: Cos β and γ for three representative mass bins, where β is the polar angle and γ is the azimuthal angle of a given ϕ in the G.J. frame.

Since these axioms strikingly agree with the data in the ϕ, J/ψ and Υ systems, and merely represent modern QCD practice, it is reasonable to consider this the discovery of glueballs.

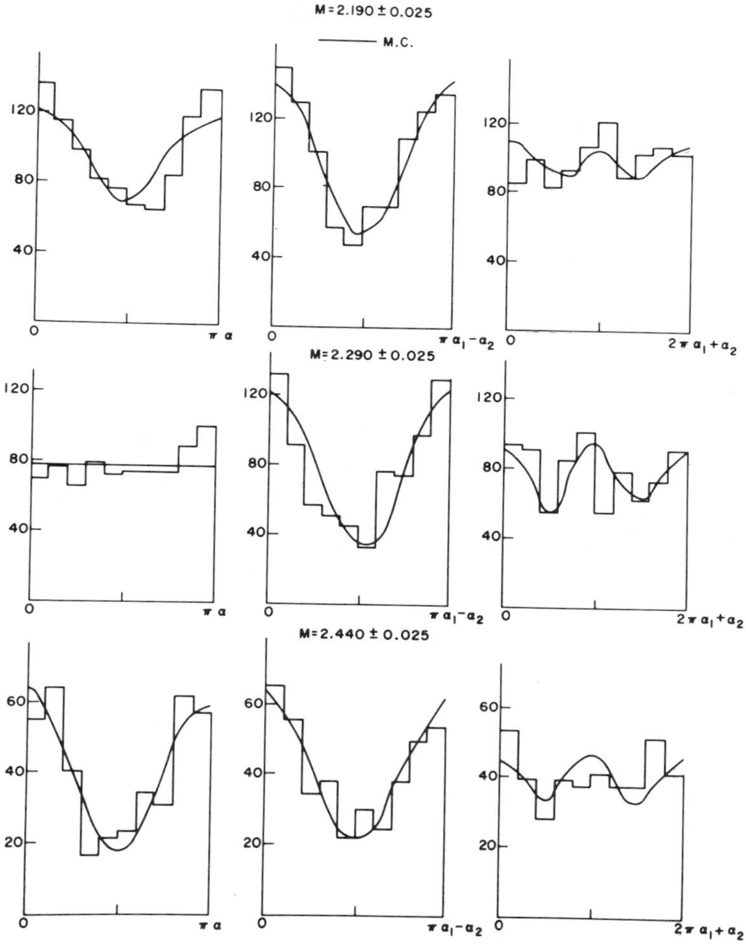

Figure 17b: α, $\alpha_1 - \alpha_2$, and $\alpha_1 + \alpha_2$ for three representative mass bins, where α is the azimuthal angle of the K^+ in the φ rest frame measured from the x-axis of the G.J. frame.

The constituent (i.e. gluon has effective mass) gluon models[23–24] would predict three low lying $J^{PC} = 2^{++}$ glueballs. The mass estimates from the MIT bag calculations and the lattice gauge groups[25–28] cover the range ≈ 1.7 - 2.5 GeV for $J^{PC} = 2^{++}$ glueballs. Thus we are clearly in the right ballpark for agreement with present phenomenological mass calculations.

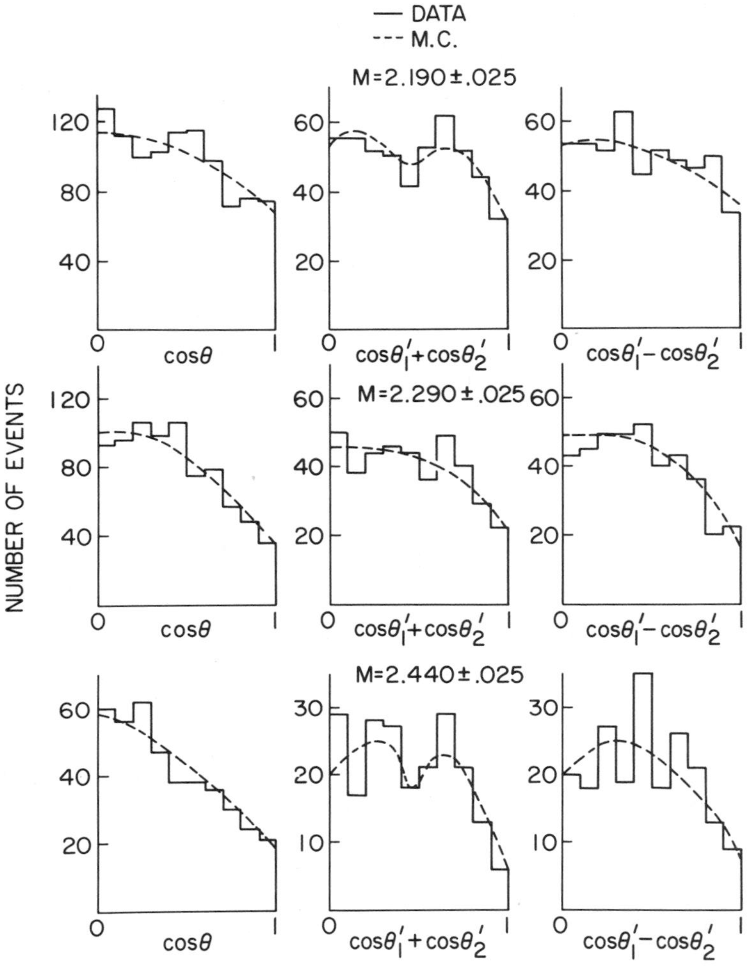

Figure 17c: Cosθ, cosθ_1' + cosθ_2', and cosθ_1' - cosθ_2' for three representative mass bins, where θ is the polar angle of the K^+ in the ϕ rest frame measured from the other ϕ as the z-axis.

T.D. Lee has analytically calculated J = 2 glueballs in the strong coupling limit[29] and obtains three glueball states which correspond to our three states. His strong coupling calculation gives the mass differences between these three states in terms of two parameters, one being essentially the effective strength of the coupling and then a mass scale parameter. In order to try to adjust his strong coupling calculation to the real world of intermediate coupling we took the mass of the 0^{++} glueball as ≈ 1 GeV from the Lattice Gauge calculations, and fit our three masses with the other parameter and found a reasonable fit.

TABLE I

Three Resonance Fit

$M_1 = 2.120^{+.020}_{-.120}$ $\Gamma_1 = .300^{+.150}_{-.050}$ ~ 40% data:

S-wave, S = 2	$\sim 30\%^{+70\%}_{-10\%}$	coupling sign (+) defined
D-wave, S = 2	$\sim 50\%^{+10\%}_{-50\%}$	coupling sign (−)
D-wave, S = 0	$\sim 20\%^{+30\%}_{-20\%}$	coupling sign (−)

$M_2 = 2.220^{+.090}_{-.020}$ $\Gamma_2 = .200 \pm .050$ ~ 40% data

S-wave, S = 2	$\sim 40\%^{+10\%}_{-20\%}$	coupling sign (+)
D-wave, S = 2	$\sim 50\%^{+20\%}_{-10\%}$	coupling sign (+)
D-wave, S = 0	$\sim 10\%^{+10\%}_{-10\%}$	coupling sign (+)

$M_3 = 2.360 \pm .020$ $\Gamma_3 = .150^{+.150}_{-.050}$ ~ 15% data

S-wave, S = 2	$\sim 25\%^{+25\%}_{-10\%}$	coupling sign (+)
D-wave, S = 2	~ 0% + 25%	coupling sign (−)
D-wave, S = 0	$\sim 75\%^{+15\%}_{-25\%}$	coupling sign (+)

ARGAND PLOT

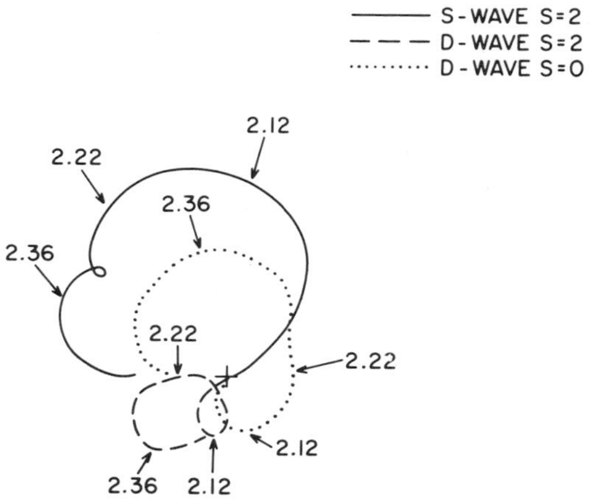

Figure 18: Argand plot from K-matrix.

I have many years ago used a similar procedure in the case of the Pauli-Dancoff strong coupling calculations of the nucleon isobars. In that case when we put in the known f^2 and a reasonable value for the cut-off, the strong coupling calculation results gave reasonable agreement with the experimental observations on nucleon isobars.

What About the Width of Glueballs?

In all hadrons, the hadronization process consists of creation of one or more $q\bar{q}$ paris. This must occur near the outer region of confinement involving strongly interacting soft glue, probably including collective interactions, if we are to have resonances decay with typical hadronic widths ($\Gamma_{hadronic} \sim 100$ to several hundred MeV).

For example the $\rho(770) \to \pi\pi$ requires production of one quark pair. The width of the $\rho(770)$ is $\Gamma_\rho = 154 \pm 5$ MeV. The $\rho'(600) \to 4\pi$ requires the production of three quark pairs. Yet $\Gamma_{\rho'} \approx 300 \pm 100$ MeV. Thus even though production of two additional quark pairs is required the $\Gamma_{hadronic}$ actually increases. This example clearly shows that hadronization easily occurs via collective soft glue effects and this is the basis of typical hadronic widths.

A glueball is nothing more than a resonating multi-gluon system. The glue-glue coupling is stronger than the quark-glue coupling and thus it would be expected, via gluon splittings before the final hadronization, to have a similar hadronization process to a $q\bar{q}$ hadron. In other words a glueball would be expected to have typical hadronic widths. This is certainly to be expected for ordinary (non-exotic) J^{PC} states. In the case of exotic J^{PC} states, this arguement may not be relevant since no one yet knows what suppresses the unobserved exotic sector. In other words, Meshkov's oddballs[23] may be narrow.

I have previously discussed[6,16b] some well-known peculiarities of the OZI rule. In particular if one introduces successive steps both of which are OZI allowed, one can on paper defeat the OZI rule.

For example, $\phi \to \rho\pi$ is OZI forbidden, but $\phi \to K^+K^+ \to \rho\pi$ represents two successive OZI allowed processes which appears to defeat the OZI rule. Similarly, $\pi^- p \to \phi n$ is OZI forbidden, but $\pi^- p \to K^+K^- n \to \phi n$ representing two successive OZI allowed processes which appears to defeat the rule. One can also introduce other complicated intermediate states or processes other than hard multi-gluons to jump the disconnected part of the diagram and also appear to defeat the rule.

Thus the OZI rule is peculiar in that you can defeat it by two-step processes or in QCD language changing the nature of the multi-gluon exchange needed in the one-step diagram to a series of the ordinary OZI allowed gluon exchanges.* Thus based on the experimental validity of the rule, Zweig's diagrams are to be taken literally as one step processes and the multi gluon exchanges needed to connect disconnected parts of the diagram are not to be tampered with.

* This may at least partly be explained by the fact that when you draw quark line diagrams for typical two-step allowed processes in a Zweig forbidden diagram, you are annihilating quark pairs after hadronization has occurred. Since annihilations occur at short distances, and hadronization as I have discussed occurs at large distances, these two-step processes are probably dynamically discriminated against. However it appears that why the OZI rule works so well in Zweig disconnected diagrams will only be understood when one has calculated the dynamics involved using QCD with intermediate and strong couplings.

If one does not accept axiom 2 and demotes the universal OZI rule to the improbable OZI accident could what we see be due to very non-ideally mixed radial excitations or 4-quark states containing $s\bar{s}$ pairs, etc.

Even in this event (for which there is no evidence) it would take a second striking accident for three $I^G J^{PC} = 0^+ 2^{++}$ resonant states and essentially nothing else to occur within the narrow high mass interval of ≈ 2120 to 2360 MeV. Since inventing enough unlikely accidents can destroy any theory I do not consider these possible explanations plausible.

OTHER GLUEBALL CANDIDATES

The radiative decay of the J/ψ is thought to occur as shown in Fig. 19 where one of the usual three gluons emitted in the annihilation of the $c\bar{c}$ pair is replaced by a photon. Thus it has been argued[30–31] that the two-gluon system could recoil from the photon and preferentially form

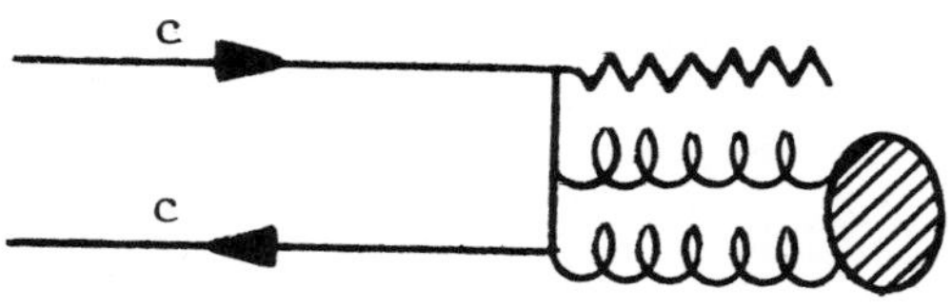

Figure 19: The dominant diagram in radiative J/ψ decay.

a glueball. The first and most discussed glueball candidate of this type is the iota (1440).[32] The status of the iota (1440) with $J^{PC} = 0^{-+}$, $M \approx 1440^{+20}_{-15}$ and $\Gamma \approx 55^{+20}_{-30}$ was recently thoroughly reviewed in the Paris Conference.[33] Some concern was expressed that the ITHEP calculations on instanton effects would move a 0^{-+} glueball up to 2.0–2.5 GeV mass region. The possibility that the iota (1440) is a radial excitation rather than a glueball has also been discussed.

Another glueball candidate of this type is the $\theta(1640)$ with $J^{PC} = 2^{++}$ favored, 95% C.L., $M \approx 1700 \pm 50$, $\Gamma \approx 160 \pm 50$. See Ref. 33 for a review of the status of these glueball candidates.

Recently at the Experimental Meson Spectroscopy Conference there were papers discussing them.[23b,34] Opinions differ strongly. The most recent and thorough review was made by Sid Meshkov.[23b] He concluded the iota (1440) and $\theta(1640)$ are not glueballs but also cited alternate explanations[34,35] in which they could be.

One can directly search for a nonet + glueball $\rightarrow$ decuplet with characteristic mixing splittings. The $g_S(1240)$ with $J^{PC} = 0^{++}$, $M = 1240 \pm 10$ MeV, and $\Gamma = 140 \pm 10$ MeV is one such a glueball candidate.[36] Of course other explanations such as the mixture of singlets from two nonets (one of which could be a radial excitation) are alternatives. The direct pattern recognition search for glueballs is a difficult and so far inconclusive program.

At the Brighton Conference just preceeding this lecture, the Mark III collaboration reported in radiative J/ψ decay,[37] new data observing the iota and the θ. For the iota, the $K^0_S K^0_S \pi^0$ mode was observed in addition to the previously seen $K^+K^-\pi^0$ and $K^0_S K^\pm \pi^\mp$ modes. The Breit-Wigner fit parameters determined were $M = 1.46 \pm 0.01$ GeV and $\Gamma = 0.097 \pm 0.0025$ GeV. In the case of the θ the Breit-Wigner parameters were determined as $M = 1.719 \pm 0.006$ GeV, $\Gamma = .117 \pm .023$ GeV. The iota and θ situation did not appear to change substantially from the prior review[33] and the only essentially new development was the evidence for a new narrow structure ($\xi(2200)$).

It should be noted the question has often been raised as to whether $\phi\phi$ states are seen in radiative decay of the ψ. The new MK III results observe $\psi \to \gamma\phi\phi$.[37] Their detection efficiency for $\phi\phi$ is very low in the mass region of the $g_T(2120)$, $g_T'(2220)$ and $g_T''(2360)$. Thus they find only ~10 events in this mass region. However if one corrects their $\phi\phi$ mass spectrum for the detection efficiency it is not inconsistent with the shape of the mass spectrum seen by BNL/CCNY. However one should note we are comparing ≈ 4,000 observed events to ~ 10. It appears that the MK III can only observe strong signal, narrow, high mass states such as the η_c and thus is not likely to be able to observe the BNL/CCNY states.

Conclusions

If you assume as input axioms:

1. QCD is correct;

2. The OZI rule is universal for weakly coupled glue in disconnected Zweig diagrams where the disconnection is due to the creation or annihilation of new flavor(s) of quark(s), then the BNL/CCNY $g_T(2120)$, $g_T'(2220)$ and $g_T(2360)$ are produced by 1-3 primary glueballs. One or two broad primary glueballs could in principle break down the OZI suppression and mix with one or two quark states which accidentally have the same quantum numbers and nearly the same mass. However the simplest explanation of the rather unusual characteristics of our data is that we have found a triplet of $J^{PC} = 2^{++}$ glueball states.

Since our input axioms are in good agreement with experiments and merely represent modern QCD practice, we have very probably discovered 1-3 $J^{PC} = 2^{++}$ glueballs.

The iota(1440) and the θ(1700) observed in J/ψ radiative decay are glueball candidates. The pros and cons of which have been discussed briefly here and more extensively in the references cited. Other recent glueball searches[38] have not yet led to candidates.

REFERENCES

1. H. Fritzch and M. Gell-Mann, XVI Int. Conf. on High Energy Physics, Chicago-Batavia, 1972, Vol. 2, pp. 135; H. Fritzch, M. Gell-Mann and H. Leutwyler, Phys. Lett. 47B, 365 (1973); S. Weinberg, Phys. Rev. Lett. 31, 49 (1973); S. Weinberg, Phys. Rev. D 8, 4482 (1973); D.J. Gross and F. Wilczek, ibid, 3633 (1973).
2. C.N. Yang and R.L. Mills, Phys. Rev. 96, 191 (1954).
3. a) Fritzch and Minkowski, Nuovo Cimento 30A, 393 (1975). b) R.P. Freund and Y. Nambu, Phys. Rev. Lett. 34, 1645 (1975). c) R. Jaffee and K. Johnson, Phys. Lett. 60B, 201 (1976). d) Kogut, Sinclair and Susskind, Nucl. Phys. B114, 199 (1975). e) D. Robson, Nucl. Phys. B130, 328 (1977). f) J. Bjorken, SLAC Pub. 2372.
4. S. Glashow, J. Ioliopulos, L. Maianai, Phys. Rev. D 2, 1285 (1970); A. Salam, Elementary Particle Theory, Nobel Symposium, Ed. N. Svartholm (Wiley Interscience 1968); S. Weinberg, Phys. Rev. Lett. 19, 1264 (1967).
5. S.J. Lindenbaum. Hadronic Production of Glueballs. Proc. 1983 Intern. Europhysics Conf. on High Energy Physics, Brighton, U.K., July 20-27, 1983, J. Guy and C. Costain, Editors (Rutherford Appleton Laboratory), p. 351-360.
6. S.J. Lindenbaum. Evidence for Explicit Glueballs from the Reaction $\pi^- p \to \phi\phi n$. Proc. of the 20th Course: Gauge Interactions; Theory and Experiment, International School of Subnuclear Physics, Erice, Trapani, Italy, August 3-14, 1982, (to be published). BNL # 32096. (Continued)

6. b) S.J. Lindenbaum, Evidence for Glueballs. Proc. 1982 APS Meeting of the Division of Particles and Fields, University of Maryland, College Park, Maryland, October 28-30, 1982. Eds. William E. Caswell and George A. Snow, AIP Conf. Proc. #98, Particles and Fields Subseries #29, pp.218-246.

7. G. Zweig, CERN REPORTS TH401 and 412 (1964).

8. S. Okubo, Phys. Letts. 5, 165 (1963); Phys. Rev. D 16, 2336 (1977).

9. J. Iizuba, Prog. Theor. Physics. Suppl. 37-38, 21 (1966); J. Iizuba, K. Okuda and O. Shito, Prog. Theor. Phys. 35, 1061 (1966).

10. A. Etkin, K.J. Foley, J.H. Goldman, W.A. Love, T.W. Morris, S. Ozaki, E.D. Platner, A.C. Saulys, C.D. Wheeler, E.H. Willen, S.J. Lindenbaum, M.A. Kramer, U. Mallik, Phys. Rev. Lett. 40, 422-425 (1978); Phys. Rev. Lett. 41, 784-787 (1978).

11. D. Cohen et al., Phys. Rev. Lett. 38, 269 (1977).

12. D.S. Ayres et al., Phys. Rev. Lett. 32, 1463 (1974).

13. H.J. Lipkin, Phys. Lett. 124B, 509 (1983).

14. S.J. Lindenbaum, Phys. Lett. 131B, 221-223 (1983).

15. S.J. Lindenbaum, C. Chan, A. Etkin, K.J. Foley, M.A. Kramer, R.S. Longacre, W.A. Love, T.W. Morris, E.D. Platner, V.A. Polychronakos, A.C. Saulys, Y. Teramoto, C.D. Wheeler. A New Higher Statistics Study of $\pi^- p \to \phi\phi n$ and Evidence for Glueballs. Proc. 21st Intern. Conf. on High Energy Physics, Paris, France, 26-31 July 1982, Journal de Physique 43, P. Petiau and M. Porneuf, Editors (Les Editions de Physique, Les Ulis, France), pp. C3-87 - C3-88.

16. a) S.J. Lindenbaum, Hadronic Physics of $q\bar{q}$ Light Quark Mesons, Quark Molecules and Glueballs. Proc. of the XVIII Course of the International School of Subnuclear Physics, July 31-August 11, 1980, Erice, Trapani, Italy, Subnuclear Series, Vol. 18, "High Energy Limit", Ed. A. Zichichi, pp. 509-562; b) S.J. Lindenbaum, Il Nuovo Cimento, 65A, 222-238 (1981).

17. T. Armstrong et al., CERN EP/82-103 (1982).

18. M. Baubillier et al., Phys. Lett. 118B, 450 (1982).

19. A. Etkin, K.J. Foley, R.S. Longacre, W.A. Love, T.W. Morris, E.D. Platner, V.A. Polychronakos, A.C. Saulys, C.D. Wheeler, C.S. Chan, M.A. Kramer, Y. Teramoto, S.J. Lindenbaum. The Reaction $\pi^- p \rightarrow \phi\phi n$ and Evidence for Glueballs. Phys. Rev. Lett. 49, 1620-1623 (1982).

20. S.J. Lindenbaum. The Discovery of Glueballs. Surveys in High Energy Physics, Vol. 4, pp. 69-126, John M. Charap, Editor (Harvard Academic Publishers, London, 1983).

21. S.J. Lindenbaum. Glueballs in the Reaction $\pi^- p \rightarrow \phi\phi n$. Invited Lecture. Proc. Hadronic Session of the Eighteenth Rencontre De Moriond, La Plagne-Savoie-France, January 23-29, 1983, Vol. I, Gluons and Heavy Flavours, J. Tran Thanh Van, Editor (Editions Frontieres, Gif sur Yvette, France), pp. 441-468.

22. R.S. Longacre. Glueballs in the Reaction $\pi^- p \rightarrow \phi\phi n$ at 22 GeV/c. Proc. of the Seventh Intern. Conf. on Experimental Meson Spectroscopy, April 14-16, 1983, Brookhaven National Laboratory, S.J. Lindenbaum, Editor (AIP Conf. Proc., Particles and Fields Subseries, to be published).

23. a) S. Meshkov. Proc. Hadronic Session of the Eighteenth Rencontre De Moriond, La Plagne-Savoie-France, January 23-29, 1983, Vol. I, Gluons and Heavy Flavours, J. Tran Thanh Van, Editor (Editions Frontieres, Gif sur Yvette, France), pp. 427-440.

23. b) S. Meshkov. Glueballs. Proc. Seventh Intern. Conf. on Experimental Meson Spectroscopy, April 14-16, 1983, Brookhaven National Laboratory, S.J. Lindenbaum, Editor (AIP Conf. Proc., Particles and Fields Subseries, to be published).
24. a) C. Carlson, J. Coyne, P. Fishbane, F. Gross, S. Meshkov, Phys. Rev. D 23, 2765 (1981). b) J. Coyne, P. Fishbane and S. Meshkov, Phys. Lett. 91B, 259 (1980); C. Carlson, J. Coyne, P. Fishbane, F. Gross and S. Meshkov, Phys. Lett. 99B, 353 (1981).
25. C.E. Carlson, T.H. Hansson and C. Peterson, Phys. Rev. D 27, 1556 (1983).
26. C. Rebbi. Proc. 21st Intern. Conf. on High Energy Physics, Paris, France, 26-31 July 1982, Journal de Physique 43, P. Petiau and M. Porneuf, Editors (Les Editions de Physique, Les Ulis, France), pp. C3-723; B. Berg, Calculations in 4d Lattice Gauge Theories (unpublished).
27. B. Berg and A. Billoire, Phys. Lett. 113B, 65 (1982); 114B, 324 (1982).
28. K. Ishikawa et al., Phys. Lett. 116B, 429 (1982); 120B, 387 (1983).
29. T.D. Lee. Time as a dynamical variable CU-TP-266; Talk at Shelter Island II Conf., June 2, 1983; also, ERICE Lectures, August 3-14, 1983 (these proceedings).
30. M. Chanowitz, Phys. Rev. Lett. 46, 981 (1981).
31. J.F. Donoghue, Experimental Meson Spectroscopy - 1980, Sixth Int. Conf., Brookhaven National Laboratory, April 25-26, 1980, Eds. S.U. Chung and S.J. Lindenbaum, AIP Conf. Prof. #67, Particles and Fields Subseries #21, pg. 104-119.
32. Edwards et al., Phys. Rev. Lett. 48, 458 (1982); 49, 259 (1982).

33. E. Bloom. Proc. 21st Intern. Conf. on High Energy Physics, Paris, France, 26-31 July 1982, Journal de Physique 43, P. Petiau and M. Porneuf, Editors (Les Editions de Physique, Les Ulis, France), pp. C3-407.

34. J. Donoghue. The η', The Iota and the Pseudoscalar Mesons. Proc. of the Seventh Intern. Conf. on Experimental Meson Spectroscopy, April 14-16, 1983, Brookhaven National Laboratory, S.J. Lindenbaum, Editor (AIP Conf. Proc., Particles and Fields Subseries, to be published).

35. W.F. Palmer and S.S. Pinski, Production, Decay and Mixing Models of the Iota Meson.

36. A. Etkin, K.J. Foley, R.S. Longacre, W.A. Love, T.W. Morris, S. Ozaki, E.D. Platner, V.A. Polychronakos, A.C. Saulys, Y. Teramoto, C.D. Wheeler, E.H. Willen, K.W. Lai, S.J. Lindenbaum, M.A. Kramer, U. Mallik, W.A. Mann, R. Merenyi, J. Marraffino, C.E. Roos, M.S. Webster, Phys. Rev. D 25, 2446 (1982).

37. K. Einsweiler. Proc. 1983 Intern. Europhysics Conf. on High Energy Physics, Brighton, U.K., July 20-27, 1983, J. Guy and C. Costain, Editors (Rutherford Appleton Laboratory), p. 348-350.

38. See Reference 5 for a brief review and references to the individual papers.

39. J.C. Pati and A. Salam, Phys. Rev. D 8, 1240 (1973); Phys. Rev. D 10, 275 (1974); H. Georgi and S.L. Glashow, Phys. Rev. Lett. 32, 438 (1974); see papers by Beg and Sugawara, XX Int. Conf. on High Energy Physics, Madison, Wisconsin, July 1980; R.N. Mohapatra and R.E. Marshak, Phys. Rev. Letts. 44, 1316 (1980).

40. Particle Data Group Tables, Phys. Letts. 111B, 12 (1982).

41. J. Lee-Franzini. Υ Spectroscopy from CUSB. Proc. of the Seventh Intern. Conf. on Experimental Meson Spectroscopy, April 14-16, 1983, Brookhaven National Laboratory, S.J. Lindenbaum, Editor (AIP Conf. Proc., Particles and Fields Subseries, to be published).

D I S C U S S I O N

Chairman: S. J. Lindenbaum

Scientific Secrataries: S. Capstick and N. Ohta

-BERNSTEIN:

This is a question from one experimentalist to another: why can't an OZI forbidden process like $\pi^- p \to \phi\phi n$ be mediated by multiple soft gluon exchange? There your argument about weak coupling does not apply, and I might think that a ϕ resonance could be made from multiple gluons.

-LINDENBAUM:

In the Zweig disconnected diagram $\pi^- p \to \phi\phi n$ you get creation of two $s\bar{s}$ pairs, and if you look at the diagram the other way around you get annihilation of two $s\bar{s}$ pairs. Those annihilations only occur at very small distances and so they emit hard gluons. The soft gluons come in the outer areas of the confinement region, and that is where you are not, in these very simple processes, creating and annihilating the quarks which make the diagram disconnected. I might add, multiple gluon creation of $q\bar{q}$ pairs would be expected to occur in allowed processes which correspond to connected diagrams. In this case the moving quark lines, as they separate (at large distances), serve as a source of soft multi-gluons, which leads to the hadronization process.

-BERNSTEIN

Is that a quantitative argument or a hope?

-LINDENBAUM:

There is no quantitative argument possible at present other than in the weak coupling region of QCD. However, the OZI forbidden processes are expected to occur at small distances and thus the quark glue coupling should be weak for these processes.

-BERNSTEIN:

Does Lipkin now agree with your assessment of his argument?

-LINDENBAUM:

I don't think so. I presented my statements on the subject at the Brighton Conference and nobody disagreed with them and a number of theorists agreed. My reply has also been accepted for publication by Physics Letters.[14]

-HOU:

If the acceptance for $\phi\phi$ is improved in the $\psi \to \chi\gamma$ experiments and your glueballs are not seen, you are still safe if they are three gluon glueballs.

Could you explain why $\psi' \to \psi\pi\pi$ and $\Upsilon' \to \Upsilon\pi\pi$ are OZI forbidden? I find both your argument and Lipkin's equally convincing. I wouldn't call it OZI suppressed because the two pions are emitted through soft gluons, with only 600 MeV available. Why is this "OZI forbidden" process dominating the decay?

-LINDENBAUM:

The hardness or softness of the gluons depends on the ratio Q^2/Λ, and Λ is not well known. If you don't attribute the narrowness of the ψ', Υ' and especially the Υ'' to the Zweig suppression, then what do you attribute it to? Lipkin argued that the $\psi'(3685)$ goes via the 2π mode 50% of the time. He then said that we can get $\pi + \psi' \to \psi + \pi$ from $\psi' \to \psi + 2\pi$ by crossing one π, which is diffractive excitation of the ψ and has a large cross section, so

-LINDENBAUM (continued):

the process is OZI allowed. He also used this so-called crossed Pomeron diagram arguement to conclude $\pi^- p \to \phi\phi n$ is $\approx$ OZI allowed, since it is related by crossing to $\phi + n \to \phi + \pi^- + p$ which is merely elastic scattering with additional production of a pion and is not expected to be suppressed. I pointed out that when you cross in that manner you get into different physical and kinematic regions, and you cannot simply relate the two processes. For example considering the kinematics alone, the crossed process corresponds to high momentum transfer and high mass diffractive dissociation and thus is expected to be negligibly small.

-HOU:

I have made a naive pole model analysis of $\psi' \to \psi\pi\pi$ and $\Upsilon' \to \Upsilon\pi\pi$ going through a 0^{++} Σ-model object. The small size of $\psi' \to \psi\pi\pi$ can be understood as partly due to phase space and partly due to the pole particle being off shell. So the narrowness cannot be used as a proof of the OZI rule.

LINDENBAUM:

This is equivalent to having two step allowed processes which replace a Zweig forbidden process, and there is no evidence for this occurring in the ϕ, J/ψ and Υ systems. However, let me reiterate, that my rigorous conclusions are based on my second axiom which states that the OZI rule cannot be broken by such processes.

-OHTA:

Is there any reason why glueballs exist only in spin-2 and possibly in spin-0 states? Also why are these spin-2 states nearly degenerate?

-LINDENBAUM:

They need not exist only in spin-0 and spin-2 states, however all theoretical calculations find the lowest lying states are 0^{++} and then 2^{++}. In our experiment we cannot probe the C = - sector because $\phi\phi$ has C = +, but our experiment has no bias to any particular J^P. It has just turned out that we have found a cluster of 2^{++} states in the mass region we are probing. Prof. Lee said that based on his strong coupling calculations he expected three J = 2 states. As described in my proceeding's paper, adjusting his strong coupling constant calculations to the real world, we find his mass formula gives agreement with the mass splittings of the states we observe. In the constituent gluon model as proposed by Meshkov, the two gluon sector contains three low-lying 2^{++} states.

-EREDITATO:

In which channel do you experimentally look for $\phi\phi$?

-LINDENBAUM:

The channel is (22 GeV/c) $\pi^- p \to \phi\phi n$. Each $\phi \to K^+K^-$ and we identify the four K's.

-CATTO:

Earlier you talked about other glueball candidates, like the $J^{PC} = 0^{-+}$ iota(1440), and you mentioned that the instanton effects would move this up to ~ 2-2.5 GeV. Do you feel this is a plausible calculation?

-LINDENBAUM:

That statement was made by the I.T.H.E.P. group and reported by E. Bloom in his rapporteur talk at the Paris Conference a year ago. He said that he was worried about the iota for that reason. The 0^- states are particularly sensitive to instanton effects.

-KLEVANSKY:

I am having a logic problem. As I understand it you wish to add to the data that exists confirming QCD by establishing the existence of glueballs. Yet your identification of glueballs relies on the assumption of the correctness of QCD. So if you identify anything under this assumption I don't see that this says anything <u>for</u> QCD. You can only disprove it.

-LINDENBAUM:

As far as I am concerned there is no theory that you can ever experimentally prove without assuming the theory. No theory can be mathematically proven without input axioms.

-KLEVANSKY:

The input axioms should not be the conclusion.

-LINDENBAUM:

But to experimentally show a theory is correct, we must take the theory and calculate or predict various things and find experimental agreement. Thus you must assume a theory to experimentally prove it.

-BATTISTON:

In your paper submitted to <u>Surveys In High Energy Physics</u> you showed the $\phi\phi$ mass distribution for $K^-p \to \phi\phi\Lambda/\Sigma$. This distribution has a mass peak at ~ 2.3 GeV. This peak is broader than your peak, but statistically significant. The reaction $K^-p \to \phi\phi\Lambda/\Sigma$ is Zweig allowed. How do you explain this peak?

-LINDENBAUM:

Our distribution peaks earlier and falls faster than that one. Because $K^-p \to \phi\phi\Lambda/\Sigma$ is OZI allowed, there is no reason to

-LINDENBAUM (continued):
expect the two distributions to be the same. That reaction will transmit things other than glueballs.

To explain the structure of the distribution we must do a partial wave analysis, but because of the poor statistics no analysis has been done.

-BATTISTON:

Can you show me the distribution of the phase space for the ϕ system with the hypothesis that no resonance is present?

-LINDENBAUM:

It would peak at higher masses and look roughly like this:

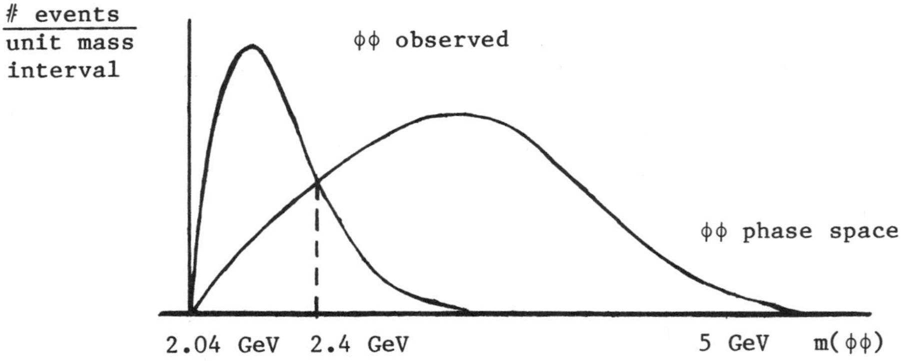

-BATTISTON:

Can you explain why your axiom of the correctness of QCD rules out the possibility of $\phi\phi$ resonance states?

-LINDENBAUM:

The second axiom rules out any breakdown of the OZI Rule except through glueballs.

-BATTISTON:

Couldn't a $\phi\phi$ resonance state increase the cross-section without introducing 2 resonant intermediate state of glue?

-LINDENBAUM:

You are ignoring the Zweig disconnected diagram argument, which I take as an axiom, and which is experimentally verified.

-KASPER:

He is not ignoring the axiom, he is assuming that it is not occurring in resonance, and that having produced the four strange quarks they resonate, and this accounts for the increased cross-section.

-LINDENBAUM:

Again you will find that you are making a two-step allowed process which changes the nature of the mulit-gluon exchange in a Zweig disconnected diagram to that corresponding to two successive allowed steps. Doing this could destroy all Zweig disconnected diagrams, which will remove the explanation for OZI suppression and change the OZI rule to the unlikely OZI accident. I made a qualitative argument for the validity of the one step diagrams, based on the fact that annihilation occurs at small distances and hadronization occurs at large distances.

-CATTO:

I noticed that some of the lattice gauge theory results are either higher or lower than the results of your experiments. For which J^{PC} states have these calculations been done?

-LINDENBAUM:

The ones I showed were all for $J^{PC} = 2^{++}$. Lattice calculations have only been done for 0^{++} and 2^{++}. The results of these

-LINDENBAUM (continued):

calculations lie been 1.7 ~ 3 GeV and they average to ~ 2 GeV.

-OGILVIE:

The lattice calculations have better figures for the 0^{++} state. Is there any hope of cleaning up the experimental status of the 0^{++} glueball?

-LINDENBAUM:

The prospects are not too good. There are too many possible explanations for the 0^{++} glueball candidates.

UNIVERSALITY FEATURES IN (pp), (e^+e^-), AND DEEP INELASTIC SCATTERING PROCESSES

A. Zichichi

CERN
Geneva
Switzerland

ABSTRACT

The use of the correct variables in (pp), (e^+e^-), and deep-inelastic scattering (DIS) processes allows universality features to be established in these - so far considered - different ways of producing multihadronic states.

INTRODUCTION

Multihadronic final states can be produced in purely hadronic interactions such as (pp), in purely electromagnetic interactions such as (e^+e^-), and in lepton-hadron deep-inelastic scattering (DIS) processes, which can be either "weak" (such as νp) or electromagnetic (such as μp).

The purpose of this paper is to review the main points needed in order to establish a common basis for a comparison between these various ways of producing multihadronic states. The interest in this study is twofold:

i) So far, these three ways of producing multiparticle systems have been considered to be basically different. Our study shows that, before reaching such a conclusion, it is imperative to use the correct variables in describing the three processes: (pp), (e^+e^-), and DIS. In fact, the use of the correct variables allows universal features to be revealed in the multihadronic final states produced in (pp), (e^+e^-), and DIS.

ii) Once these - so far considered - different ways of producing multihadronic final states are brought within the correct framework, the comparison can be made at a deeper level, and basic differences can thus be studied, if they exist. These differences must be at a level which is below our present one, where we have established striking common features.

We will see that (pp) interactions can be studied à la (e^+e^-) and à la DIS. However, the DIS way turns out to be incorrect when comparing DIS data with those obtained in (e^+e^-). The equivalence found between (pp) and (e^+e^-) data allows the "differences" reported between DIS and (e^+e^-) data to be understood. These differences are reproduced if (pp) data are analyzed à la DIS, and disappear when the correct variables are used.

In Section 2 we introduce the correct variables. In Section 3 we report all results obtained so far when comparing (pp), (e^+e^-), and DIS. In Section 4 we extrapolate our findings to the multihadronic systems to be studied at collider energies, such as at the CERN $(p\bar{p})$ machine. Section 5 gives the conclusions.

Notice that the universality features discovered are a zero-parameter fit to the various properties of the multihadronic systems produced in (pp), (e^+e^-), and DIS.

2. THE IDENTIFICATION OF THE CORRECT VARIABLES

The identification of the correct variables for describing hadron production in (pp) interactions, (e^+e^-) annihilation, and DIS processes is the basic starting point for putting these three ways of producing multiparticle hadronic systems on an equal footing. In this section we show how this can be done.

2.1 <u>e^+e^- Annihilation</u> is illustrated in Figure 1, where q_1^{inc} and q_2^{inc} are the four-momenta of the incident electron e^- and positron e^+; q^h is the four-momentum of a hadron produced in the final state, whose total energy is

$$(\sqrt{s})_{e^+e^-} = \sqrt{(q_1^{inc} + q_2^{inc})^2} = 2E_{beam} \tag{1}$$

(when the colliding beams have the same energy).

As we will see later,

$$\begin{aligned} q_1^{inc} &= q_1^{had} , \\ q_2^{inc} &= q_2^{had} , \end{aligned} \tag{2}$$

where $q_{1,2}^{had}$ are the four-momenta available in a (pp) collision for the production of a final state with total hadronic energy

$$\sqrt{(q_1^{had} + q_2^{had})^2} = \sqrt{(q_{tot}^{had})^2} . \tag{3}$$

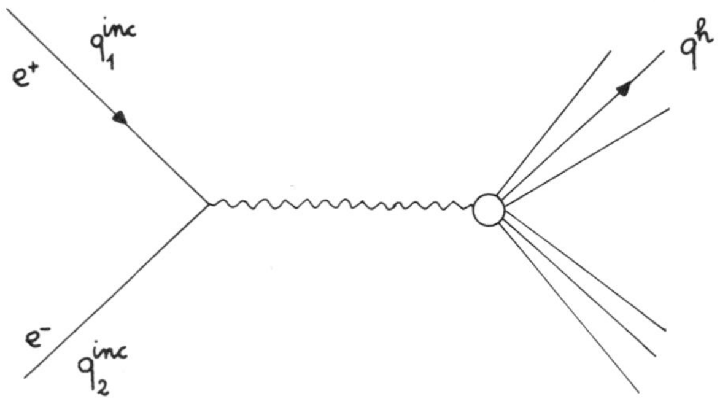

Fig. 1. Schematic diagram for the (e^+e^-) annihilation

It is this quantity $\sqrt{(q_{tot}^{had})^2}$ which should be used in the comparison with (e^+e^-) annihilation, and therefore with

$$(\sqrt{s})_{e^+e^-} \quad . \tag{4}$$

This means that

$$(\sqrt{s})_{e^+e^-} = \sqrt{(q_{tot}^{had})^2} \quad . \tag{5}$$

Moreover, the fractional energy of a hadron produced in the final state of an (e^+e^-) annihilation is given by

$$(x)_{e^+e^-} = 2\,\frac{q^h \cdot q_{tot}^{had}}{q_{tot}^{had} \cdot q_{tot}^{had}} = 2\,\frac{E^h}{(\sqrt{s})_{e^+e^-}} \tag{6}$$

where the dots indicate the scalar product, and E^h is the energy of the hadron "h" measured in the (e^+e^-) c.m. system. Notice that the four-momentum q_{tot}^{had} has no space-like part:

$$q_{tot}^{had} \equiv \left[\, i\vec{0};\ (\sqrt{s})_{e^+e^-} \right] \quad . \tag{7}$$

2.2 DIS processes are illustrated in Figure 2, where q_1^{inc} and $q_1^{leading}$ are the four-momenta of the initial- and final-state leptons, respectively; q_2^{inc} is the four-momentum of the target nucleon; q_1^{had} is the four-momentum transferred from the leptonic to the hadronic vertex, whose time-like component is usually indicated as ν:

$$q_1^{had} \equiv (i\vec{p}_1^{\,had};\ \nu \equiv E_1^{had}) \quad . \tag{8}$$

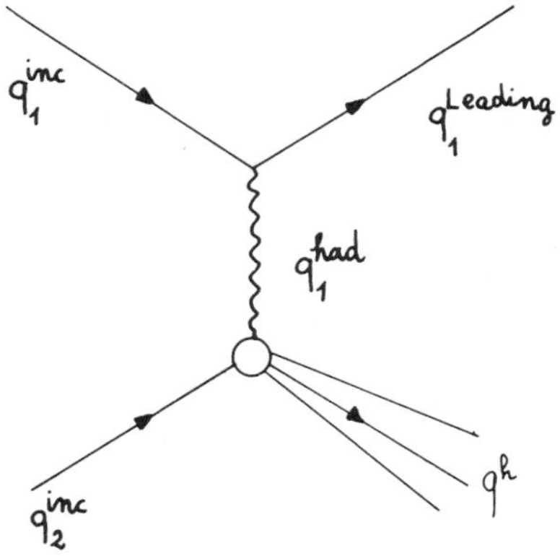

Fig. 2. Schematic diagram for the DIS processes.

Notice that in order to easily identify the equivalent variables in (pp) interactions, we have introduced a notation in terms of E_1^{had} and $\vec{p}_1^{\,had}$.

A basic quantity in DIS is the total hadronic mass

$$(W^2)_{DIS} = (q_1^{had} + q_2^{inc})^2 , \qquad (9)$$

and the fractional energy is

$$(z)_{DIS} = \frac{q^h \cdot q_2^{inc}}{q_1^{had} \cdot q_2^{inc}} , \qquad (10)$$

where again the dots between the four-momenta indicate their scalar product.

2.3 <u>(pp) interactions</u> are illustrated in Figure 3, where $q_{1,2}^{inc}$ are the four-momenta of the two incident protons; $q_{1,2}^{leading}$ are the four-momenta of the two leading protons; $q_{1,2}^{had}$ are the space-like four-momenta emitted by the two proton vertices; q^h is the four-momentum of a hadron produced in the final state.

Now, attention! A (pp) collision can be analyzed in such a way as to produce the key quantities proper to (e^+e^-) annihilation and DIS processes.

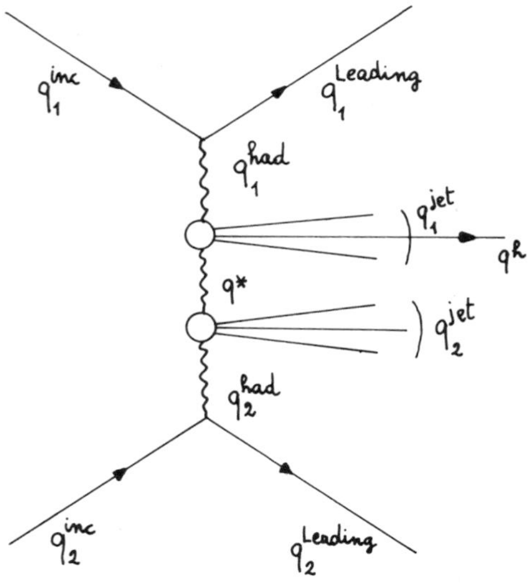

Fig. 3. Schematic diagram for the (pp) interactions.

In fact, from Figure 3 we can work out the following quantities, which are needed if we want to compare (pp) physics with (e^+e^-), i.e.

$$(q_{tot}^{had})_{pp} = (q_1^{had} + q_2^{had})_{pp} \quad ; \tag{11}$$

and [see formula (5)]

$$\sqrt{(q_{tot}^{had})_{pp}^2} = (\sqrt{s})_{e^+e^-} \,. \tag{12}$$

Moreover,

$$(x)_{pp}^{had} = 2\,\frac{q^h \cdot q_{tot}^{had}}{q_{tot}^{had} \cdot q_{tot}^{had}} \,, \tag{13}$$

to be compared with

$$(x)^{had}_{e^+e^-} = 2\,\frac{q^{h} \cdot (q^{had}_{tot})_{e^+e^-}}{(q^{had}_{tot})_{e^+e^-} \cdot (q^{had}_{tot})_{e^+e^-}} . \tag{14}$$

The subscripts (e^+e^-) in q^{had}_{tot} are there to make it clear that these quantities are measured in (e^+e^-) collisions and are the quantities equivalent to q^{had}_{tot} measured in (pp) interactions.

The same (pp) diagram (Figure 3) can be used to work out the key quantities needed when we want to compare (pp) physics with DIS. In this case we have

$$(W^2)^{had}_{pp} = (q^{had}_1 + q^{inc}_2)^2 \tag{15}$$

and

$$(z)^{had}_{pp} = \frac{q^{h} \cdot q^{inc}_2}{q^{had}_1 \cdot q^{inc}_2} . \tag{16}$$

Note the in W^2 the leading proton No. 2 is not subtracted. This is the reason for the differences found in the comparison between DIS data and e^+e^- (see Section 3 and Reference 15). In fact (W^2) is not the effective total energy available for particle production, owing to the presence there of the leading proton.

3. EXPERIMENTAL RESULTS

A series of experimental results, where (pp) interactions have been analyzed à la e^+e^- and à la DIS, have given impressive analogies in the multiparticle systems produced in these - so far considered - basically different processes: (pp), (e^+e^-), DIS.

The experimental data where (pp) interactions are compared with (e^+e^-) are shown in Figures 4-11 and References [1-14].

The experimental data where (pp) interactions are compared with DIS are shown in Figures 12-14 and References [15-17].

These comparisons show striking analogies with respect to the following quantities:

i) the inclusive fractional energy distribution of the produced particles[1,2,9,11] (see Figures 4,5,14);

ii) the average charged-particle multiplicities[3,8,12,15,16] (see Figures 6,12,13);

iii) the ratio of the average energy associated with the charged particles over the total energy available for particle production[5] (see Figure 7);

iv) the inclusive transverse momentum distribution of the produced particles[7,10] (see Figures 8-10);

v) the correlation functions in rapidity[14] (see Figure 11).

Notice the power of the (pp) interaction. Once this is analyzed in the correct way it produces results equivalent to (e^+e^-) and DIS.

This means that there is an important universality in these ways of producing multihadronic systems.

4. EXTRAPOLATIONS OF COLLIDER PHYSICS

From the above analysis we can conclude that:

i) the leading effects must be subtracted and the correct variables have to be used if we want to compare purely hadronic interactions with (e^+e^-) and DIS;

ii) the old myth, based on the belief that in order to compare (pp) with (e^+e^-) and DIS you need high-p_T (pp) interactions, is dead. In fact we have proved that low-p_T (pp) interactions produce results in excellent analogy with (e^+e^-) annihilation and DIS

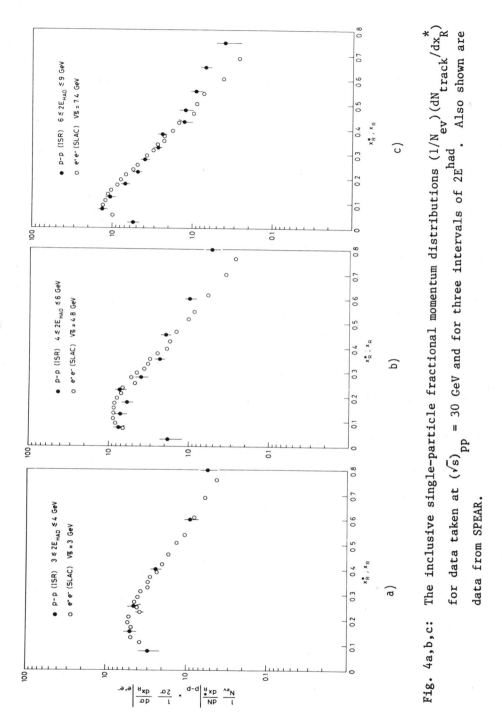

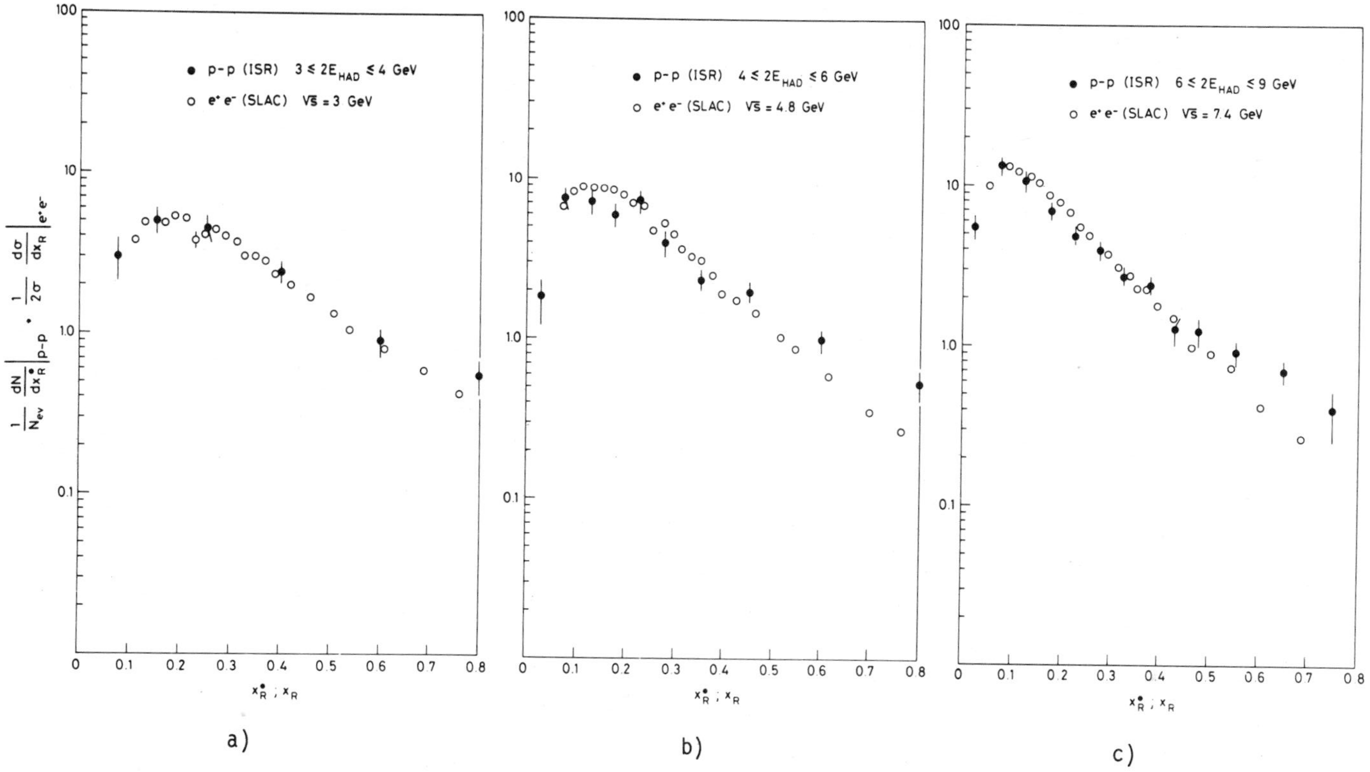

Fig. 4a,b,c: The inclusive single-particle fractional momentum distributions $(1/N_{ev})(dN_{track}/dx_R^*)$ for data taken at $(\sqrt{s})_{pp} = 30$ GeV and for three intervals of $2E^{had}$. Also shown are data from SPEAR.

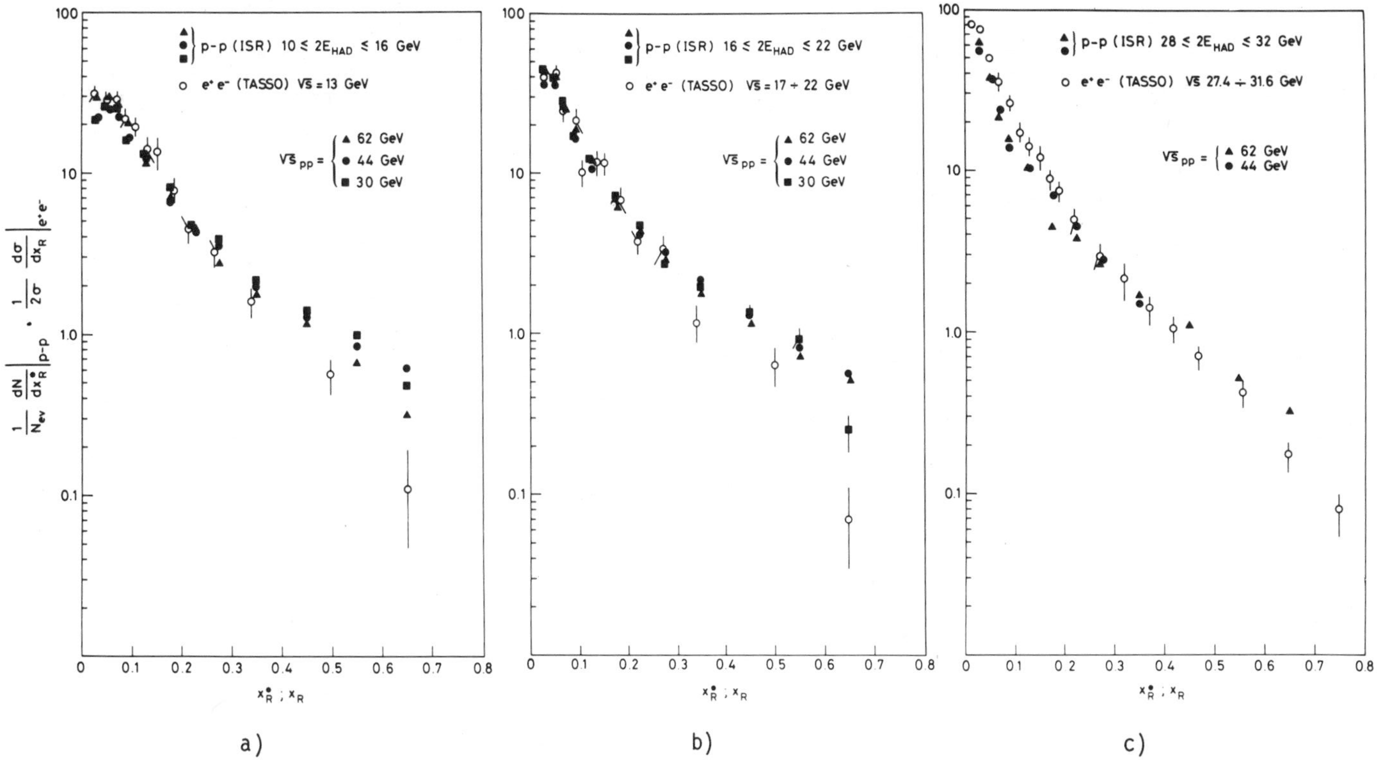

Fig. 5a,b,c: The inclusive single-particle fractional momentum distributions $(1/N_{ev})(dN_{track}/dx^*_R)$ in the same $2E^{had}$ interval, but different $(\sqrt{s})_{pp}$. Also shown are data from TASSO at PETRA.

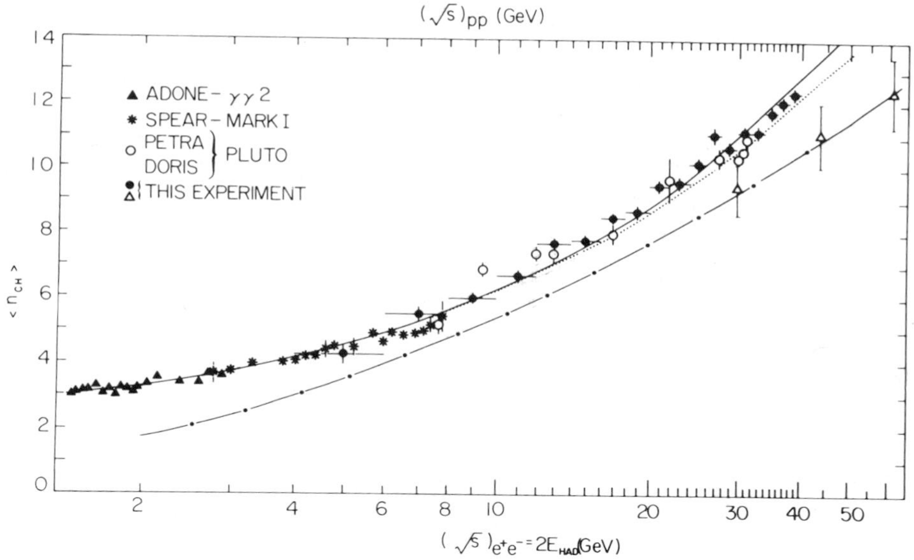

Fig. 6. Mean charged-particle multiplicity [averaged over different $(\sqrt{s})_{pp}$] versus $2E^{had}$, compared with (e^+e^-) data. The continuous line is the best fit to our data according to the formula $\langle n_{ch}\rangle = a + b \exp[c\sqrt{\ln (s/\Lambda^2)}]$. The dotted line is the best fit using PLUTO data. The dashed-dotted line is the standard (pp) total charged-particle multiplicity with, superimposed, our data as open triangular points.

processes, the basic parameter in (pp) interactions being $\sqrt{(q_{tot}^{had})^2}$ for a comparison with (e^+e^-), and $\sqrt{W^2_{(pp)}}$ for a comparison with DIS.

The existence of high-p_T events means that point-like constituents exist inside the nucleon. But low-p_T events contain the same amount of basic information as high-p_T events. The only difference is expected in $\langle p_t\rangle$. In fact our analysis of the inclusive transverse momentum distribution, in terms of the renormalized variable $p_t/\langle p_t\rangle$ [notice that p_t indicates the transverse momentum of the particles produced with respect to the jet axis, and p_T with

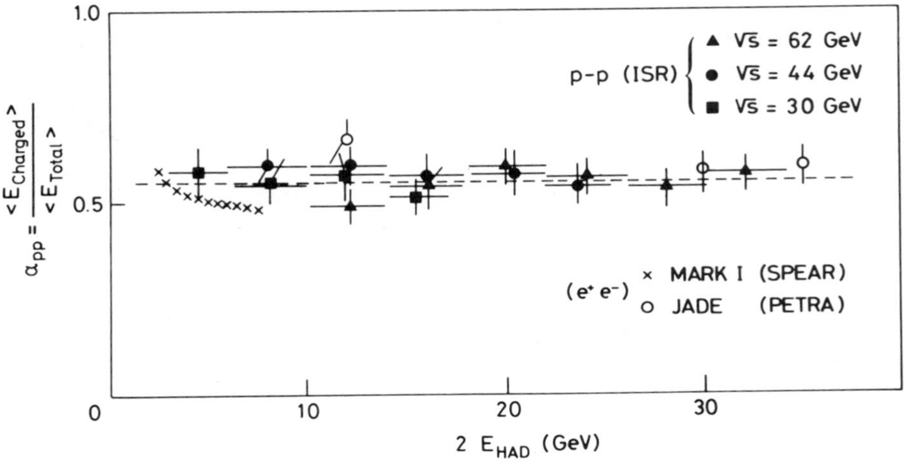

Fig. 7. The charged-to-total energy ratio obtained in (pp) collisions α_{pp}, plotted versus $2E^{had}$ and compared with (e^+e^-) obtained at SPEAR and PETRA.

respect to the colliding (pp) or ($p\bar{p}$) axis], is suggestive of a very interesting possibility: multiparticle systems produced at high p_T could show, at equivalent $\sqrt{(q_{tot}^{had})^2}$ high values of $\langle p_t \rangle$. This should be the only difference between multiparticle systems with the same $\sqrt{(q_{tot}^{had})^2}$ produced at low-p_T and high-p_T.

There are two ways of producing $\sqrt{(q_{tot}^{had})^2}$:

i) one is at low p_T, and we have seen what happens;

ii) the other is at high p_T: we have not been able to compare, at constant values of $\sqrt{(q_{tot}^{had})^2}$, the multiparticle systems produced in (pp) interactions at high p_T and low p_T, the reason being the lack of CERN ISR time. However, as mentioned above, the agreement between our data and (e^+e^-) data in the variable $(p_t/\langle p_t \rangle)$, makes it possible to foresee what should change between low-p_T and high-p_T multiparticle jets.

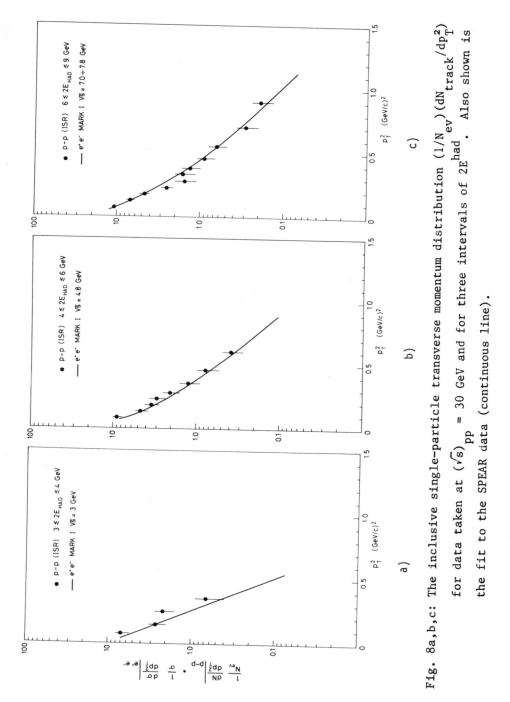

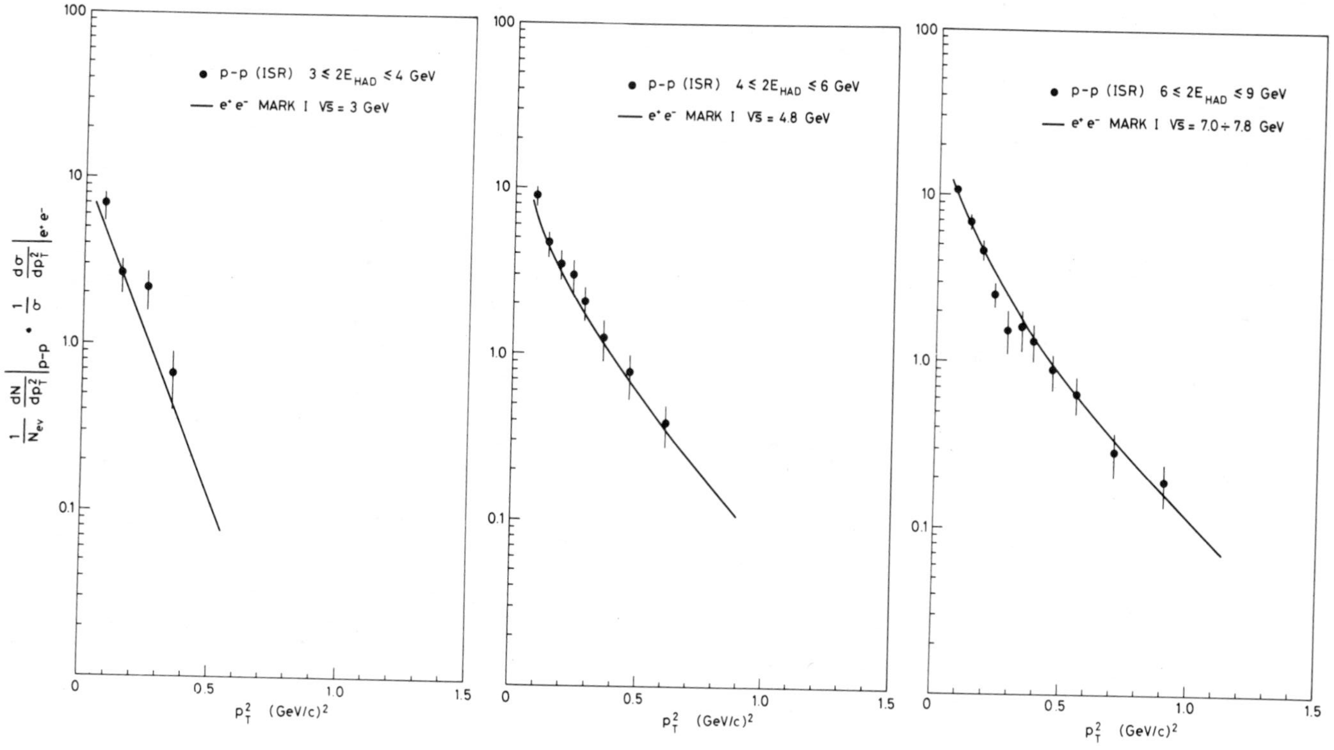

a) b) c)

Fig. 8a,b,c: The inclusive single-particle transverse momentum distribution $(1/N_{ev})(dN_{track}/dp_T^2)$ for data taken at $(\sqrt{s})_{pp} = 30$ GeV and for three intervals of $2E^{had}$. Also shown is the fit to the SPEAR data (continuous line).

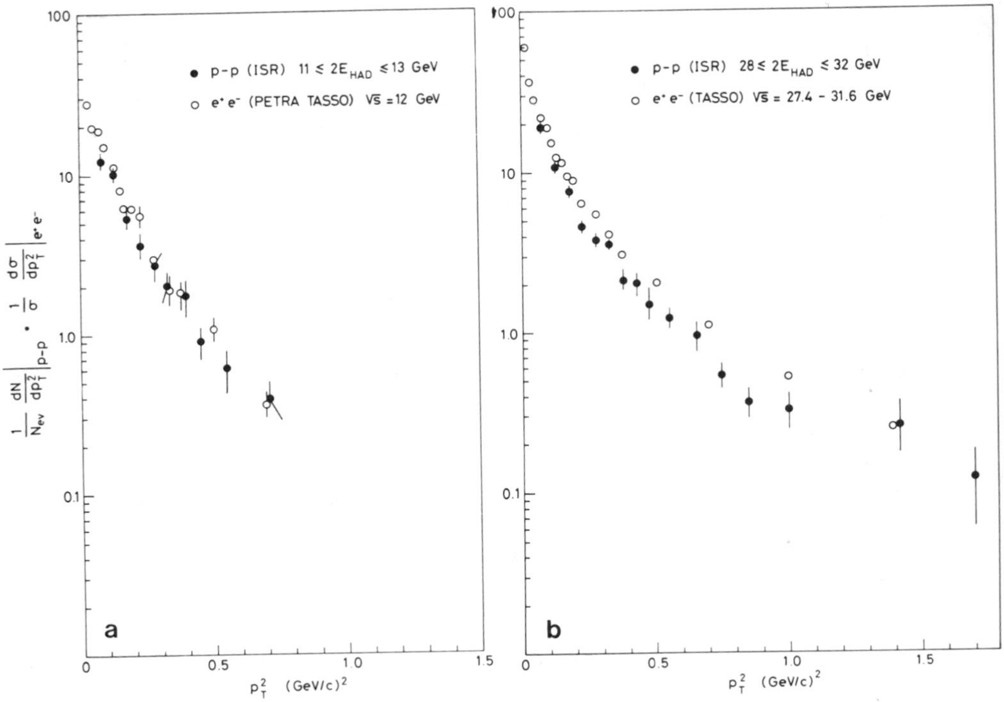

Fig. 9a,b: The inclusive single-particle transverse momentum distributions $(1/N_{ev})(dN_{track}/dp_T^2)$ for two E^{had} range. Also shown are data from TASSO at PETRA.

Now we come to the extrapolation: The extrapolation of our method to the CERN $p\bar{p}$ Collider[18,19] would allow a large energy jump and could produce clear evidence for or against our prediction. Let us give an example. If two jets at the $(p\bar{p})$ Collider are produced back-to-tack with the same transverse energy E_T, then we have

$$\sqrt{(q_{tot}^{had})^2} \cong 2E_T \ .$$

Suppose that we are at

$$2E_T = 100 \text{ GeV} \ .$$

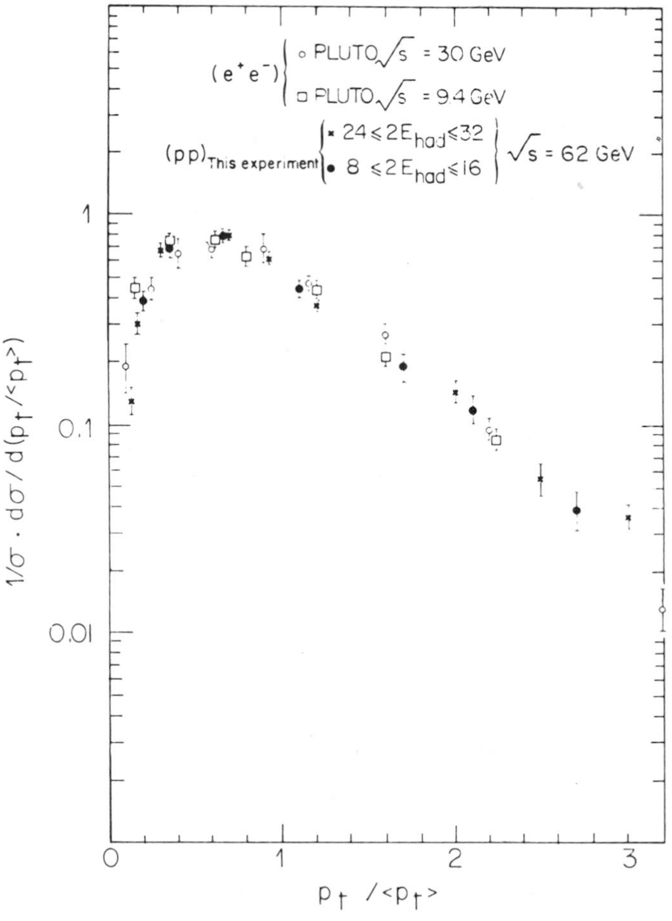

Fig. 10a,b: The differential cross-section $(1/\sigma)[d\sigma/d(p_t/<p_t>)]$ versus the "reduced" variable $p_t/<p_t>$. These distributions allow a comparison of the multiparticle systems produced in (e^+e^-) annihilation and in (pp) interactions in terms of the renormalized transverse momentum properties.

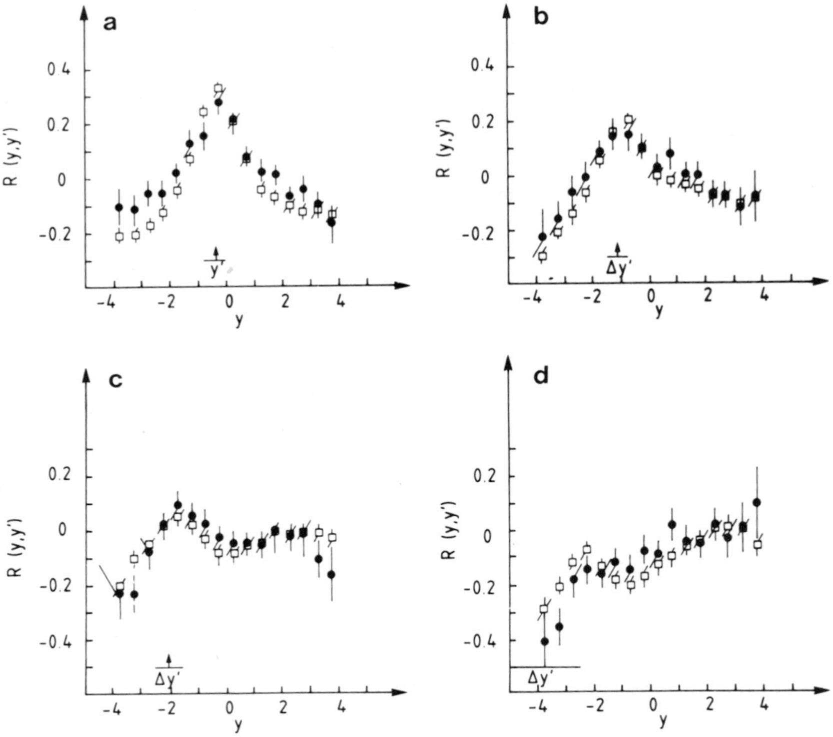

Fig. 11. Two-particle correlation in rapidity space: R(y,y'), for different y' intervals, as measured in the present experiment after leading proton subtraction in the $\sqrt{(q_{tot}^{had})^2}$ range 25 to 36 GeV (black points), compared with the results by the TASSO Collaboration at $(\sqrt{s})_{e^+e^-}$ between 27 and 35 GeV (open squares).

This system, according to our extrapolation, should be like a multiparticle state produced by $(\sqrt{s})_{e^+e^-}$ =100 GeV.

However, there is a very important check to make using collider data, without the need for the (e^+e^-) data.

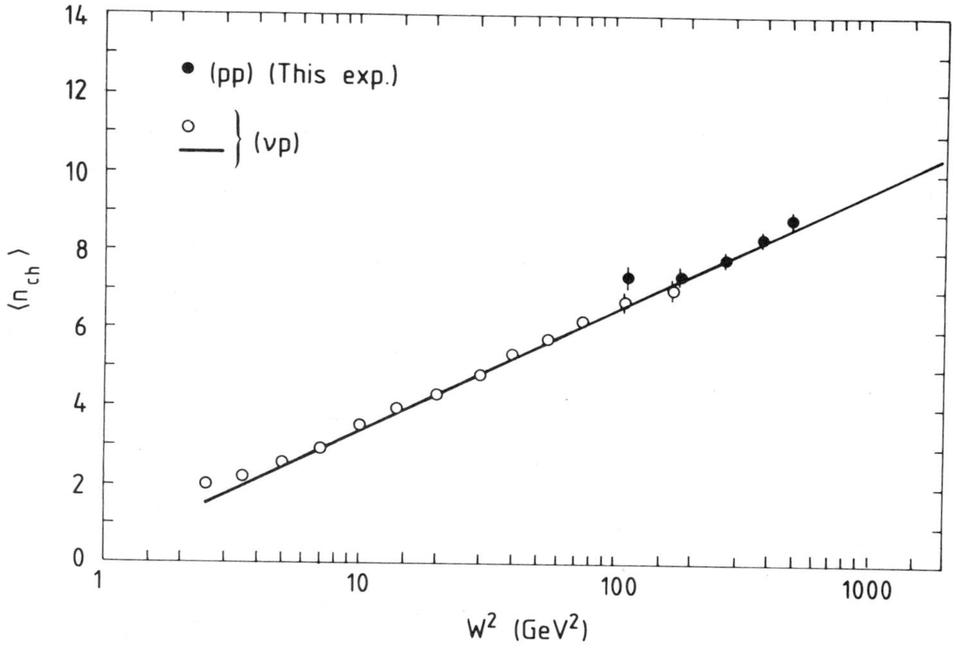

Fig. 12. The average charged-particle multiplicities $\langle n_{ch} \rangle$ measured in (pp) at $(\sqrt{s})_{pp} = 30$ GeV, using a DIS-like analysis, are plotted versus W^2 (black points). The open points are the (νp) data and the continuous line is their best fit.

The key point is to see if, at the CERN $p\bar{p}$ Collider, a multiparticle system produced at low p_T but with

$$\sqrt{(q_{tot}^{had})^2} = 100 \text{ GeV}$$

looks like the one produced at high E_T. The main difference we can expect is the value of $\langle p_t \rangle$.

To check these points is another important contribution to understanding hadron production at extreme energies.

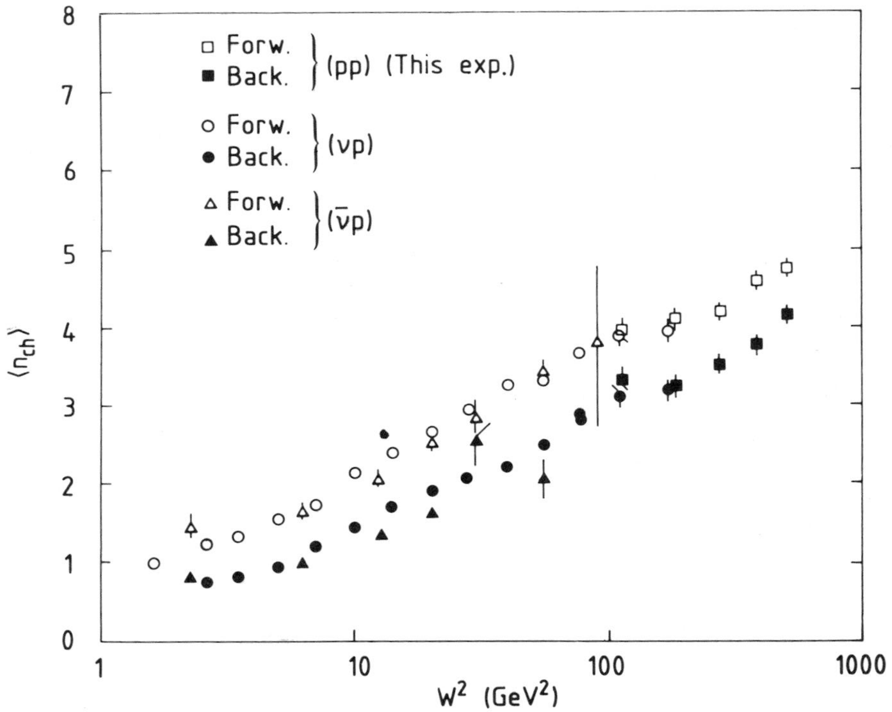

Fig. 13. The mean charged-particle multiplicities $\langle n_{ch} \rangle_{F,B}$ in the forward and backward hemispheres versus W^2, in (νp), ($\bar{\nu}$p), and (pp) interactions.

5. CONCLUSIONS

The new method of studying (pp) and (p$\bar{\text{p}}$) collisions - based on the subtraction of the "leading" effects and the use of correct variables - allows us to put on equal footing the multiparticle systems that are produced in purely hadronic interactions, in (e^+e^-) annihilation, and in DIS processes.

- Purely hadronic interactions means using machines such as the CERN Intersecting Storage Ring (ISR), the CERN p$\bar{\text{p}}$ Collider, the BNL-CBA collider, and the FNAL p$\bar{\text{p}}$ Collider.

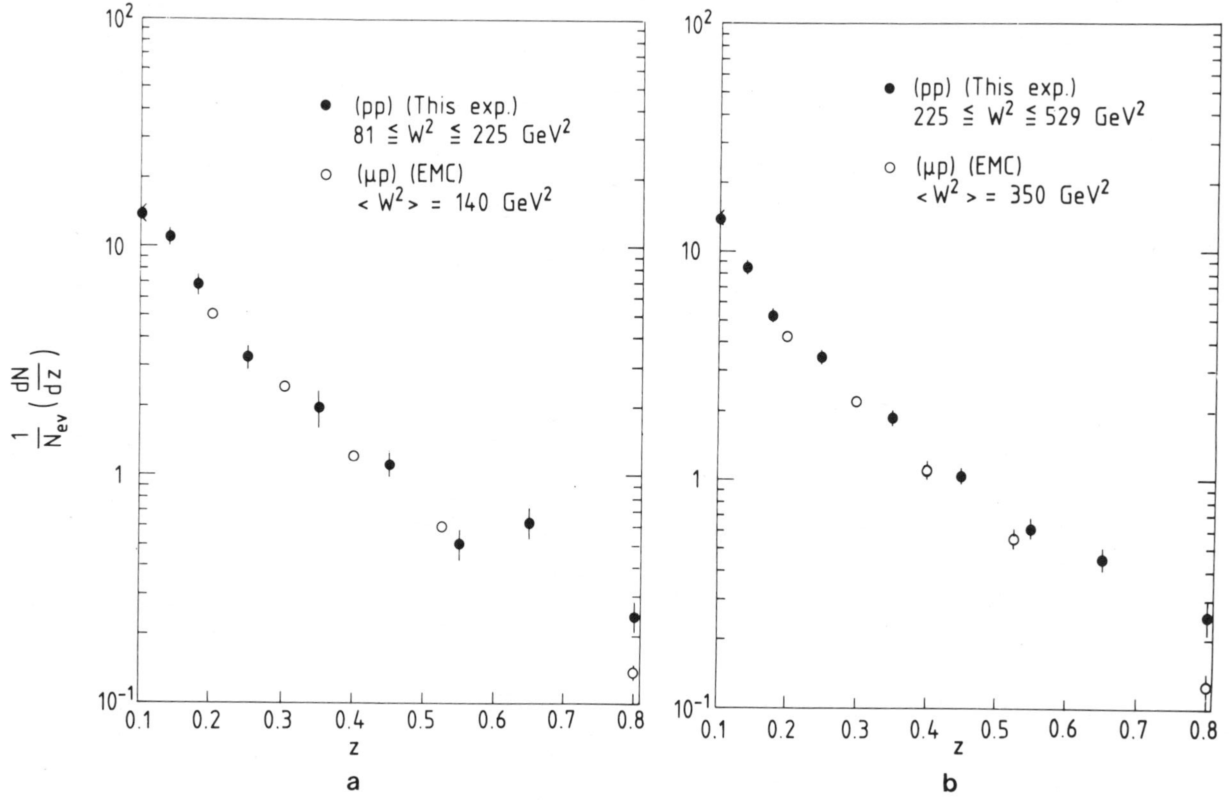

Fig. 14a,b: The inclusive distribution of the fractional energy z for (pp) reactions: a) in the energy interval ($81 \leq W^2 \leq 225$) GeV^2 compared with the data from (μp) reactions at $\langle W^2 \rangle = 140$ GeV; b) in the energy interval ($225 \leq W^2 \leq 529$) GeV^2, compared with the data from (μp) reactions at $\langle W^2 \rangle = 350$ GeV^2.

- (e^+e^-) annihilation means using machines such as LEP and its possible developments.
- DIS processes means using machines such as HERA.

The "leading" subtraction and the use of the correct variables allow us to show that a universal feature is at work, in the production of multibody final states in (pp), (e^+e^-), and DIS.

REFERENCES

1. M. Basile et al., Phys.Lett., 92B, 367 (1980).
2. M. Basile et al., Nuovo Cimento, 58A, 193 (1980).
3. M. Basile et al., Phys.Lett., 95B, 311 (1980).
4. M. Basile et al., Nuovo Cimento, 29, 491 (1980).
5. M. Basile et al., Phys. Lett., 99B, 247 (1981).
6. M. Basile et al., Nuovo Cimento Lett., 30, 389 (1981).
7. M. Basile et al., Nuovo Cimento Lett., 31, 273 (1981).
8. M. Basile et al., Nuovo Cimento, 65A, 400 (1981).
9. M. Basile et al., Nuovo Cimento, 65A, 414 (1981).
10. M. Basile et al., Nuovo Cimento Lett., 32, 210 (1981).
11. M. Basile et al., Nuovo Cimento, 67A, 53 (1982).
12. M. Basile et al., Nuovo Cimento, 67A, 244 (1982).
13. M. Basile et al., Nuovo Cimento, 73, 329 (1983).
14. G. Bonvicini et al., preprint CERN-EP/83-29 (1983), submitted to Nuovo Cimento Letters.
15. M. Basile et al., Nuovo Cimento Lett., 36, 303 (1983).
16. G. Bonvicini et al., Nuovo Cimento Lett., 36, 555 (1983).
17. G. Bonvicini et al., preprint CERN-EP/83-42 (1983).
18. UA2 Collaboration, Phys. Lett., B118, 203 (1982).
19. UA1 Collaboration, preprint CERN-EP/83-02 (1983).

D I S C U S S I O N

CHAIRMAN: A. ZICHICHI

Scientific Secretary: J. Bagger

- *ROTELLI:*

A fair percentage of the time, what comes out of a hadronic process is not a leading hadron but a leading hadron resonance, or a 'spectator jet'. Is it possible to identify the pions which come from the decay of this leading jet? According to your prescription, these pions should be subtracted along with the leading particle.

- *ZICHICHI:*

What you say is not correct. If one studies $d\sigma/dx|_{inclusive}$ in pp interactions, one sees that the probability of finding a proton is independent of x. We have made a world analysis of the various processes, and we have established how the leading effect varies when the leading particle in the final state is not the same as the particle in the initial state. For example, we have measured how much $\Delta(3/2, 3/2)$ resonance is in our final state, and we find only 7%. The presence of a 'leading' jet cannot be mistaken for a leading proton. In fact, the jet is made with many particles: each will have an x_F much lower than the x_F of the leading proton.

- *KLEVANSKY:*

How do you identify the leading proton?

- *ZICHICHI:*

Thanks to nature, the fastest positive track in a (pp) interaction is for $x_F > 0.5$, in a large fraction of cases, the leading proton. At low x_F the identification becomes harder because the pion contamination increases. This may be seen in the following figure:

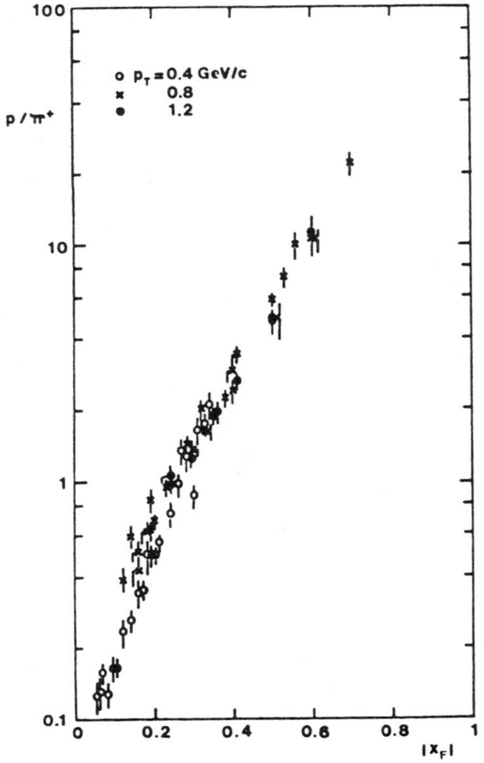

We have restricted our analysis to the region $0.3 \lesssim |x_F| \lesssim 0.8$, where identification of the leading proton is straightforward.

- *EKSPONG:*

Today you demonstrated close similarities between pp and e^+e^- reactions if the leading particles are subtracted. However, your explanation of the forward/backward multiplicity correlation was too condensed for me to understand. Could you elaborate on this?

- *ZICHICHI:*

Until now the forward/backward multiplicity correlation was explained in terms of long-range correlations. In our new approach there exists a simple explanation. The correlation is due to the fact that selecting a multiplicity in one hemisphere is equivalent to picking up a certain amount of effective energy. In fact, on the average, the number of particles in one hemisphere (call it 'forward') will be ($\frac{1}{2}$) of the total number of particles produced:

$$\langle n_{ch} \rangle_F = \frac{1}{2} \langle n_{ch} \rangle^{total} .$$

In the other hemisphere you will be left with the remaining half of the total multiplicity, i.e.

$$\langle n_{ch} \rangle_B = \frac{1}{2} \langle n_{ch} \rangle^{total} \ .$$

The correlation is a straightforward consequence.

- *EREDITATO:*

What are the systematic errors caused by misidentifying the leading proton?

- *ZICHICHI:*

The systematic errors change with x_F in a well-known way. This gives us good control over them. One need only worry about systematics when there is no parameter to vary. In our case, the systematics are no problem, and the results follow our expectations.

- *EREDITATO:*

How do you account for the momentum carried by neutrinos?

- *ZICHICHI:*

The cross-section for producing neutrinos is enormously suppressed compared with the cross-section for hadronic processes.

- *LEURER:*

What is the diagrammatic argument which explains the similarity between the hadronic states produced in e^+e^- annihilation and pp interactions, when analysed in your way?

- *ZICHICHI:*

The picture is as follows: in e^+e^- annihilation the incoming particles annihilate to form one virtual photon of mass

$$\sqrt{(q_1 + q_2)^2} \equiv (\sqrt{s})_{e^+e^-} \ .$$

This virtual photon then produces a quark-antiquark pair. Each one hadronizes into a jet.

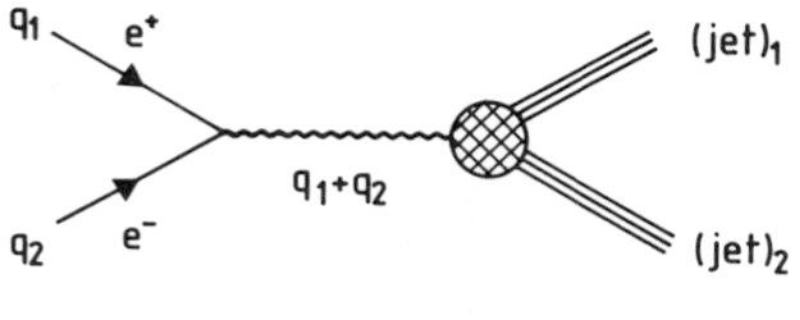

In pp-interactions, on the other hand, one must first subtract the four-momentum carried by the leading protons.

The correct effective energy is

$$\sqrt{(q_1^{had} + q_2^{had})^2} \equiv \sqrt{(q_{tot}^{had})^2} \ .$$

It is *this* energy which must be compared with $(\sqrt{s})_{e^+e^-}$ to recover the same results.

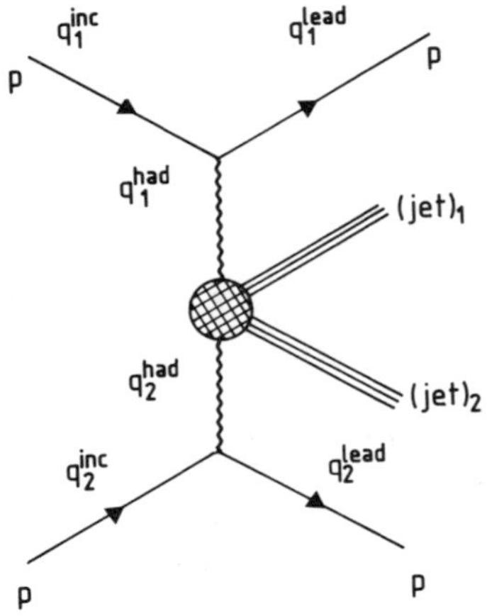

- *CHEN:*

There is a slight complication in the analysis of pp or $p\bar{p}$ interactions. In these reactions, more than 50% of the time one has quark-gluon or gluon-gluon scattering, and the gluons are connected to the leading particles by colour lines. At some stage this should spoil the universality between e^+e^- and pp interactions.

- *ZICHICHI:*

Our analysis is the starting point for investigating at what level the universality is violated. So far, our results are valid to 20%. We are studying where and how to find possible features that will distinguish (e^+e^-) and (pp) produced multiparticle systems.

- *WEIDBERG:*

Can you use your model to make predictions for mean multiplicities at collider energies?

- *ZICHICHI*

This could be done, and we are working on it.

- *WEIDBERG:*

Have you compared the shape of the multiplicity distributions in pp and e^+e^- interactions?

- *ZICHICHI:*

We are actually studying it.

- *CHEN:*

In the spirit of your search for the perfect variables to describe e^+e^- and pp interactions, I would like to say that $(\sqrt{s})_{e^+e^-}$ is not the best variable for describing e^+e^-. There is a radiative correction to $\sqrt{s}$ which lowers the true effective energy.

- *ZICHICHI:*

This is a well-expected correction. It does not mean that $\sqrt{s}$ is not the best variable. It is the only one we know to assign a given energy to the (e^+e^-) annihilation. This is not the case for $(\sqrt{s})_{pp}$, as I have reported in my lecture.

EVENTS, LAWS OF NATURE, AND INVARIANCE PRINCIPLES

E.P. Wigner

Princeton University
Princeton, NJ

INTRODUCTION

I have often discussed this and related subjects but I still find it difficult to call attention to all the relevant points - points which we all should remember. Actually, in the first part of the discussion, I would prefer to replace in the title, so kindly suggested by Dr. Zichichi, the "Events" by "Initial Conditions".

The sharp distinction between Initial Conditions and Laws of Nature was initiated by Isaac Newton and I consider this to be one of his most important, if not the most important, accomplishment. Before Newton there was no sharp separation between the two concepts. Kepler, to whom we owe the three precise laws of planetary motion, tried to explain also the size of the planetary orbits, and their periods. After Newton's time the sharp separation of initial conditions and laws of nature was taken for granted and rarely even mentioned. Of course, the first ones are quite arbitrary and their properties are hardly parts of physics while the recognition of the latter ones are the prime purpose of our science. Whether the sharp separation of the two will stay with us permanently is, of course, as uncertain as is all future development but this question will be further discussed later. Perhaps it should be mentioned here that the permanency of the validity of our deterministic laws of nature became questionable as a result of the realization, due initially to D. Zeh, that the states of macroscopic bodies are always under the influence of their environment; in our world they can not be kept separated from it.

As to the invariances, they can be characterized as laws which the laws of nature have to obey. If a certain behavior follows from a set of initial conditions, from similar but displaced initial conditions, the same behavior but equally displaced behavior follows.

The three concepts, initial conditions, laws of nature, and invariances, and their changes in the course of the development of physics, will be discussed next in a bit more detail.

INITIAL CONDITIONS

The initial conditions of a system, together with the laws of nature, are supposed to determine the behavior of the system as long as this remains isolated, i.e. is not subject to the influence of other systems. They are supposed to be independent from each other. The initial conditions of Newton's mechanics are most easily defined in this connection: they are the positions and velocities of all the bodies of the system. This means, of course, that the "bodies" are all point like, i.e. have no variable inner coordinates. This is essentially true for the planets to which Newton applied his theory principally, it could be, but is not generally, true of atoms. But Newton's mechanics was greatly generalized rather soon after its establishment and the basic idea remained unchanged. It may be of some interest to observe in this connection that Aristotle's initial conditions involved only the positions - he assumed that the velocities are proportional to the forces which the body experiences.

It may be worth reminding here already that Newton's laws are, and were already known by him, to be valid only in a coordinate system which is "at rest", or is in uniform straight motion. And the "rest" or "uniform motion" of the coordinate system can not be defined abstractly - the only simple and general definition is that the coordinate system is such that Newton's laws are valid in connection therewith. This may appear to make his laws meaningless but they become meaningful by the postulate that there are such coordinate systems and the validity of Newton's laws with respect to it extends to the whole universe.

Perhaps I should mention also that even though the definition of the initial conditions for Newton's mechanics is very simple, their measurement was very difficult even for the system for which its application was ground-breaking: the motion of the planets around the sun. The full determination of the positions of the planets was difficult, principally because we see only the projection of their three coordinates on the plane perpendicular to the direction at which we see it. But the full determination of the positions is quite old - it seems to have been first carried out by the Greeks -

Ptolemy's work is most often cited. But, of course, the coordinate system in which he specified the position was not the one at rest with the sun but at rest with respect to the Earth.

The next fundamental change in the definition of the initial conditions, together with a fundamental extension of the area of physics, came with Maxwell's equations. The initial conditions for these are more complex mathematically and also, in principle, more difficult to measure - they are <u>functions</u> of the three dimensions of space. Only an infinity of numbers can identify a function, even if this is arbitrarily differentiable and the determination of the initial conditions is, therefore, in this case infinitely more difficult in principle than for the Newtonian system. But states with definite properties can be, and have been, produced and the Maxwell equations are well confirmed experimentally. It is perhaps good to remark here that Newton's mechanics was also extended to continuous materials (liquids, gases, also solid bodies) and the same observation applies to these expensions as was made to the electromagnetic field equations.

One other remark should be made concerning the initial conditions of Maxwell's equations. They can be given in terms of the three components of the electric and of the magnetic field strengths at a given time - but the magnetic field's divergence is zero, so these are not free variables. The other description, by means of the four components of the vector potential and their time derivatives, is even less perfect because an arbitrary four dimensional gradient can be added to the vector potential without changing its physical content, i.e. without changing the electric and magnetic field strengths derivable from them. If it is postulated that the divergence of the vector potential be zero, the situation is not better than with the description by means of the electric and magnetic field strength. These are epistemological difficulties but, in practice, do not impair the usefulness and applicability of the equations.

The last fundamental change in the description of the state of a system, and hence of the initial conditions, came with the establishment of quantum mechanics. As you know, this also increased the area of physics, even more than the establishment of Maxwell's equations. The states of quantum mechanics are described by complex so called state vectors in the infinite dimensional Hilbert space - which does not give a greater complexity than the electromagnetic field specifications of Maxwell's theory. In fact the vectors in Hilbert space can be replaced, in most practical applications, by "wave functions" which are complex functions in 3n dimensional space where n is the number of particles. The correspondence between the physical stat and the state vector describing it is almost one-to-one - two state vectors characterize the same state only if all their components differ by the same factor, i.e. if they have the same direction in Hilbert space.

In this regard, the description of the states, hence also of the initial conditions, of quantum mechanical theory comes closer to the ideal described originally than that of Maxwell's theory. But whereas the determination of the state can be done at least approximately for the electromagnetic field, the microscopic state, that is the state vector, of the quantum state can not be determined from the outside even approximately. Every measurement undertaken on the system is very likely to change its state fundamentally. It is true, on the other hand, that many of the possible states of a microscopic system can be produced, i.e. that, particularly systems consisting only of one, two, or perhaps three or even more particles can be pushed into a variety of states. Not into all states which can be described by a state vector - some of these can be proved to be absolutely unproducable (superselection rules). Many, in fact most, can be produced only approximately by a finite apparatus. But, of course, the apparatus producing it has to be large only from a microscopic point of view - not necessarily large as compared with our own bodies.

In sum, the true initial conditions, describing the full microscopic structure of a system, are very difficult to determine - difficult even for microscopic systems. Some such states can be produced and the theory can be tested on them but we must admit that states of most "state vectors" are impossible to create. Just the same, the confirmation of the theory for the obtainable state vectors convinces us of the validity of the theory and its actual significance is, of course, overwhelming. I trust it is unnecessary for me to give evidence for this.

Let me just summarize by admitting that the nature of the initial conditions, that is also the description of the states of the systems in which physics is interested, has changed fundamentally from the old theory, where it was straightforward and simple, to Maxwell's field theory where it remained straightforward but ceased to be simple, then again to quantum mechanics for which it is neither simple nor straightforward. It is remarkable that relativity theory did not introduce such drastic changes - this is perhaps because it did not fundamentally expand the area of physics which, as you know, is now able to describe the properties of all common bodies and indeed contains, in principle, all of chemistry. But, of course, only in principle.

Let me mention, finally, one effect which the theory of relativity should have introduced into the description of the initial conditions and perhaps also into the description of all states. The state vectors, as we use them, describe the properties of the states which they assume at a definite time. But these are unobservable - we can not get signals instantaneously from a distance. It would be, therefore, more reasonable for the state vector, or the wave function, to describe the state on the negative light cone from the points of which signals may reach the observer. I have proposed this and tried to do it also but with very little success, at least so far. Let me now go over to the discussion of the laws of nature.

LAWS OF NATURE

It is not necessary to explain the nature of the laws of nature to this audience. If the initial conditions are given for a system, they predict its future as long as the system is isolated and does not receive signals from outside its negative light cone. The formulation of the laws of nature depends naturally on the type of initial conditions that they receive - in fact one can say that they give the initial conditions which describe the system at times later than the time of the original initial conditions.

Naturally, all our laws of nature are approximate. But there are conditions under which their accuracy is marvelous. Elementary quantum mechanics gives the energy levels of H or He with an accuracy of about 10^{-6} and the present theory improves this further. Newton's law of gravitation, as applied to the planets, gives their positions with a truly surprising accuracy - in the worst case, that of the Mercury, the error is less than 1/20.000 after one circulation around the sun. As you know, even this deviation from Newton's laws was eliminated by the general theory of relativity and we hope that other deviations from the present laws of nature will be also eliminated, or at least decreased, by future laws of nature. And that the validity of the present laws of nature will be further extended, surely to high energy phenomena, but perhaps also to the phenomenon of life. But it is truly wonderful that we are provided with situations in which the present laws of nature have such high accuracy. It would have been much more difficult to create science if these situations were not available.

The laws of nature we know are both very interesting and most useful. But, we must admit, they are not perfect, in many cases not consistent. In particular the macroscopic and microscopic theories are not united - we can measure times and distances almost solely macroscopically and we use macroscopic instruments also for quantum mechanical measurements. In fact, the Russian physicist V. Fock declared that the measuring apparata must be described classically, that is not quantum - mechanically, the systems on which the measurement is undertaken are quantum objects. S. Ludwig's defense of quantum mechanics is based on the same demand: measuring and registering apparata must be treated classically but - I must admit - he does not admit that this is a limitation of the validity of quantum mechanics, the hiding of its inability to describe macroscopic objects. But surely, we must realize that the quantum mechanical measurement process, the outcome of which is probabilistic, is not describable by the present quantum mechanical equations which are deterministic. It is natural to try to attribute the probabilistic outcome of the measurement to the uncertainty of the initial state of the measuring apparatus but it is easy to prove that this is impossible, impossible at least if there is no a priori relation betweem the pre-measurement states of system and measuring apparatus.

Perhaps I should mention that this suggested to me a modification of the quantum mechanical equations as applied to macroscopic bodies. This was stimulated by the article of D. Zeh (1970) according to which a macroscopic body can not be isolated from the interaction of the environment - not even in intergalactic space. This means that the present deterministic equations of quantum mechanics do not strictly apply to it and I suggested (in 1979) the addition of a term to these equations which should take care of the interaction with the environment. This interaction renders the microscopic state (i.e. the wave function) of macroscopic bodies to be subject to probability laws and would account also for the probabilistic outcome of the measurement process assuming that the measuring apparatus is macroscopic. Although the proposed equation is surely not the final one, it is surely true that the present deterministic equations of quantum mechanics are not strictly valid for macroscopic bodies.

Another conflict between the present microscopic and macroscopic theories, that is between quantum mechanics and general relativity, stems from general relativity's fundamental concept: the space-time distances between space-time points - the basis of the g_{ik}. A point in space-time can be defined only as the crossing point of two world lines, that is by the position and time of a collision. But quantum mechanic's collision theory does not specify the position of the collision precisely and also attributes an extended time-interval thereto. Hence, according to the microscopic point of view space-time points can not be defined with absolute accuracy. Of course, the actual applications of the general theory of relativity do not require the measurement of truly microscopic distances - the distances of the planets from the sun are not microscopic. But the virtual non-existence of such distances shows that the general theory of relativity, that is our present theory of gravitation, is not in harmony with quantum mechanics.

Actually, the postulate of the definability of infinitely small space-time points underlies also the quantum field theories and, in the opinion of some of us, is responsible for the infinities and for the need of renormalization. There are, therefore, attempts to abandon the concept of the classical and absolutely precise space-time point concept and Dr. T.D. Lee presented to us such an attempt which I hope will be fully successful. My own attempts in that direction were not and we must admit that the departure from the ideas of standard geometry would be as fundamental as Einstein's departure from the idea of absolute simulataneity.

The last part of my discussion points to the lack of absolute consistency of the present laws of physics and their weaknesses. I could have mentioned in this connection also that they do not yet give a truly satisfactory description of nuclear structure and even less of the very high energy phenomena which we heard about a great deal. What bothers me even more is that they make absolutely no reference to the phenomenon of life except for the vague admission that there is an observer behind each measuring apparatus. But we should not forget how amazingly successful they are in the description of certain phenomena, how much further they went than Heisenberg's original postulate of giving the energy levels of atoms and the transition probabilities between then. Their influence was not only scientific, they also had technical applications and changed our mode of life greatly.

Let me mention finally the wonderful property of the laws of nature, not postulated by their definition which is, in Einstein's words that they are "simple and mathematically beautiful". They are given by equations which use only mathematical symbols (though we must admit that some of these were originated by physicists) and use them in a simple way. This is truly remarkable - it could be very, very different and this would reduce the power of man, and his interest in science. We admit that our present laws of nature are approximate and have validity only under certain conditions but these conditions were furnished to us and we recognized at least approximate laws, and this is wonderful. Whether man will ever recognize the true laws of nature is questionable - I hope he will come closer and closer to them but always remain very far. If he recognized them, science could not be developed further and this would deprive man of an important source of pleasure and a deflection from the quest for power. I hope this will not happen.

I will give now a short description of the invariance principles and this will be followed by some reservations which, I hope, describe the limitations of the discussions preceding it.

INVARIANCE PRINCIPLES

As was mentioned before, the invariance principles are laws which the laws of nature must obey. Surely, if the laws of nature were different at different locations, or if they depended irregularly on time, it would not be possible to recognize them and, in fact, they would not be laws of nature. That the laws of nature are the same in coordinate systems which are uniformly moving with respect to the one defined at the discussion of Newtonian mechanics' initial conditions - and hence uniformly moving with respect to each other - is much more surprising and even some of the greatest scientists originally did not believe it - they questioned Einstein's theory of special relativity. Aristole's theory also contradicts it, as was implied before.

Of course, when we speak about the invariance of the laws we should not forget that the laws apply only to isolated systems and, here on the Earth we are subject to its gravitational attraction. This removes most invariances from everyday experience. It was, therefore, not so easy to recognize all the invariances. But the magnitude of the outside influences can often be judged and their effect taken into account or, under better conditions, such as the Earth's gravitational attraction on fast moving particles, neglected.

As was said before, the initial and most important function of the principles of invariance is to serve as necessary conditions on the validity of laws of nature and hence in establishing, or at least proposing, such laws. But the invariance principles do have other, less obvious applications. The oldest one of these is the establishment of conservation laws. This is usually credited to E. Noether but was proposed even earlier by Hamel. It is far most easily obtained from the quantum mechanical formalism: the commutability of the spatial displacement operators, and of the infinitesimal operators of rotation with the time displacement operator leads at once to the conservation laws of momentum and angular momentum. The commutability of the infinitesimal time displacement operator with the finite such operators establishes the energy conservation law. In pre-quantum mechanical theories the connection is much less obvious and the works of Hamel and Noether do deserve respect.

But the invariance principles have also other, and even more effective consequences in quantum mechanics. If we consider a system at rest - originally an atom was the subject of this consideration - the discrete energy levels' states' state vectors (or wave functions) are all linear combinations of a finite number of state vectors. Naturally it is best to choose these orthogonal to each other. If we subject these states to a rotation, the resulting state vectors can be expressed as linear combinations of the original ones. The matrix which transforms the original state vectors into the rotated ones is, apart from a constant factor, uniquely determined. The matrices so obtained for the various rotations form, apart from a constant, a "representation" of the rotation group, i.e. the product of two such matrices gives a matrix which corresponds to the product, that is successive application, of the two rotations. It follows that a representation of the rotation group corresponds to each discrete energy level. These representations describe then several properties of the underlying states - they give, for instance, their total angular momentum. But they give many other properties of these states, the selection rules for optical transitions, the ratios of these transition probabilities between the various states of one energy level to another one, and so on. This is an application of the symmetry principles which is available only in quantum mechanics and is based on the linear nature of the description of the states in that theory. It was first established by the present writer and given in a book in 1931.

Perhaps even more interesting consequences can be derived if one uses an approximate Hamiltonian. In the case of atomic spectra it is natural to assume that the velocities of the electrons are small, hence the magnetic forces also small and that this applies also to the spin interaction. This leads to the L-S coupling theory of spectra. Another assumption leads to the J-J coupling theory in nuclear physics and we have heard of many other approximate symmetry applications in high energy physics. Perhaps it is good if I mention that the weak interaction is neglected as a rule and this leads to the assumption of reflection symmetry and hence to the parity concept.

In summary, in quantum mechanics the symmetry principles lead to an easy establishment of several consequences of the theory and are very useful in this connection also. But the prime function of the symmetry principles remains in my opinion to help in the establishment of the laws of nature and to control the validity of proposed laws.

I will now try to give a critical, very critical, review of the concepts considered to be fully valid in the preceding discussion and allude to the consequences of their possible violations. According to the great philosopher-physicist E. Mach all laws of nature and all concepts have only limited validity and he may be right in this regard.

RESERVATIONS

As to initial conditions, it is not clear that they can be obtained in the microscopic sense for all systems. A macroscopic body's energy levels are so close to each other that it appears infinitely difficult to determine its exact state. In addition, as will be mentioned next, the difficulty is increased by its vigorous interaction with the environment. If the initial state can not be determined, not even in principle, the concept of "initial state", as we now use it, loses universal validity.

The same is true of the "law of nature" concept if it is impossible to retain the system in an isolated state. As was mentioned before, the difficulty to maintain the isolated state of a macroscopic body was first pointed out by D. Zeh in 1970 and his argument was further strengthened by the present writer around 1979. It was pointed out, in particular, that even if a macroscopic body, such as a cubic centimeter of tungsten, even if it is taken out into intergalactic space, far from all bodies which could influence it much more, the cosmic radiation would affect it so much that it would remain "isolated" only for about a thousandth of a second. Of course, this fact does not reduce the practical usefulness of solid state physics, even though this treats even macroscopic systems as isolated, but it does attack the basic principles of our physics.

Another point which was emphasized by various colleagues is the probability that the coordinate system which is at rest with respect to the center of mass of the universe has simpler laws of nature than moving coordinate systems, that is that the Galileo-Newton theory of the invariance with respect to a uniform motion of the coordinate system is not exactly valid. We have, at present, no clear evidence for this, but the arguments denying the absolute validity of this invariance, which is incorporated into the Lorentz transformation and the theory of relativity, are quite reasonable.

What the "reservations" just made mean is that our physics will be fundamentally modified also in the future, as it was several times in the past. And we may hope that it will be also extended further, perhaps even to the phenomenon of life which is not described by our present physics. To finish with an optimistic remark, let me mention one extension of the area of physics which describes properties of objects such as solids - or liquids and gases. The extension resulted from the acceptance of the atomic and molecular structure of bodies - an acceptance opposed by the first physics book I read. It said "atoms and molecules may exist but this is irrelevant from the point of view of physics". It now seems relevant! Physics developed and extended its area greatly in our century! We can hope that it will extend it further.

D I S C U S S I O N

CHAIRMAN : E.P. WIGNER

Scientific Secretaries : G. Heise and F. Lamarche

- *MILOTTI* :

At the end of your talk you said that life is substantially different from ordinary matter.

- *WIGNER* :

The most obvious evidence for this is that quantum mechanics is based on the theory of observations, and observations the outcome of the observer is indeterminant, i.e. subject to probability laws.

It can be thought that this is attributable to the fact that the original state of the observer is undetermined, and that being in some states he finds one result, and that if in another state he finds another result, so that the probabilistic nature of the outcome of the observation is due to the uncertainty of the original state of the observer. But it is easily proved that this is not a valid explanation, no matter in what state Ψ the observer was originally, if the object is in a state ϕ_k in which the outcome of the measurement is uniquely given, we have

$$\phi_k \times \Psi \to \phi_k \times \Psi_k \qquad \text{(observation of a definite state)}$$

where Φ_k is the state in which the observation result is k. It then follows from the linearity of the equation for the interaction that if the original state of the object was a linear combination $\Sigma\alpha_{k\phi k}$ then

$$\sum^{k} a_k\phi_k \times \Psi \to \sum^{k} a_k\,(\phi_k \times \Psi_k) \qquad \text{(observation of a linear comb.)}$$

From the linearity, and the assumption that the state of the observer is independent of that of the object, it follows that, after the measurement, the observer is in a combination of states, which is obviously unreasonable. If I look at the Stern-Gerlach experiment, I see the flash either up or down, but I am not in a superposition of seeing it up and down. This shows that the process of observation cannot be described by present quantum mechanics. Macroscopic objects are subject to probabilistic laws, they cannot be given a wave function, for they cannot be isolated. This is particularly evidenced in the process of measurement.

- *MILOTTI* :

It seems to me that if you adopt Bohr's interpretation of quantum mechanics, there is no such contradiction.

- *WIGNER* :

I doubt that the process of observation can be described by linear quantum mechanics. Many other people have published articles on this.

- *LAMARCHE* :

Do you think that "initial conditions" are all located in the past or that they may be spread in space and time, i.e., do you believe in determinism ?

- *WIGNER* :

The concept of determinism is not simple. I don't believe that man is able to give a wave function to a macroscopic object, for such an object cannot be isolated, as illustrated by my example of the cube of tungsten in intergalactic vacuum. If you could obtain a wave function for the whole universe, perhaps the deterministic interpretation would be all right, but it does not seem possible to do so.

- *LEE* :

Maybe a very cold neutron star could be an isolated system?

- *WIGNER* :

The neutron star also receives cosmic radiation. To give a wave function to a star is much beyond my imagination. It would need a configuration space of such dimension that I could hardly write it out in decimal system in this room.

- *TAVANI* :

Have you comments to make on the "EPR-paradox" and the property of non-locality emerging from J.S. Bell's inequalities ?

- *WIGNER* :

In my opinion the so-called EPR-paradox is not really a paradox. Similar phenomena can be created also within the range of classical theory.

As to the J.S. Bell inequalities, I consider them truly important, inasmuch as they prove that in the case considered by him, one cannot define a non-negative probability function which describes the state of his system in the classical sense, i.e., gives non-negative probabilities for all possible events. He considers two particles in a certain quantum mechanical state, and the probability that the spin of one has a given direction, and the measurement of the spin component of the other in two other directions has a positive or negative value. He shows that if none of these probabilities is negative, the probabilities of the outcomes of quantum mechanically possible measurements contradict the postulates of quantum mechanics which have been, incidentally, confirmed experimentally. It follows that the true states of the system considered by him, cannot be described classically. It is not possible to give a single distribution function which gives the correct probabilities for the outcomes of all possible measurements.

This is a very interesting and very important observation and it is truly surprising that it has not been made before. Perhaps some of those truly interested in the epistemology of quantum mechanics took it for granted but they did not demonstrate it.

CLOSING CEREMONY

The Closing Ceremony took place on Saturday, 13 August 1983. The Director of the School presented the Prizes and Scholarships as specified below.

PRIZES AND SCHOLARSHIPS

Prize for Best Student - awarded to Jonathan A. BAGGER, Princeton University, USA

The Scholarships open for competition among the participants were awarded as follows:

Jun John Sakurai Scholarship - Jonathan A. BAGGER, Princeton University, USA

Patrick M.S. Blackett Scholarship - Robert H. BERNSTEIN, University of Chicago, USA

James Chadwick Scholarship - Cornelis P. VAN DEN DOEL, University of California, Santa Cruz, USA

Amos-de-Shalit Scholarship - Miriam LEURER, The Weizmann Institute of Science, Israel

Gunnar Källen Scholarship - Sandra P. KLEVANSKY, University of Frankfurt, FRG

André Lagarrigue Scholarship - Marco TAVANI, University of Rome, Italy

Giulio Racah Scholarship - Clifford BURGESS, University of Texas, USA

Benjamin Wu Lee Scholarship - Sultan CATTO, Yale University, USA

Giorgio Ghigo Scholarship - George HOU, UCLA, USA

Enrico Persico Scholarship - Petar SIMIC, Rockefeller University, New York, USA

Peter Preiswerk Scholarship - Pierre SALATI, LAPP, Annecy, France

Prize for Best Scientific Secretary - awarded to Sultan CATTO, Yale University, USA

The following students received *honorary mentions* for their contributions to the activity of the School:

Seiju AMI	- Freie Universität Berlin, FRG
Richard BALL	- University of Cambridge, UK
Giuseppe BALLOCCHI	- University of Bologna, Italy
Roberto BATTISTON	- University of Perugia, Italy
Chafik BENCHOUK	- Centre de Physique des Particules, Marseille, France
Javier BERDUGO	- DESY, Hamburg, FRG
Massimo CAMPOSTRINI	- Scuola Normale Superiore, Pisa, Italy
Giancarlo D'AMBROSIO	- University of Naples, Italy
Ghanasmyan DATE	- The Weizmann Institute of Science, Israel

Marcel DEMARTEAU	- DESY, Hamburg, FRG
Graciela GELMINI	- CERN, Geneva, Switzerland
Gerhard HEISE	- University of Köln, FRG
Randy A. JOHNSON	- Brookhaven Nat. Lab., USA
Stephen KAYE	- SLAC, Stanford, USA
Carlos MANA	- DESY, Hamburg, FRG
Tukashi MARUYAMA	- SLAC, Stanford, USA
Edoardo MILOTTI	- University of Trieste, Italy
Richard P. MOUNT	- CALTEC, Pasadena, USA
Dennis NEMESCHANSKY	- Institute for Advanced Study, Princeton, USA
Michael C. OGILVIE	- Brookhaven Nat. Lab., USA
Nobuyoshi OHTA	- INFN, Rome, Italy
Frank Peter POSCHMANN	- DESY, Hamburg, FRG
Pietro ROTELLI	- University of Lecce, Italy
Antony WEIDBERG	- CERN, Geneva, Switzerland
Dieter ZEPPENFELD	- Max-Planck-Institute, München, FRG

The following participants gave their collaboration in the scientific secretarial work:

Jonathan BAGGER
Richard BALL
Giuseppe BALLOCCHI
Javier BERDUGO
Robert H. BERNSTEIN
Clifford BURGESS
Simon CAPSTICK
Sultan CATTO
Giancarlo D'AMBROSIO
Marcel DEMARTEAU
Gerhard HEISE
George HOU
Sandra P. KLEVANSKY
Francois LAMARCHE
Miriam LEURER
Carlos MANA

Barbara MELE
Edoardo MILOTTI
Dennis NEMESCHANSKY
Michael C. OGILVIE
Nobuyoshi OHTA
Frank Peter POSCHMANN
Pierre SALATI
Petar SIMIC
Marco TAVANI
Cornelis Pieter VAN DEN DOEL

PARTICIPANTS

Guido ALTARELLI	Istituto di Fisica dell'Università Piazzale Aldo Moro, 5 00185 ROMA, Italy
Seiju AMI	Freie Universität Berlin FB 20 WE 4 Arnimalle 3 1000 BERLIN, FRG
Michele ARNEODO	Istituto di Fisica dell'Università C. Massimo d'Azeglio, 46 10125 TORINO, Italy
Jonathan A. BAGGER	Princeton University Department of Physics PRINCETON, NJ 08544, USA
David C. BAILEY	CERN EP Division 1211 Geneva 23, Switzerland
Richard BALL	University of Cambridge Department of Applied Mathematics and Theoretical Physics Silver Street CAMBRIDGE CB3 9EW, UK

Giuseppe BALLOCCHI

Istituto di Fisica dell'Università
Via Irnerio, 46
40126 BOLOGNA, Italy

Wolfgang BANZHAF

Universität Karlsruhe
Institut für Theoretische Physik
Kaiserstrasse 12
Physikhochhaus
7500 KARLSRUHE, FRG

Roberto BATTISTON

Istituto di Fisica dell'Università
Via Elce di Sotto, 10
06100 PERUGIA, Italy

Chafik BENCHOUK

Centre de Physique des Particules
70, route Léon Lachamp
Case 907
13288 MARSEILLE Cedex 2, France

Javier BERDUGO

DESY/Group F13
Notkestrasse 85
2000 HAMBURG 52, FRG

Robert H. BERNSTEIN

University of Chicago
Enrico Fermi Institute
5640 South Ellis Avenue
CHICAGO, IL 60637, USA

Graziano BRUNI

Istituto di Fisica dell'Università
Via Irnerio, 46
40126 BOLOGNA, Italy

Clifford BURGESS

University of Texas
Department of Physics
AUSTIN, TX 78712, USA

Massimo CAMPOSTRINI

Scuola Normale Superiore
Piazza dei Cavalieri, 7
56100 PISA, Italy

Simon CAPSTICK | University of Toronto
Department of Physics
TORONTO, M5S 1A7, Canada

Clara CASTOLDI | Istituto Nazionale di Fisica Nucleare
Via A. Bassi, 6
27100 PAVIA, Italy

Stefano CATANI | Istituto Nazionale di Fisica Nucleare
Largo E. Fermi,2
50125 FIRENZE, Italy

Sultan CATTO | Yale University
Physics Department
NEW HAVEN, CT 06520, USA

Min CHEN | Massachusetts Institute of Technology
MIT
Department of Physics
CAMBRIDGE, MA 02139, USA

Guido COGNOLA | Libera Università degli Studi
di Trento
Dipartimento di Fisica
38050 POVO (Trento),Italy

Giancarlo D'AMBROSIO | Università di Napoli
Istituto di Fisica Teorica
Mostra d'Oltremare, Pad. 19
80125 NAPOLI, Italy

Ghanashyam DATE | Weizmann Institute of Science
Department of Nuclear Physics
76100 REHOVOT, Israel

Paolo DEL GIUDICE | Istituto di Fisica dell'Università
Piazzale Aldo Moro, 2
00185 ROMA, Italy

Marcel DEMARTEAU — DESY/F13
Notkestrasse, 85
2000 HAMBURG 52, FRG

Alberto DEVOTO — Florida State University
Physics Department
TALLAHASSEE, FL 32306, USA

Gosta EKSPONG — University of Stockholm
Institute of Physics
Vanasdisvägen 9
11346 STOCKHOLM, Sweden

Antonio EREDITATO — Istituto di Fisica Sperimentale
dell'Università
Mostra d'Oltremare, Pad. 20, 3
80138 NAPOLI, Italy

Louis FAYARD — CERN
EP Division
1211 GENEVA 23, Switzerland

Sergio FERRARA — CERN
TH Division
1211 Geneva 23, Switzerland

Enrico FRANCO — Istituto di Fisica dell'Università
Piazzale Aldo Moro, 2
00185 ROMA, Italy

Harald FRITZSCH — Max-Planck-Institut für Physik
und Astrophysik (MPI)
Föhringer Ring 6
Postfach 40 12 12
8000 MUNCHEN 40, FRG

Peter GAIGG
Institut für Theoretische Physik
Technische Universität Wien
Karlsplatz 13
1040 WIEN, Austria

Ugo GASPARINI
Istituto di Fisica dell'Università
Via Marzolo 8
35100 PADOVA, Italy

Yannick GIRAUD HERAUD
Laboratoire de Physique Corpusculaire
Collège de France
11, Place Marcellin-Berthelot
75231 PARIS CEDEX 05, France

Sheldon L. GLASHOW
Harvard University
Jefferson Physical Laboratory
CAMBRIDGE, MA 02138, USA

Gerhard HEISE
Am Flutgraben 17
5000 KOLN 80, FRG

George HOU
University of California
Department of Physics
LOS ANGELES, CA 90024, USA

Stephen HWANG
University of Goteborg
Chalmers University of Technology
Institute of Theoretical Physics
41296 GOTEBORG, Sweden

Randy A. JOHNSON
Brookhaven National Laboratory
Associated Universities, Inc.
Physics Department
UPTON, L.I., NY 11973, USA

Gordon KANE | University of Michigan
Physics Department
ANN ARBOR, MI 48104, USA

Peter KASPER | Rutherford Appleton Laboratory
Chilton
DIDCOT, Oxon, OX11 0QX, UK

Stephen KAYE | Stanford Linear Accelerator Center
Bin 94
P.O. Box 4349
STANFORD, CA 94305, USA

Sandra P. KLEVANSKY | University of Frankfurt
Institut für Theoretische Physik
Postfach 111932
6000 FRANKFURT/M 11, FRG

François LAMARCHE | University of Ottawa
Department of Physics
OTTAWA, Ontario K1N 6N5, Canada

Elliot LEADER | Westfield College
Physics Department
Kidderpore Avenue
LONDON NW3 7ST, UK

Tsung Dao LEE | Columbia University
Department of Physics
NEW YORK, NY 10027, USA

Miriam LEURER | Weizmann Institute of Science
Department of Nuclear Physics
P.O. Box 26
REHOVOT, Israel

Sam J. LINDENBAUM — Brookhaven National Laboratory
Associated Universities, Inc.
Department of Physics
UPTON, L.I., NY 11973, USA

City College of New York
Physics Department
Convent Ave at 138th St.
NEW YORK, NY 10031, USA

Carlos MANA — DESY/F13
Notkestrasse 85
2000 HAMBURG 52, FRG

Tukashi MARUYAMA — Stanford Linear Accelerator Center
P.O. Box 4349
STANFORD, CA 95305, USA

Riccardo MEGNA — Istituto di Fisica dell'Università
C. M. D'Azeglio, 46
10125 TORINO, Italy

Barbara MELE — Istituto di Fisica dell'Università
Piazzale Aldo Moro, 2
00185 ROMA, Italy

Edoardo MILOTTI — Istituto di Fisica dell'Università
Via Valerio, 2
34127 TRIESTE, Italy

Richard Philip MOUNT — California Institute of Technology
Department of Physics, 256-48
PASADENA, CA 91125, USA

Dennis NEMESCHANSKY	Institute for Advanced Study School of Natural Sciences PRINCETON, NJ 08540, USA
Michael C. OGILVIE	Brookhaven National Laboratory Associated Universities, Inc. Department of Physics UPTON, L.I., NY 11973, USA
Nobuyoshi OHTA	Istituto Nazionale di Fisica Nucleare Piazzale Aldo Moro, 2 00185 ROMA, Italy
Stefan OLEJNIK	Slovak Academy of Science Institute of Physics, EPRC Dùbravskà cesta 842 28 BRATISLAVA, Czechoslovakia
Fabrizio PALUMBO	Istituto Nazionale di Fisica Nucleare Laboratori Nazionali di Frascati P.O. Box 13 00044 FRASCATI (Roma), Italy
M. Andrew PARKER	CERN EP Division 1211 GENEVA 23, Switzerland
Monica PEPE'	Istituto Nazionale di Fisica Nucleare Viale Benedetto XV, 5 16132 GENOVA, Italy
Martin POHL	DESY Notkestrasse 85 2000 HAMBURG 52, FRG

Frank Peter POSCHMANN	DESY/F13 Mark-J Experiment Notkestrasse 85 2000 HAMBURG 52, FRG
Peter N. RATOFF	California Institute of Technology Lauritsen Laboratory High Energy Physics, 356-48 PASADENA, CA 91125, USA
Richard S. RAYMOND	Brookhaven National Laboratory Associated Universities, Inc. AGS Experiment 748 U/M UPTON, L.I., NY 11973, USA
Claudio REBBI	Brookhaven National Laboratory Physics Department UPTON, L.I., NY 11973, USA
Pàl RIBARICS	Hungarian Academy of Science Central Research Institute for Physics P.O. Box 49 1525 BUDAPEST, Hungary
Pietro ROTELLI	Dipartimento di Fisica dell'Università Via Arnesano 73100 LECCE, Italy
Carlo RUBBIA	CERN EP Division 1211 Geneva 23, Switzerland
Gunnar RYDNELL	University of Goteborg Chalmers University of Technology Institute of Theoretical Physics 41296 GOTEBORG, Sweden

Karel SAFARIK	Joint Institute of Nuclear Research Laboratory of Nuclear Problems Head Post Office P.O. Box 79 101000 MOSCOW, USSR
Pierre SALATI	Laboratoire de Physique des Particules B.P. 909 74019 ANNECY-LE-VIEUX, France
Pascal SCHIRATO	Université de Genève Département de Physique Nucléaire et Corpusculaire 32 Boulevard d'Yvoy 1211 GENEVA 23, Switzerland
Petar SIMIC	Rockefeller University Physics Department 1230 York Avenue NEW YORK CITY, NY 10021, USA
Iuliu STUMER	CERN EP Division 1211 GENEVA 23, Switzerland
Karlheinz STUPPERICH	Universität Gesamthoschchule Siegen Postfach 210209 5900 SIEGEN 21, FRG
Marco TAVANI	Istituto di Fisica dell'Università Piazzale Aldo Moro, 2 00185 ROMA, Italy
Roberto TENCHINI	Istituto Nazionale di Fisica Nucleare Via Livornese 56010 S. PIERO A GRADO(Pisa), Italy

Cornelis Pieter VAN DEN DOEL | University of California
Department of Physics
SANTA CRUZ, CA 95064, USA

E. VAN DER SPUY | Senior Consultant
Theoretical Physics
Nuclear Development Corporation
Private Bag X 256
PRETORIA 0001, South Africa

Matteo VILLANI | Istituto di Fisica dell'Università
Via Amendola 173
70126 BARI, Italy

Antony WEIDBERG | CERN
EP Division
1211 GENEVA 23, Switzerland

Eugene P. WIGNER | Princeton University
Department of Physics
P.O. Box 708
PRINCETON, NJ 08544, USA

Frank WILCZEK | University of California
Institute for Theoretical Physics
SANTA BARBARA, CA 93106, USA

Dieter ZEPPENFELD | Max-Planck-Institut
für Physik
Föhringer Ring 6
Postfach 401212
8000 MUNCHEN, FRG

Fabio ZWIRNER | Istituto di Fisica dell'Università
Via Marzolo, 8
35100 PADOVA, Italy

INDEX